AF520243

WEED SEED BIOLOGY

ABOUT THE AUTHORS

Dr. K. Vanangamudi Former Dean (Agriculture) is now working as a Professor of Seed Science and Technology and Co. Principal Investigator of the ICAR NAIP scheme on "Development of e-course materials for B.Sc (Ag) degree programe". He is Principal Investigator of GOI schemes NIMI on "Development of Instructional materials for ITI students in Agriculture sector" and "Development of Instructional materials for COE both BBBT & advance modules". He was the former Dean (Agri), Agricultural College & Research Institute, TNAU, Coimbatore from March 2006 to July 2009, former Dean of AP Agricultural College, TNAU, Kalavai, Vellore district from September 2003 to February, 2006 and former Professor & Head of Dept. of Seed Science & Technology, TNAU, CBE 3 from March 2001 to August 2003. During his tenure as Dean (Agri) he started three new UG degree progarmmes namely B.Tech (Bioinformatics) in 2006-07, B.Tech (Agricultural Information Technology) in 2006-08 and BS (Agribusiness management) in 2007-08. As a faculty Dean of TNAU, Dr. Vanangamudi is having a rich experience in e-education. Introduced e-learning, e-teaching, e-class, e-registration, e-attendance, e-communication, e-assignment, online examination and e-evaluation in UG education of TNAU. This is implemented in 12 UG degrees being offered at 10 constituent and 4 affiliated colleges.

Dr. M. Bhaskaran, Ph.D., is now positioned as Special Officer (Seeds), Seed Centre, Tamil Nadu Agricultural University, Coimbatore. He is also holding the additional charge of Professor and Head, Dept. of Seed Science & Technology, TNAU, Coimbatore, since 01.06.2011. He obtained Ph.D. in Seed Science and Technology from Tamil Nadu Agricultural University in 1995. He is having 21 years of experience in seed planning, production and quality control, teaching, research, extension and administration. He established and operationalised National Seed Research & Training Centre & Central Seed testing laboratory with all infrastructures as founder Director and obtained ISTA member laboratory status. He was the Principal Investigator of three schemes funded by Ministry of Environment and Forestry, Neyveli Lignite Corporation under Ministry of Coal & Mines (GOI). He is a member in central Seed Committee and Central Sub Committee on variety release and Notification, GOI, technical expert for SAARC project of Ministry of External Affairs on Seed Quality Regulation aspects, technical expert for OECD Seed Schemes of Government of India and member in Research Advisory Council of Directorate of Seed Research of Indian Council of Agricultural Research (ICAR). He was awarded with R.V. Swaminathan best Ph.D. scholar award (1995), TNAU best Teacher Award (2003) and MASU award on book writing in English for the book entitled, "Principles of seed production and quality control" (2003). He has published 25 research papers and one textbook "Principles of seed production and quality control". He has visited Brazil, Italy, USA, France, Afghanistan, Sri Lanka, Bhutan, Bangladesh, Nepal, South Africa and New Zealand in connection with seed science research and seed testing.

Ms. S. Balavidhya has completed B.Sc. (Horti) during 2007 and M.Sc. (Ag.) majoring in Horticulture during 2009 from faculty of Agriculture, Annamalai University, Annamalainagar, Tamil Nadu. For her thesis, she worked on "Studies of micropropagation of Rose" under the guidance of Dr. Arumugam Shakila. She has received "Vallalar Endowment Award" for the year of 2003-2007 and she is an awarder of "Srilochani Varadharajalu Endowment" for the topper in her MSc. She is now working as Senior Research Fellow in (GOI) NIMI scheme "Development of Instructional materials for COE both BBBT & advance modules". She has published two research papers in Asian Journal of Horticulture 2010 and a review paper on "Seed Biopriming" in Agrobios Research Journal in 2012.

Dr. P.Murali Arthanari, Assistant Professor (Agronomy) graduated from Faculty of Agriculture, Annamalai University, obtained post graduate and doctoral degree in Agronomy from Coimbatore campus of Tamil Nadu Agricultural University, excelling in all the courses of study. He is currently working at Directorate of Weed Science Research Centre, Department of Agronomy, TNAU, Coimbatore and handling research projects on weed management in field crops and involving UG and PG teaching on weed science. He has to his credit more than 35 research publications in National and International journals. He is recipient of international award on food security, T. Kondareddy award for outstanding research in Agronomy and young scientist award from ISEP, Gorakhpur.

WEED SEED BIOLOGY

EDITORS

K. VANANGAMUDI **M. BHASKARAN**

S. BALAVIDHYA **P. MURALI ARTHANARI**

DEPARTMENT OF SEED SCIENCE AND TECHNOLOGY
TAMIL NADU AGRICULTURAL UNIVERSITY
COIMBATORE-641 003

Published by:

Scientific Publishers (India)
5 A, New Pali Road, P.O. Box 91
Jodhpur 342 001 (India)

Branch Office
Scientific Publishers (India)
4806/24, Ansari Road, Daryaganj
New Delhi - 110 002 (India)

E-mail: info@scientificpub.com
Website: www.scientificpub.com

ISBN: 978-81-7233-827-5

Typeset by Rajesh Ojha
Printed in India

PREFACE

Weeds have been existing on the earth ever since the man started domesticating/cultivating plants and animals around 10,000 B.C. Out of 2,50,000 plant species, weeds constitute about 250 species, which are prominent in agricultural and non-agricultural system. In the world, there are 30,000 weed species and out of these 18,000 species cause damage to the crops. Weeds are the most severe and widespread biological constraint to crop production and cause invisible damage till the crop is harvested.

Weeds are the most underestimated crop pests in tropical agriculture although they cause maximum reduction/loss in the yields of crops than other pests and diseases. They decrease quantity and quality of produce/food, fibre, oil, forage/ fodder, animal products (meat and milk) and cause health hazards for humans and animals.

Contamination of food grains with poisonous weed seeds fetches low price. Foundation or certified seed is rejected, if weed seeds exceed 2% and also the market value of seed is drastically reduced. Thus, knowledge of weed seed biology, seed dormancy and germination, allelopathic effects on germination inhibition and seed production, will be helpful to adopt strategic plan in weed management in seed production, so that good quality seeds could be produced free from weed seed contamination.

Keeping in mind the importance of weed seed, this book on "Weed Seed Biology" provides information on origin, distribution, habit, habitat, floral biology (Flower type and fruit structure), flowering, fruiting, pollination, propagation, seed dispersal, life cycle, seed dormancy and germination, seed storage including seed burial and longevity, effects on crop production, control, uses and past work done on Seed Science and Technology for 42 dryland weeds, 33 gardenland weeds and 21 wetland weeds.

The book on "Weed Seed Biology" is first of its kind in India. We hope this publication would be extremely helpful to the Students, Academicians, Extension functionaries, Seed certification officials, Seed producers, Seed growers and those who are actively engaged in agriculture research and in particular Seed Science Technology.

Authors

CONTENTS

PART - I
SEED BIOLOGY OF DRYLAND WEEDS

D. Mariselvam[1] and S. Lakshmi[2]

1. PG scholar, 2. Assistant professor,
Department of Seed Science and Technology,
Tamil Nadu Agricultural University, Coimbatore – 641 003,
E-mail: maris.tnau@gmail / lakku_seed@yahoo.com

INDIAN MALLOW (*Abutilon indicum*)

Common name: Indian Mallow, Country Mallow, Abutilon, Indian Abutilon

Botanical name: Abutilon indicum

Tamil name : Paniyaratutti, Tutti

Hindi name : Kanghi

Family: Malvaceae (Mallow family)

Origin and distribution : Native to tropic and subtropical regions

Botany

Habitat : A common weed in waste places.

Habit : A diffusely much-branched shrub; growing to a height of 60 to 150cm (Fig. 1).

Fig. 1. Plant

Stem : Round with smooth close tomentum often tinged with purple.

Leaf : Broad, cordate, long-petioled, dentate, lower surfaces silky.

Floral biology

Flower : Solitary, axillary, yellow sometimes with a purple centre; floral parts as in other members of Malvaceae; carpels 10-20, syncarpous, individual carpels pointed or mucronatae with thin tomentum and scattered tuft of still hairs; Pedicel 5cm long and jointed at the top, calyx is tomentose, cupular below with 5 lobes, corolla is yellow sometimes with purple centre, petals 5 obovate and rounded with prominent veins (Fig. 2).

Fig. 2. Flower

Flowering and fruiting : September-April

Fruit : Fruit is quite interesting - it is circular in shape, consisting of 11-20 radiating hairy carpels, brown when dry; each carpel flattened, somewhat boatshaped (Fig. 3).

Fig. 3. Fruit

Seed biology

Seed : Seeds are kidney-shaped (Fig. 4).

Fig. 4. Seed

Propagation : By seed.

Seed dispersal : Seeds are mainly distributed through winds.

Medicinal uses : The plant is very much used in Siddha medicines. In fact, the root, bark, flowers, leaves and seeds are all used for medicinal purposes by Tamils. The leaves are used as adjunct to medicines used for pile complaints. The flowers are used to increase semen in men.

STARBURR (*Acanthospermum hispidum*)

Common name : Starburr

Botanical name : Acanthospermum hispidum

Tamil name : Multhulasi

Family : Compositae (Asteraceae).

Origin and distribution : This weed has become naturalized in Africa, the Hawaiian Islands, India, Australia and the West Indies.

Botany

Habitat : Common weed of dry lands, roadsides and wastelands.

Habit : Erect branched, hairy herb, upto 50cm height, annual (Fig. 1).

Fig. 1. Plant

Stem : Round, striate and rough with a dense clothing of spreading white hairs.

Leaf : Opposite, simple ovate, entire or serrate, acute, up to 4 x 2.2 cm.

Floral biology

Flower : Heads in the forks of dichotomous branches, heterogamous, rayed; corolla 3-lobed in female, 5-lobed in bisexual, lobes acute; stamens 5, anther base obtuse, not tailed (Fig. 2).

Fig. 2. Flower

Fruit : The fruits are achenes, long, with a triangular form, and covered with irregular hairs that lend the plant other popular names, including "carrapicho-decarneiro" (sheep's-thistle) or "carrapicho-de-tres-pontas" (three-pointed-thistle) (Fig. 3) (Aranha *et al.*, 1972; Correa, 1926; Lorenzi, 1982; Coimbra, 1994).

Fig. 3. Fruit

Seed biology

Seed : Seed burs are wedge-shaped, strongly compressed, 4-7 mm long, ribs bearing 1-2 rows of hooked prickles 1-2 mm long, 2 apical prickles stout, curved or straight, 3-4 mm long, protected by hairs and thorns (Fig. 4).

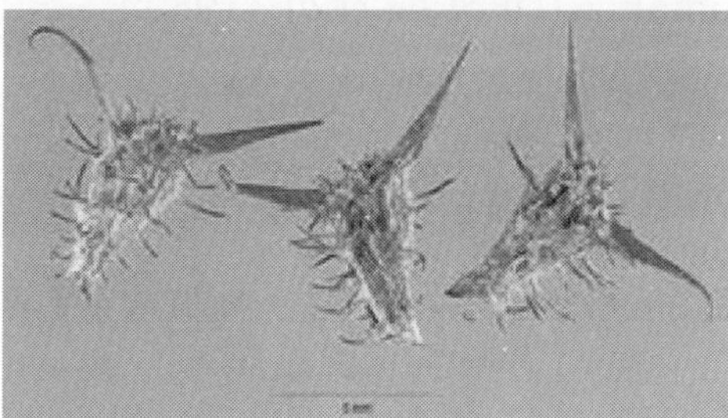

Fig. 4. Seed

Seed production : This noxious weed is a prolific seed producer. Plants with 25 cm in height are capable of producing mature seeds. The seeds are produced in abundance until the plants freeze in the fall. Average yield is about 4500 kg ha^{-1}. Seeds are shed from the plant very soon after maturing and most of them that remain on the plant at harvest are immature and result in low germination.

Propagation : By seeds

Seed dispersal : Seeds are spread by vehicles, machinery, animals, birds, humans, pasture seed, grain and water.

Seed dormancy and germination : The germination of harvested seeds is around 10 per cent, but soaking in water or KNO_3 increases germination to 60 per cent and when coupled with chilling, germination is increased to 80 per cent. Soaking seeds with scarified seed coats improves germination by 80 to 90 per cent.

Scarification by pre-freezing, acid or sandpaper reduced germination by 50 to 80 per cent. Immature embryo lead to poor germination (36%). Thus, the dormancy of these seeds seems to be due to the combination of following factors: immature embryo, impermeable seed coat and substances inhibiting germination.

Acanthospermum hispidum will germinate on a wide variety of soils, from sandy to clay (Miranda, 1996).

In general, the seeds of *A. hispidum* remain dormant in the soil for periods of upto eight years, reappearing at each new preparation of the soil (Schwerzel and Thomas, 1979) according to the rainfall and temperature regime.

Garcia (1992) examined seed dormancy and related it to climatic factors, noting that dormancy could be broken by incubation in the dark for 10-30 days.

Seed storage and burial : Seeds buried in soil below 7.5 cm have been shown to loose all viability after three years. Elemar Voll *et al.* (1997) reported 10.9 years longevity from seed burial study.

Control measures : A. hispidum has shown to resistant to a number of different types of herbicides (Reddy, 1965; Luo and Matsuomoto, 2002) and has showed an insecticidal activity against the larvae of a species of lady-bug (*Epilachna varivestes*) (Kraus *et al.*, 1994).

Uses : Useful for hemorrhage, weakness and diarrhoea. External formation, abortion and visceral anomalies (Di Stasi *et al.*, 1989.)

REFERENCES

Aranha, C., O. Bacchio, and H.F. Leitão Filho. (1972). *Plantas Invasoras de Culturas.* São Paulo: HUCITEC/*Min. da Agricultura/Agiplan* - BID, v. 2.

Coimbra, R. (1994). Manual de Fitoterapia. Belém: CEJUP.

Correa, M.P. (1926). Dicionário das Plantas Úteis do Brasil: Rio de Janeiro: Editora Nacional.

Di Stasi, L.C., E.M.G. Santos, C.M. Santos and C.A. Hiruma. (1989). Plantas Medicinais na Amazônia. São Paulo: *Editora da Universidade Estadual Paulista.*

Elemar Voll, L.P. Dionisio Gazziero and Decio Karam. (1997). Population dynamics of *Acanthospermum hispidum* dc. under soil and herbicide management practices, *Plant Physiology*, 32(2).

Garcia, Q.S. (1992). Dormencia e germinacao em aquenio de *Acanthospermum hispidum.* Campinas. *Dissertação de Mestrado em Biologia Vegetal na Universidade Estadual de Campinas*, SP. 100.

Kraus, W, M. Koll-Weber and R. Maile. (1994). Biologically active constituents of tropical and subtropical plants. *Pure Appl. Chem.*, 66: 2347-2352.

Lorenzi, H. (1982). Plantas daninhas do brasil. Nova Odessa/SP: *Instituto Plantarum.*

Luo X-Y and H. Matsuomoto. (2002). Susceptibility of a broadleaved weed, *Acanthospermum hispidum* to the grass herbicide fl uazifop-butyl. *Weed Biol. Manag.*, 2: 98-103.

Miranda, L.F.P. (1996). Análise comparativa da composição química de *Acanthospermum hispidum* DC em suas diferentes fenofases. Salvador, *Dissertação de Mestrado, Instituto de Biologia da Universidade Federal da Bahia.*78

Reddy, D. B. (1965). Control of *Acanthospermum hispidum* with weedicides. *Plant Prot Bull* 13: 67-68.

Schwerzel, P.J and P.E.L. Thomas. (1979). Effect of cultivation frequency on the survival of seeds of six weeds commonly found in Zimbabwe Rhodesia. *Rhod Agric. J. Zimb.*, 5: 195-199.

PRICKLY CHAFF FLOWER *(Achyranthus aspera)*

Common name : Prickly chaff flower

Botanical name : Achyranthes aspera (L.)

Tamil name : Naiurivi

Hindi name : Chirchita, Apang.

Family : Amaranthaceae

Origin and distribution : The plant is widespread in the world as a weed, in Baluchistan, Ceylon, Tropical Asia, Africa, Australia and America. In the Northern part of India, it is known as a medicinal plant in different systems of folk medicine

Botany

Habitat : Common in waste lands and on bunds in dry lands.

Habit : Erect herb about 90 cm high; sparingly branched, biennial or perennial (Fig. 1).

Stem : Round, striate and hairy.

Leaf : Leaves thick, orbicular-obvate or elliptic, obtuse, pubescent tomentose or velvety, rarely glabrous, petiole short.

Fig. 1. Plant

Fig. 2. Flower

Floral biology

Flower : Spike long, about 25 cm long in fruit; bracts membranous, spinescent, persistent; bracteoles spinescent; perianth 4-5, rigid, lanceolate, aristatem strongly ribbed, shining and hard; stamens 2-5, filaments connate at the base and alternate with staminodes which are toothed lacerate or with a toothed scale at the back; anthers two celled; ovary subcompressed, oblong, one celled; ovule solitary, pendulous, style filiform, stigma cap-like (Fig. 2).

Fruit : An utricle, oblong or ovoid, indehiscent, rounded or areolate at apex; seed inverse, oblong (Fig. 3).

Fig. 3. Fruit

Seed biology

Seed : Subcylinderic, truncate at apex, reddish brown (Fig. 4)

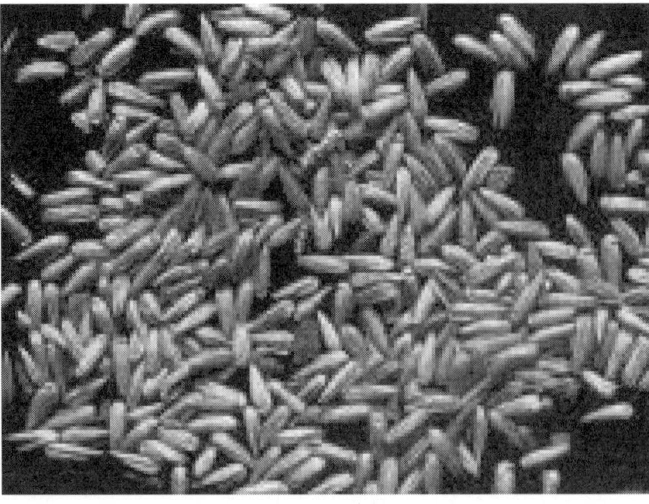

Fig. 4. Seeds

Propagation : By seeds

Seed dispersal : Seeds dispersed by natural herbivores

Life cycle : Life span less than 1 year to up to 3 years

Seed dormancy and germination : A 7 month period of after ripening in dry storage at room temperature increased the germination of *Achyranthus aspera* utricles from 0 to 50%; however when fresh utricles were washed in running water for 24 hr seeds germinated to 100% (Pandya and Pathak, 1980).

Achyranthus aspera having physiological dormancy (Pandya and Pathak, 1980).

Effects on crop production : Its allopathic effects can be persistent in the soil and affect other crops

Uses : Useful for reclamation of wastelands. Leaf is consumed as potherb. Seeds rich in protein, cooked and eaten. Used in religious ceremonies in India.

REFERENCES

Pandya, S.M and V.S. Pathak. (1980). Seed dormancy imposed by covering structures in *Achyranthes aspera*, *Geobios*, (Jodhpur) 7:74-76.

BUIKALLAN (*Aerva lanata*)

Common name : Buikallan or Boor

Botanical name : Aerva lanata

Tamil name : Sirukkan poolai

Hindi name : Gorakhbuti or Kapuri jadi

Family : Amaranthaceae

Origin and distribution : Originated from Africa and distributed throughout the world.

Botany

Habitat : A perennial, common dry land weed.

Habit : Branched under shrub, erect or prostrate, hairy tomentose, perennial growing to a height of 30-60 cm (Fig. 1).

Stem : Striate, round and softly woody tomentose.

Leaf : Small, woolly or glabrate, petiolate, elliptic, orbicular-obovate or orbicular-obtuse or acute.

Fig. 1. Plant

Floral biology

Flower : Small in solitary or panicled spikes; bracts and bracteoles small; perianth calycine, membranous 4–5 lobed, all or the inner woolly; stamens 4-5, filaments subulate, connate below with interposed linear staminodes into a cup, anthers 2-celled; ovary ovoid or subglobose, style long or short, stigma cap-like or bifid, ovule one, pendulous (Fig. 2).

Flowering and fruiting : Flowering and fruiting throughout the year

Fruit : Indehiscent utricle or circumsessile capsule; seeds inverse (Fig. 3).

Fig. 2. Flower

Fig. 3. Fruit

Seed biology

Seed : Kidney shaped with shining black coriaceous testa. Size of seed is 0.6 cm & 0.4 cm and weight of seed is 0.14 mg/seed.

Propagation : By seeds

Seed dispersal : Seed dispersed through animals.

Seed dormancy and germination : Dehusking (removal of seed covering structures) affected seed germination highly significantly, promoting germination in *Aerva lanata*.

Uses : **Leaves:** Leaf-decoction is prepared as a gargle for treating sore-throat and used in various complex treatments against guinea-worm. Babies that have become unconscious during an attack of malaria or of some other disease are washed with a leaf decoction, at the same time smoke from the burning plant is inhaled. The leaf-sap is also used for eye-complaints. An infusion is given to cure diarrhoea and in an unspecified manner at childbirth, and on sores.

Root: The root is used in snake-bite treatment.

Flowers: For pains in the lower part of the back, leaves and flowers are reduced to ash which is rubbed into cuts on the back.

Spiritual: It gives protection against evil spirits, is a good-luck talisman for hunters, and safeguards the well-being of widows.

JAVANESE WOOLPLANT *(Aerva tomentosa)*

Common name : Javanese Woolplant

Botanical name : *Aerva tomentosa*

Tamil name : Poolai keerai

Hindi name : Bui

Family : Amaranthaceae

Origin and distribution : Tropical Asia and Africa

Botany

Habitat : Common dryland weed with characteristic appearance.

Habit : White, tomentose, conspicuous, erect, herb, growing upto a height of 1.0 m.

Stem : Round, striate and thickly white tomentose.

Leaf : Leaves alternate, linear oblong or spathulate, obtuse or acute, densely tomentose.

Floral biology

Flower : Flowers dioecious in densely white woolly spikes forming terminal panicles; perianth calycine 5-lobed, membranous, woolly; stamens 5, alternating with staminodes, connate at the base; ovary superior, single celled with one ovule, style as long as the stigma, stigmas 2 in number (Fig. 1).

Fig. 1. Flower

Flowering and fruiting : Flowering period for this shrub is January, February, March, April, May, June and December.

Fruit : An utricle, membranous; seeds coriaceous.

Seed biology

Seed : Seeds lenticular, reniform, subglobose, or shortly cylindric, smooth or verruculose.

Propagation : By seeds and cuttings.

Seed dispersal : Seeds are mainly dispersed by animals.

Uses : The shrub is suitable for pots, shrubbery and herbaceous border, propagated from seeds or cuttings. Flowers and seeds are used against swelling, headache and rheumatism.

ALYCE CLOVER (*Alysicarpus glumaceus*)

Common name : Alyce clover (Foster).

Botanical name : *Alysicarpus glumaceus*

Tamil name : Namappoondu

Hindi name : Mordi

Family : Papaloidenae(Faboideae)

Origin and distribution : Alysicarpus is native to tropical Africa, Madagascar, Afghanistan, India, Pakistan, Vietnam and Indonesia. It is widespread throughout the tropics.

Botany

Habitat : A common dry land weed predominantly in black soils.

Habit : Diffusely branching, densely pubescent, or subglabrous, ascending annual herb and attain a length upto 45 cm (Fig. 1).

Fig. 1. Plant

Fig. 2. Flower

Stem : Round, densely covered with long spreading hairs.

Leaf : 1-3 foliate leaflets subglabrous, oblong, obtuse, linear-lanceolate, and variable.

Floral biology

Flower : Flowers purple, small, in pairs adpressed to a subglabrous axis, pedicels short; calyx ciliate, longer than the first joint of the pod, teeth imbricate, lanceolate; corolla papilonaceous, not exerted; stamens diadelphous. Ovary superior, single celled, many ovuled; style filiform, incurved at the tip (Fig. 2).

Flowering and fruiting : September- November.

Fruit : Fruits are indehiscent oblong pods, 18-25 mm long. Pod moniliform, joints strongly and closely ribbed transversely, turgid.

Seed biology

Seed : Seeds are oblong, ellipsoid, 2.5 mm × 1.5 mm × 1.2 mm (Mannetje, 2002) (Fig. 3).

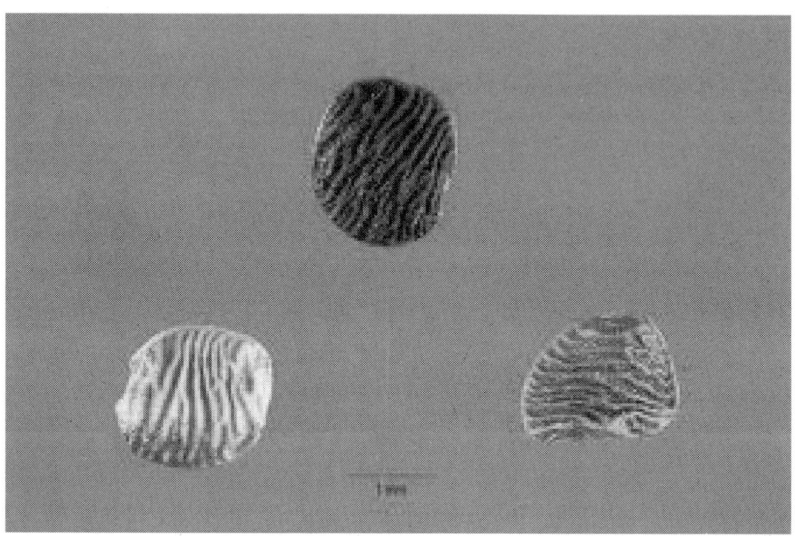

Fig. 3. Seed

Propagation : By seeds

Life cycle : It is available during the rainy season and at early stages of drought periods (FAO, 2010).

Seed dispersal : Seeds are dispersed by winds and animals.

Uses : Alysicarpus is an N-fixing legume (Mannetje, 2002).

REFERENCES

FAO (2010). *Grassland Index.* A searchable catalogue of grass and forage legumes. FAO.

Mannetje, LT. (2002). *Alysicarpus ovalifolius.* **In:** PROTA (Plant Resources of Tropical Africa/Ressources vegetales de l'Afrique tropicale. Ed: Oyen, L.P.A. and R.H.M.J. Lemmens, Wageningen, Netherlands.

SCARLET PIMPERNEL (*Anagallis arvensis*)

Common name : Scarlet pimpernel

Botanical name : Anagallis arvensis

Tamil name : Anicham

Hindi name : Neel, Dharti Dhak, Krishnaneel

Family : Myrsinaceae

Origin and distribution : Native to Europe, but spread throughout the temperate zone in Africa, Asia, Europe, North America, and South America

Botany

Habitat : Cultivated areas, wastelands, roadsides.

Habit : Herbs annual, biennial, or perennial, usually glabrous (Fig. 1).

Stem : Stems quadrangular, short winged on ridges, usually branched at base.

Leaf : Leaves opposite, occasionally in whorls of 3, sessile, ovate to narrowly ovate.

Floral biology

Flower : Calyx lobes linear-lanceolate, 3.5–6 mm, hyaline margined, apex long acuminate, costate. Corolla blue or red, rotate, 4–6 mm, parted nearly too base lobes. Obovate-elliptic, 2.7–3 mm wide, margin entire to denticulate. Stamens 1/3–1/2 as long as corolla; filaments pubescent. Style is 1.5 mm (Fig. 2).

Fig. 1. Plant

Fig. 2. Flower

Flowering : Spring, summer, autumn

Fruit : Capsule 3.5 mm in diameter globose, circumscissile, many seeded (Fig. 3).

Seed biology

Seed : The seeds are tiny brown balls. There are several seeds in a seed pod. The seed pod is a ball, opening round the middle (Fig. 4). The seeds of the plant, which are very numerous and enclosed in small capsule are brown in colour (Fig. 4).

Fig. 3. Fruit

The average seed number per plant is around 900, but large specimen may yield as many as 1200 seeds (Salisbury, 1962).

The weed produces many thousand seeds per plant. The seeds generally remain viable in the soil for about 10 years.

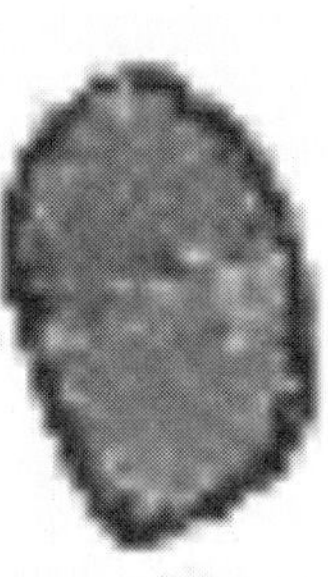

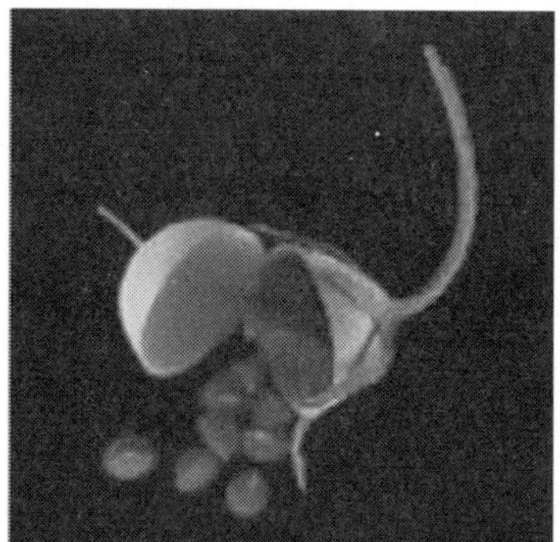

Fig. 4. Seed

Propagation : Propagation is entirely by seeds, as the plant is an annual, completely dying at the end of each season, both above and below ground.

Seed dispersal : As the autumn comes on, the fruit in the centre of each flower swells and ripens. It is in the form of a little urn or capsule, full of tiny seeds. When the latter are quite ripe, the urn splits round its circumference into two halves - the upper half lifts up like a lid and the seeds are shaken out with every movement of the wind.

The stiff persistent style at the apex may assist in the removal of the lid as the capsule is touched by a passing animal or brushed against other plants when shaken by the wind.

The seeds of the plant, which are very numerous, and enclosed in small capsules, are much eaten by birds.

Seed dormancy and germination : *Anagallis arvensis* seeds having post harvest dormancy and water soluble inhibitor.

It was observed that freshly collected seeds gave very low germination percentage, even when moisture, low temperature and light were provided. But, when those seeds were washed in running water, under same experimental conditions, more than 80 per cent germination occurred. These observations suggested the occurrence of post harvest dormancy and the presence of some water soluble inhibitor in the seeds of A. arvensis.

Grant Lipp and Ballard (1963) have pointed out that, apart from being genetically controlled, the duration of dormancy in *Anagallis arvensis* depends on the temperature prevailing during seed formation and maturation. They have shown that lower temperature during seed maturation cause prolonged dormancy.

Climatic factors, like temperature and rainfall, to which the seed *Anagallis arvensis* are exposed after maturation, seem to be of great importance in controlling the duration of dormancy. Higher temperature reduces the period of dormancy. Similarly, washing of seeds artificially or by natural rainfall immediately removes dormancy. Under Indian conditions, seeds of *A. arvensis* mature in March-April. Afterwards, they are subjected to high temperatures (upto 45°C) during summer (April-June), which is followed by moderately high temperature (30-35°C) and plenty of moisture during the rainy season (July-September). Later the temperature drops to about 15-20°C by the end of October, when germination of *A.arvensis* is observed in the fields (Singh, 1967).

Seed storage and burial : Seed is able to remain viable in soil for at least 10 years, while in dry storage remain viable for 8 years. Seed recovered from excavations and dated at more than 30 years old is reported to have germinated.

Control measures : Hand pull or post-emergence application of 2, 4-D at 500 g ha^{-1} or metsulfuron at 4 g ha^{-1}.

Uses : Diuretic, diaphoretic and expectorant. The ancient reputation of Scarlet Pimpernel has survived to the present day, especially in dealing with diseases of the brain. Doctors have considered the herb remedial in melancholy and in allied forms of mental disease, the decoction or a tincture being employed.

Past work done on Seed Science and Technology

Seed dormancy, germination and storage

Singh (1967) observed that duration of dormancy varies in different seed collections. After four months, seeds from Jhansi, Gwalior and Allahabad were almost out of dormancy, while those from Agra and Banda remained dormant to a certain extent, even after a period of eight months in storage (Table 1).

Table 1. Percentage germination of seeds from different localities during storage at room temperature

Localities		Storage period in months		
		4	8	12
Agra	....	56.0 ± 7.0	70.6 ± 4.0	82.3 ± 3.0
Etawah	....	75.6 ± 3.9	84.0 ± 1.6	90.9 ± 0.9
Jhansi	....	88.8 ± 2.3	94.3 ± 3.6	96.6 ± 1.8
Gwalior	....	86.3 ± 2.1	89.0 ± 3.7	95.9 ± 2.1
Kanpur	....	74.6 ± 2.6	82.0 ± 1.6	89.9 ± 1.1
Banda	....	49.0 ± 2.4	60.0 ± 2.9	75.3 ± 1.9
Allahabad	....	85.6 ± 1.6	89.0 ± 1.4	94.6 ± 2.0

(Mean ± standard deviation)

According to him, seeds stored at 40°C had become non-dormant, whereas those stored at lower temperatures were still partially dormant, after 3 months of storage (Table 2)

Table 2. Percentage germination of seeds stored at different temperatures

Storage temperature °C		Storage period in months	
		3	8
0	....	63.3 ± 3.7	63.3 ± 3.7
10	....	81.0 ± 0.8	85.6 ± 2.9
20	....	83.3 ± 2.7	90.3 ± 1.8
30	....	86.3 ± 0.8	80.3 ± 1.2
40	....	92.0 ± 1.6	— —

The seeds lying in the surface layer of soil (10 cm) become completely non-dormant within 15 days, presumably due to greater leaching activity. On the other hand, seeds buried at greater depths show slightly lower percentage germination. But in general, seeds enclosed in nylon bags gave higher percentage germination than those kept in sealed tubes. Washing of seeds increases percentage germination due to removal of water-soluble inhibitor (Table 3).

Table 3. Effect of washing on germination

Washing duration hours		Germination (%)
0 (control)	....	12.2 ± 2.2
24	....	80.5 ± 2.4
48	....	82.0 ± 1.9
72	....	74.0 ± 3.2

It is evident that as the seeds age, the percentage as well as the rate of germination increases. Fifteen days old seeds gave 17 per cent germination in 18 days. After eight months, 91.5 per cent germination was obtained within 12 days (Fig. 5) (Singh, 1967).

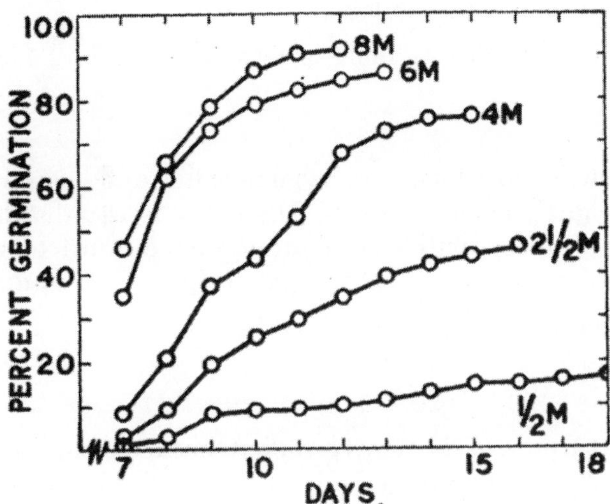

Fig. 5. Effect of age (months) on the rate and the percentage germination of seeds of *Anagallis arvensis*

REFERENCES

Grant Lipp, A.E. and L.A.T. Ballard. (1963). Germination pattern shown by light sensitive seeds of *Anagallis arvensis*. *Aust. J. Biol. Sci.,* 16:72-84.

Singh, K.P. (1967). Distribution and growth performance of *Anagallis arvensis* L plants raised from seed of different localities. In *PL 480 Annual Report*, 96-105.

Salisbury, E. (1962). The Biology of Garden Weeds. Part 1. *J. Royal Hort. Soc.*, 87:338-350, 390-404.

KARIYAT (*Andrographis serpyllifolia*)

Common name : Kariyat

Botanical name : Andrographis serpyllifolia

Family : Acanthaceae

Tamil name : Kattuppooraan Kodi, Siyan kodi

Hindi name : Birbhubat

Origin and distribution : Native to India distributed to China, Japan and Bangladesh.

Botany

Habitat : Common dry land weed.

Habit : Trailing procumbent herb.

Stem : Rootstock woody; stem villous.

Leaf : Sessile, orbicular, obtuse, more or less hairy, up to 2.5 cm diameter.

Floral biology

Flower : The flowers are bisexual, zygomorphic and usually associated with conspicuous, often brightly colored bracts. The calyx is usually deeply 4-5 lobed or sometimes is highly reduced with more numerous minute teeth. The corolla is sympetalous, usually pentamerous, mostly zygomorphic, and commonly 2 lipped.

Fruit : Capsule glabrous, acute at both ends.

Seed biology

Seed : The mature seed is nearly naked because the seed coat is almost completely digested. The seed stalk or funiculus of each seed is modified into a hook shaped jaculator or retinaculum that functions in flinging out the seeds during dehiscence. Seeds are deeply rugose and spoon shaped.

Propagation : By seeds

Uses : Internally for malaria, influenza, syphilis, liver, digestive problems and dysentery. To prevent and treat cold, flue, allergies, constipation, pain, tonsilitis, and sinusitis; as an antiseptic.

PRICKLY POPPY (*Argemone mexicana*)

Common name : Prickly poppy

Botanical name : Argemone mexicana L.

Tamil name : Brahmadandu

Hindi name : Pharangi dattura, Bhatbhamt

Family : Papavaraceae

Origin and distribution : Argemone mexicana is native of Mexico and West Indies, but has become pan tropical after accidental introduction or introduction as an ornamental. It is naturalized in most African countries, from Cape Verde, East to Somalia, and South to South Africa.

Botany

Habitat : A troublesome weed commonly growing in dry lands.

Habit : An erect branching prickly annual herb, with yellow latex, grows to a height of 60 -120cm (Fig. 1).

Stem : Round, glaucous, usually very prickly but sometimes nearly glaburous and full of yellow juice.

Leaf : Sessile, pinnately dissected.

Floral biology

Flower : Flowers axillary, solitary, bright yellow; sepals 2-3; petals 4-6; stamens indefinite, yellow; ovary superior, one celled with many ovules on parietal placentation, style very short, stigma lobed (Fig. 2).

Fig. 1. Plant

Fig. 2. Flower

Flowering and pollination : Anthesis of *Argemone maxicana* occurs over three consecutive days with flowers opening each morning and closing in the late afternoon. This stigma is receptive in early morning for first-day flowers and by late morning, numerous stamens begin to dehisce, releasing large quantities of pollen. By the end of the second day, anthers shrivels and the stigma becomes non-receptive. During the third day, the petals and stamens abscise. Hymenopterans (*Dialictus* and *Halictus*) and Coleopterans (*Euphoria* and *Lytta*) are the most abundant and effective pollinators Hymenopterans are effective cross-pollinators while Coleopterans facilitate self-pollination. The overall floral architecture as well as the floral behaviour are adaptations which facilitate the pollination of *Argemone* by beetles.

In the tropics, *Argemone mexicana* flowers and fruits throughout the year. The flowers open early in the morning, and last for 2–3 days. Small stingless bees are the main pollinators, but *Argemone mexicana* is predominantly self-pollinated.

Fruit : Fruit is a capsule, spiny, 2.5-5 cm long and 2 cm wide, with 4-6 valves opening at the tip to release numerous seeds (Fig. 3).

Fig. 3. Fruit

Seed biology

Seed : Seeds are brownish-black, nearly spherical, about 1 mm in diameter, covered in a fine network of veins, oily. The seed is light, has a waxy coat and is pitted (Fig. 4).

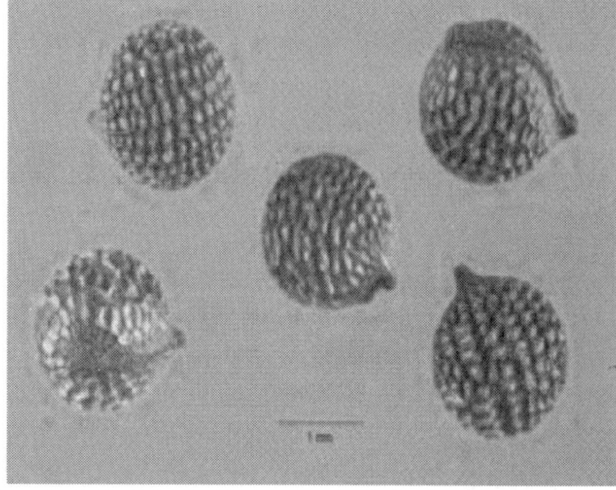

Fig. 4. Seed

Propagation : By seeds

Life cycle : A. mexicana is a long-blooming plant. In the temperate zone, flowering occurs in summer - autumn. In South India (Western Ghats), the plant flowers and fruits through the year (Matthew, 1999).

Seed production : Seed production can be 18, 000–36,000 seeds per plant. Seeds are readily eaten by a number of bird species in Puerto Rico as indicated by the presence of many seeds of the species in birds' stomachs (Barnés, 1946).

Mauritius reports the greatest seed production with an average of 60 to 90 capsules per plant with 300 to 400 seeds in each capsule (Holm *et al.*, 1977).

Seed dispersal : Seed may be dispersed by wind and water and is known to spread quickly in irrigation schemes. Dispersal also occurs by soil adhering to farm machinery and by man and livestock.

Seed dormancy and germination : The seeds have a type of underdeveloped embryo and therefore some kind of morphological dormancy according to Nikolaeva (1977) and Baskin and Baskin (1998), i.e. the embryo needs to grow inside the seed before germination. It germinates in 3-4 weeks at 15°C.

Seeds can remain dormant for many years. Seeds germinate best in moist soil with a temperature of upto 25°C. In some regions, they germinate throughout the year if enough moisture is available.

Seeds are dormant when shed and have an after-ripening period of several weeks or months. In Eastern Africa and Australia, seeds germinate at any time of year if moisture is available but, in Mauritius, germination only occurs in the cooler months (Parsons and Cuthbertson, 1992). In India also, it has been noted that germination occurs late in the season, with low night temperatures (Ambasht, 1992).

Seed storage and burial : In Ethiopia, most seeds do not normally germinate the year after shedding. Instead they enter the seed bank and seedlings establish, even in well maintained field, probably for many years (Karlsson *et al.*, 2003).

Effects on crop production : A. mexicana is a principal weed of beans and maize in Tanzania, cereals in Australia and India, cotton in Nicaragua, potatoes in India, tobacco in Argentina and Puerto Rico, and wheat in Pakistan (Holm *et al.*, 1977).

A. mexicana has an inhibitory effect on germination and seedling growth of vegetables (Hazarika and Sannigrahi, 2001) and weed residues may affect growth and development of bambara groundnut (*Vigna subterranea*) and sorghum (*Sorghum bicolor*) because of the inhibitory effects of allelochemicals present (Karikari *et al.*, 2000).

Control measures : Hand harvesting of field crops can be painful in the presence of *Argemone mexicana.* A biological control programme has been initiated in Australia. In Mexico, several predatory insects are identified including an extremely damaging species of root and leaf-feeding weevil.

Uses : Throughout the tropics, *Argemone mexicana* is widely used as a medicinal plant. It is considered a painkiller, diuretic, cholagogue and anti-inflammatory. The seed oil is used as a purgative and as pomade. Both the seed oil and leaf infusions are drunk to relieve cough. Root and leaf decoctions are applied to the skin to cure oedema, inflammation, muscle pain, ulcers, and yaws, to remove warts, to kill Guinea worm and to promote wound healing.

Past work done on Seed Science and Technology

Karlsson *et al.* (2003) stated that seeds are highly dormant, with very little initial germination and no substantial germination (upto 20%) after-ripening effect, any response to various combinations of incubation environments. This would suggest that very few seeds would germinate in the season following seed dispersal. Substantial germination occurred only in response to gibberellic acid (Fig. 5). Because of the clear germination response to gibberellic acid and the conti-nuous embryo growth and germination when part of the seeds coat was removed (Fig. 5), *A. mexicana* probably do not exhibit deep simple double morphophysiolo-gical dormancy. The seed coats of *A. mexicana* do not inhibit imbibition and therefore the species do not have physical dormancy. Further, seeds of both species did germinate, without any mechanical treatment of the seeds, in response to gibberellic acid. Therefore, the species do not have mechanical dormancy. The germination in response to removal of part of the seed coat might be explained by a hypothetical germination-inhibiting substance being present, and that this substance is more easily leached out of the seeds after the treatment. However, this seems less likely since there was no response to the different times with water treatment in the leaching experiment, and therefore, at least *A. ochroleuca* do not have chemical dormancy.

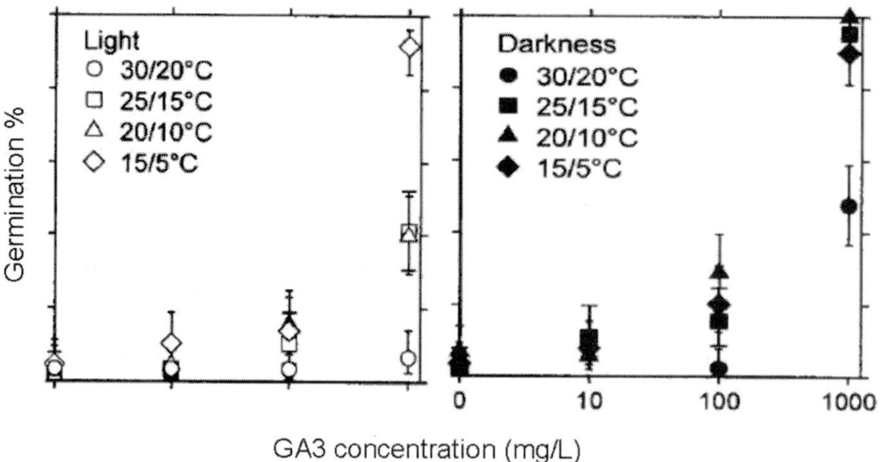

Fig. 5. Germination of *A. mexicana* in various concentrations of gibberellic acid and under various light and temperature conditions. Germination with 95% confidence interval, based on statistics for binomial distribution, is shown.

REFERENCES

Ambasht, R.S. (1982). **In**: Biology and Ecology of Weeds. The Hague, Netherlands Exell A.W, Wild H. eds: W. Junk Publisher. anaerobiosis: A question about the "inhibitor hypothesis" of seed dormancy. *Planta*, 129:109-112.

Barnes, V. (1946). The birds of Mona Island, Puerto Rico. *Auk*, 63:318-327.

Baskin, C.C. and J.M. Baskin. (1998). Seeds – ecology, biogeography and evolution of dormancy and germination. Academic Press, San Diego.

Hazarika. B, and A.K. Sannigrahi. (2001). Allelopathic research in vegetable production - A Review. *Environment and Ecology*, 19:799-806.

Holm, L.G., L. Plucknelt, D. Pancho and J. Herberger. (1977). World's worst weeds. Distribution and Biology. *University press of Hawaii*, Honolulu: 280–284.

Karikari, S.K., C. Bagai and A. Segwagwe. (2000). Allelopathic activity of five botswana weed species on Bambara groundnut [*Vigna subterranea* (L.) Verdc] and sorghum [*Sorghum bicolor* (L.) Moench]. *Crop Research (Hisar)*, 20(3): 397-406.

Karlsson L.M., T. Tamado and P. Milberg. (2003). Seed dormancy pattern of the annuals *Argemone ochroleuca* and *Argeemone mexicana* (Papaveraceae). *Flora*, 198:329-339.

Matthew, K.M. (1999). The flora of the Palni Hills, South India. Part 1, 'Ranunculaceae' 'Alangiaceae'. Tiruchirapalli: The Rapinat Herbarium.

Nikolaeva, M.G. (1977). Factors controlling the seed dormancy pattern. **In**: The physiology and biochemistry of seed dormancy and germination. Ed: A.A. Khan – *North-Holland Publishing Company*, Amsterdam. pp. 51-74.

Parsons, W.T. and E.G. Cuthbertson. (1992). Noxious Weeds of Australia. Melbourne, Australia : *Inkata Press*.

WORM KILLER (*Aristolochia bracteata*)

Common name : Worm killer

Botanical name : Aristolochia bracteata

Tamil name : Aduthinnapalai

Hindi name : Kidamari

Family : Aristolochiaceae

Origin and distribution : Australia is the primary origin and it is distributed from India to Papua New Guinea

Botany

Habitat : A very troublesome weed in dry lands, especially in black soils.

Habit : Slender glabrous, perennial herb, with weak prostrate or ascending branches 9-18 inches long (Fig. 1).

Stem : Striate, glabrous and leafy. The underground stems are pale yellow or white with numerous vegetative buds.

Leaf : Alternate, glaucous, broadly ovate, cordate at base.

Floral biology

Flower : Flowers solitary, bracteate, bract large, orbicular; perianth tube greenish, inflated at the base, contracted into a cylindric neck, hairy within, lip dark purple, 2.5 to 5 cm long, linear with revolute margins; stamens 6, adnate to the style; ovary inferior, about 6 - celled, style column thick, stigmatic lobes glandular, hairy, 3-6 in number (Fig. 2).

Fig. 1. Plant **Fig. 2.** Flower

Flowering and pollination: Flowers are often pollinated by flies. The Aristolochiaceae have co-evolved with swallowtail butterflies (Papilionidae), with the butterfly larvae feeding exclusively on members of the family. Larvae tolerate the aristolochic acids produced by the plants, but the poisonous chemicals deter predators.

Fruit: Capsule oblong-ellipsoidal; seeds covered with glands on one side (Fig. 3).

Fig. 3. Fruit

Seed biology

Seed : Seeds compressed, mostly winged all round or thick with an elaiosome. Numerous, usually flat, with copious hard or fleshy endosperm and minute embryo (Fig. 4).

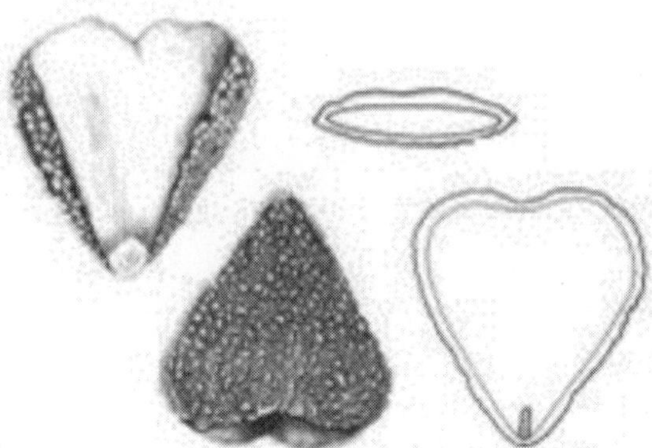

Fig. 4. Seed

Propagation : Seeds and underground stems.

Seed dispersal : Seeds are easily dispersed by winds.

Seed dormancy and germination : Germination tests were conducted at 30°C under three nitrate concentrations (1, 10 and 20 mM), fluctuating temperature (27/20°C) and light and dark conditions. It was found that seeds have under-developed embryos, which take about 301±178 h (±SD) to achieve seed coat rupture, another 205±126 h to reach radicle protrusion and 176±76 h more to the final stage of cotyledon emergence. Germinability was above 52% in all treatments, except in the dark (15%). For all treatments, average germination time was above 290±123 h. Potassium nitrate increased germinability to >87%. No particular treatment was required for embryo development, but seeds in the population that continued to germinate after 1 month were probably in various states of non-deep, simple morphophysiological dormancy. Increased germinability in nitrate treat-ments and light requirement for germination could prevent germination under unsuitable environmental conditions (Alves-Da-Silva *et al.*, 2011)

Soak the seeds for 24 - 48 hours in normal water, changing water a couple of times to wash out potential germination inhibitors. Sow the seeds over poor substrate with good drainage (containing sand or similar + 50%), cover with sand (2 mm.) and keep at cool temperatures of about 12 - 15°C. The germination starts after about two weeks. When germination has begun, the temperature may be raised to about 20°C. These plants are not tropical, excessive heat and humidity may kill them.

Uses : Aristolochia bracteata is used in traditional medicine as a gastric stimulant and in the treatment of cancer, lung inflammation, dysentery and snake bites (Negi *et al.*, 2003).

REFERENCES

Alves-Da-Silva. D., F. Borghetti, K. Thompson, H. Pritchard and J.P. Grime. (2011). Under-developed embryos and germination in *Aristolochia galeata* seeds. *Plant Biology,* 13: 104–108.

Negi, P.S., C. Anandharamakrishnan and G.K. Jayaprakasha. (2003). Antibacterial activity of *Aristolochia bracteata* roots extracts. *J. Med. Food.,* 6(4): 401-403.

SHAGGY BUTTON PLANT (*Borreria articularis*)

Common name : Shaggy button plant

Botanical name : Borreria articularis (Syn. *B. hispida*)

Tamil name : Noatha churi

Hindi name : Vasuka

Family : Rubiaceae

Origin and distribution : Growing wild in many parts of India. Distributed to China and Malaya.

Botany

Habitat : A common dry land weed associated with ragi, groundnut and pulses. A weed with many biotypes capable of adjusting to the drought conditions.

Habit : A hispid procumbent annual herb with long internodes (Fig. 1).

Fig. 1. Plant

Stem : Long, prostrate or procumbent branches usually dark pink in colour.

Leaf : Opposite, almost sessile, ovate, stipules connate with marginal bristles.

Floral biology

Flower : Flowers small, axillary or terminal fascicles with long narrow corolla tube, white or pink; stamens 4; gynoecium inferior, bicarpellary, syncarpus, bilocular with only one ovule in a locule on axile placena.

Fruit : Fruit two-seeded mericarp; seeds pale red, oblong with a ventral groove (Fig. 2).

Seed biology

Seed : Seeds are oblong; granulate, opaque, unusually variable, 3 mm or less in length (Fig. 3).

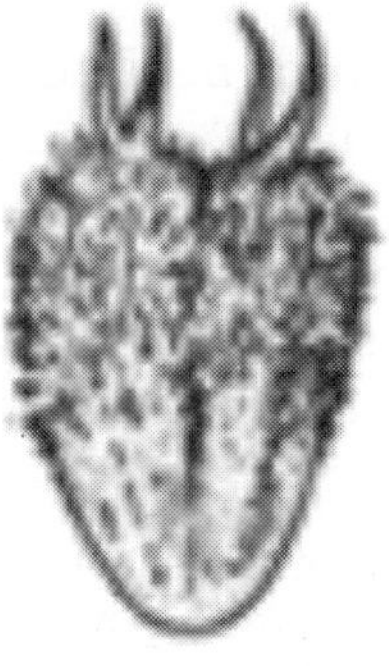

Fig. 2. Fruit

Fig. 3. Seed

Propagation : By seeds

Seed dispersal : Mature seeds are dispersed by wind, rain water, cultivation, soil on tools or even grazing.

Effect on crop production : Borreria commonly found in sorghum and groundnut field. It affects the growth and development of sorghum and groundnut crop.

Uses : Plant pacifies vitiated vata, pitta, arthritis, pain, muscle ache, edema, trauma, indigestion, colic, skin disease, menorrhagia and leucorrhea. Seeds are used to control demulcent, diarrhoea, dysentery.

MADAR (*Calotropis gigantea*)

Common name : Gigantic Swallow-wort; Madar

Botanical name : Calotropis gigantea

Tamil names : Erukkam, yerukku

Hindi name : Aak, Alarkh, Madar

Family : Asclepiadaceae

Origin and distribution : Calotrop is native to Cambodia, Indonesia, Malaysia, Philippines, Thailand, Sri Lanka, India and China (Forster *et al.*, 1996).

Botany

Habitat : Common in all plains, waste lands, dry lands, especially in black soil.

Habit : Large shrub, trunk stout, pale, shoots woolly, perennial (Fig. 1).

Stem : Erect, stout and round.

Leaf : Sessile, opposite, ovate or obovate, covered with soft white wool.

Floral biology

Flower : Flowers large, crowded, pale in lateral pedunculate cymes, bracts oblong; calyx 5-lobed, lobes ovate, glandular within; corolla pale purple, broadly campanulate, lobes spreading, ovate, lanceolate; corona scales 5, narrow, shorter than the staminal column, spur upcurved, involute; stamens inserted at the base of the corolla; anthers short, inflexed over the style tip, pollen masses solitary, pendulous, attached to the carriers by slender caudicles; ovary 2-celled, stigma depressed, 5-angled; follicles large, recurved (Fig. 2).

Flowering and pollination : Calotropis is an example of entomophilous pollination (pollination by insects) and pollination is achieved with the help of bees. The pollen are arranged in a structure named pollinia which is attached to a glandular, adhesive disc at the stigmatic angle (Translator mechanism). These sticky discs get attached to the legs of visiting bees so that pollinia are pulled out when the bee moves away. When such a bee visits another flower, that flower gets pollinated by the sticky pollinia.

Fruit : The fruit is a follicle (Fig. 3).

Fig. 1. Plant

Fig. 2. Flower

Seed biology

Seed : Seeds broadly ovate, comose (Fig. 4).

Propagation : By seed

Life cycle : This plant plays host to a variety of insects and butterflies.

Seed dispersal : When dry, seed dispersal is by wind. Seeds spread by wind and water over large distances (Smith, 2002).

Fig. 3. Fruit

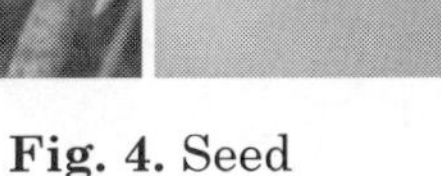

Fig. 4. Seed

Seed dormancy and germination : Older seed will not require the cold period to break its dormancy.

Uses : Flowers last long, and in Thailand they are used in various floral arrangements.

REFERENCES

Forster, P.I., D.J. Liddle and A. Nicholas. (1996). Flora of Australia. Asclepiadaceae, *Australian Government Publishing Service*. 28: 214 Canberra.

Smith, Nicholas M. (2002). Weeds of the wet/dry tropics of Australia - a field guide. Environment Centre NT, Inc. pp. 28-29.

KING'S CROWN (*Calotropis procera*)

Common name : King's crown

Botanical name : *Calotropis procera*

Tamil name : Vel yerukku

Hindi name : Madar, Akada, Akdo, Aak

Family : Asclepiadaceae

Origin and distribution : Giant milkweed is native to West Africa and distributed throught the world

Botany

Habitat : Common in all plains, waste lands, dry lands.

Habit : Large shrub, trunk stout, pale, shoots woolly (Fig. 1).

Fig. 1. Plant

Stem : Erect stout and round.

Leaf : Sessile, opposite, ovate or obovate, covered with soft white wool.

Floral biology

Flower : The flower clusters are umbelliform cymes that grow at or near the ends of twigs. The flowers are shallowly campanulate with five sepals that are 4 to 5 mm long, fleshy and variable in colour from white to pink, often spotted or tinged with purple (Fig. 2).

Fig. 2. Flower

Flowering and pollination : Flowering and fruiting takes place throughout the year (Little *et al.*, 1974). Hundreds to thousands of seeds may be produced per plant each year. Highly cross-pollinated through insects such as monarch butterflies.

Fruit : The fruits are inflated, obliquely ovoid follicles that split and invert when mature to release flat, brown seeds with a tuft of white hairs at one end (Fig. 3). (Howard, 1989; Liogier, 1995; Little *et al.*, 1974).

Seed biology

Seed : Seeds broadly ovate, comose. Seeds with a pappus of silky hairs (Fig. 4). The seeds in a Puerto Rican collection averaged 0.0095 + 0.0027 g seed^{-1} or about 100,000 seeds kg $^{-1}$.

Propagation : By seeds

Fig. 3. Fruit

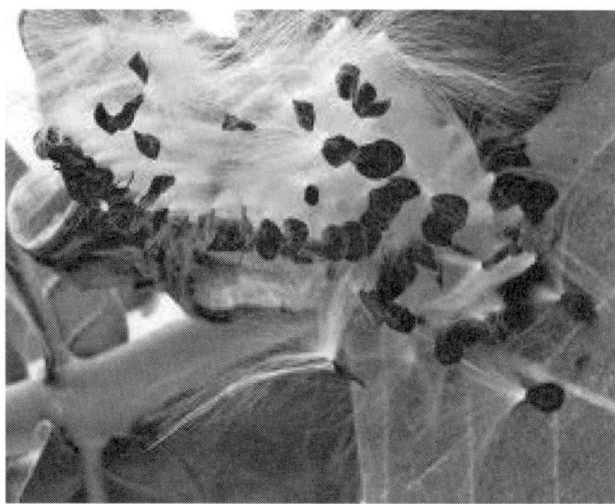

Fig. 4. Seed

Seed dispersal : Half the seed weight was found in the wing (silk). The seeds are dispersed by wind and may fly for several hundred yards in gentle breezes. Seedlings may arise in abundance after rainy periods, but only a few survive the first season.

Seed collection and extraction : Mature fruits should be collected just before dehiscence and the hairs should be removed from seeds after collection (Ashwani Kumar, 2009).

Seed dormancy and germination : No seed dormancy is recorded and seeds readily germinated upon sowing. Eighty-nine percent germination took place in potting mix between 7 and 64 days after sowing.

Seedlings may arise in abundance after rainy periods, but only a few survive during the first season. Using the reserves in its large taproot, giant milkweed can resprout year after year when burnt or cut (Ashwani Kumar, 2009).

Seed storage : Seed storage behaviour is orthodox. It is beneficial to dry the seeds in sunshine for 3-4 days before storage to avoid the fungal infection and other moisture related problems.

Seeds of Calotropis can be stored for one year and no reduction in percent seed germination was recorded after one year of storage (Ashwani Kumar, 2009).

Effects on crop production : Giant milkweed does not form dense stands, normally occurring as scattered individuals. It can be a serious weed in pastures, overgrazed rangelands, and poorly managed hay fields.

Other uses : Fruit fibres and seed hairs may be used for filling cushions and for making rope.

Soil improver: A source of green manure. The plant can help improve soil water conditions and also acts as a soil binder.

Pollution control: C. procera is an ideal plant for monitoring sulphur dioxide emissions in the air.

Other uses: A suitable indicator of exhausted soil.

The flowers are considered to be sweet, bitter, digestive, stomachic, tonic and useful in cough, asthma and catarrh. Flowers are also employed in skin diseases, worms, mucus, haemorrhage caused by the mutual and simultaneous vitiation of rakta and pitta (Aiyer and Kolammal, 1953; Nadkarni, 1954; Sharma, 1969).

Past work done on Seed Science and Technology

Studies on seed germination of *Calotropis procera*

Ashwani Kumar (2009) reported that germination studies have been carried out on seeds collected from Jaisalmer, Jaipur, Banthala, Nagaur and Dausa. 100 seed weight varied significantly in seeds collected from different places. Maximum 100 seed weight (0.84 g) was recorded for seeds collected from Jaisalmer which was followed by Nagaur (0.51 g), Lalsot, Dausa (0.450 g), Jaipur (0.49 g), Banthala (0.49 g). Likewise percent germination also varied in the seeds collected from Jaisalmer (98%), which was followed, by seeds collected from Nagaur (96%), Jaipur (94%), Bantahala (88%), and Lalsot (86%). Seed size was also variable for seeds collected from different locations.

REFERENCES

Aiyer K.N. and M. Kolammal. (1953). Editors. Pharmacognosy of Ayurvedic Drugs, Series 1. 1st ed. Trivandrum: The Central Research Institute;

Ashwani Kumar. (2009). *Calotropis procera* : a potential plant for hydrocarbons. *Earth Sciences.*

Howard, R.A. (1989). Flora of the lesser antilles, leeward and windward islands. Dicotyledoneae. *Arnold Arboretum, Harvard University*, Jamaica Plain, 6: 658.

Liogier, H.A. (1995). Descriptive flora of Puerto Rico and adjacent islands. *Editorial de la universidad de puerto rico*, San Juan, 4: 617.

Little, E.L., Jr., R.O. Woodbury and F.H. Wadsworth. (1974). Trees of Puerto Rico and the virgin islands. *Agriculture Hand book* 449. U.S. Department of Agriculture, Washingtion, 2:1,024.

Nadkarni, A.K. (1954). *Indian Materia Medica*, Bombay. pp. 408, 1476.

Sharma, B.M. (1968). Root systems of some desert plants in Churu, Rajasthan. *Indian Forester* 94(3): 240-246.

Websites

http://www.science20.com/humboldt_fellow_and_science/blog/calotropis_procera_potential_plant_hydrocarbons

TANNER'S CASSIA (*Cassia auriculata*)

Common name : Tanner's cassia

Botanical name : Cassia auriculata

Tamil name : Aavaarai

Hindi name : Tarvar, Tarval

Family : Caesalpinioideae (Leguminosae)

Origin and distribution : Plant occur throughout India and Srilanka. Distribution of the plant is pan-tropical. It is a small bush, which grows wild in South India with flowers and pods throughout the year (Matthew, 1983).

Botany

Habitat : A gregarious weed on black cotton soil, dry, stony and hilly land. It thrives well along roadside, in degraded forests and on wastelands.

Habit : Tall, branching shrub; branches and underside of leaves grayish (Fig. 1).

Fig. 1. Plant

Stem : Branchlets finely pubescent.

Leaf : Nearly sessile, pinnate, with subulate glands between all the leaflets, leaflets 8-12 pairs, elliptic obtuse, rigidly subcoriaceous; stipules cordate, foliaceous.

Floral biology

Flower : Flowers large, yellow, showy, in dense corymbs; calyx 5-lobed, tube short; corolla imbricate, petals 5, subequal; stamens 10, 6-7 fertile, rest staminodes; ovary superior, one celled, many ovuled, style incurved, stigma truncate (Fig. 2).

Fruit : Pod flat, papery, obtuse at each end, stalked, few seeded, ligulate, straight, glabrous, flexible with a distinct space between the uniseriate seeds (Fig. 3).

Seed biology

Seed : Seeds are minute, quadrangular bluntly pointed with 5-6 shallow transverse furrows. The average weight of the seed is 0.13 g (Fig. 4).

Fig. 2. Flower

Fig. 3. Fruit

Fig. 4. Seed

Propagation : By seed and stem cuttings.

Seed dispersal : Seeds are dispersed when the legume fruit splits open.

Seed dormancy and germination : For quick germination, seeds are scarified and held in running water. The seedlings are fairly resistant to desiccation. Stem cuttings are planted 5 – 12.5 cm apart in rows.

In most Leguminosae, the inner ovular integument is completely destroyed during seed development and does not take part in seed coat formation (Prakash, 1987).

Uses : The flowers and seeds of *Cassia auriculata* has been reported to show a very significant antidiabetic effect. It is the source of yellow coloured dye, obtained from its flowers and seeds (Chandramouli, 1995).

It also possesses medicinal properties: the bark is astringent, leaves and fruits anthelmintic, seeds used in eye troubles and root employed in skin diseases. It has been used for the treatment of ulcers, leprosy and liver disease (Kumar *et al.*, 2002).

REFERENCES

Chandramouli, K.V. (1995). Sources of natural dyes in India. A compendium with regional names. *PPST Foundation*, Adayar, Madras Tamil Nadu, India.

Kumar R.S., M. Ponmozhi and M. Nalini. (2002). Effect of *Cassia auriculata* leaf extract on lipids in rats with alcoholic liver injury. *Asia Pacific J. Clin. Nutr.*, 11(2): 157-163.

Matthew, K.M. (1983). The Flora of the Tamil Nadu Carnatic. *The Diocesan Press*, Madras, Tamil Nadu, India.

Prakash, N. (1987). Embryology of the Leguminosae. *In*: Advances in legume systematic. [Ed.] Stirton C.H. Royal Botanical Gardens, Kew, UK. 241 – 278.

COFFEE SENNA (*Cassia occidentalis*)

Common name : Coffee senna

Botanical name : Cassia occidentalis L.

Tamil name : Peyavarai, Thagarai

Hindi name : Kasaundi, Agauthi

Family : Caesalpinioideae(Leguminosae)

Origin and distribution : Naturalized in Eastern USA, distributed around the world.

Botany

Habitat : An introduced weed from West Indies, now wild in waste lands, along roadsides and dry places.

Habit : Diffuse, sub glabrous, annual under shrub, growing 2-5 feet high (Fig. 1).

Fig. 1. Plant

Stem : Stem is four sided when young becoming round with age.

Leaf : Leaves pinnate, leaflets 6-10, ovate to oblong, acuminate, gland single, just above the base of the petiole, none between the leaflets.

Floral biology

Flower : Flowers yellow, in short peduncle, few flowered, corymbs or panicles; bracts thin, ovate acuminate, caduceus; calyx tube short, sepals obtuse, glabrous, petals 5, conspicuously veined; stamens 10, 6 – 7 fertile, others staminodal; ovary superior, one celled, many ovuled (Fig. 2).

Flowering and pollination : Coffee senna flowers and fruits throughout the year or seasonally, depending upon the rainfall and cold seasons.

Flowers of the Cassia usually do not secrete nectar as the pollen is shed through short slits or pores in the anthers and the pollen is released through the vibration of the flowers by bees during 'buzz pollination' (Gottsberger and Silberbauer-Gottsberger, 1988).

Fruit : A long pod (legume), straight to slightly curved, from 3 to 5 inches in length, without hairs (glabrous) or minutely pubescent, The seed pods are dark brown and curve slightly upward; The seeds, which are in long pods, can be roasted and made into a coffee like drink (Fig. 3).

Seed biology

Seed : The seeds are sub-cylindrical, greenish brown, smooth glabrous and flattened on both ends (Fig.4).

Fig. 2. Flower

Fig. 3. Fruit

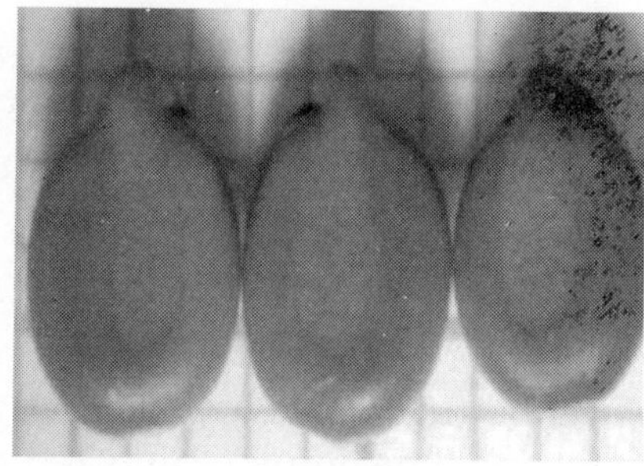

Fig. 4. Seed

Propagation : By seed.

Seed dispersal : Long distance seed dispersal in nature is mostly by stream flow, water movements over the soil surface or in mud attached to the feet and fur of animals. In weedy situations seed can be moved in mulch, in mud on machinery and vehicles and on footwear. Cattle, horses (Anning *et al.*, 1989) and goats will nibble the pods of Cassia and ingest the seed, some of which survive passage through the gut and are spread in the dung (Anning *et al.*, 1989, Anon., 1989). The seeds are dispersed by grazing animals (Sánchez and Uranga, 1993).

Life cycle : In seasonally cold or dry climates, the life cycle of coffee senna is completed in 6 to 9 months, but in warm, continually moist areas, plants may last a full year or grow through the second year.

Seed dormancy and germination : Scarification is necessary for good seed germination. Mechanical scarification, acid treatment, and immersing in boiling water all worked well, giving 82 to 100 percent germination (Demel and Teketay, 1996). The seeds are dispersed by grazing animals (Sánchez and Uranga, 1993).

Germination characteristics were similar in the range of 15 to 36°C. Germination of coffee senna was drastically reduced at 39°C by Rhizophus sp. Seedling growth was maximum from 30 to 36°C. It has hard seed coat. Abrasion for 5, 10, or 15 s, puncturing with a needle, and scarification using sulphuric acid were all effective in initiating germination. It emerged from a soil depth of 12.5 cm; however, emergence of coffee senna was 63% and emerging from 2.5 cm after only 3 days (Teem *et al.,* 1980).

Seeds germinated equally with or without light, and pretreatment with red or far-red light did not affect germination. Optimum temperature for germination was 25°C, and a high germination percentage (>70%) occurred from 12.5 to 30°C. The low temperature threshold for germination was between 10 and 12.5°C, whereas the upper threshold was near 45°C. Coffee senna germination in response to moisture stress and solution pH differed at 15 and 30°C. At 0.4 MPa, no germination occurred at 15°C, whereas 15% germination occurred at 30°C.

Untreated seed of foetid cassia species has negligible germination (2%), but acid scarified seed gave almost total germination (Cock and Evans, 1984).

Fire scarifies the seed and in Australia has caused mass emergences of seedlings following good rain (Anning *et al.*, 1989).

During seed development, the nucellar tissue mainly disappears. A few nucellar layers were observed between the cellular endosperm and the testa, both in mature seeds of Senna.

Effect of crop production : While coffee senna can be a weed of cultivated fields and plantations, it is principally a problem because it accumulates in heavily grazed pastures. Poisoning of pigs fed coffee senna seeds resulted in muscle necrosis (Timm and Riet-Correa, 1997).

Uses : Seed of coffee senna are roasted and used as a coffee substitute (Guzmán 1975). Ingestion of large amounts of seeds has been implicated in deaths of cows, horses, and goats (Raintree 2002).

Past work done on Seed Science and Technology

Germination of *Cassia occidentalis* seed at different osmotic potential levels

According to Maria Elena Aparecida Delachiave and Sheila Zambello de Pinho, (2003), the effect of osmotic potential and salinity on the germination of seeds of *Senna occidentalis*, with and without a change of solutions. The percentage and rates of germination decreased according to decrease of potential, being more drastic when the substitutions of solutions were not made. The largest reductions were observed with the PEG solutions, with and without substitutions.

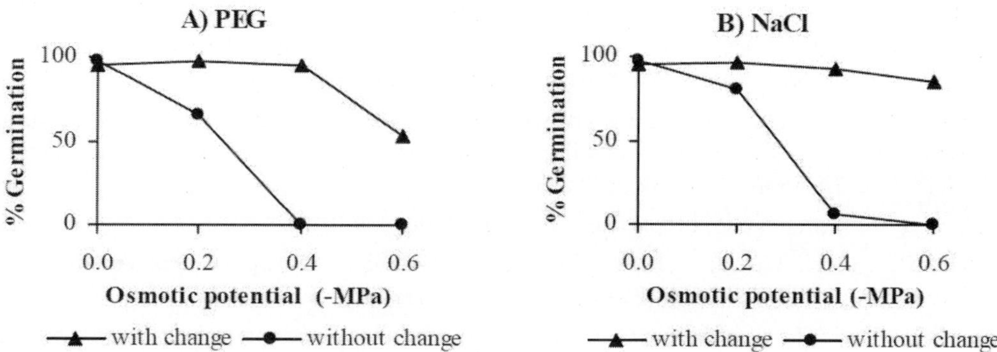

Fig. 5. Average percentage of germination of seeds of *Cassia occidentalis*, submitted to different osmotic potentials with and without a change of solution, induced by PEG (A) and NaCl (B),under light, at 25°C.

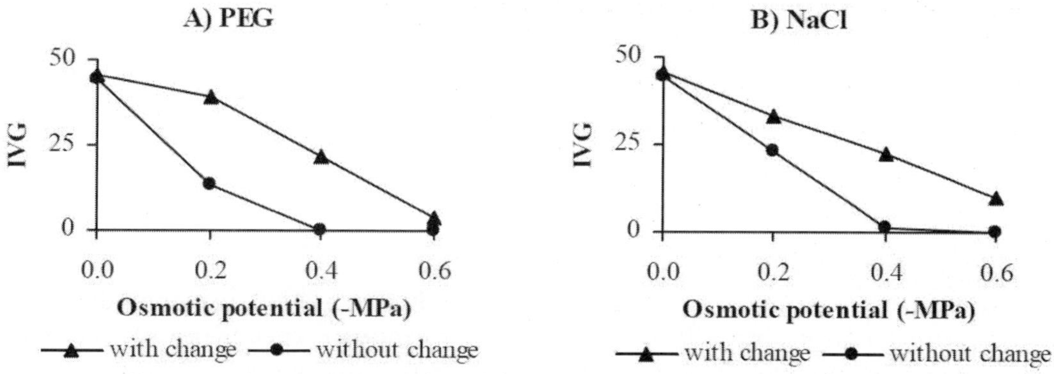

Fig. 6. Average velocity of germination (IVG) of *Cassia occidentalis* seeds, submitted to different osmotic potentials with and without change of solution, induced by PEG (A) and NaCl (B),under light, at 25°C.

REFERENCES

Anning, P., H.G. Bishop, G. Lambert and G. Sutherland. (1989). Sicklepod replaces overgrazed tropical pastures. *Queensland Agricultural Journal,* 115: 188-192.

Anon. (1989). Sicklepod - another problem weed for the tropics. *BSES Bulletin* 27: 18-19.

Cock, M.J.W. and H.C. Evans. (1984). Possibilities for biological control of *Cassia tora* and *Cassia obtusifolia. Tropical Pest Management* 30: 339-350.

Demel Teketay. (1996). The effect of different pre-sowing seed treatments, temperature and light on the germination of five Senna species from Ethiopia. *Biomedical and Life Sciences.* 11(2): 155-171

Gottsberger, G. and I. Silberbauer-Gottsberger. (1988). Evolution of flower structures and pollination in neotropical Cassiinae (Caesalpiniaceae) species. *Phyton (Horn)* 28: 293-320.

Guzman, D.J. (1975). Especies utiles de la flora salvadoreña. *Ministerio de Educación, Dirección de Publicaciones,* San Salvador, El Salvador. 703.

Maria Elena Aparecida Delachiave and Sheila Zambello de Pinho. (2003). Germination of *Cassia occidentalis* Link: Seed at different Osmotic Potential levels. *Brazilian Archives of Biology and Technology.* 46: (2):163-166

Rain Tree (2002). Fedegoso. http://rain-tree.com/ fedegosa.html.

Sanchez, P. and H. Uranga. (1993). Plantas indeseables de importancia economica en los cultivos tropicales. *Editorial Científico-Técnica,* Havana, Cuba. 166.

Teem, D.H., C.S. Hoveland and G.A. Buchanan. (1980). Sicklepod (*Cassia obtusifolia*) and coffee senna (*Cassia occidentalis*) geographic distribution, germination and emergence. *Weed Sci.,* 28: 68-71.

Timm, C.D. and F. Riet-Correa. (1997). Plants toxic to pigs. *Ciencia Rural* 27(3): 521-552.

LAMB'S QUARTERS (*Chenopodium album*)

Common name : Lamb's quarters

Botanical name : Chenopodium album

Tamil name : Paruppukkirai

Hindi name : Bathua or Bathuwa

Family : Amaranthaceae

Origin and distribution : Plants native in Eastern Asia are included under *Chenopodium album*, but often differ from European specimens. It is widely introduced elsewhere, e.g. Africa, Australasia, North America, and Oceania.

Botany

Habitat : Fields, pastures, agronomic and vegetable croplands, gardens, orchards, vineyards, landscaped areas, roadsides, and other disturbed locations.

Habit : It tends to grow upright at first, reaching heights of 10 - 150 cm (rarely to 3 m), but typically becomes recumbent after flowering (due to the weight of the foliage and seeds) unless supported by other plants (Fig. 1).

Stem : The stems are stout, angular, and variably coloured.

Leaf : The leaves are alternate and can be varied in appearance. 1 - 5 cm long and 0.4 - 2 cm broad; they are waxy-coated, unwettable and mealy in appearance, with a whitish coat on the underside.

Floral biology

Flower : The small flowers are radially symmetrical and grow in small cymes on a dense branched inflorescence 10 - 40 cm long. Tiny, green, stalkless flowers are packed in dense clusters at the tips of the main stem and branches. The tiny flowers lack petals, and like the leaves, are also covered in a white powdery coating (Fig. 2).

Fig. 1. Plant

Fig. 2. Flower

Flowering : Flowers bloom from May to November. Tiny, green, stalkless flowers are packed in dense clusters at the tips of the main stem and branches. The tiny flowers lack petals, and like the leaves, are also covered in a white powdery coating.

Pollination : Flowers are wind-pollinated; they are visited by few insects.

Fruit : Fruits are enclosing with flower. 1-1.5 mm diameter; wall ± rough at 20X. Adherent to seed (Fig. 3).

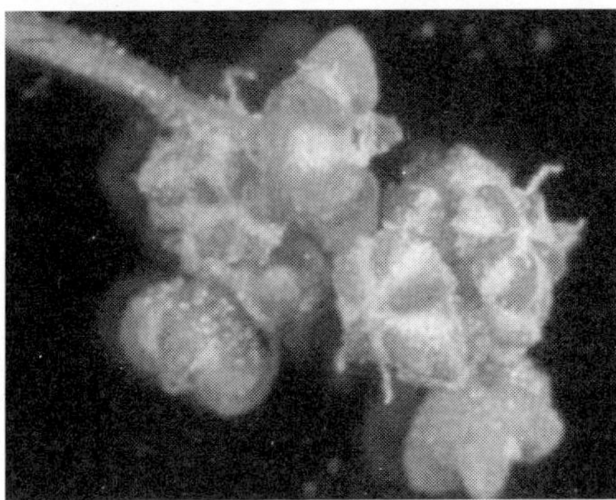

Fig. 3. Fruit

Seed biology

Seed : Each plant may produce thousands of tiny, black, shiny seeds with rounded edges. If they seem dullish brown rather than shiny and black, it is due to the papery covering. Seeds are tiny and plants produce two types: smaller black seeds and larger brown ones. The seeds are enclosed by a membranous pericarp that is easily removed. The seed is ovate, with convex sides and acute, keeled margins. The testa has faint radial striations but is otherwise smooth. Seeds range from 1.1-1.5 mm in diameter (Fig. 4) (Williams, 1963).

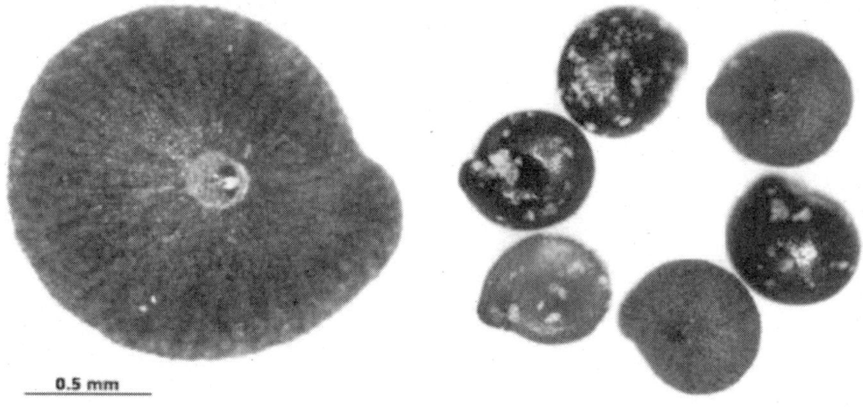

Fig. 4. Seed

Propagation : By seed

Seed dispersal : Seed dispersal through birds and squirrels.

Seed dormancy and germination : According to Bozena Sera *et al.* (2008), the seeds show considerable polymorphisms. There are two different seed forms on one plant: brown seed and black seed. The brown seed has a thin seed coat and can germinate after the harvest, while the black seed has a thick seed coat and is dormant. Brown seeds are larger than black seeds. The surface of the black seeds is either smooth or reticulate.

The length of primary dormancy of *C. album* seeds differs between various populations and can also vary between individuals of the same population (Holm *et al.*, 1977).

Seed longevity : The seeds can remain viable in the soil for 40 years. *C. album* seeds remaining in the soil for 20 years may still have a 23% viability (Harrison, 1990)

Effects on crop production : *C. album* is vulnerable to leaf miners, making it a useful trap crop as a companion plant. Growing near other plants, it attracts leaf miners which might otherwise have attacked the crop to be protected. It is a host plant for the beet leaf hopper, an insect which transmits curly top virus to beet crops.

Uses : The leaves and young shoots may be eaten as a leaf vegetable; either steamed in its entirety or cooked like spinach, but should be eaten in moderation due to high levels of oxalic acid. Seeds are high in protein, vitamin A, calcium, phosphorus, and potassium.

Past work done on Seed Science and Technology

Dormancy breaking treatments

Jursik *et al.* (2003) expressed that germination in the light, *C. album* was 67%, whereas in the dark it was only 12%. *C. album* germinated in a wide temperature range (5–33°C). At 5°C, 4% germination was found, but the increase of germination with increasing temperature was only slow. Significantly higher germination was registered at about 14°C (39%). The highest germination value (75%) was achieved at 18°C. An additional temperature increase caused a decrease of germination at 24°C. Only 13% germination was recorded at 30°C. The seeds of *C. album* showed low germination energy at 15 and 24°C. Only at 33°C, the germination energy was high, but the total germination was only 26.5%. The first seeds germinated within 24 hours at 24°C and 33°C (3.5% and 6.5%, respectively). At 24°C, 66.25% of seeds germinated within six days. The following increase of germination was not significant ($\alpha = 0.05$). A very low germination energy was recorded at 15°C. The first seeds (3%) started to germinate on the third day. The increase of germination was very low and a higher increase occurred only after 7 days. 58.5% of seeds germinated after 11 days and further increase was insignificant ($\alpha = 0.05$).

Duration and strength of seed dormancy

The seeds of *C. album* matured in August and October showed a significant differentiation in the duration and strength of dormancy (Fig. 5). The earliest ripened seeds (August) revealed a strong primary dormancy (50-100 days). The following increase of germination was slow with an average germination of 27%, 150 days after ripening. The seeds of *C. album* ripened in October were partially dormant after the harvest (germination 14.5%) but 50 days after the harvest, germination of these seeds achieved 72%. Such germination level was preserved till the end of the tests in both years.

Bozena Sera *et al.* (2008) reported that the seeds under plasma irradiation of 12 - 48 minutes started to germinate before the first data monitoring. These seeds started to germinate during the first 24 hours of the germination test. The seeds with 6 minutes of plasma irradiation and the seeds in the control conditions started to germinate before the second data monitoring (between 24 hours and 48 hours of the test). On the fourth day of the germination test a significant difference between the plasma treated seeds and the control ones was observed (Fig. 6). The major differences between the control and plasma treated seeds were found also on the sixth day. Only 15% of the control seeds germinated while at least 55% of the plasma treated, for 30 minutes and 48 minutes, germinated, i.e., three times more. Germination of the seeds also depended on the duration of plasma exposure.

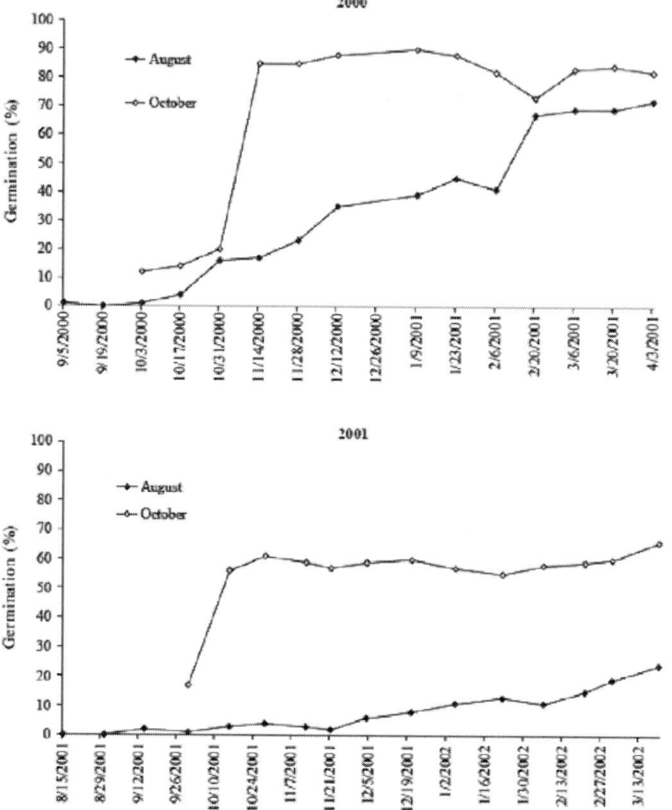

Fig. 5. Dormancy in seeds of *C. album* from different collections

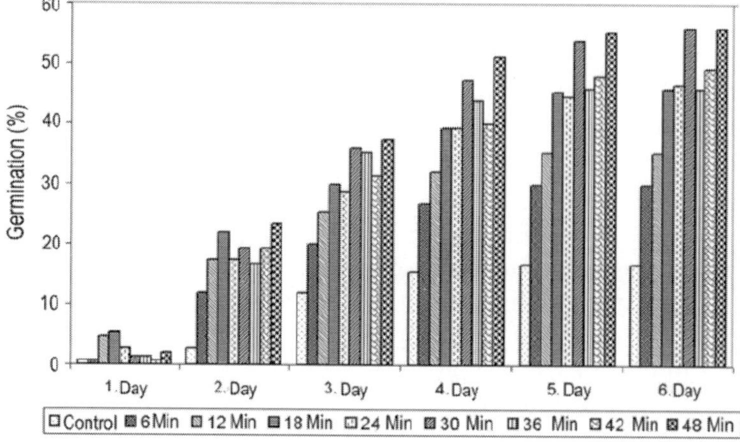

Fig. 6. Percentage of the germinated seeds of *Chenopodium album* treated with plasma for different durations.

The major differences in the sprout accretion were found only on the first day of seed germination (Fig. 7). Significant differences between the control and the seeds with plasma irradiation for 6 minutes to 48 minutes were observed. The greatest difference was found between the seeds treated for 48 minutes (6.4 mm) and the control (3.1 mm). The mean sprout accretion was the most obvious from all sprout characteristics. The mean sprout accretion significantly differed between the control and the seeds treated by plasma for 12 - 48 minutes (on the first day) and for 18 - 48 minutes (from the second to the fourth day cultivation). The highest difference was observed between the control and the 48- minute plasma treated seeds (13.9 mm, while the control was 4.6 mm) on the fifth day. The sprout growth process is shown in Fig. 8.

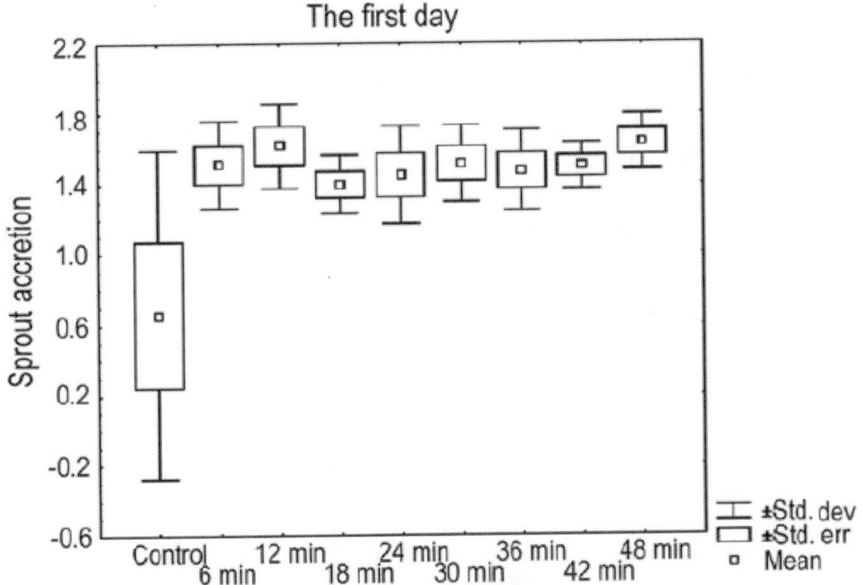

Fig. 7. Sprout accretion of growing seeds of *Chenopodium album* treated with plasma on the first day of cultivation. Standard deviation and standard error of the mean are shown. Significant differences between the control and treated seeds were registered (Tukey HSD test; for details, see the text). The arcsin transformation is used before the figure creation.

The SEM of seed coats of control and treated seed showed that the plasma treated seeds have an etched/eroded surface. It was suggested that the seeds with an etched/eroded seed coat would germinate faster than healthy ones (or than those with a thick seed coat). Changed seed surface in plasma treated seeds could probably enhance and accelerate water uptake. The seeds can also change their dormancy and germination processes. The process of etching has an analogous effect to the seed scarification (e.g., in the digestive tract of the bird, the seeds with a thick seed coat).

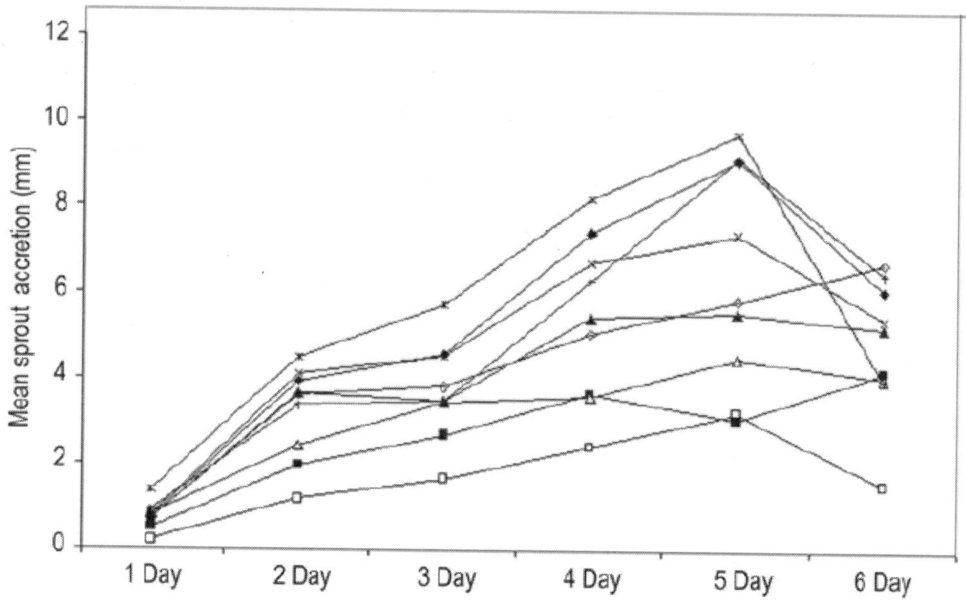

Fig. 8. Mean sprout accretion of growing seeds of *Chenopodium album* treated with plasma for different times of exposure. Standard deviation and mean are shown

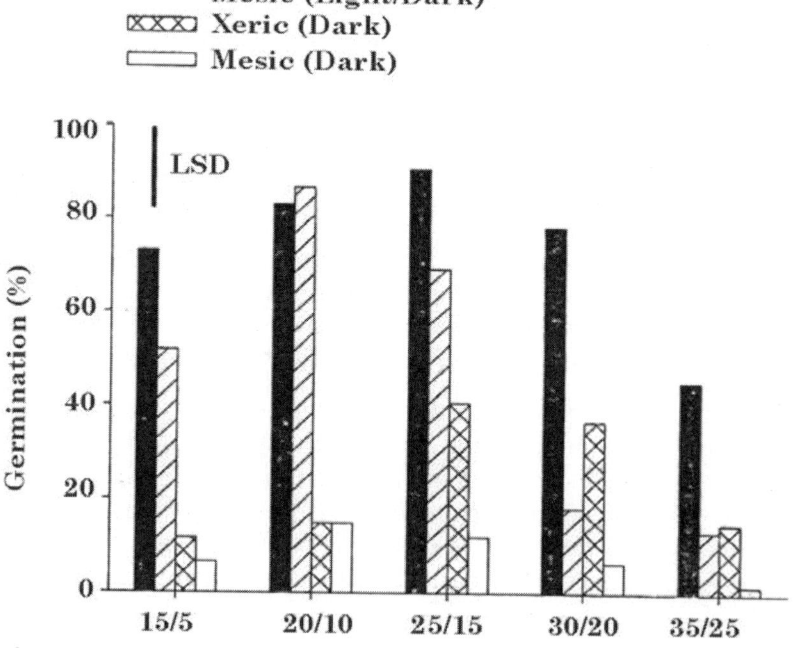

Fig. 9. Effect of temperature and light on seed germination of common lambsquarters xeric and mesic populations

Eslami (2011) reported that germinability was significantly greater under light/dark compared to continuous darkness in both populations (Fig. 9). Germinability of the xeric population was greater than that of the mesic population in almost all experimental scenarios except 20/10°C, which indicated that this temperature regime (20/10°C) provided the best conditions for germination of the mesic population. Xeric and mesic populations were best germinated at 25/15°C and 20/10°C, respectively. Germination of both populations declined at temperatures above 25/15°C. However, the reduction was more severe in mesic than in xeric population, indicating the greater susceptibility of mesic population germination to high temperatures. This may be because mesic populations are typically not exposed to high temperatures during spring and early summer.

Salinity significantly decreased the germinability of both populations; however, the magnitude of its impact depended on population and salinity level. For both populations, the three-parameter logistic model provided a satisfactory fit for the response of seed germination to NaCl concentration (Fig. 10). Germination of xeric population was about 80% up to the salinity level of 20 dS m^{-1} and its germinability was >40% even at 30 dS m^{-1}. Increasing salinity level to 35 dS m^{-1} sharply decreased xeric population germination (<2%) and germination was completely ceased at 40 dS m^{-1}. Mesic population germinability was lower than that of xeric population at all salinity levels. The mesic population exhibited a germination percentage greater than 35% at salinity level of 20 dS m^{-1}. Further increase of salinity to 25 dS m^{-1} dropped mesic population germinability to less than 2% and its germinability was completely hindered bu a salinity level of 35 dS m^{-1}. The parameter x_{50} that represents the salinity level for 50% inhibition of the maximum germination was 29.6 and 13.8 dS m^{-1} for xeric and mesic populations, respectively. The x_{50} values demonstrate the greater salt tolerance in xeric compared with mesic population at germination stage.

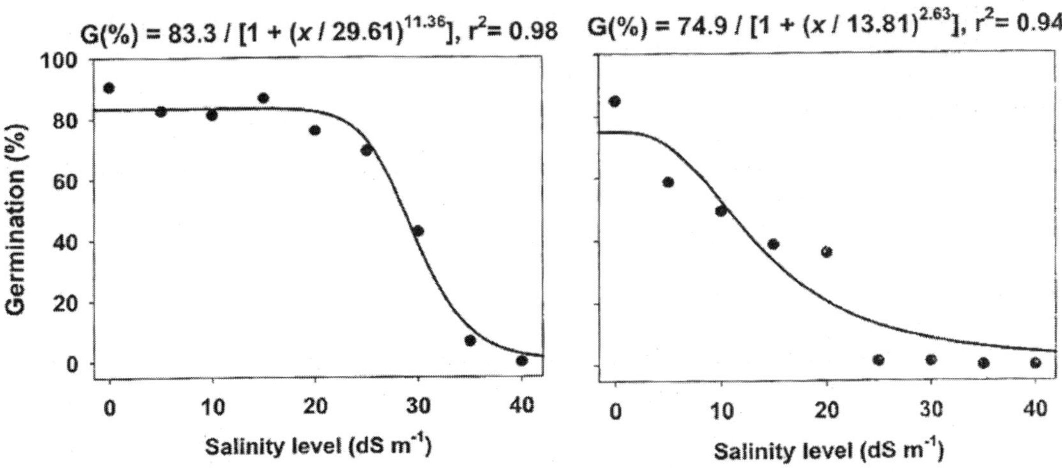

Fig. 10. Effect of NaCl concentration on seed germination of common lambsquaters xeric (left) and mesic (right) populations.

Depth of burial strongly limited seedling emergence of both populations. Germination was noted the sixth day after establishing the experiment for xeric population seeds placed on the soil surface (27%) as well as xeric population seeds sown at 0.5 cm (21%). Increasing the seeding depth decreased emergence percentage of both populations (Fig. 11). The greatest emergence of xeric (77%) and mesic (70%) populations occurred for seeds placed on the soil surface and no seedlings emerged from burial depth 3 cm. Mesic population emergence percentage, however, was greater than that of xeric population from depths deeper than 0.5 cm. This might be due to the larger size of seeds of mesic population compared with that of xeric population. Emergence of weeds is affected by the amount of food reserves in seed and the depth of burial in the soil. The sigmoidal decline model was best fitted to the data of xeric and mesic population emergence in relation to seeding depth (Fig. 11). According to the model, the seeding depth that decreased xeric and mesic population emergence by 50% was 0.64 and 0.93 cm, respectively.

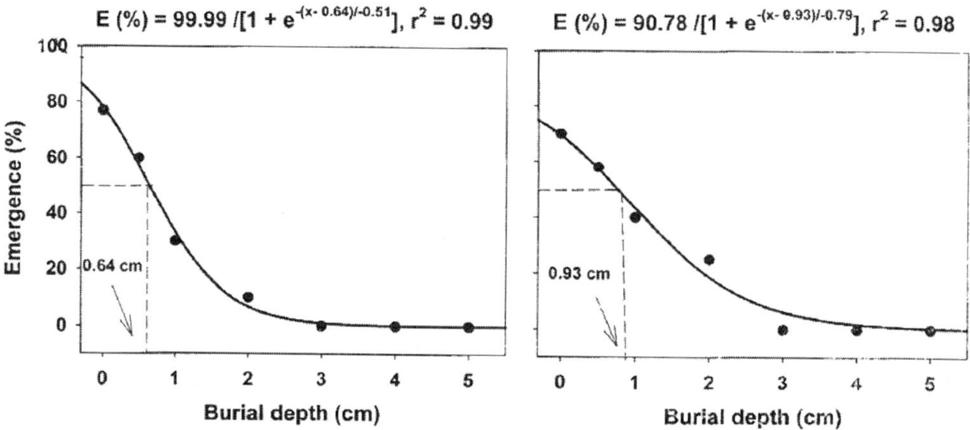

Fig. 11. Effect of seed burial depth on seedling emergence (%) of common lambsquarters xeric (left) and mesic (right) populations.

Effect of moisture stress

Sauernorn *et al.* (1988) found that germination of several tropical weeds decreased with a decreasing osmotic potential, but *Chenopodium album* seed were able to germinate at -1.0 MPa (Wei *et al.*, 2009).

REFERENCES

Bozena Sera, Vitezslav Stranak, Michal Sery, Milan Tichý and Petr Spatenka. (2008). Germination of *Chenopodium album* in response to microwave plasma treatment. *Sci. Technol.*, 10: 506.

Eslami, S.V. (2011). Comparative germination and emergence ecology of two populations of common lambsquarters (*Chenopodium album*) from Iran and Denmark. *Weed Sci.*, 59(1): 90-97.

Harrison, S.K. (1990). Interference and seed production by common lambsquarters (*Chenopodium album*) in soybeans (*Glycine max*). *Weed Sci.*, 38:113–118.

Holm, L.G., L.D. Plucknelt and Pancho J. Herberger (1977) World's worst weeds. Distribution and Biology. *University press of Hawaii*, Honolulu: 280–284.

Jursík, M., J. Soukup, V. Venclová and J. Holec. (2003). Seed dormancy and germination of shaggy soldier (*Galinsoga ciliata* Blake.) and Common lambsquarter (*Chenopodium album* L.). *Plant Soil Environ.*, 49 (11): 511–518.

Sauerborn, J., W. Koch and J. Krage. (1988). On the influence of light, temperature, depth of burial and water stress on the germination of selected weed species. *Z.pflanzenk and Pflanz,* 11: 47-53.

Wei, S., C. Zhang, X. Li, H. Cui, H. Huang, B. Sui, QinghuiMeng and H. Zhangm. (2009). Factors affecting buffalobur (*Solanum rostratum*) seed germination and seedling emergence. *Weed Sci.*, 57: 31-35.

Williams, J.T. (1963). Biological flora of the British Isles. *Chenopodium album* L. *J. of Eco.*, 51: 711-725.

SWOLLEN FINGERGRASS (*Chloris inflata*)

Common name : Swollen fingergrass

Botanical name : Chloris inflata Link (Syn. *Chloris barbata* Sw.)

Tamil name : Mayil kondai pul

Family : Poaceae

Origin and distribution : Originating in Tropical America and distributed in Indian continent, China, S. E. Asia and Java.

Botany

Habitat : This is a good fodder grass, also known to serve as alternate host for shoot fly of sorghum. A common dry land grass.

Habit : A tufted perennial, commonly found as a weed in dry cultivated fields and waste places (Fig. 1).

Stem : Culms up to 60 cm high.

Leaf : Leaves 7.5 to 12.5 cm long.

Floral biology

Flower : Inflorescence composed of raceme. Inflorescence with 6 to 15 digitately arranged spikes; spikelets usually purplish in colour (Fig. 2).

Fig. 1. Plant

Fig. 2. Flower

Flowering : Flowering and fruiting throughout the year.

Seed biology

Fruit / Seed : Caryopsis with adherent pericarp, fusiform, dorsally compressed, trigonous, estipitate, without sulcus, 1.2 mm long, smooth, apex unappendaged. Embryo 0.75 length of caryopsis. Hilum is punctiform. Endosperm farinose (Fig. 2). The ovary of a grass develops into a one-seeded grain (*caryopsis*). Brown, oval, 2 mm long by 0.6 mm wide. The seed weight is about 0.5 mg or 2 million seeds kg^{-1} (Fig. 3).

Fig. 3. Fruit (Seed)

Propagation : Reproduction by seeds and vegetatively. The reproductive stage of grass species *Chloris barbata* is during September-November. Post rainy months favoured robust fibrous root production in all the grass species.

Seed dispersal : Dispersal of seeds in the fur of large herbivores (epizoochory)

Life cycle : Perennial or occasionally annual. Seed germinates 1-7 days after planting or in spring and it grows rapidly over summer. Seed heads are formed in summer. Stolons may form in the first and subsequent summers to spread the plant. It makes little growth over the cooler winter period. In some environments with long dry periods or cold winters, it may act as an annual.

Seed dormancy and germination : One study was to evaluate light influence on the germination of *Chloris barbata* (L.) Sw. seeds. Thus, two experiments were carried out in BOD incubators at 30°C. In the first one, the experimental design was in a 3 × 2 factorial arrangement, i.e. three light regimes (dark, dark with two 10 min periods of exposure to green light and dark with 10 min periods of exposure to white light), combined with two substrate moistening substances (water and KNO3 solution). In the second experiment, design was in a 2×2×2 factorial arrangement, consisting of two light conditions (present and absent), two dispersal units (spikelets and caryopsis) and two moistening substances (water and KNO_3 solution).

The germination percentage of chloris seeds kept in the dark on water-moistened substrate was 4%, whereas the exposure to two white light fluxes increased germination to 12% in the water-moistened substrate and to 23% when moistened with KNO_3. Non-germinated seeds kept their viability, since 98 to 100% of them germinated up to the seventh day after transference to an environment with daily eight hours of light.

In the second experiment, irrespective of the dispersal unit, the final germination percentage was higher in the presence of light (spikelet and caryopsis), peaking up to 100%; in the dark, 8 and 30% maximum values were obtained for seeds moistened with water and KNO_3, respectively. Also in the dark, germination of caryopsis was 7% higher than that of spikelets. The presence of caryopsis coats delays the speed of the process, but the final germination is not changed. Green light does not activate the seed germination process and can be used as safety light in assays involving dark treatments; under such a condition, KNO_3, stimulates the partial germination of seeds (Jose luiz da Silva *et al.*, 2009)

Control measures : Fire and grazing are not very effective on established stands.

REFERENCE

Jose luiz da Silva, Sebastiao Carneiro Guimaraes and E. Oscar Mitsuo Yamashita. (2009). Germination of *Chloris barbata* (L.) Sw. Seeds under light. Revista de Revista de Ciências Agro-Ambientais, *Alta Floresta*, 7(1): 23- 34.

SURYAVARTI (*Chrozophora rottleri*)

Common name : Suryavarti, Rottler's Chrozophora

Botanical name : *Chrozophora rottleri*

Synonyms : *Chrozophora plicata*

Tamil name : Purapirakkai

Hindi name : Neel Kanthi, Shahdevi

Family : Euphorbiaceae

Origin and distribution : Chrozophora plicata occurs from Senegal East to Somalia and South throughout East Africa to Northern South Africa. It also occurs from Egypt and Saudi Arabia East to Tropical Asia.

Botany

Habitat : Gregarious in habit and a common black soil weed particularly in dry areas and road sides.

Habit : Erect annual herb, covered with sessile, stellate hairs, growing to a height of 75cm (Fig. 1).

Fig. 1. Plant

Stem : Ascending branches trichotomously forked from base.

Leaf : Leaves simple, alternate, cordate, base unequal, margin wavy, 3-lobed, both surfaces covered with stellate hairs; monoecious.

Floral biology

Flower : Flowers in axillary racemes, male crowded near the tip, female below, pedicelled; calyx 5 lobed, globes, valvate; petals yellow 5, free, alternate with sepals, short; disc of 5 glands, alternate with petals; stamens 5-15, filaments connate below;

ovary superior , 3 carpelled, syncarpous, 3 celled, ovule one in each cell, styles 3, each bifid, stigma red (Fig. 2).

Fig. 2. Flower

Flowering and fruiting : September – February

Fruit : Capsule reddish with dense stellate hairs giving a blue colour when wet (Fig. 3).

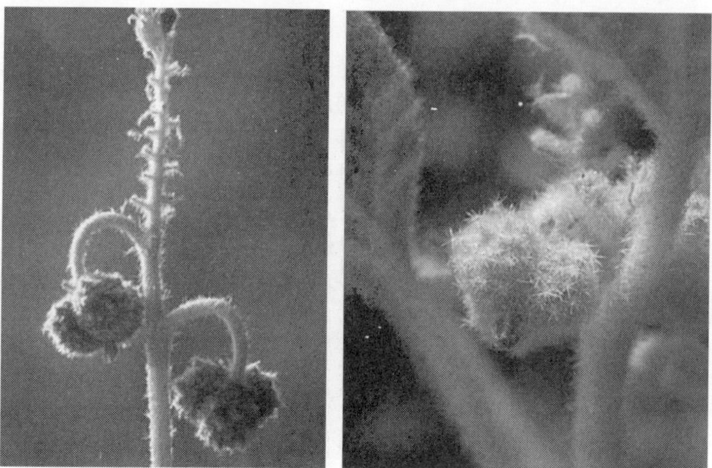

Fig. 3. Fruit

Seed biology

Seed : Seeds ovoid, 3.5 mm long, smooth or minutely dotted, pale or dark brown to blackish.

Propagation : By seeds.

Uses : In Sudan, pounded stems or whole plants are applied to wounds to improve healing. In Ethiopia, an infusion of the seeds and leaves is taken as a laxative. The plant is also used medicinally in Saudi Arabia, Pakistan and India, e.g. against jaundice and to purify blood.

In Senegal, the plant is not browsed by most stock, except occasionally by sheep and goats, as it causes vomiting and diarrhoea. In Kenya camels graze it. The fruits yield a purplish blue dye, which is used in East Africa to dye mats.

WILD JUTE (*Corchorus trilocularis*)

Common name : Wild Jute, African Jute

Botanical name : Corchorus trilocularis L.

Tamil name : Punnakku poondu

Hindi name : Kadukosta, Kadvapat, Hardikaket

Family : Tiliaceae

Origin and distribution : The primary center of origin and diversity of the wild taxa of Corchorus appears to be Africa.

Botany

Habitat : A common waste land weed.

Habit : An erect or semispreading annual fibrous herb differing from *Corchorus aestuans* in having hairy, pink, stem (Fig. 1).

Fig. 1. Plant

Stem : Fibrous, round, pinkish and covered with sparsely distributed tubercle - based hairs.

Leaf : Shining leaf surface; 3-4 angled, 3-4 valved.

Floral biology

Flower: Inflorescence called cymes. Flowers are yellow, opposite the leaves, either solitary in pairs. Bracts are filiform and subulate. The buds are ovoid-ovate and yellow. Stamens are numerous, yellow, inserted on a cushion-like structure called the torus. Ovary is superior, 3-4 celled and many ovuled (Fig. 2).

Flowering : A day for maximum flowering is 85-90 days.

Fruit : Fruit is an elongate, 3-4 valved, scabrous capsule 2-3 inches long, with a short beak (Fig. 3).

Seed biology

Seed : Seeds are small, trigonous, oblong – ovoid and black colour. Upto 1 mm in diameter, Seeds separated by septum, 100 seed weight (mg) 66.0 ± 1.00 (Fig.4).

Fig. 2. Flower

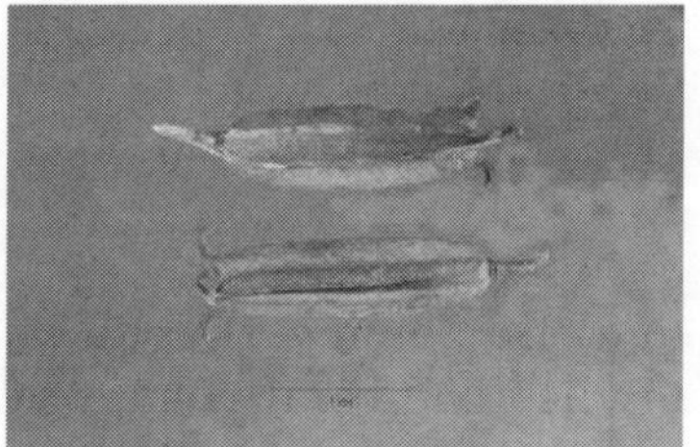

Fig. 3. Fruit

Fig. 4. Seed

Propagation : Seeds

Seed dispersal : Seed dispersal takes place through birds and animals.

Seed storage : The viability of *Corchorus trilocularis* ranges from 2 to 3 years. Since lower moisture content at storage may lead to hard seededness, which ultimately hinders germination, the accessions of wild *Corchorus* species should not be conserved below 10% moisture during medium to long term storage (Sinha *et al.*, 2011).

Effects on crop production : In pot culture, the weed residues incorporated at 100 g kg^{-1} of soil inhibited the germination and growth of onion. The weed residues reduced the plant height, root length and shoot and root dry weight in onion.

Uses : Stem fibres are used for rope making. To cure syphilis, root decoction is used for several days, while seeds are powdered.

REFERENCE

Sinha, M.K., C.S. Kar, T. Ramasubramanian, A. Kundu, S. Mahapatra and C. Kole. (2011). (Ed.), Wild Crop Relatives: Genomic and Breeding Resources, Industrial Crops, *Springer-Verlag* Berlin Heidelberg.

CYANOTIS (*Cyanotis cucullata*)

Common name : Cyanotis

Botanical name : Cyanotis cucullata Kunth.

Tamil name : Nayini-poondu

Family : Commelinaceae

Origin and distribution : Found in upland forest and rocky areas in Africa, Asia, India, S. E. Asia, Malaysia and Tropical Australia.

Botany

Habitat : Common weed in dry and garden lands, it thrives well under varied conditions of moisture.

Habit : Fleshy, stout annual herb; stem succulent, glabrous or sparsely hairy, lateral branches prostrate, trailing, rooting at the nodes (Fig. 1).

Fig. 1. Plant

Fig. 2. Flower

Stem : Round, green or pinkish, glabrous except for a longitidunial line of short hairs, and is full of slimy juicy.

Leaf : Leaves linear, sheathing, acuminate, sheaths ciliate, base broad or narrowed.

Floral biology

Flower : Flowers blue, clustered in the sheaths, axillary; bracteoles small, slender, linear-lanceolate, glabrous or minutely ciliate; sepals 3, subequal; petals 3, often united in a tube below, limb fimbriate; stamens 6, all fertile, filaments nearly naked, glabrous; ovary 3-celled, ovules 2 in each cell, style inflated below the apex (Fig. 2).

Flowering : Flowers from July to December in Java; flowers open only once for a few hours.

Fruit : Capsule 3 horned at the apex, depressed between the horns; seeds subquadrate, smooth or obscurely pitted.

Seed biology

Seed : Seeds 1 or 2 per valve, cylindric or quadrangular, reticulate or pitted; hilum at adjoining end of 2 seeds.

Propagation : Reproduced by seed and fragment of stem (stolons).

Seed dispersal : Seeds are dispersed through winds.

Uses : This species is used to treat boils and ascites. A plant is decoction of the whole plant is reportedly used in swellings above the abdomen. The plant is used as food for pigs.

INDIAN GOOSEGRASS (*Eleusine indica*)

Common name : Indian goosegrass

Botanical name : Eleusine indica

Tamil name : Thippa ragi, Kevuru

Hindi name : Jangali marua, jhingari

Family : Poaceae

Origin and distribution : Eleusine indica is of African origin and it is distributed throughout the warmer areas of the world to about 50 degrees latitude.

Botany

Habitat : It is widely cultivated throughout South India.

Habit : Goose grass grows in clumps from a central tap root that has a fibrous root system (Fig. 1).

Fig. 1. Plant

Stem : Hollow stems called culms, which are plugged (solid) at intervals called nodes, the points along the culm at which leaves arise.

Leaf : Leaves are a distinct white or silver color at the base, folded or flattened in shape, and smooth to the touch; very few hairs are present on the foliage.

Floral biology

Flower : Raceme. Seed heads composed of 2-13 spikes each 1 1/2 to 6 inches long, 3-7 mm wide, in clusters at the top of stems. Two rows of flattened spikelet occur along each spike (Fig. 2).

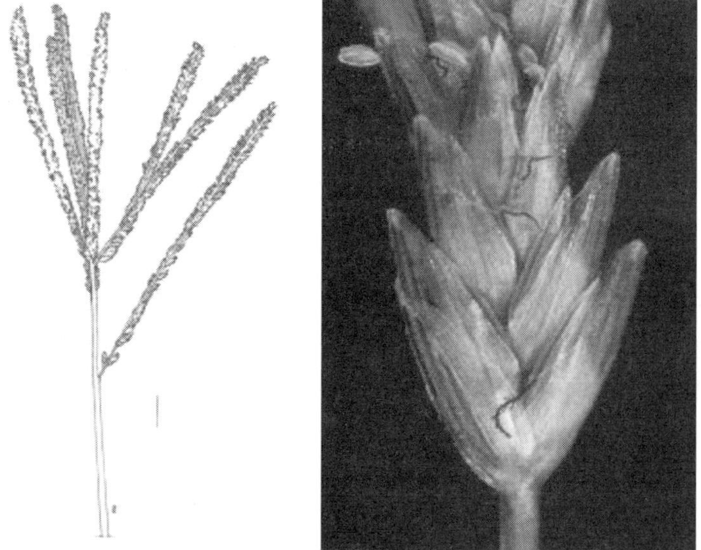

Fig. 2. Flower

Fruit : Each spikelet contains 3-6 light brown to black seeds that are 1-2 mm long (Fig. 3).

Fig. 3. Fruit

Seed biology

Seed : Seeds are dark brown. Seeds are arranged in a distinctive herringbone pattern along each spike (Fig. 4). Single plant can produce upto 50,000 seeds. As a result, failure to control goose grass can increase the quantity of goose grass seed in the soil seed bank. The seeds are eaten by wild and domestic animals and are occasionally grown for grain in Africa and India, but *Eleusine coracana*, finger millet, with somewhat larger seeds is far better for this purpose.

Fig. 4. Seeds

Propagation : By seed

Life cycle : Germination of goose grass seed will generally start sometime in late April and will continue throughout the summer. Time line for goose grass special control is given below.

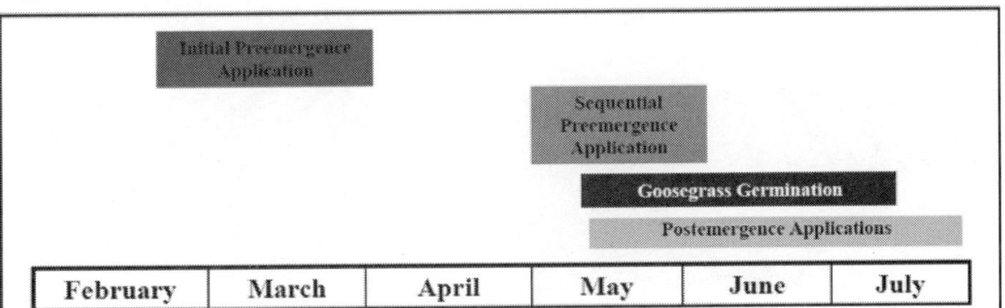

Seed dispersal : *E. indica* grows and flowers well in all seasons and a single plant may produce more than 50000 small seeds, which move readily by wind, in mud on the feet of animals and in the tread of machinery.

Seed dormancy and germination : Goose grass germination is suppressed by a lack of light and low temperatures.

Effect of crop production : The seed heads may contain high levels of cyanogenic glycosides and are believed to be responsible for occasional cases of stock poisoning.

Control measures : Chlordane (on clay) gave excellent control of goose grass when applied prior to germination. Goose grass can be suppressed by late season chemical treatments. DSMA appeared to be the most effective of the several chemicals used.

Chemical treatments applied in mid September gave upto 90 per cent reductions in viable seed set. This shows the fallacy in curtailing treatment of goose grass even though cool weather has slowed growth of this weed.

ASTHMA WEED (*Euphorbia hirta*)

Common name : Snakeweed, Asthma weed

Botanical name : Euphorbia hirta L.

Tamil name : Amman paccharisi.

Hindi name : Dudhiya

Family : Euphorbiaceae

Origin and distribution : Euphorbia hirta is native to Central America and a very common weed of the tropics and subtropics; it occurs throughout tropical Africa and also in South Africa.

Botany

Habitat : Most common dry land weed, flowering and fruiting all round the year on roadsides and waste lands.

Habit : Straggling, ascending annual hispid herb with copious latex; branches with yellow hair (Fig. 1).

Fig. 1. Plant

Fig. 2. Flower

Stem : Round and covered yellow crisp hairs; it is usually solitary and bent backwards at the top in characteristic way.

Leaf : Leaves obliquely elliptic or lanceolate, acute, serrulate.

Floral biology

Flower : Involucres in terminal and axillary cymes; flowers monoecious; male several surrounding a single female in each cyathium; limb of stalked glands minute; stamen 1; ovary superior, tricarpellary syncarpous; cocci appressed, pubescent (Fig. 2).

Fruit : Fruit is berry (Fig. 3).

Seed biology

Seed : Seeds reddish with shallow furrows. Seeds are minute, quadrangular bluntly pointed with 5-6 shallow transverse furrows. The average weight of the seed is 0.13 g (Fig. 4).

Fig. 3. Fruit

Fig. 4. Seed

Propagation : By seeds

Euphorbia hirta produces up to 3000 seeds per plant. Regeneration can also be achieved from leaf explants or stem segments.

Seed dispersal : It is observed that maximum number of seeds are dispersed to a distance of about 5 to 25 cm around the plant. In nature, the seeds dispersed by the bursting of the capsules and further be carried away by wind currents.

Ants may be dispersal agents of *Euphorbia hirta* seeds.

Seed dormancy and germination : Maximum germination of 48 percent is obtained in diffused daylight. In continuous darkness germination percentage is low and in continuous light, it is very poor.

Euphorbia hirta shows a germination rate of upto 92%. Seeds can be dried without damage to low moisture contents, usually much lower levels than normally reached in nature.

Seed longevity : Over a wide range of storage environments, their longevity increases with reduction in both moisture content and temperature.

Effects on crop production : Euphorbia hirta is a host to many fungal pathogens and may act as a reservoir of pathogenic fungi, which can infect nearby susceptible crops.

Trypanosomatid flagellates (*Phytomonas* spp.) are detected in *Euphorbia hirta* plants in coconut plantations.

Euphorbia hirta also acts as a host to several insect vectors, including the aphid *Aphis craccivora*, a vector of the rosette virus disease of groundnut and *Aphis gossypii*.

Uses : Euphorbia hirta is an important medicinal herb used throughout its distribution area, including Tropical Africa. It is used, as a decoction or infusion to treat gastrointestinal disorders, including intestinal parasites, diarrhoea, peptic ulcers, heartburn, vomiting and amoebic dysentery.

In Java and India, the tender shoots serve as famine food, raw or steamed, but they may cause intestinal complaints. In West Africa, *Euphorbia hirta* is used as a livestock fodder.

Past work done on Seed Science and Technology

Development of embryo and seed

Kajale (1954) in *Euphorbia hirta* reported that fertilization is porogamous. The pollen tube passes in between the wall of the embryo sac and of the synergids. Generally one and sometimes both the synergids are destroyed. The polar nuclei increase in size prior to triple fusion. Probably syngamy precedes triple fusion.

The proembryo consists of three to four cells and is T-shaped. Embryo development conforms to the *Euphorbia* variation of the Onagrad Type. The small suspensor is multiseriate towards the embryo side and persists in the mature seed. The structure of the mature embryo and the distribution of starch are described.

The primary endosperm nucleus divides earlier than the oospore. The free nuclear endosperm later becomes cellular. Wall formation starts at the periphery and extends towards the centre. One to four nucleoli are present in the nuclei of the endosperm cells. The endosperm surrounds the embryo except at the tip of the root cap.

The hypostase persists in the mature seed as perisperm and serves as an organ of storage.

The testa consists of four layers of cells. The innermost layer representing the outer epidermis of the inner integuments becomes brittle and stony. Next come three more layers derived from the outer integument. Granular deposits are present in the epidermal cells of the testa which have a finely warty appearance. There is a difference in the size and shape of the cells of the different layers.

REFERENCE

Kajale, L.B. (1954). A contribution to the embryology of the Phytolaccaceae II. Fertilization and the development of embryo, seed and fruit in *Rivina humilis* Linnaeus and *Phytolacca dioica* Linnaeus. *J. Indian Bot. Soc.*, 33: 206-225.

SPURGE (*Euphorbia hypercifolia*)

Common name : Spurge, diamond frost

Botanical name : Euphorbia hypercifolia L.

Family : Euphorbiaceae

Origin and distribution : A native of the New World Tropics, introduced into many parts of the Old World Tropics and subtropics; humus-rich soils; near sea-level.

Botany

Habitat : A dryland weed common up to 1000 m on slopes.

Habit : Erect, slender, annual, glabrous or sparsely pubescent herb (Fig. 1).

Fig. 1. Plant

Stem : Stem sparsely branched.

Leaf : Leaves opposite, obliquely oblong, obtuse, serrulate, nerves distinct, stipulate, stipules minute, setaceous.

Floral biology

Flower : Flowers monoecious, involucrate, in axillary and terminal peduncled cymes; bracts many, subulate, subtended by 2 floral leaves; involucre very minute, turbinate, glabrous, glands very shortly stipitate, rounded, white, lobes lanceolate projecting above the glands, limb of the gland white or pale pink; stamens solitary; ovary superior, tricarpellary, syncarpous, 3 celled with one ovule in each cell, style very short (Fig. 2).

Fig. 2. Flower

Fruit : Capsule subglobose, pubescent or glabrous; seeds reddish smooth or with shallow transverse pits (Fig. 3).

Seed biology

Seed : The seeds are 4 angled, oval or spherical, and in some species have a caruncle (Fig. 4).

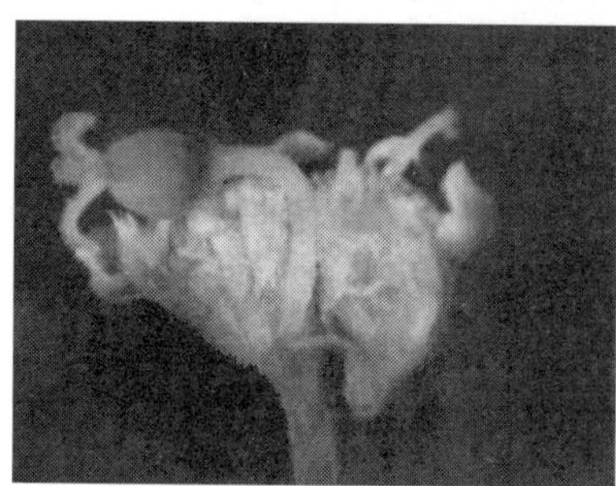

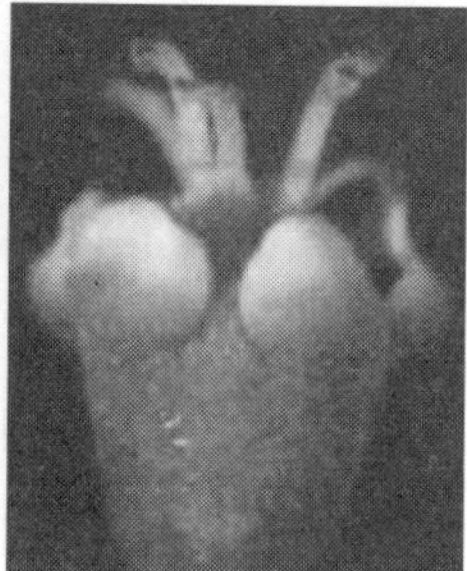

Fig. 3. Fruit

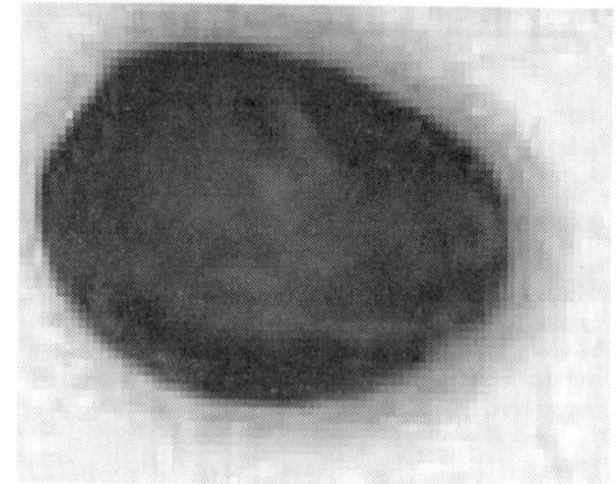

Fig. 4. Seed

Propagation : By seed

Seed dispersal : Ants as a group have a dual role of predation and dispersal of Euphorbia seeds.

Uses : Abundant bloomer for landscape areas, border fronts, hanging baskets and containers. May be grown as a houseplant.

DWARF MORNING GLORY (*Evolvulus alisinoides*)

Common name : Dwarf morning-glory, Shankhpushpi, Shankhahuli.

Botanical name : Evolvulus alisinoides L.

Tamil name : Vishnukirandi

Hindi name : Vishnu krantha, Shyamakrantha

Family : Convolvulaceae

Origin and distribution : China, Bangladesh, Cambodia, India, Indonesia, Japan (Ryukyu Islands), Laos, Malaysia, Myanmar, Nepal, Pakistan, Philippines, Thailand, Vietnam; Africa, Australia (Western Australia; Northen Territory; Queensland; New South Wales), North America, Pacific Islands, South America.

Botany

Habitat : Common weed in dry lands, waste places, pastures and scrub jungles.

Habit : Prostrate, perennial hairy herb with numerous branches, growing upto 60cm (Fig. 1).

Fig. 1. Plant

Stem : Round and densely clothed with white appressed and long spreading hairs.

Leaf : Leaves small, variable in size, entire, lanceolate, oblong-ovate or sub-orbicular.

Floral biology

Flower : The flowers are funnel shaped, purple and white in colour and arranged in solitary or aggregated inflorescence. The complete bisexual flowers consist of accessory as well as reproductive parts. The sepals are persistant in nature. However, petals are white and purple. The bicarpellary ovary has two ovule in each locule with a long style and a short bifurcated stigma. The size of flowers is not much varied but stamens are dimorphic (two types long as well as short) in nature. The anthers are small showing longitudinal slits dehiscence (Fig. 2).

Fig. 2. Flower

Flowering, pollination and fruit set : Flowers are self as well as cross-pollinated as confirmed by various hand pollination experiments. They were visited by different floral visitors like, honey bees i.e., (*Apis dorsata, Apis cerana indica*), small bees (*Melliopona* Sps.) bumble bees, black ant (*Componotus compestris*), beetle (*Coccinella punctata*), butterflies (*Colias earytheme*), common wasps (*Polisters herbraeus*). They visit flowers between 7.00-8.00 a.m. It is interesting to note that all of them are not pollen-carriers. Honey bees, bumble bees and butterflies are pollen-carriers as evident from the presence of pollen on their body parts. On the basis of their visitation rate and pollen load on their body parts, the bees are found to be most efficient pollinators. Pollen load on the bees is distinctly observed on the lower side of the abdomen of honey bees, a higher amount of pollen load than that of other insects. The fruit set percentage is calculated as 65%.

Fruit : The fruits are 6.00 mm in diameter, they are oval and spherical in shape. The type of fruit is capsule non fleshy, non-hairy, dehiscent, a capsule long loculicidal. Fruits are usually 2 seeded. It is clear from the present observation that the average number of fruit/inflorescence is 6-8. Unripe fruits are yellow green blotched in colour which becomes brownish yellow at the time of maturity (Fig. 3).

Seed biology

Seed : The total number of seeds/fruit is 2-4 and seed set percentage is 96.4%. Seeds are smooth, reniform, glabrous and black in colour with 0.16 cm length. These seeds are hard under normal conditions. The weight of 100 seeds measured 1.2 g.

Propagation : By seed

Seed dispersal : Seed dispersal takes place through birds and winds.

Fig. 3. Fruit

Seed dormancy and germination : It germinates at the beginning of the rainy season, and starts flowering after 8 weeks.

Uses : The whole plant is used medicinally and possesses antidysentric and antiseptic properties. Shankhpushpi is also known as brain tonic, increasing memory, reducing nervous debility and so on (Chermexcil, 1992). Its paste is applied externally in skin diseases. The whole plant juice is traditionally used in various mental disorders. In epilepsy and hysteria, the fresh plant juice is recommended with kustha powder and honey. It promotes intelligence, enhances memory and retention especially of school going children.

REFERENCE

Chermaxcil. (1992). Selected medicinal plants of India: A monograph of identity, safety and clinical usage. Pharmaceutical and cosmetics export promotion council, *Bombay*: 151-153.

LOTUS SWEET JUICE (*Glinus lotoides*)

Common name : Lotus sweet juice

Botanical name : *Glinus lotoides*

Tamil names : Cheruppadai, Ceruppati

Hindi name : Gandhi-buti, Bakada, dholakani, Hata

Family : Molluginaceae

Origin and distribution : Glinus lotoides is native to Eurasia and Africa and has become widespread in tropical, subtropical and warm-temperate areas worldwide.

Botany

Habitat : Occurring in various types of woodland, scrub, riverine fringes, grassland and along roadsides, often in sandy soils

Habit : Annual procumbent herb upto 40 cm long, with stellate woolly parts (Fig. 1).

Fig. 1. Plant

Stem : Stems upto 45 cm long

Leaf : Leaves sub-verticillate, 0.6-2.0 cm long, 0.5-1.8 cm broad, orbiculate or more or less cuneate, often mucronate; petiole 2-8 mm long.

Floral biology

Flower : Flowers in clusters of 3-15; sepals lanceolate, 3.5-4.1 × 0.4-2 mm, stellate-pubescent abaxially, glabrous adaxially, apex rounded to acute or slightly mucronate; stamens 3-5 (Fig. 2).

Flowering and fruiting : Flowering late summer-fall.

Fig. 2. Flower

Fruit : Capsule sub-globose or oblong, 6 mm long, membranous, enclosed in the sepals (Fig. 3).

Fig. 3. Fruit

Seed biology

Seed : Seeds 10-25 per locule, orange-brown, 0.4-0.6 × 0.3-0.4 mm, papillate, somewhat glossy or dull; papillae sometimes black. $2n = 36$. Seeds many, less than 1 mm long, tuberculate, strophiolate, with a curved linear appendage. 3/4 the size of the seed.

Propagation : Seeds as well as vegetative propagation

Seed dispersal : Seeds are dispersed through ants.

Uses : In India, the species is used as treatment for diarrhoea, boils and abdominal diseases, as well as weakness in children (Kirtikar and Basu, 1935).

REFERENCE

Kirtikar K.R. and B.D. Basu. (1935). *Indian Medicinal Plants.* Vol.4, Allahabad.

INDIAN HELIOTROPE (*Heliotropium indicum*)

Common name : Indian Heliotrope

Botanical name : Heliotropium indicum L.

Tamil name : Thel-kodukku.

Hindi name : Siriyari, Hatishura

Family : Boraginaceae

Origin and distribution : Native to Asia or perhaps South America. Established in all tropical and subtropical countries on the planet.

Botany

Habitat : Dry land weed, common on roadside and waste places.

Habit : Annual hirsute herb (Fig. 1).

Stem : Erect, deeply grooved and sparsely roughly hairy.

Leaf : Leaves alternate or subopposite, hispid, ovate, obtuse, serrate, cordate at the base, decurrent on long petioles.

Floral biology

Flower : Flowers pale violet in long scorpioid cymes, leaf opposed; calyx 5 lobed, lobes narrow, linear; corolla 5 lobed, tubular, tube narrow cylindric, lobes spreading, small rounded, crenate; stamens 5 on the corolla tube, acuminate; ovary 4 celled, ovules solitary in each cell, style terminal, stigma obtuse, above a lobed ring (Fig. 2).

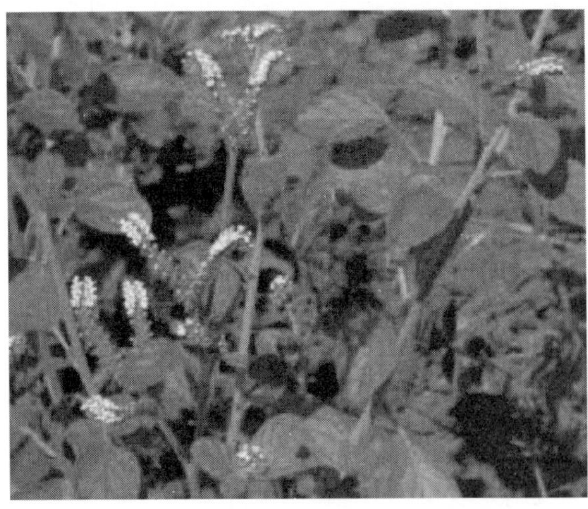

Fig. 1. Plant

Fig. 2. Flower

Flowering : July - November.

Seed biology

Fruit / Seed : Nutlets combined in pairs, beaked, angled, separating later into pyrenes, each without an empty cavity Fruit ribbed, 3-3.5 mm, glabrous or nearly so, deeply 2-cleft into mericarps each divided into 2 1-seeded pyrenes; mericarps longitudinally ribbed (Fig. 3).

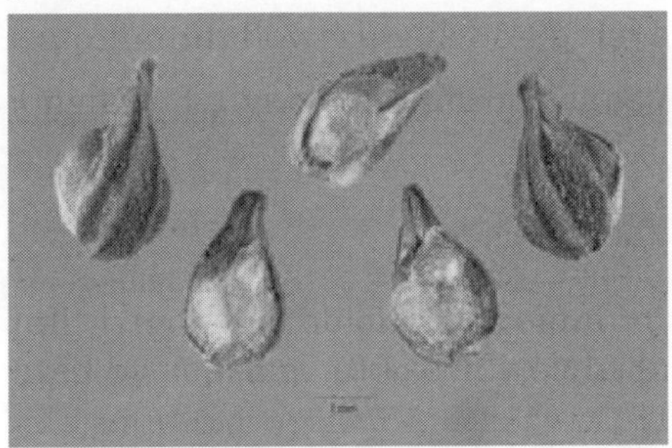

Fig. 3. Fruit/seed (Nutlet)

Propagation : By seed

Seed dispersal : Seeds are dispersed by wind and water.

Life cycle : Perennial. Seeds germinate in summer, often with a flush in autumn. A rosette of leaves is formed, then several, long, low lying stems emerge. Flower stems arise from these in spring and flowering starts in November and will continue all summer in moist areas. Older plants are semi dormant in winter and produce a flush of growth in spring and autumn. Mature plants tend to flower in spring and autumn.

Seed dormancy and germination : Seeds of Indian heliotrope germinated at a range of alternating temperatures (25/15, 30/20 and 35/25°C day/night) and the final result show the germination of Indian heliotrope was always greater in light and seed germination was not affected by a high level of salt and a range of pH between 5 and 9, but was sensitive to high degrees of water stress. Seedling emergence of Indian heliotrope (70%) was optimal when seeds were sown in the top 0.2-cm soil layer. A burial depth of 2cm completely inhibited emergence of Indian heliotrope (Chauhan and Johnson, 2008).

Uses : The leaves are used for the treatment of ophthalmic disorders, erysipelas, pharyngodynia and anti-inflammatory, anti-tumor. The roots are used as astringent, expectorant and febrifuge. The extract of leaves was proved to be active against Schwart's leukemia, wound healing and anti-inflammatory activities.

REFERENCE

Bhagirath S. Chauhan and David E. Johnson. (2008). Seed germination ecology of purple-leaf button weed (*Borreria ocymoides*) and Indian heliotrope (*Heliotropium indicum*): Two common weeds of rain-fed rice. *Weed Sci.*, 56(5): 670-675.

INDIAN SARSAPARILLA (*Hemisdesmus indicus*)

Common name : Indian Sarsaparilla

Botanical name : Hemidesmus indicus (L.) R. Br.

Tamil name : Nannari

Hindi name : Anantamul

Family : Asclepiadaceae

Origin and distribution : It occurs over the greater part of India, from the upper Gangetic plain Eastwards to Assam and in some places in Central, Western and South India.

Botany

Habitat : Dry land weed, common on roadside and waste places.

Habit : Straggler; latex milky (Fig. 1).

Stem : Stem wiry

Leaf : Leaves highly vaiable, linear lanceolate to orbicular, opposite.

Floral biology

Flower : Flowers 5-15, in axillary fasciles. Calyx-lobes imbricate, herbaceous to thick, glandular within. Corolla yellow/purple, (sub) rotate; lobes valvate, thick, rugose within. Stamens connivent around styles; filaments incurved, distinct; pollinia suberect; pollinial bags spherical, closely appressed to the brownish, sheathy caudicle. Corona single, corolline. (Fig. 2).

Fruit : Ovaries conic. Follicles terete (Fig. 3).

Fig. 1. Plant

Fig. 2. Flower

Fig. 3. Fruit

Seed biology

Seed : The seeds are coma brownish white, oblong and covered in tiny silvery white hairs.

Propagation : By seeds and tubers.

Seed dormancy and germination : Collect fresh seeds any time of the year (since fruiting is observed throughout the year). Soak them in water for 24 hours and sowing raised beds at the onset of monsoon for better growth. 60% of the seeds germinate within fortnight.

Uses : It is tonic, diuretic, demulcent, disphoretic and blood purifier. Employed in nutritional disorders, syphilis, chronic rheumatism, gravel and other urinary diseases and skin afections. It is also employed as a vehicle for potassium iodine.

FALSE WATERWILLOW (*Indoneesiella echioides*)

Common name : False Waterwillow

Botanical name : Indoneesiella echioides (Synonyms: *Andrographis echioides)*

Tamil name : Gopurathaangi

Hindi name : Ranchimani

Family : Acanthaceae

Origin and distribution : False Waterwillow is found in India and Sri Lanka.

Botany

Habitat : It is common in all the dry districts.

Habit : An erect, annual hairy herb, simple or slightly branched with oblong, obtuse leaves. Branchlets to 50 cm (Fig. 1).

Stem : Stem hairy.

Leaf : Leaves oblong to oblanceolate, 3-5 x 0.7-1cm. Racemes to 5 cm, usually not exceeding the leaves, scarcely branched.

Floral biology

Flower : Flowers are borne in spike-like racemes, up to 2 cm long. The stalk carrying the raceme is densely hairy. Flowers are erect. Sepal tube is 2 mm long, with thread-like sepals up to 9 mm long (Fig. 2).

Flowering and fruiting : March-June, October-December.

Fruit : Capsule 1 x 0.5 cm.

Seed biology

Seed : Seed is modified into a hook shaped jaculator or retinaculum that functions in flinging out the seeds during dehiscence (Fig. 3).

Fig. 1. Plant

Fig. 2. Flower

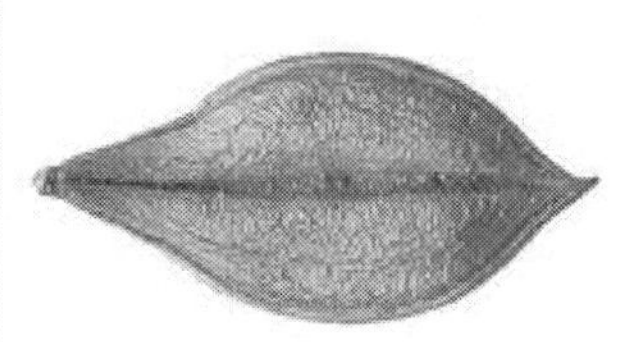

Fig. 3. Seed

Propagation : By seeds.

Seed dispersal : Seeds are dispersed by winds.

Uses : Leaf juice is very bitter to taste and administered for its anthelmintic activity. Leaf powder consumed along with rice water for snake bite and for eczema.

LANTANA (*Lantana camera* var. *aculeata*)

Common name : Lantana

Botanical name : Lantana camara L. var. *aculeata*

Tamil name : Unni

Hindi name : Raimuniya, Chaturang, Khaneri

Family : Verbenaceae

Origin and distribution : Naturalized in approximately 60 countries or island groups between 350°N and 350°S latitudes. Occurs widely in the Asia-Pacific region, Australia, New Zealand, Central and South America, West Indies and Africa. The plant is still widening its range.

Botany

Habitat : A wildly growing weed especially in waste lands, difficult to control.

Habit : A prickly shrub, often strongly smelling (Fig. 1).

Fig. 1. Plant

Stem : Stem much branched, prickled.

Leaf : Leaves opposite, ovate, lanceolate with toothed edges, simple, cuneate, cordate, rugose, 2-10 cm long and 2-8 cm wide.

Floral biology

Flower : Flowers form in dense clusters and vary in colour from red-yellow, orange-pink and white; depending on the type, maturity and location (Fig. 2). Sub umbellate spikes; calyx small toothed; corolla tube slender, limb salver-shaped lobes 4 or 5 spreading; stamens 4, didynamous anthers oblong; ovary bilocular, one ovule in each locule (Rod Ensbey, 2008).

Fig. 2. Flower

Flowering, pollination and fruit set : *L. camara* flowers prolifically. It can greatly alter fire regimes in natural systems (Day *et al.*, 2003) and is reported to proliferate well after fires. It usually flowers early in the first growing season and is capable of flowering all year round, if adequate moisture and light are available (Day *et al.*, 2003; Duggin and Gentle, 1998). Flowers are efficiently pollinated by butterflies, thrips and sunbirds (in India), and humming birds (in Brazil) resulting in 85% fruit set (Hilje, 1985).

Fruit : Succulent, one-seeded drupe or berry about 6-8 mm in diameter. These develop in clusters and consist of aggregate, fleshy segments that are green at first and then turn a shiny, dark purple to black when ripe (Fig. 3) (Rod Ensbey, 2008).

Fig. 3. Fruit

Seed biology

Seed : Seeds exalbuminous (Fig. 4). Seed setting takes place between September to May with 1 - 20 seeds on each flower head. Mature plants produce upto 12,000 seeds annually.

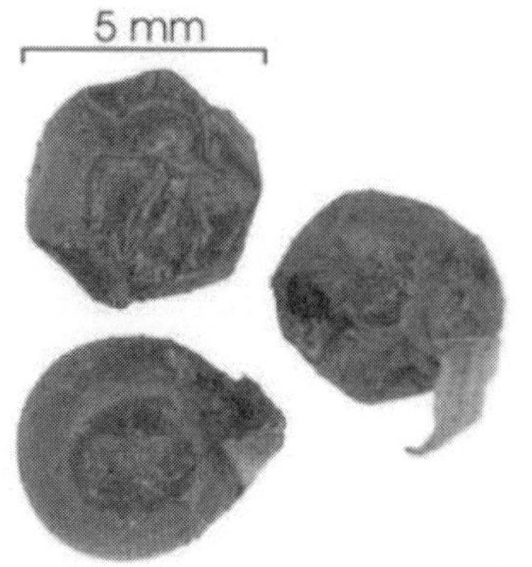

Fig. 4. Seed

Propagation : Seed as well as vegetative propagation.

Vegetative propagation

Lantana has a shallow root system forming a dense root mat that grows mainly within the top 10–30 cm of soil. The plant does not sucker from damaged or broken roots but will regrow vigorously from the base of stems. Horizontal stems are able to take root when they contact moist soil. Lantana will also shoot from the base of vertical stems.

Seed dispersal : Seeds are dispersed by birds (Swarbrick *et al.,* 1998; Day *et al.,* 2003). Seed dispersal is through frugivorous birds, fox and rodents.

Movement of water, contaminated soil and machinery, deliberate planting and poorly disposed garden waste can help lantana spread.

Fruit eating birds are the main cause of lantana spread. Birds and some mammals eat the fruit and the seed in their droppings is viable (Rod Ensbey, 2008).

Life cycle

Germination can be year round, but peaks after summer rains. Plants mature in one year, but they must complete one whole season before seeding. Vegetative spread can produce dense thickets (Rod Ensbey, 2008).

Seed dormancy and germination : Germination rate of fresh seed is generally low, but the germinability gets improved when the seed passes through the digestive system of birds and animals. High light intensity and soil temperature will stimulate germination of seeds which means that clearing of forest areas, inappropriate burning and other disturbances will help spread of the weed. Seeds are capable of surviving the hottest fires.

Seed germination occurs when sufficient moisture is present; germination is reduced by low light conditions.

Chemicals present in smoke may help to enhance water absorption due to enhanced permeability and may lead to greater seed germination (Raizada and Raghubanshi, 2010).

Seeds are capable of germinating any time of the year if sufficient soil moisture, light, and warmth are available (Duggin and Gentle, 1998; Parsons and Cuthbertson, 2001). Although seed germination rate in *L. camara* is low (4-45%) due to seed dormancy, low seed viability, and meiotic instability (Duggin and Gentle, 1998; Sahu and Panda, 1998), low seedling mortality (Sahu and Panda, 1998) and fast vegetative spread compensate for this deficit.

L. camara is known to proliferate and germinate well after disturbances such as grazing, cutting, and fire (Duggin and Gentle 1988; Gentle and Duggin, 1997; Hiremath and Sundaram, 2005).

Seed storage : Some studies have shown that 50 per cent of seed remain viable in dry conditions for upto two years; unpublished studies have shown seed to last five years. The length of time is likely to depend on variety, soil type and soil moisture levels (Rod Ensbey, 2008).

Effects on crop production : Lantana threatens natural habitats and native flora and fauna. In Australia, nineteen endangered and threatened species are under threat due to the weed. It infests pastures, grazing lands, orchards and crops like, tea, coffee, oil palm, coconut and cotton, and reduces the economic viability of the crops. The allelopathic qualities of lantana reduce the vigour of native plant species and limit their productivity.

Control and management : Integrated weed management uses a range of control methods in order to achieve more effective control. Lantana can be controlled by using a combination of manual control (hand grubbing, hand cutting, hand pulling), mechanical control (bulldozing or slashing), herbicides, fire, pasture improvement, and grazing management techniques (Rod Ensbey, 2008).

Uses : Lantana is mainly used as a herbal medicine and in some areas as fire-wood and mulch. In some countries, it is planted as a hedge to contain or keep out livestock. Leaf extracts of lantana exhibit antimicrobial, fungicidal, insecticidal and nematicidal properties.

Past work done on Seed Science and Technology

Raizada and Raghubanshi (2010) reported that *L. camara* seeds are of medium size and they tend to get buried deep in the litter layer. When there is a fire, seeds in deeper litter remain protected from fire and are exposed only to smoke while seeds that come into direct contact with fire are burnt and destroyed.

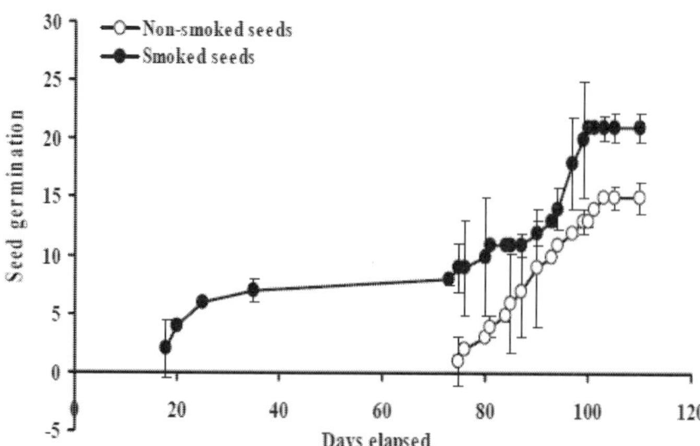

Fig. 5. Cumulative numbers of germinated *L. camara* seeds as a function of days after exposure to smoke

According to them, smoke significantly enhanced seed germination in *L. camara*. Out of 100 seeds, 16 ± 1.3 and 21 ± 1.2 seedlings were recruited in non-smoked and smoked conditions, respectively. In the case of smoked seeds, a third of the 21 seedlings were the result of recruitment of double seedlings from a single seed. Smoke exposure hastened seedling recruitment. Smoke exposed seeds started germinating within a few weeks, while in non smoked seeds germination only started after 2 months (Fig. 5).

The faster germination in smoked seeds led to a significantly higher value of the germination velocity index compared to seeds in the control treatment. Further, the seedlings recruited from smoked seeds also had significantly higher values of the vigour index. The healthier seedlings emerging from the smoked seeds led to lower seedling mortality, as would be expected, although seedling mortality did not differ significant between treatments (Fig. 6).

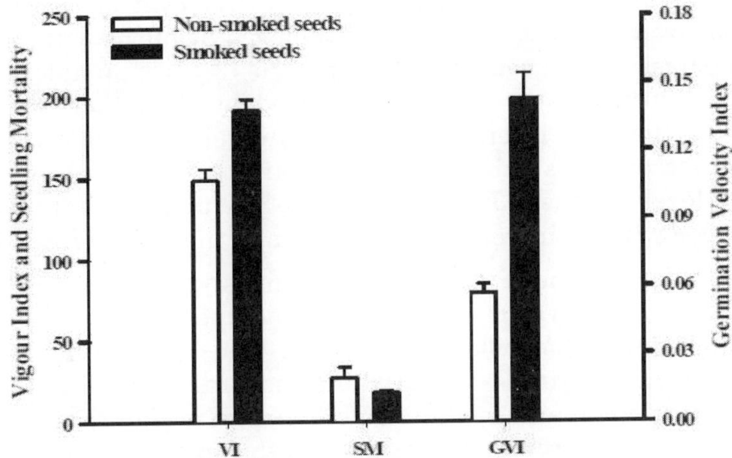

Fig. 6. Germination vigour index (GVI), seedling mortality (SM), and vigour index (VI) of smoked and non-smoked *L. camara* seeds

Vivian-Smith *et al.* (2006) reported that the seed damaging fly, *Ophiomyia lantanae* is removed fewer damaged fruits than undamaged fruits, suggesting that frugivores select against damaged fruit. In germination experiments, these differential responses were primarily due to significantly reduced germination of damaged PER seeds. Seedling emergence from pots of the PER biotype was also lower in damaged seeds (27.5%) than undamaged seeds (56.1%). In contrast, emergence increased with seed damage in the pink biotype, from 28.2 to 41.1%. *O. lantanae* damage significantly reduced seed mass only in the smaller seeded PER biotype. In the larger seeded pink biotype, it may have promoted seedling emergence by interfering with dormancy mechanisms. Mean viable seed densities ranged from 78.6 to 402.8 seeds m^{-2}. Of these, non-dormant seeds comprised 24.6 to 98.2 seeds m^{-2}, suggesting that recruitment is unlikely to be seed limited. They conclude that while *O. lantanae* influences recruitment and dispersal processes, the magnitude of the responses measured are unlikely to greatly influence plant densities in south-east Queensland infestations.

Duggin and Gentle (1998) evaluated the invasion of *Lantana camara* into dry rainforest–open forest ecotones was evaluated over a range of disturbance intensities namely wildfire (unburnt, low, and high intensity), overstorey damage (different levels of biomass removal), and nutrient addition (NPK). Seedling establishment occurred in control treatments but growth was weak (averaging 0.85 g plant^{-1} year^{-1}) and they failed to flower during the 27 months experimental period. High levels of nutrient addition (60–80 g m^{-2}) without other forms of disturbance significantly increased growth (to 0.95 g plant^{-1} year^{-1}) and germination (41.6–44.5%), while survival increased modestly (14.1–15.9 plants plot^{-1}). Biomass increased significantly to 1.78 g plant^{-1} year^{-1} when the shrub layer was removed or burnt by a low intensity wildfire while germination and survival increased by 13.5 and 22.0% respectively over the control. Invasion increased further when the overstorey was physically removed or burnt with a high intensity wildfire as biomass increased significantly to 2.42 g plant^{-1} year^{-1}, while germination and survival significantly increased by a further 35.6% and 40.7% respectively over the low intensity wildfire and shrub removal. Invasion was positively increased with disturbance intensity and increased resource availability. Light at ground level increased from 21.3 to 30.5% of ambient light when the shrub layer was damaged while it increased to 84.3% when the overstorey was damaged.

Vivian-Smith and Dane Panetta (2009) investigated germination, seedling emergence, and seed survival of two lantana biotypes (Pink and pink-edged red [PER]) in southeastern Queensland, Australia. The treatments comprising two differing environmental regimes (irrigated and natural rainfall) and sowing depths (0 and 2 cm). Seed survival and seedling emergence were significantly affected by all factors (time, biotype, environment, sowing depth, and control) ($P < 0.001$). Seed dormancy varied with treatment (environment, sowing depth, biotype, and control) ($P < 0.001$), but declined rapidly after 6 months. Significant differential responses by the two biotypes to sowing depth and environment were detected for both seed survival and seedling emergence ($P < 0.001$). Seed mass was consistently lower in the PER biotype at the population level ($P < 0.001$), but this variation did not adequately explain the differential responses. Moreover, under natural rainfall the magnitude of the biotype effect was unlikely to result in ecologically significant differences. Seed survival after 36 months under natural rainfall ranged from 6.8 to 21.3%. Seed survival projections from the regression model were greatest for buried seeds under natural rainfall (11 yr) and least under irrigation (3 yr).

REFERENCES

Day, M.D., C.J. Wiley, J. Playford and M.P. Zalucki. (2003). Lantana: current management status and future prospects for international agricultural research. Australian Centre, Canberra.

Duggin, J.A. and C.B. Gentle. (1998). Experimental evidence on the importance of disturbance intensity for invasion of *Lantana camara* L. in dry rainforest-open forest ecotones in north-eastern NSW, Australia. *Forest Ecology and Management.* 109:279-292.

Gentle, C.B. and J.A. Duggin. (1997). *Lantana camara* L. invasions in dry rainforest-open forest ecotones: the role of disturbances associated with fire and cattle grazing. *Australian Journal of Ecology* 22: 298-306.

Hilje, L. (1985). Insectos visitadores y efficiencia reproductive de *Lantana camara L.* (Verbenaceae). Brenesia 23: 293-300.

Hiremath, A.J. and B. Sundaram. (2005). The fire-lantana cycle hypothesis in Indian forests. *Conservation and Society* 3: 26-42.

Parsons, W.T. and E.G. Cuthbertson. (2001). *Noxious Weeds of Australia.* CSIRO Publications.

Purnima raizada and Akhilesh S. raghubanshi. (2010). Seed germination behaviour of *Lantana camara* in response to smoke, *Tropical Ecology,* 51(2S): 347-352.

Rod Ensbey (2008). Lantana, *Primefact 673,* NSW Department of Primary Industries .

Sahu, A.K. and S. Panda. (1998). Population dynamics of a few dominant plant species around industrial complexes in West Bengal, India. *Bombay National History Society Journal* 95: 15-18.

Swarbrick, J.T., B.W. Willson and M.A. Hannan-Jones. (1998). *Lantana camara* L. in Panetta, F.D., R.H. Groves and R.C.H. Shepherd (Eds.). *The Biology of Australian Weeds.* Melbourne. pp. 119–136.

Vivian-Smith, G. and F. Dane Panetta. (2009). Lantana (*Lantana camara*) seed bank dynamics: seedling emergence and seed survival, *Invasive plant science and management* 2(2):141-150.

Vivian-Smith, G., C.R. Gosper, A. Wilson and K. Hoad. (2006). *Lantana camara* and the fruit- and seed-damaging fly, *Ophiomyia lantanae* (Agromyzidae): seed predator, recruitment promoter or dispersal disrupter, *Biol. Control,* 36:247–257.

COMMON LEUCAS (*Leucas aspera*)

Common name : Common Leucas

Botanical name : Leucas aspera

Tamil name : Thumbai

Hindi name : Chhota halkusa

Family : Lamiaceae (Labiatae)

Origin and distribution : Distributed throughout India from the Himalayas down to Ceylon.

Botany

Habitat : A very common dry land weed. An erect, slender, fragrant annual, which grows gregariously, and is seen commonly in harvested fields and on roadsides.

Habit : A small annual herb, stem quadrangular, pubescent (Fig. 1).

Fig. 1. Plant

Stem : Square, grooved on all sides and covered with coarse hairs.

Leaf : Leaves linear or oblong - lanceolate, opposite.

Floral biology

Flower : Flowers occurring in verticellaster inflorescence, white, bilipped; calyx tube ribbed and hispid; corolla 4, fused; stamens 4, didynamous, epipetalous; ovary superior, bicarpellary, syncarpous, 4 - chambered with one ovule in each locule (Fig. 2).

Flowering : July - November

Fruit : Fruit 4 nutlets, 2.5 mm long, oblong, brown, smooth, inner face angular and outer face rounded (Fig. 3).

Seed biology

Seed : Nutlets 0.2 x 0.1 cm, obovoid-oblong, angular on inner surface, rounded on outer, smooth, brownish black. (Fig. 4).

Fig. 2. Flower

Fig. 3. Fruit

Fig. 4. Seed

Propagation : By seeds

Seed dispersal : Mature seeds are dispersed by wind, rainwater, cultivation, soil on tools, or even grazing.

Life cycle : An erect, slender, fragrant annual, which grows gregariously, and is seen commonly in harvested fields and on roadsides. Bees pollinate the flowers.

Uses : Used as pot herb. Leaves are used in traditional medicine. The juice of the leaves is used as local application for psoriasis, chronic skin eruptions and chronic rheumatism (Kirtikar *et al.,* 1991). *L. aspera* leaves are also used as insecticides and mosquito repellent in rural area (Reddy *et al.*, 1993).

REFERENCES

Kirtikar K.R. and B.D. Basu. (1991). Indian medicinal plants. Bidtter, E., J.F. Caius and K.S. Mhaskar (Ed). *Periodical Experts Book Company,* 20:19.

Reddy, M.K., S. Viswanathan, P. Thiruganeshansambantham and L. Kameswaran. (1993). Effect of *Leucas aspera* on snake venom poisoning in mice and its possible mechanism of action. *Fitoterapia,* 64: 442-446.

THE HOARY BASIL (*Ocimum canum*)

Common name : The Hoary Basil

Botanical name : Ocimum canum

Tamil name : Nai Thulasi

Hindi name : Kali tulsi, Manjari, Ram tulsi.

Family : Labiatae (Lamiaceae)

Origin and distribution : Genus *Ocimum* is widespread over Asia, Africa and Central and Southern America; it appears to have its centre of diversity in Africa. Basil was probably first put to cultivation in India.

Today, basil is cultivated in many Asian and Mediterranean countries; main exporters (for the European market) are France, Italy, Morocco and Egypt. There is also significant basil production in California.

Botany

Habitat : A common weed in dry lands.

Habit : Erect much branched aromatic herb, growing up to 60 cm (Fig. 1).

Stem : Fourangled, grooved, hairy when young becomes glabrous with age.

Leaf : Leaves elliptic lanceolate, slightly serrate, pubescent, long-petioled.

Floral biology

Flower : Flowers white, pink or purplish arranged in closed whorl in a raceme; calyx small, persistent, villous and pilose; corolla bilipped; stamens and ovary as in Leucas (Fig. 2).

Fig. 1. Plant

Fig. 2. Flower

Flowering and fruiting : They produce flowers over a long period, a few flowers being produced each day. The open flowers offer nectar and pollen as rewards and are visited and sternotribically pollinated by day-flying animals. The blossoms are flag-shaped, and the reproductive organs are close to the lower corolla lip. The stamens and stigma show movements immediately after anthesis and remain for 20–30 minutes and this may cause self-pollination. Both species reproduce primarily through autogamy. Flowers are open during 5:30–13:30h in *O. americanum*. Certain bee species like *Apis florea, A. cerana indica, Amegilla* sp., and *Pseudapis oxybeloies*, and the butterfly *Surandra queretroum* are the most frequent and consistent visitors and can be pollinators for both plant species (Raju, 1989).

Fruit : Consists of four black ellipsoidal nutlets that become mucilaginous when wetted (Fig. 3).

Seed biology

Seed : Seeds are tiny in nature.

Fig. 3. Fruit

Fig. 4. Seed

Propagation : By seed.

Seed dispersal : Seed dispersal mainly through wind and animals.

Seed dormancy and germination : A brief light exposure is essential for the germination of dark imbibed seeds of *Ocimum americanum* Linn. Seeds with testa broken at the radicular end were independent of light requirement for germination. Red and far-red light were found to affect germination in a fashion typical of other photomorphogenic reactions mediated through a phytochrome system. The stimulatory effect of light was not constant if the imbibed seeds were maintained in darkness for long periods. Imbibed seeds entered a secondary dormancy (skotodormancy) under prolonged darkness. This was relieved by cold treatment.

Uses : The seeds of basil have some use as thickening agent in Thailand, but do not share the leaves' fragrance.

Fresh basil leaves have a strong and characteristic aroma, not comparable to any other spice, although there is a hint of cloves traceable.

All basil varieties have in common that their dried leaves is much less aromatic than fresh ones; deep-freezing the herb is the best method of preservation.

REFERENCE

Raju, A.J.S. (1989). Reproductive ecology of *Ocimum americanum* L. and *O. basilicum* L. (Lamiaceae) in India. *Plant Species Biol.*, 4: 107–116.

CONGRESS GRASS (*Parthenium hysterophorus*)

Common name : Congress grass

Botanical name : *Parthenium hysterophorus* (synonym - *Parthenium lobatum* Buckl.)

Tamil name : Visha poondu.

Hindi name : Gajar ghas, Chatak Chandani

Family : Compositeae (Asteraceae)

Origin and distribution : *Parthenium hysterophorus* occurs throughout the tropical and sub-tropical Americas from Florida to Southern Brazil and Northern Argentina (Dale, 1981).

It became the major noxious weed for over the last 25 years in India (Mahadevappa, 1997) and Australia (Navie *et al.,* 1996).

Botany

Habitat : Parthenium weed is capable of growing in most of the soil types, but becomes most dominant in alkaline, clay loam soils, cultivated fields and road sides.

Habit : A fast maturing branching annual herb (Fig. 1).

Fig. 1. Plant

Stem : An erect stem upto 2 m high becoming woody with age, slender hairy

Leaf : Leaves alternate, sessile, bipinnate, upto 20 x 10 cm, pubescent, lobes entire, acute.

Floral biology

Flower : Small, creamy white flowers occur on the tips of the numerous stems. Heads many, peduncled in panicles, florets white, outer florets 5, female, supported by two hyaline wings at base; inner few male florets; corolla in female an obovate, entire ligule; in male tubular, 5 lobed; anthers obtuse below (Fig. 2).

Fig. 2. Flower

Flowering and pollination : Parthenium weed normally germinates in spring and early summer, produces flowers and seed throughout its short life and dies in late autumn. However, with the right conditions (rain, available moisture, mild soil and air temperatures); parthenium weed can grow and produce flowers at any time of the year. In a good season, four or five generations may emerge. In summer, if plants are stressed (e.g. due to lack of water), parthenium weed can complete its life cycle in four weeks (Fig. 3).

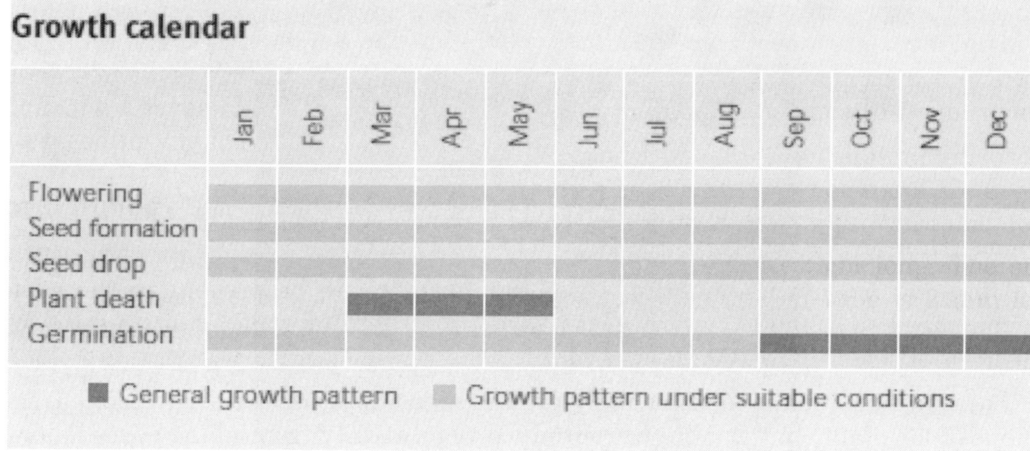

Fig. 3. Life cycle

Fruit : An achene, dorsally compressed, narrowed below; pappus of 2 lateral reflexed awn (Fig. 4).

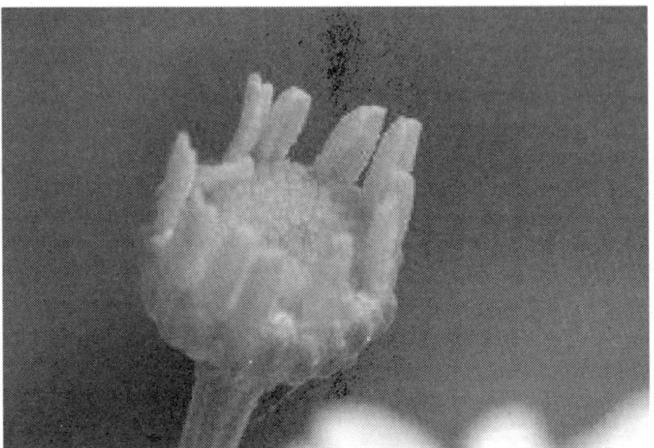

Fig. 4. Fruit

Seed biology

Seed : Seeds are black in colour and wedge shaped that are 2 mm long with thin white scales (Fig. 5). Each flower head produces four or five black achenes weighing 0.2 mg to 0.4 mg. The seeds are contained and dispersed within the 'achene complex', which consists of a single seed crowned by a pappus of two or three orbicular scales, two attached sterile disc florets and a subtending involucral bract (Navie *et al.,* 1996). The achenes are shed gradually or retained on the inflorescence until stem senescence.

Fig. 5. Seed

Seed production : A typical mature plant can produce from 15000 to 25000 achenes (Haseler, 1976; Joshi, 1991).

Upto 340 million parthenium seeds per hectare can be present in the surface soil compared to 120,000 native grass seeds. Parthenium weed can produce large quantities of seed, up to 100,000 per plant.

Parthenium weed is an extremely prolific seed producer, with upto 25,000 seeds (achenes) per plant, but unable to reproduce vegetatively from plant parts (Navie *et al.*, 1996), and with an enormous seed bank, estimated as 200,000 seeds per m^2 (Joshi, 1991)

Propagation : By seeds.

Seed dispersal : The seed is easily spread by vehicles, machinery, animals, pasture seed, stock feed and water. Most long distance spread is in produce, vehicles and farm machinery. It can also be spread by flooding and blown by winds (Auld *et al.*, 1983; Gupta and Sharma, 1977).

Proper cleaning of farm equipment, sowing of uncontaminated seed and a short-term quarantine of livestock in parthenium infested area will reduce the risk of spreading the weed (Tamado *et al.*, 2002).

Seed extraction : The seeds are air-dried for 3 days and threshed by hand (Tamado *et al.*, 2002).

Germination test : An adequate time for germination of *P. hysterophorus* seeds is 2 weeks according to Williams and Groves (1980), Pandey and Dubey (1989) and Navie *et al.*, 1998.

Seed dormancy and germination : Navie *et al.* (1998) reported an average germination percentage of 51% of 'fresh' seeds without any significant difference between light and dark conditions (25/20°C).

Williams and Groves (1980) concluded from their experiment that only temperatures higher than 35°C limit germination of this species.

They also reported maximum germination (88%) of the seed in dark, under a day/night temperature regime of 21/16°C. The percentage of the germination decreased as the day/night tempera-ture differential was increased.

Buried seeds can remain dormant for many years and seed close to the soil surface will germinate readily. Most seeds germinate within 2 years, if conditions are suitable, although upto 12% of buried seeds may be viable after 2 years (Butler, 1984).

Its large and persistent soil seed bank, fast germination rate and ability to undergo dormancy make it well adapted to semi-arid environments. It also releases chemicals during rain that inhibit the germination and growth of pasture grasses and other plants.

Several internal and external factors prevent seed germination. Among the internal factors, the presence of biochemical inhibitor in the seed and immature embryo are common. The commonest external factors are soil water content and temperature (Fernandez-Quiatanilla *et al.*, 1991).

Butler (1984) and McFadyen (1992) reported absence of a primary dormancy mechanism in the seeds, but the study by Navie *et al.* (1998) showed initial inhibition of germination in freshly shed seeds.

According to Ramaswami (1997), non dormancy in seeds of parthenium is one of the major factors which help its extensive spread and establishment.

The conditions under which seeds mature on the mother plant can determine the level of seed germinability and dormancy and consequently affect time of seed germination and fate of next generation (Meyer and Allen, 1999).

Parthenium weed achieves highest germination rates at temperatures ranging from 12°C to 27°C and the optimum germination temperature being 22–25°C (John Chamberlain and Amanda Gittens, 2004).

Seed storage and burial : Butler (1984) reported that germination declines from 66% after 1 week of burial to 29% after burial for 2 years. However, Navie *et al.*(1998) reported 74% seed viability after 2 years burial and the predicted half life of the seeds to be about 6 years.

The longevity of seed represents a major mechanism of survival for weed species. It leads to a continuous source of emergency (Carvalho and Favoretto, 1995). However, the longevity of seed in soil varies according to the characteristics, burial depth, and climatic conditions of seeds (Carmona, 1992).

Parthenium seed is viable for an average of two years and upto 20 years. Parthenium seeds can be stored in fresh water for three to five years and it can germinate (Monaco *et al.,* 2001).

The seedlings of *P. hysterophorus* emerged from shallow buried (< 0.5 cm) seeds and none from more than 5 cm depth possibly due to exhaustion of seedling reserves before emergence or an induced dormancy (Tamado *et al.*, 2002).

Parthenium seed banks are persistent with seed viability greater than 50 per cent after more than two years in the soil. Seed near the soil surface is rarely viable beyond two years. However, undisturbed, buried seed will stay dormant for a longer period. It has been recorded as viable for upto six years and even longer. In climates where rainfall is irregular, dormancy mechanisms prevent untimely germination. The size and persistence of the seed bank has implications several years beyond a reduction in parthenium weed populations (John Chamberlain and Amanda Gittens, 2004).

Buried seeds have been found to last much longer than seed on the soil surface, and a significant proportion can still germinate after eight to ten years.

Persistence tests demonstrated that more than 70% of parthenium seeds buried at 5 cm below the soil surface survived for at least 2 years, whereas seeds on the soil surface did not survive for longer than 6 months.

Effects on crop production : In India, according to Khosla and Sobti (1981), about 40% sorghum (*Sorghum bicolor* L.) yield reduction due to parthenium weed was recorded.

Channappagoudar *et al.* (1990) also reported that the presence of parthenium weed in irrigated sorghum in India reduced grain yield from 6.47 to 4.25 tons/ha (34.3%) and decreased grain weight by 30%.

However, its overall impact on the production system is multifaceted, both direct and indirect, thus making it difficult to quantify losses (Evans, 1997).

Impact : The weed adversely affects human and animal health and can cause

- Respiratory problems
- Severe dermatitis
- Tainted meat, if stock eat the plant within one month of slaughter
- Tainted milk.

Control

Type of infestation	Chemical	Biological	Physical	Pasture management	Mechanical
Light – few plants over a small area	Spot spray before seeds set.	Not suitable.	Hand pulling is not recomm-ended, because of the health risks associated with parthenium weed.	Maintain good pasture cover by not overgrazing	Some land hold-ers have achie-ved success by ploughing in parthenium weed in the rose-tte stage before it sets seeds, but this must be followed up by sowing a crop or direct seeding of perennial pasture.
Medium – plants over a medium area		Release biological control agents			
Heavy – Large number of plants		Establish a nursery site for biological control agents, if possible	Use strategic fencing to sepa-rate different land types and improve grazing management.		

Past work done on Seed Science and Technology

Navie *et al.* (1998) observed that the majority (73.7%) of the seeds were still viable after 24 months of burial. The remainder could not be recovered (18.0%) or were no longer viable (8.3%). There was a log-linear decline in persistence of germinable seeds over time, which indicated a constant rate of loss and a half-life of about 6 years. Although there was considerable rainfall (31 mm), seedlings did not emerge during the first month of this experiment. In the succeeding 3 months, there was substantial seedling emergence after rainfall, and 51.4% of seeds had germinated by the end of the fourth month. After 5 months, further seedling emergence was not detected, and intact seeds could not be located. These findings suggest that seed incorporation into the soil is important to the long-term persistence of *P. hysterophorus* seeds. In an initial test of germination, unburied seeds from the same seed lot exhibited a degree of innate dormancy, and this may explain the delayed germination observed in the surface-sown seeds. In the seed burial and recovery experiment, innate dormancy was lost after 2 months of burial in the field, although *in situ* germination of buried seed remained low for at least 24 months. Therefore, it appears that more than one dormancy mechanism may contribute to the persistence of *P. hysterophorus* seeds.

Tamado *et al.* (2002) observed that the mean germination percentage of the freshly harvested seeds of *P. hysterophorus* just before burial in the soil was only 5.3 ± 1.8% in light and 0% in darkness at room temperature. After 1 month of seed burial in the soil, however, the germination percentage was substantially higher both in light and darkness (Fig. 6). Therefore, it seems that *P. hysterophorus* has seeds that are initially dormant.

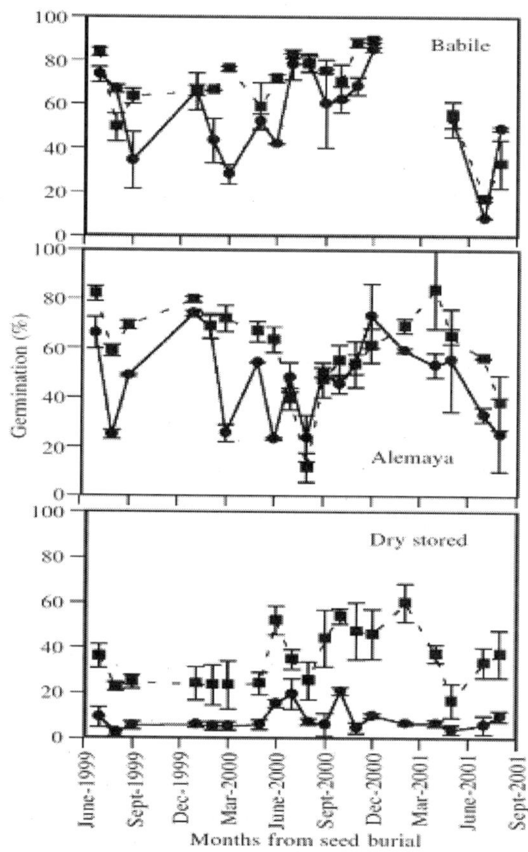

Fig. 6. Temporal pattern in germination of *Parthenium hysterophorus* seeds buried at Babile and Alemaya and dry stored in the laboratory. Data from Babile are missing from Feb. and April 2001. ■ dashed lines indicate germination in light. ● solid line show germination in darkness. Bars indicate ± SE.

After 26 months of seed burial, more than 50% of the buried seeds remained viable (Fig. 7). The loss in viability (Y) with period of burial months (X) was greater at Babile (Y = 95.8 - 1.33 X; t 14 df = 3.90; $P = 0.0014$) than at Alemaya (Y = 98.4 - 0.99 X; t 15 df = 3.73; $P = 0.0018$) possibly because of higher temperature at the former site. The regression predicted a 'half-life' of seeds of approximately 34 months and 48 months at Babile and Alemaya, respectively. These results suggest eradication of a population of *P. hysterophorus* to be a very difficult task because of its long-lived seeds.

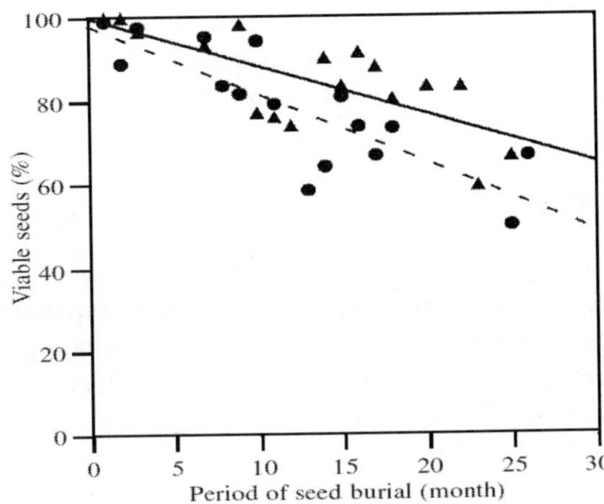

Fig. 7. Temporal variation in viability of *Parthenium hysterophorus* seeds buried under field conditions at Babile (● dashed line) and Alemaya (▲ solid line). The lines are fitted using linear regression. Each point represents 300 seeds.

Germinability of seeds dry-stored in the laboratory was generally lower than seeds buried in the field suggesting that dormancy in buried seeds is reduced by alternating temperature, moisture, chemical or biological factors operating in the field. Seeds exposed to light showed higher germination than those in darkness, but with time this difference tended to become smaller, especially in seeds that were buried in the field.

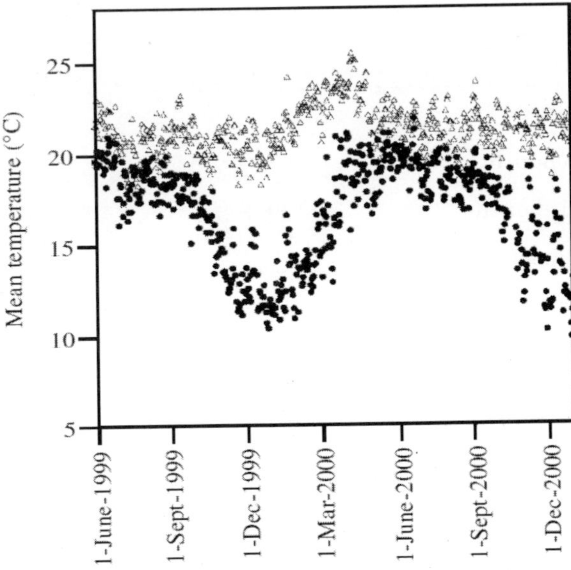

Fig. 8. Daily mean air temperature during the periods of Parthenium hysterophorus seed burial experiment at Babile (Δ) and Alemaya (●).

Absence of large seasonal changes in temperature at the sites where the seeds were buried (Fig. 8) could be the possible reason for absent or only weak seasonal variations in dormancy as these are believed to be governed by seasonal changes in temperature.

Seed age, temperature and light

Freshly harvested seeds collected in 1998 germinated only in light but, as seeds aged in dry storage, a minor fraction lost its light requirement and germinated in darkness. At all four test occasions, however, germination was substantially higher in light than in darkness (Table 1). The seeds collected in 2000 also showed an increased germination at 25°C with time (Table 2).

Table 1. Germination percentage (±SE) of *Parthenium hysterophorus* seeds stored dry for different periods of time at room temperature

Seed age (months)	Light	Darkness
0	6.7 ± 2.4	0
6	55.3±10.9	18.0 ± 13.3
12	46.7±7.0	15.3 ± 4.4
24	53.3±4.7	10.7±1.3

Table 2. Germination percentage (±SE) of *Parthenium hysterophorus* seeds stored dry for different periods of time at constant temperatures of 10°C and 25°C

	10°C	25°C
Light		
1 month	30.3± 2.6	47.0 ± 4.9
11 months	29.3±6.5	67.2 ± 5.5
Darkness		
1 month	0.9±0.9	3.9 ± 2.0
11 months	0.5± 0.5	21.7±4.9

Parthenium hysterophorus seeds germinated well over a wide range of fluctuating temperatures between 12/2°C to 35/25°C (Table 3) and only the lowest temperature impeded germination markedly.

Williams and Groves (1980) concluded from their experiment that only temperatures higher than 35°C or lower than 5°C limit germination of this species.

Table 3. Germination percentage (± SE) of *Parthenium hysterophorus* seeds tested at different daily flucutuating temperatures

Temperatures (°C)	Germination (%)
12/2	28.6 ± 4.5
15/5	62.7±3.7
20/10	71.3±4.2
25/15	71.2±2.6
30/20	68.0±3.0
35/25	73.9±7.6

Enhanced germination of *Parthenium hysterophorus* was reported under dark conditions (Dhawan and Dhawan, 1994).

Simulated moisture stress

Germination of both *P. hysterophorus* and sorghum decreased with decreasing osmotic potential and there was also a remarkable delay in germination at high PEG concentrations. The effect of moisture stress on germination of *P. hysterophorus* was more severe at higher temperature (Fig. 9) where no germination occurred at PEG concentrations greater than 210 g litre^{-1} (ys < -0.52 MPa). At 20°C, however, some *P. hysterophorus* seeds could germinate up to PEG concentration of 261 g litre^{-1} (ys = -0.86).

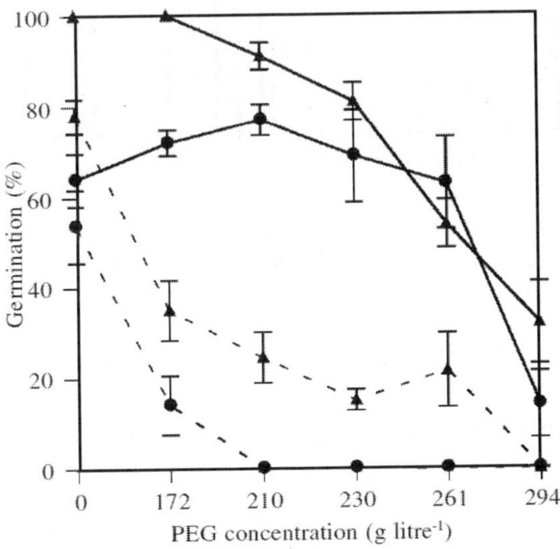

Fig. 9. Germination of *Parthenium hysterophorus* seeds (dashed lines) sorghum grain (solid lines) at 20°C and 27°C (•) in different levels of moisture stress created by PEG. Bars indicate ±SE.

Thus, if a sorghum crop is sown early in the field, it can take advantage of the limited available moisture and get an earlier start and thereby a competitive advantage towards *P. hysterophorus* and other late-germinating weed species in eastern Ethiopia.

Depth of seed burial on emergence

Tamado *et al.* (2002) reported that emergence from both non-pregerminated and pregerminated *P. hysterophorus* seeds depended on depth of burial. Seedlings from some pregerminated seeds emerged from 5 cm, but none from greater depths (Fig. 10).

Shallow burial (0.5 cm) gave the highest emergence percentage and the lower emergence from surface-placed seeds could be due to loss of seeds by predation or due to poorer soil contact and thereby diminished moisture availability. Decreasing emergence with depth of burial may be caused by exhaustion of seedling reserves before emergence.

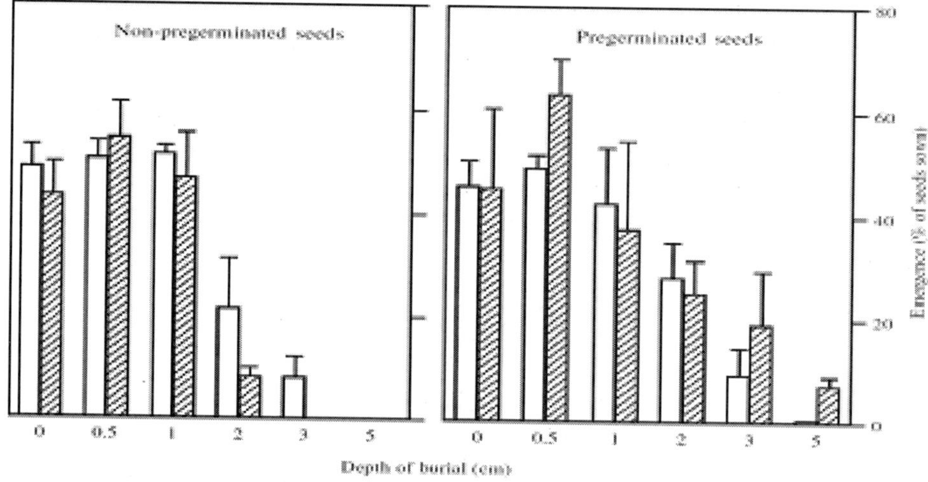

Fig. 10. Influence of depth of seed burial on emergence of *Parthenium hysterophorus* seedlings in sandy loam (unshaded bar) and loamy sand (diagonally striped bars) soils. Bars indicate SE. No seedlings emerged from 7.0 cm and 10.0 cm depth.

According to Pandey and Palni (2005), light exposure in imbibed state increased germination of freshly harvested *Parthenium hysterophorus* L. seeds to 68% from 8% in the dark. Dormancy breaking chemicals including GA_3, ascorbic acid, ethrel, thiourea, β-mercaptoethanol, kinetin, indole acetic acid (IAA), potassium nitrate, sodium azide and chlorine increased or resulted in maximum germination of the seeds in the dark substituting for illumination. However, with time the dark germination increased by 3 months to maximum by 7 months.

Nigatu *et al.* (2010) reported that in the soil seed bank, the viable seed density for *P. hysterophorus* was 68.5%. Similarly, the soil seed bank under the low, medium and high infestation sites was dominated by *P. hysterophorus*, which contributed 25.1, 65.4 and 87.4% of the viable seed bank, respectively.

Javaid and Shafique (2010) showed a great variation in final seed germination percentage of seeds collected in different months of the year 2006. Environmental conditions especially temperature and solar radiation during the formation of seeds seem to be the major contributing factor of parthenium seed dormancy. There was highly significant negative correlation between final germination percentage of parthenium seeds and prevailing environmental temperature conditions during the seed formation (Table 4). Maximum germination was recorded in the seeds collected during the coldest months of January and December when the mean daily temperature dropped to 14.1 and 15.1°C, respectively. Germination was slightly decreased to 97 and 87% during November and October respectively with the rise of temperature to 21.1 and 26.6°C. Lowest germination of 0 – 7% was recorded in seeds collected during summer months of April to September when the mean daily temperature raised from 29.2 to 33.6°C (Fig. 11). Final germination percentage was also negatively and significantly correlated with solar radiation (Table 4). Seeds formed under high solar radiation (SR) of 18.74 – 23.25 MJ/MF/day during April to September showed dormancy, while those under low solar radiation of 9.28 – 16.83 MJ/MF/day during October to December and January do not (Fig. 12). The final germination percentage exhibited an insignificant negative correlation with wind velocity, total precipitation, number of rainy days and relative humidity (Table 4).

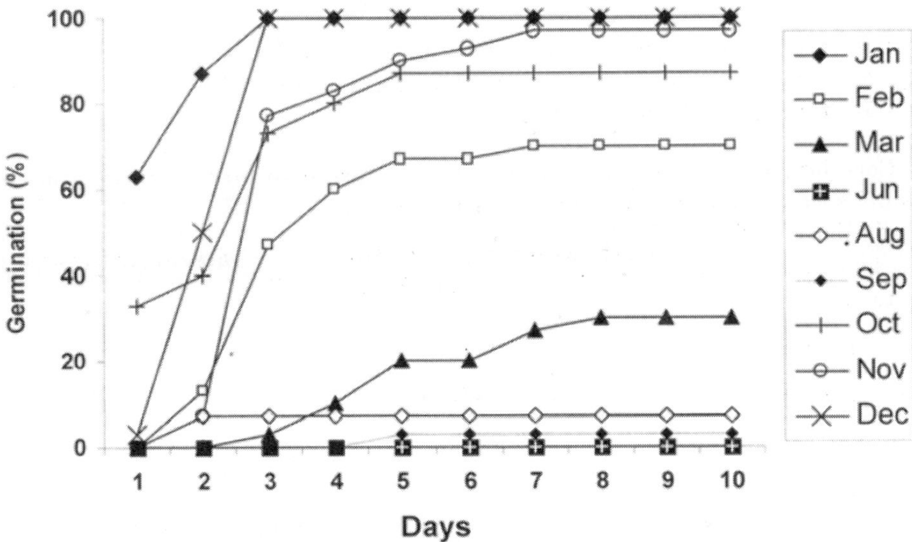

Fig. 11. Daily germination percentage of parthenium seeds collected in different months of the year 2006. Germination of seeds collected in April to July was zero.

There was a significant variation in rate of germination of parthenium seeds in different months. Highest germination rate of 33.3% was recorded in seeds collected during December and January. Germination rate during February, October and

November was 10, 17.4 and 13.8%, respectively. Seeds collected during rest of the months showed very low germination rate of 0–3.8% (Fig. 12). Similar to that of final germination percentage, germination rate also showed a highly significant negative correlation with mean daily, mean minimum and mean maximum environmental temperatures and a significant correlation with solar radiation during the seed formation period.

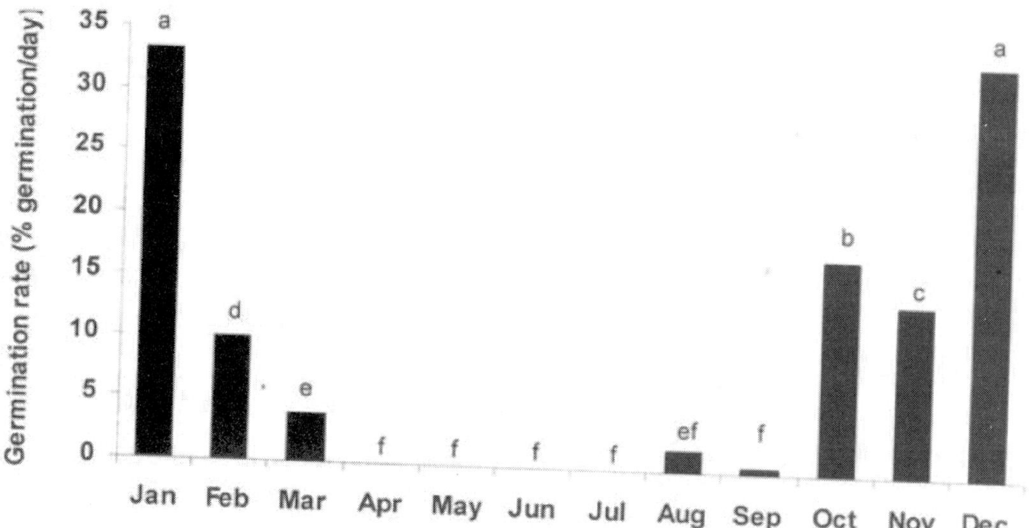

Fig. 12. Percentage germination rate of parthenium seeds collected in different months of the year 2006. Bars with different letters show significant difference as determined by Duncan's Multiple Range Test at $p<0.05$.

Table 4. Correlation of maximum germination percentage and germination rate of parthenium seeds with different environmental factors.

	Maximum germination	Germination rate
Mean daily temperature	– 0.86*	– 0.87**
Mean maximum temperature	– 0.86*	– 0.87**
Mean minimum temperature	– 0.85*	– 0.86*
Precipitation	– 0.55	– 0.44
No. of rainy days	– 0.59	– 0.43
Relative humidity	– 0.44	– 0.41
Bright sunshine duration	– 0.56	– 0.53
Solar radiation	– 0.90**	– 0.86*
Wind speed	– 0.58	– 0.48

REFERENCES

Auld, B.A., J. Hosking and R.E. McFadyen. (1983). Analysis of the spread of tiger pear and parthenium weed in Australia. *Australian Weeds,* 2: 56-60.

Butler, J.E. (1984). Longevity of *Parthenium hysterophorus* L. seed in the soil. *Australian Weeds* 3:6.

Carmona, R. (1992). Poblematica e manejo de bancos de sementes de invasoras em solos. *Planta Daninha,* 10: 5-16.

Carvalho, P.C. and V. Favoretto. (1995). Impacto das reserves de sementes no *Solosobre dinamica* populacional das pastogens Informativo Abrates, 5(1): 87-108.

Channappagoudar, B.B., Y.C. Panchal, S. Manjunath and R.V. Koti. (1990). Studies on influence of *Parthenium* on sorghum growth under irrigated conditions. *Farming Systems* 6: 102-104.

Dale, I.J. (1981). Parthenium weed in America. Report on the ecology of *Parthenium hysterophorus* in South Central and North America. *Australian Weeds,* 1:8-14.

Dhawan, S. R and P. Dhawan. (1994). Congress grass: effect of temperature and light on seed germination. Adv. Pl. Sci., 7: 177-178.

Evans, H.C. (1997). *Parthenium hysterophorus* L. A Review of its weed status and the possibility of biological control. *Bio-control News and Information,* 18:89-98.

Fernandez-Quintanilla, C., Soavedra, and M. S. Molas Hierbas. (1991). Conceptos Generals. **In:** Garcia Torre, L., Fernandez-Quintanilla, C. Fundamentos Sobre Malas hierbas Y herbicides. Madrids: Mundi Prensa, pp. 26-48.

Gupta, O.P and J.J Sharma. (1977). *Parthenium menace* in India and possible control measure. FAO, *Plant Protection Bulletin,* 25: 112-117.

Haseler, W.H. (1976). *Parthenium hysterophorus* L. in Australia. Pest Articles & News Summaries (PANS). 22: 515-517.

Javaid, A.S. and S. Shafique. (2010). Seasonal pattern of seed dormancy in *Parthenium hysterophorus* L. *Pak. J. Bot.,* 42(1): 497-503.

Joshi, S. (1991). Interference effects of *Cassia uniflora* Mill. on *Parthenium hysterophorus* L. *Plant and Soil.,* 132: 215-223.

John Chamberlain and Amanda Gittens. (2004). Department of Natural Resources, Mines and Energy, Queensland NSW Agriculture.

Khosla, S.N and S.N. Sobti. (1981). Effective control of *Parthenium hysterophorus* L. *Pesticides,* 15:18-19.

Lisanework Nigatu, Asresie Hassen, Janmejai Sharma and Steve W. Adkins. (2010). Impact of *Parthenium hysterophorus* on grazing land communities in north-eastern Ethiopia. *Weed Biology and Management.* Volume 10, Issue 3, pages 143–152.

Mahadevappa, M. (1997). Ecology, distribution, menace and management of parthenium. **In:** Mahadevappa, M. and Patil, V. C. (eds), Proceedings of the 1st International Conference on Parthenium Management, University of Agricultural Sciences, Dharwad, India, 1-12.

McFadyen, R.C. (1992). Biological control of parthenium weed in Australia. *Crop protection* 11: 400-407.

Meyer, S.E. and P.S. Allen (1999). Ecological genetics of seed germination regulation in *Bromus tectorum* L. I. Phenotypic variance among and within populations. *Oecologia,* 120: 27-34.

Monaco, J. Thomas, S.C. Weller and F.M. Ashton (2001). Weed biology and ecology. **In**: *Weed Sci.,* Principles and Practices 4th ed., pp. 13-31. *Academic Publisher*, USA.

Navie S.C., F.D. Panetta, R.E. McFadyen and S.W. Adkins (1998). Behaviour of buried and surface sown seeds of *Parthenium hysterophorus. Weed Res.,* 38: 335-341.

Navie, S.C., R.E. McFadyen, F.D. Panetta and S.W. Adkins (1996). The Biology of Australian Weeds 27. *Parthenium hysterophorus* L. *Plant Protection Quarterly,* 11:76-88.

Nigatu, L., A. Hassen, J. Sharma and S. W. Adkins (2010). "Impact of *Parthenium hysterophorus* on Grazing Land Communities in North-Eastern Ethiopia." *Weed Biology and Management* 10 (3): 143-152.

Pandey, H.N and S.K. Dubey (1989). Growth and population dynamics of an exotic weed *Parthenium hysterophorus* L. *Proceedings of the Indian Academy of Science (Plant Sciences),* 99:51-58.

Pandey, D.K. and L.M.S. Palni (2005). Germination of *Parthenium hysterophorus* L. seeds under the influence of light and germination promoting chemicals, *Seed Science and Technology,* 33(2): 485-491.

Ramaswami, P.P. (1997). Potential uses of Parthenium. **In:** *Proc. First Int. Conf. Parthenium Manage,* 77-80.

Tamado, Schütz and Milberg (2002). Germination ecology of the weed *Parthenium hysterophorus* in eastern Ethiopia. *Ann. appl. Biol.,* 140:263-270.

Tamado, T., L. Ohlander and P. Milberg (2002). Interference by the weed *Parthenium hysterophorus* L. with grain sorghum: influence of weed density and duration of competition. *International Journal of Pest Management,* 48.

Williams, J.D. and R.H. Groves (1980). The influence of temperature and photoperiod on growth and development of *Parthenium hysterophorus* L. *Weed Research* 20:47-52.

AFRICAN FOXTAIL (*Pennisetunm cenchroides*)

Common name : African foxtail, Buffel grass, Dhaman grass, Anjan grass

Botanical name : Pennisetunm cenchroides

Tamil name : Kozhukattai pul

Hindi name : Dhaman

Family : Gramineae (Poaceae)

Origin and distribution : Buffel grass is a bunch grass native to the hotter and drier areas of India, the Mediterranean basin and Tropical and Southern Africa.

Botany

Habitat : Grown in all kind of soil and grows even when the soil is dry.

Habit : Perennial grass (Fig.1).

Stem : It consists of aerial branches and under-ground rhizomiferous stems, bearing thick fibrous roots and numerous buds covered by scarious sheaths. The aerial branches are tufted, erect or decumbent and geniculately ascending.

Leaf : Leaf sheath is slightly compressed; keeled, with scattered long hairs outside, shorter than the internodes. The ligule is a short thin membrane fringed with hairs. Leaf blade is linear, tapering to a very fine point.

Floral biology

Flower : Inflorescence is a raceme of spikes, with spikes mostly densely arranged, though occasionally distant and not close-set, on a long, slender, puberulous or scaberulous peduncle; rachis is flexuous, flattened, grooved and scaberulous. The spikes have involucels, consisting of two series of bristles. Spikelets oblong-lanceolate, one to three in a spike and sessile. Four glumes. Anthers are three, yellow, stigmas white, feathery and the styles shortly united at the base (Fig. 2).

Fig. 1. Plant

Fig. 2. Flower

Flowering and fruiting : An early flowering short day plant with an optimal photoperiod of 12 hours. Flowering time varies among ecotypes.

Fruit : Lodicules are not present. Fascicle type of fruit.

Seed biology

Seed : 330,000-550,000 seed units/kg, or 900,000-2,000,000 caryopsis/kg.

Propagation : Seeds and stem bits.

Seed dispersal : Spreads readily by seed, which is well adapted to dispersal by wind or water movement. Seed also moved by livestock through adhesion to fur, or through ingestion and defecation. No natural spread on soils of pH <7, even though establishment with cultivation is possible on these soils.

Seed dormancy and germination : Fresh seed often has high level of dormancy. Germination rate can be improved with storage of 6-18 months after harvest or by separating caryopses.

Effect on crop production : It spreads very quickly and will often kill local native plants such as palo verdes by taking away nearby water. This plant has a very low ignition threshold and can burst into a raging flame even during the peak growing season. Its flammability and quick regrowth allow it to compete successfully against almost all vegetation in this region.

Control measures : Can be controlled using a combination of glyphosate and ammonium sulphate, possibly in repeat applications. Seedlings can be controlled using the grass -selective herbicide, fluazifop-p-butyl or dicamba, 2,4-D, 3,6-dichloropicolinic acid, triclopyr, tebuthiuron, or hexazinone. Older stands, particularly freshly cut material can be at least reduced using hexazinone or tebuthiuron.

Uses : Mainly used as a permanent pasture, but can be used for hay or silage. Not suited to short-term pasture because too difficult to remove and binds nutrient.

MADRAS LEAF-FLOWER (*Phyllanthus maderaspatensis*)

Common name : Madras leaf-flower

Botanical name : Phyllanthus maderaspatensis L.

Tamil names : Melaanelli

Hindi name : Hajarmani

Family : Euphorbiaceae

Origin and distribution : Phyllanthus maderaspatensis is widespread in tropical Africa and the Indian Ocean Islands. It occurs throughout the Old World tropics and subtropics.

Botany

Habitat : Grows in flood plain areas on heavy soils and may rely on appropriate and intermittent rainfall and flooding events for its survival. The species is described as being a summer-growing annual and is thus dependent on seasonal conditions. Often associated with open grasslands and eucalypt woodlands in or near creek beds, and grassy flats and lives near water sources. Plants are usually infrequent in abundance but have been recorded as common in disturbed areas.

Habit : Monoecious, annual or perennial, erect to spreading, unbranched, glabrous herb up to 90 - 120 cm tall; branches angular, red or brown tinged (Fig. 1).

Leaf : Leaves glabrous, subcoriaceous, scattered, subsessile, cuneatly obovate, glaucous beneath, stipulate; stipules peltate, lanceolate, membranous

Fig. 1. Plant

Floral biology

Flower : Flowers very small, monoecious, male facicled, female shortly pedicelled, solitary; sepals broad, green with white margins rounded or obcuneate; disk glandular in both; stamens 3, almost sessile on the column, erect, apiculate; ovary superior, tricarpellary, syncarpous, styles free, minute, two lobed (Fig. 2).

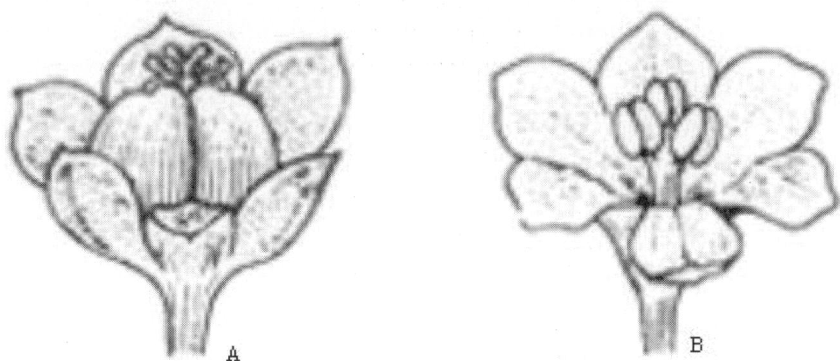

Fig. 2. A. Male flower, B. Female flower

Fruit : Globose, schizocarpic; seeds striated, trigonous (Fig. 3).

Seed biology

Seed : Seeds striated trigonous, 1.5 mm long, dark brown, shiny (Fig. 4).

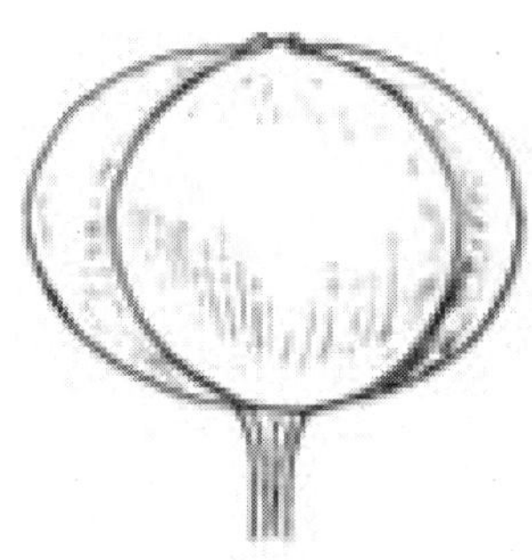

Fig. 3. Fruit

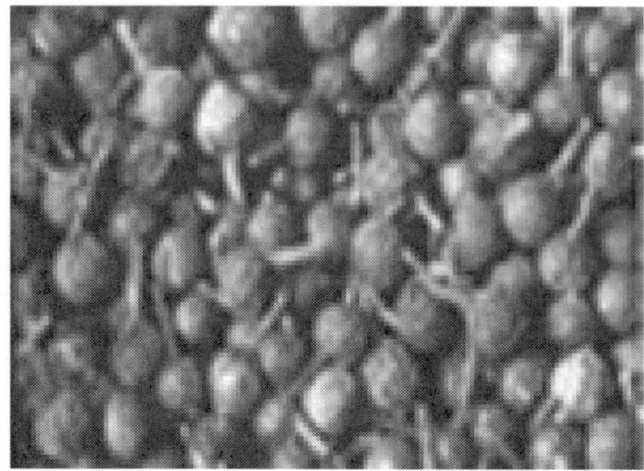

Fig. 4. Seed

Propagation : By seed. Vegetative propagation by budding, grafting, cutting and root sprouting is possible.

Seed dispersal : Seed dispersal through winds and animals.

Life cycle : Flowering time is spring to summer (May – July) and the species is a summer growing annual. Fruiting is from August to November. Seeding is recorded in March. Occurs after summer rains and readily drops its leaves as it dries off.

Seed dormancy and germination: Seeds of *Phyllanthus maderaspatensis* require light to germinate. Germination rates increase linearly with temperature from 15–35°C and decline from 36°C. Below 15°C and above 40°C, germination is very poor. Germination is poor under moisture stress. *Phyllanthus maderaspatensis* is generally grown from seed.

Uses : The plant sap and leaf decoction are credited with emetic and purgative activities.

In Tanzania, the whole plant is pounded and the solution applied to scabies. A root decoction is taken to cure constipation, diarrhoea, lack of appetite, intestinal pain, menstrual problems, gastrointestinal disorders, testicular swelling, chest complaints and snakebites.

Plant sap is used as nose drops to treat toothache. Ground leaves are rubbed on the skin with lemon juice as treatment for rheumatism.

In Niger, the plant is used as an aphrodisiac.

In Somalia, *Phyllanthus maderaspatensis* is considered poisonous.

In India, *Phyllanthus maderaspatensis* is widely medicinally used to treat headache, bronchitis, earache and ophthalmia. Powder from dried plant material mixed with milk is drunk to treat jaundice.

In Kenya, smoke from the burning plants is used to kill caterpillars in maize. Cattle will browse *Phyllanthus maderaspatensis* but only when green.

PRICKLY SIDA (*Sida spinosa*)

Common name : Sickle pod, Prickly sida

Botanical name : Sida spinosa

Tamil name : Mayir manikkam

Hindi name : Bariyara, Gulasakari

Family : Malvacaeae

Origin and distribution : Tropical American weed naturalized in Southern United States.

Botany

Habitat : An annual weed. The plant needs high temperatures for its development but may also grow in mountains upto 2.500 m height.

Habit : It is a subshrub, forb/herb and grows upto 20-90 cm tall depending on growing conditions (Fig. 1).

Stem : The stem is tough, upright, with many branches, covered with hairs. The main nodes carry 2-3 thorns.

Leaf : Leaves are alternate, elongated, with toothed margins, 2-4 cm long, petiolate.

Floral biology

Flower : Flowers are axillary, situated at the end of short pedicels, sometimes in small clusters. The corolla is composed of 5 light yellow petals. There is a single thin taproot (Fig. 2).

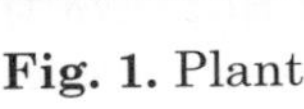

Fig. 1. Plant

Fig. 2. Flower

Fruit : The fruit is a ring of 5 one seeded segments, each with 2 sharp spines at the tip. The ring breaks up maturity, releasing 5 seeds (Fig. 3).

Fig. 3. Fruit

Seed biology

Seed : The seeds are ovate, light brown to grey brown, 2-3 mm long (Fig. 4).

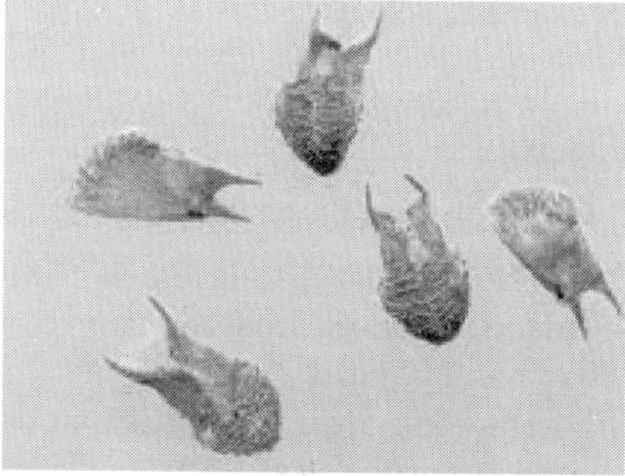

Fig. 4. Seed

Propagation : By seeds

Seed dispersal : Seeds are also carried with harvested seeds of many crops, grain and hay, but also with other plant products. Animals help to distribute the seeds as the spines on the segments (Wagner *et al.*, 1999).

Seed dormancy and germination : Fresh seeds do not imbibe water and germinate because the seed coat is impermeable to water. Seeds imbibe water when a portion of the coat in the chalazal area separates from an underlying layer of subpalisade cells. Thin walled subpalisade cells are limited to the area beneath the chalazal area. Palisade cells in the coat region that become permeable are tall, only lightly lignified, contain abundant hemicellulose and have large cell lumena. Palisade cells in regions of the coat that remain impermeable to water are short, have heavily lignified thick walls, small cell lumena and are tightly packed. The seed coat dormancy is released when moisture eventually reaches the hygroscopic hemicellulose deposits in cell lumena in the lightly lignified palisade cells of the chalazal region. The subsequent expansion of the palisade cells causes the thin-walled subpalisade cells to break, resulting in separation of palisade from subpalisade cells and free passage of water through the exposed surface to the embryo, culminating in germination (Egley *et al.,* 1986).

Freshly produced, mature prickly seeds (18 to 21 days after anthesis, < 20% water content) were dormant and neither imbibed water nor germinated when incubated for upto 4 weeks under several light and temperature conditions. Over 80% of the freshly-produced, immature seeds (12 to 16 days after anthesis, > 20% water content) germinated when removed from the mother-plant before dehydration and incubated in alternating temperature of 20° to 30°C for 4 weeks. The onset of dormancy coincided with the later stages of seed dehydration and coat-hardening. Over 90% of the mature seeds imbibed water and germinated when incubated at 35°C after 4 month dry storage at 25°C. A puncture through the seed coat, either over the radicle or cotyledons, permitted water imbibition by all mature seeds, but the puncture over the radicle was significantly more effective in inducing germination. Seed coat impermeability was important, but was not the only factor responsible for prickly sida dormancy.

Sickle pod have a hard seed coat. Abrasion for 5, 10, or 15 s, puncturing with a needle, and sulphuric treatment were all effective in initiating germination. It is emerged from a soil depth of 12.5 cm and the emergence of sickle pod is 63%.

Various environmental factors were tested under laboratory conditions to determine their effects on germination of seeds of prickly sida. Neither freezing and thawing nor moist chilling at 5°C promoted seed germination. However, increasing the incubation temperature and subjecting seeds to wet-dry cycles enhanced germination; high temperatures were more effective than alternate wetting and drying. Shifting seeds from a lower to a higher temperature regime increased germination. Seeds shifted from 15/6, 20/10, 25/15, or 30/15°C to higher regimes of 20/10, 25/15, 30/15, 35/20, or 40/25°C germinated to greater percentages than did seeds kept continuously at the lower thermo periods. With an increase in length of time, seeds were at a lower temperature, there was an increase in the percentage that germinated after they were moved to a higher regime.

Seed burial and longevity : Prickly sida emergence was optimal at a planting depth of 0.5 cm, and declined rapidly at deeper planting depths (Smith *et al.,* 1992).

Seeds of buried for 3.5 years showed 21% viability (Egley and Chandler, 1983).

Effects on crop production : All agricultural cultivated land may be infested by *Sida spinosa*, especially cotton and soybean fields, as well as pastures, meadows and uncultivated land.

Sida spinosa is an annual weed, reproducing by seeds, which strongly competes with crops for water and nutrients. It can seriously reduce the yield and quality of harvested cereals and other field crops. In pastures, the weed replaces other herbaceous plants, reducing the quality of pasturage.

Past work done in Seed Science and Technology

Effect of temperature on germination

Singh and Singh (2009a) recorded sicklepod germination was recorded at all temperature regimes from 15 to 45°C; though germination was reduced at both the extremes. A reduction of 67 and 72% in germination was recorded at 15 and 45°C, respectively, compared to the highest germination of 54% at 30°C. Germination of sicklepod was similar between temperature ranges of 20 to 40°C.

Higher germination of freshly harvested dormant seed of prickly sida was recorded at 35 than 15 or 25°C under dark conditions (Egley, 1976). Germination of 9 month stored seeds at 25°C was similar under dark or light conditions at 35°C.

According to Singh and Singh (2009a) a temperature range of 25 to 40°C was ideal for the germination of prickly sida; germination was reduced by 50% at 45 or 20°C compared to 30°C. Germination was greatly reduced at 15°C as only 4% seeds were able to germinate compared to 79% at 35°C.

Alternating higher temperature and wet/dry cycle have been found to promote prickly sida germination by breaking its dormancy (Baskin and Baskin, 1984).

Effect of pH on germination

Singh and Singh (2009a) stated that in case of prickly sida, highest germination was recorded at pH 9 and germination decreased both with increase or decrease in pH from 9. No germination of prickly sida was recorded at pH 3; germination was similar at pH 7 and 11, but higher than pH 5.

Effect of water stress on germination

An osmotic stress of -0.2 Mpa reduced Sida spinosa germination (Smith *et al.*, 1992).

Singh and Singh (2009) stated that a water potential of -0.1 had no effect on the germination of prickly sida, but further lowering the water potential to -0.2 and -0.4 Mpa registered 32 and 97% reduction in germination (Table 1).

Table 1. Effect of osmotic potential and duration on percent germination of prickly sida seed*

Duration (DAT)	Control	-0.1 Mpa	-0.2 Mpa	-0.4 Mpa	-0.8 Mpa	-1.2 MPa
7	79	45	24	1	0	0
14	79	71	50	1	0	0
21	79	78	54	3	0	0

*Means followed by same letter within species do not significantly differ using Student-Newman-Keuls (P=0.5). DAT=Days after transplanting.

Hoveland and Buchanan (1973) reported that under artificial drought *Ipomoea lacunosa* was the most tolerant weed species followed by *Sida spinosa*.

Singh and Singh (2009b) expressed that increasing the temperature from 15 to 20 and 30°C increased germination of weed species. Germination was reduced drastically at -0.1 and -0.5 osmotic potential compared to control treatments. There was no germination of weed species at -1.0 Mpa osmotic potential. Significant positive correlation was observed in prickly sida as germination increased with increase in temperature and water stress. Germination was 66% at 30°C compared to only 31% at 20°C when -0.1 Mpa water stress was imposed (Fig. 5). Only 3% germination of prickly sida was recorded at -0.5 Mpa at 30°C and no germination at lower temperatures.

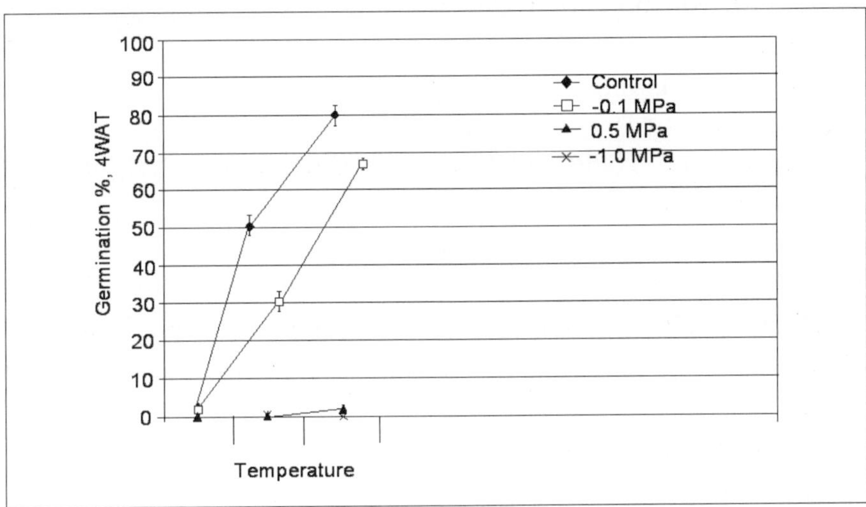

Fig. 5. Effect of temperature and water potential on germination of prickly sida. Bars indicate standard error of means (SEm).

REFERENCES

Baskin, J.M. and C.C. Baskin. (1984). Environmental conditions required for germination of prickly sida (*Sida spinosa*). *Weed Sci.,* 32 : 786-791.

Egley, G. H. (1976). Germination of developing prickly sida seeds. *Weed Sci.,* 24 : 239-243.

Egley, G.H. and J.M. Chandler. (1983). Longevity of weed seeds after 5.5 years in the Stoneville 50-year buried seed study. *Weed Sci.,* 31:264 – 270.

Egley, G.H., R.N. Paul and A.R. Lax. (1986). Seed coat imposed dormancy: Histochemistry of the region controlling onset of water entry into *Sida spinosa* seeds. *Physiologia Plantarum,* 67: 320–327.

Singh, S. and M. Singh. (2009a). Effect of Temperature, Light and pH on Germination of Twelve Weed Species. *Indian J. Weed Sci.,* 41(3&4): 113-126.

Singh, S. and M. Singh. (2009b). Effect of Temperature and Water Potential on Germination of Twelve Weed Species. *Indian J. Weed Sci.,* 41(3&4): 135-145.

Smith, C.A., D.R. Shaw and J. Newsom. (1992). Arrow leaf sida (*Sida rhombifolia*) and prickly sida (*Sida spinosa*): germination and emergence. *Weed Res.*, 32: 103–109.

Wagner, W.L., D.R. Herbst and S.H. Sohmer. (1999). Colocasia. In: Manual of the Flowering Plants of Hawaii. *University of Hawaii Press*, Honolulu, Hawai'i, pp. 1356-1357.

SILVERLEAF NIGHTSHADE (*Solanum elaeagnifolium*)

Common name : Silverleaf nightshade

Botanical name : Solanum elaeagnifolium

Tamil names : Kattu kandan-kattiri.

Family : Solanaceae.

Origin and distribution : Solanum elaeagnifolium is native to North-East Mexico and South-West USA. It is also considered native to Argentina. In India, it is a common weed in Tamil Nadu and Karnataka (Boyd *et al.,* 1984; Roche, 1991).

Botany

Habitat : Throughout India, in roadsides and waste places

Habit : Armed, diffuse subshrub, prickles numerous, straight to 2 cm (Fig. 1).

Stem : The stem is somewhat round and covered with sparsely distributed, star like hairs.

Leaf : Leaves lacerate, 3.5-8 x 1.5-5 cm, prickly along midnerve.

Floral biology

Flower : The inflorescence is a solitary cyme of 1-7 flowers, with long peduncules (5–20 mm). The calyx measures 5 - 7 mm, with 2- 4 linear lobes. The corolla is 25-35 mm large and is orbicular, generally bright blue to purple but sometimes white with yellow anthers of 7-9 mm (Fig. 2).

Fig. 1. Plant

Fig. 2. Flower

Flowering and pollination: Pollination is believed to occur *via* insects. Individual plants produce upto 200 berries per growing season, approximating to 1500–7200 seeds per plant.

Fruit : The fruit is an irregularly dehiscent berry, initially spherical, green (with white patches) and fleshy, drying and becoming yellow to orange (10 -15 mm in diameter) at maturity (Fig. 3). A single plant generally produces 40–60 fruits, each containing 60–120 seeds, smooth, flat, greenish-brown, 2–3 mm in diameter, closely resembling those of tomatoes.

Fig. 3. Fruit

Seed biology

Seed : Seeds yellowish brown to dark yellow brown, 2.5 - 3.5 mm long, 1.8 - 2.5 mm wide, smooth (finely granular with magnification), semi glossy (Fig. 4).

Individual berries produce 24 to 149 seeds (Boyd and Murray, 1982), which can add upto 5 million to 100 million seeds per acre (Cooley and Smith 1971).

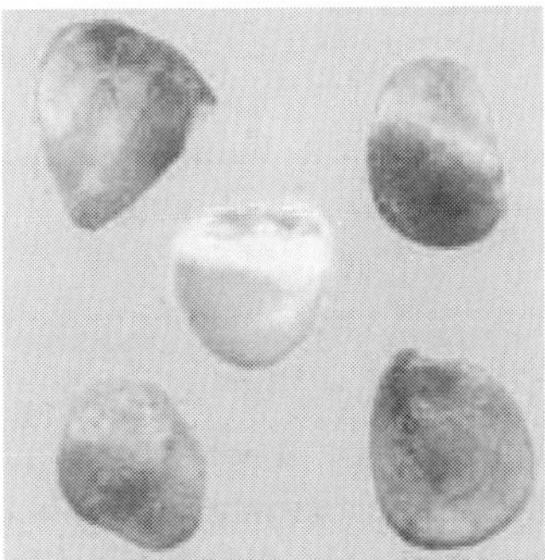

Fig. 4. Seed

Propagation : Solanum elaeagnifolium can be spread by seed, rhizomes, and/or root fragments (Boyd and Murray, 1982).

Reproduction by seed is secondary, though seeds are highly viable and last at least 10 years in soil. High levels of dormancy and infrequent germination can lead to the establishment of extensive viable seed banks.

S. elaeagnifolium has a highly regenerative capacity from rootstocks, separated shoots, rhizomes and roots but the rhizome is the principal means and has more regenerative capacity than the other organs probably rhizomes contain more carbohydrate.

Seed dispersal : Seeds may be dispersed by wind, water, machinery, agricultural produce or animal feces; studies indicate that 10% of seed is still viable after passing through sheep. Dried plants may also blow like tumbleweeds, spreading seed along the way (Boyd *et al.*, 1984).

Life cycle : The life cycle of the plant is composed of 5 phases: vegetative regeneration and germination during spring; vegetative development, the duration of which depends on the biotope; flowering from spring to the end of summer and fructification from the end of spring until autumn (Fig. 5).

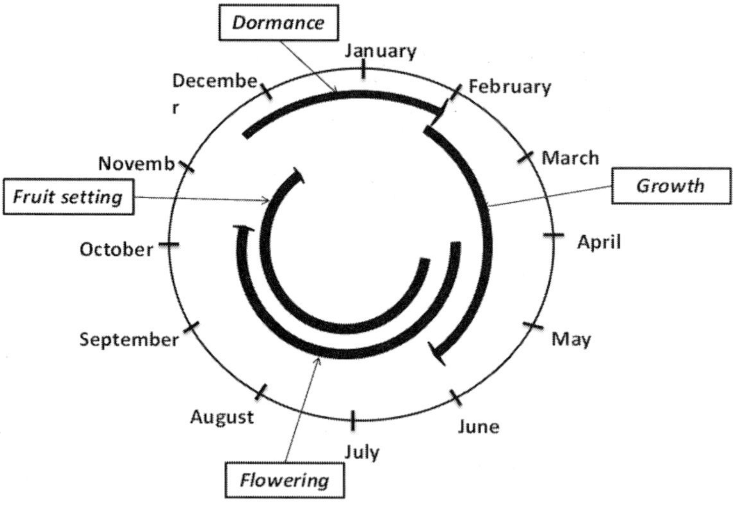

Fig. 5. Life cycle

The plant grows vegetatively during the beginning of the growth period and flowers later in the same period, after fruit maturity the plant goes dormant in winter as the herbaceous above ground shoots die.

Seed dormancy and Germination : Seeds require fluctuating temperatures to germinate. Boyd and Murray (1982) obtained a maximum germination rate of 57% when they germinated seeds at 20°C for 16 hours and 30°C for 8 hours; light and dark had no effect. It requires high temperatures (20–34°C) for normal germination and growth. Germination is advanced under favourable moisture conditions, notably heavy rains, and by alternating temperatures. They also found that a pH of 6 or 7 was optimal for germination. Rutherford (1978) indicated that immersing seeds in running water for relatively long periods may improve germination rates.

Hairy nightshade seeds were in an innate dormant state when initially separated from the berries. Moist compared with dry storage was more effective for breaking dormancy at 4°C, but dry storage was more effective at 17°C (Jingkai Zhou *et al.*, 2005).

Studies indicate that 10% of seed is still viable after passing through the digestive tract of sheep (Washington State Noxious Weed Control Board, 2006). Seeds require fluctuating temperatures to germinate. Boyd and Murray (1982) obtained a maximum germination rate of 57% when they germinated seeds at 20°C for 16 h and at 30°C for 8 h; light and darkness had no effect. They also found that a pH of 6 or 7 was optimal for germination. Rutherford (1978) indicated that immersing seeds in running water for relatively long periods may improve germination rates. Factors controlling seed germination are poorly understood, however, the reproductive biology of *S. elaeagnifolium* is comprehensively reviewed by Heap and Carter (1999), Parsons (1981) and Richardson and McKenzie (1981).

Seed storage and burial : Seeds are highly viable and last at least 10 years in soil. High levels of dormancy and infrequent germination can lead to the establishment of extensive viable seed banks.

Effect of crop production : S. elaeagnifolium competes for moisture and nutrients with many crops. The most affected crops are cotton, sorghum, wheat etc., Infestations are more serious in drylands, although irrigated croplands are also very prone to invasion and many other horticultural and orchard crops are affected.

In Morocco, losses of upto 64% in maize without treatment (Baye and Bouhache, 2007) and 78% in cotton have been reported. In Australia, wheat losses varied from 12 to 50% (Cuthbertson *et al.,* 1976) according to climatic conditions and weed density, but were highest at dry, sandy sites or during low rainfall years (Heap and Carter, 1999).

Uses : Fruits eaten as an anthelmintic and for indigestion. Root is an expectorant, used in Ayurvedic medicine for cough, asthma and chest pain. Also used for flatulence, sore throat, and toothache. Has high concentration of solasodine, a starting material for the manufacture of cortisone and sex hormones. It cures asthma, cough, bronchspasm, sore throat, constipation, an effective expectorant and diuretic.

Past work done on Seed Science and Technology

Biology of *Solanum elaeagnifolium* Cavi. (Solanaceae): Growth and survival (Abdessatar Omezine and Fethia Skhiri-harzalla, 2009).

1. Effect of rhizome storage on sprouting and viability

The environmental conditions surrounding the rhizome fragments during storage, either in refrigerator or ambient temperature affect the longevity. Drying into ambient temperature to less than 40% of the original moisture content was lethal to all rhizome fragments. This moisture content was obtained after 2 days of rhizome exposition to ambient temperature (Fig. 6). The storage in refrigerator (cold temperature) also affects the longevity; however the longevity in cold temperature was longer than that in the ambient temperature that is 40 days. This result indicated that these rhizomes were very sensitive to desiccation.

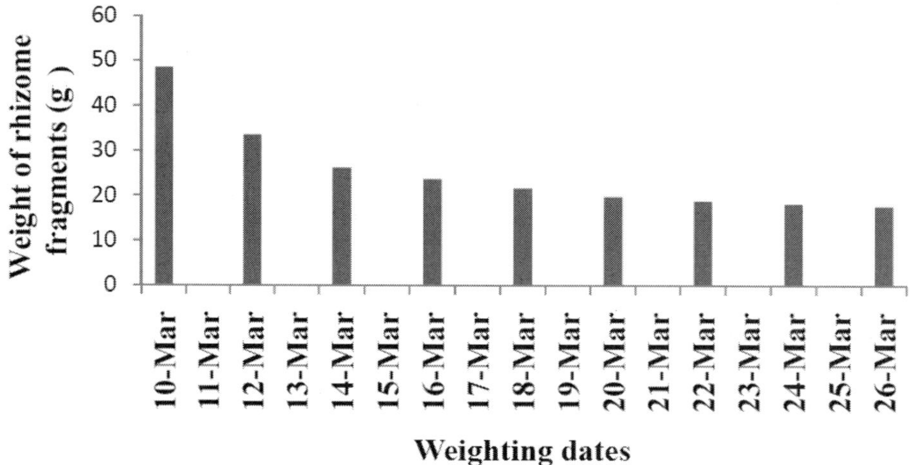

Fig. 6. Effect of storage of *S. elaeagnifolium* rhizome fragments in ambient temperature (27°± 2°C).

2. Effect of burial and duration of depth on emergence and viability of rhizomes

The study of the effect on planting depths showed that there was no difference between the effects of planting depths on the ability of *S. elaeagnifolium* rhizome fragments to sprout. But, a net difference was observed between the planting depth and emergence (Fig. 7). The result showed that deeper rhizome fragments took more time to emerge. The effect of burial depth exhibits an inverse relationship between depth and its emergence. Rhizome fragments emerged readily from a depth of as little as 2 cm, but no emergence was observed deeper than 30 cm. Emergence dropped off rather sharply with increases in depth. It takes 5 days to emerge from 2 cm depth and 45 days to emerge from 30 cm depth. This failure of emergence from deeper than 30 cm depth can be explained by either the low quantity of carbohydrate content, large quantity of carbon oxide content in soil depth or exhibition of dormancy of rhizome buds or decay of rhizomes. The dormancy duration for rhizome fragments buried 60 cm deep in soil was calculated to be seven (7) months. The effect of burial depth of rhizome fragments exhibits a direct relationship between the depth to which rhizome fragments were buried into the soil and their retention of viability (Fig. 8). Moreover, the result from the viability test of excavated rhizome fragments from the rootstock indicated that the deeper rootstock showed a decreased viability probably it was caused by the absence of viable buds or exhausting buds from repeated growth and competition. The relationship between depth of excavated rhizome fragments and their retention of viability is shown in Fig. 7. *Solanum elaeagnifolium* showed a short lived nature of their rhizome fragments.

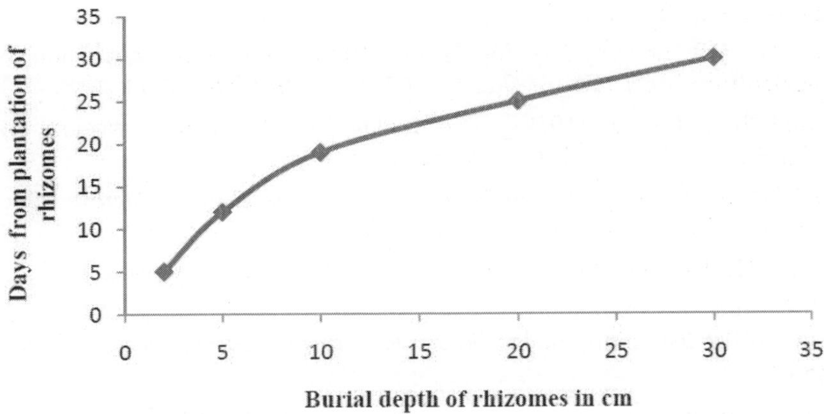

Fig.7. Number of days to emerge from the indicated burial depth of *S. elaeagnifolium* rhizome fragments.

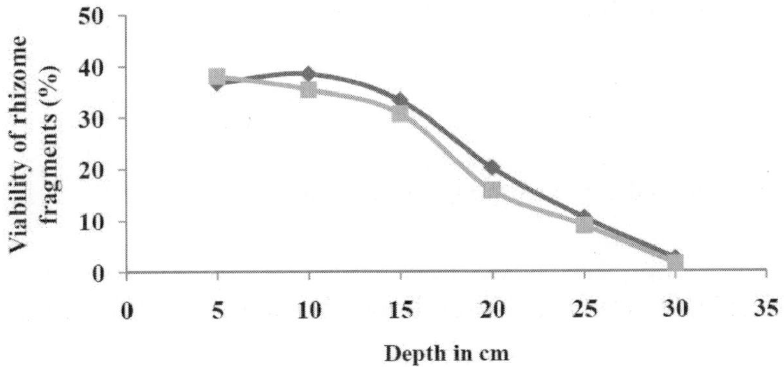

Fig. 8. Viability of rhizome fragments of *S. elaeagnifolium* (%) either excavated (black) from or buried (gray) in the soil at the indicated depth (cm).

REFERENCES

Abdessatar Omezine and Fethia Skhiri-Harzalla. (2009). Biology of *Solanum elaeagnifolium* Cav. (Solanaceae): Growth and survival, *Pak. J. Weed Sci. Res.*, 15(4): 251-268,

Baye, Y. and M. Bouhache. (2007). Etude de la compétition entre la morelle jaune (*Solanum elaeagnifolium* Cav.) et le maïs de printemps (*Zea mays* L.) *Bulletin OEPP/EPPO Bulletin*, 37: 129–131.

Boyd, J.W, D.S. Murray, and R.J. Tyri (1984). Silverleaf nightshade, *Solanum elaeagnifolium*, origin, distribution and relation to man. *Eco. Bot.,* 38:210-217.

Boyd, J.W. and D. S. Murray. (1982). Effects of shade on silverleaf nightshade (*Solanum elaeagnifolium*). *Weed Science,* 30:264-269.

Cooley, A.W and D.T. Smith. (1971). Silver leaf night shade (white weed) establishment from seed and clipped seedlings. Progress Report of Texas A&M University, Texas Agricultural Experiment Station, PR- 3198.

Cuthbertson, E.G., A.R. Leys and G. McMaster (1976). Silverleaf nightshade – a potential threat to agriculture. *Agricultural Gazette of New South Wales* 87:11–13.

Heap, J.W and R.J. Carter (1999). The biology of Australian weeds. 35. *Solanum elaeagnifolium* Cav. *Plant Protection Quarterly,* 14: 2–12.

Jingkai Zhou, Edward L. Deckard and William H. Ahrens. (2005). Factors affecting germination of hairy nightshade (*Solanum sarrachoides*) seeds. *Weed Science, Vol.* 53(1): 41-45.

Parsons, W.T. (1981). Noxious Weeds of Victoria. *Inkata Press*, Melbourne (AU).

Richardson, R.G. and D.N. Mckenzie. (1981). Regeneration of, and toxicity of 2,4-D to root fragments of silver leaf nightshade (*Solanum elaeagnifolium* Cav.). *J. Austr. Inst. Agri. Sci.*, 47: 48-50.

Roche, C. (1991). Silver leaf nightshade (*Solanum elaeagnifolium* Cav.) Pacific Northwest Extension Publications, 365.

Rutherford, P.A. (1978). Effect of time of immersion in running and still water on the germination of silver-leaf nightshade (*Solanum elaeagnifolium*). *Proceedings of the conference of the Council of Australian Weed Science Societies* 1:372-378.

Washington State Noxious Weed Control Board (2006). Silverleaf Nightshade (*Solanum elaeagnifolium* Cav.). Olympia, Washington (US).

Website

http:// www.nwcb.wa.gov/weed_info/Solanum_elaeagnifolium.html

YELLOW-BERRIED NIGHTSHADE (*Solanum xanthocarpum*)

Common name : Yellow-berried Nightshade

Botanical name : Solanum xanthocarpum

Tamil name : Kandan kattiri.

Hindi name : Kateli

Family : Solanaceae.

Origin and distribution : It occurs throughout India, in dry situations as a weed along the roadsides and wastelands. It is also distributed in Ceylon, Asia, Malaya, Tropical, Auastrana and Polynessia

Botany

Habitat : Roadsides and waste places

Habit : Armed, diffuse subshrub, prickles numerous, straight, to 2 cm (Fig. 1).

Stem : The stem is somewhat round and covered with sparsely distributed, star-like hairs

Leaf : Leaves lacerate, 3.5-8 x 1.5-5 cm, prickly along midnerve

Floral biology

Flower : Cymes 4-5-fold. Extra-axillary; peduncle to 3 cm. Calyx-lobes almost free, prickly, acuminate. Corolla purple, 2.5 cm across. The purple flowers, that are 2 cm long with five petals, can be seen in small bunches, sometimes opposite to the leaves. (Fig. 2).

Fig. 1. Plant

Fig. 2. Flower

Flowering and fruiting : Flowers and fruits appear during summer

Fruit : Berry green with white stripes, ripening yellow, globose, 1.5 cm across; seeds smooth, circular (Fig. 3).

Fig. 3. Fruit

Seed biology

Seed : Seeds are 2.5 mm in diameter and glabrous (Fig. 4).

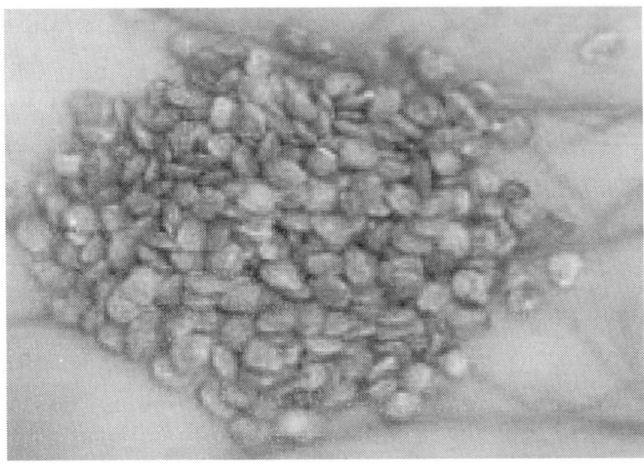

Fig. 4. Seed

Propagation : It is naturally propagated by seed in waste lands.

Seed dispersal : Seed dispersal takes place through birds (77%), crab-eating fox (19%) and lizards (4%).

Life cycle : The flowering time is August – September. It can be propagated through seeds. This weed is easily dispersed through animals as the fruits have hooked bristles and two strong hooked beaks.

Seed dormancy and germination : Germination rates of seeds collected directly from fruit (control), bird droppings, the faeces of crab-eating foxes and lizards were, respectively, 64, 64, 53, and 80%.

Uses : Useful in treating worms, cold, hoarseness of voice, fever, dysuria, enlargement of the liver, muscular pain, spleen and stone in the urinary bladder.

Roots and seeds are used as an expectorant in asthma, cough and pain in chest.

Its stem, flowers and fruits, being bitter and carminative, are used for relieving burning sensation in the feet.

WILD INDIGO (*Tephrosia purpurea*)

Common name : Wild Indigo

Botanical name : Tephrosia purpurea

Tamil name : Kolinji.

Hindi name : Sharpunkha

Family : Leguminosae

Origin and distribution : Originated from India and it is distributed in tropical, subtropical and warm temperate regions of the World.

Botany

Habitat : A common waste land weed also found in drylands and roadsides.

Habit : Erect perennial, copiously branched pubescent undershrub (Fig. 1).

Stem : Stem is woody at the base, round and pubescent in young parts that become glabrous with age.

Leaf : Leaves short petioled imparipinnate, leaflets 11-13, small; stipules linear subulate, ascending or reflexed.

Floral biology

Flower : Flowers bluish purple, bracteate, in extra axillary leaf opposed racemes; bracts minute; calyx closely silky, teeth linear subulate, as long at the tube; corolla thinly silky, papilionaceous; stamens diadelphous; ovary superior, single celled, style subglabrous, penicillate at the tip (Fig. 2).

Fig. 1. Plant

Fig. 2. Flower

Fruit : Pod flat, linear, 2-4.5 cm x 3-5 mm, somewhat up-curved towards the end, convex around the seeds, flattened between, margins thickened, dehiscent with twisted valves, 2-8/10 seeded (Fig. 3).

Fig. 3. Fruit (Pod)

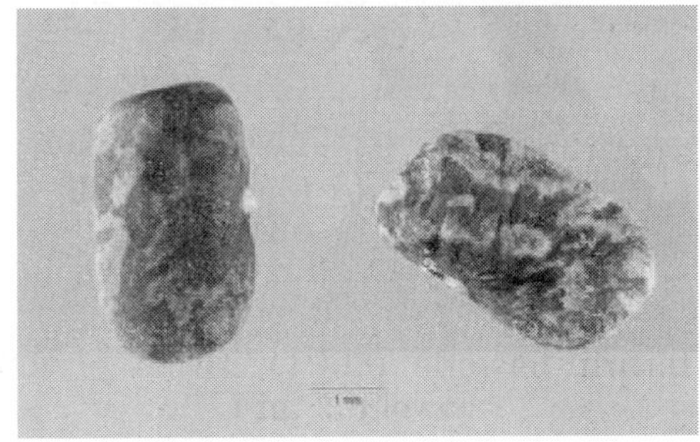

Fig. 4. Seed

Seed biology

Seed : Seed rectangular to transversely ellipsoid, 2.5-5 mm x 1.8-3 mm, light to dark brown to black, sometimes mottled (Fig. 4).

Propagation : By seed.

Seed dispersal : Seeds are dispersed through animals.

Seed dormancy and germination : The seed have a waxy, impermeable hard seed coat and do not quickly germinate. To hasten germination, the seeds are to be abraded with sand or steeped in hot water at 55°C for two to three minutes.

Uses : A decoction of the roots is given in dyspepsia, diarrhoea, rheum-atism, asthma and urinary disorders. The root powder is salutary for brushing the teeth, where it is said to quickly relieve dental pains and stop bleeding.

T. purpurea is used as green manure for vegetables, rice, coconut and banana, especially in India and Sri Lanka, and on a more limited scale in Indonesia, Malaysia and southern China. When grown as a green manure on saline-sodic soils in Rajastan (India), it is most successful in reducing soil salinity and lowering the pH (Orwa *et al.,* 2009).

Past work done on Seed Science and Technology

SEM study on seed structure

Al-Ghamdi and Zahrani (2010) studied the seed morphology of *Tephrosia purpurea* by SEM and reported that the seeds are reniform, dark brown, flat with black ornamentation and hilum is sub central (Fig. 5). The length and width of seed is 4.25 and 3.00 m, respectively

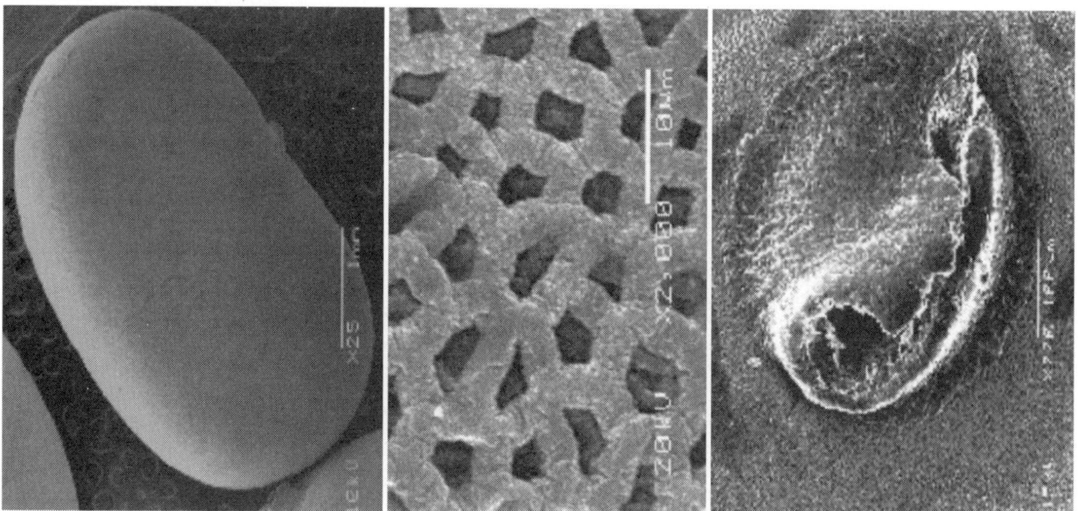

Fig. 5. *Tephrosia purpurea* (L.) entire seed; enlargement of seed coat; hilum shape

Proline accumulation in *Tephrosia purpurea* seeds

Erakar and Murumkar (1995) recorded maximum proline content in young seeds followed by pod wall. In mature seeds, proline content and also moisture content was lowest (Fig. 6). Tephrosia seeds are rich in nitrogen and ultimately in proteins; moreover, they posses significant amount of proline. This may help to survive under unfavorable conditions during germination.

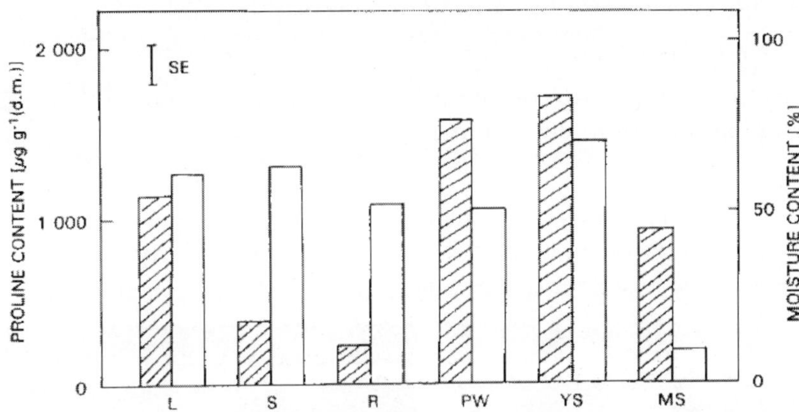

Fig. 6. Proline (hatched columns) accumulation and moisture content (open columns) in various plant parts (L - leaves, S - stem, R - roots, PW - pod wall, YS - young seeds, MS - mature seeds) of *Tephrosia purpurea* Pers.

REFERENCES

Al-Ghamdi. F.A. and R.M. Zahrani. (2010). Seed morphology of some *Tephrosia* species, *Feddes Report*, Weinheim, 121: 1–2.

Erakar, S.R. and C.V. Murumkar. (1995). Proline accumulation in *Tephrosia purpurea* pers. *Biologia plantarum,* 37 (2): 301-304,

Orwa. C, A. Mutua, R. Kindt , R. Jamnadass and A. Simons (2009). Agroforestree database: A tree reference and selection guide version 4.0 (http://www.worldagroforestry.org/af/treedb/)

GROUND BUR-NUT (*Tribulus terrestris*)

Common name : Ground Bur-nut; Small Calltrops.

Botanical name : Tribulus terrestris L.

Tamil name : Nerinji.

Hindi name : Gokhru

Family : Zygophyllaceae.

Origin and distribution : Tribulus terrestris is native to Southern Europe (Parker, 1972), Africa, temperate and tropical Asia. Plant is widely distributed in different parts of India up to 3000 m altitude.

Botany

Habitat : A common dry land weed.

Habit : An herbaceous annual with a deep root (Fig. 1).

Stem : The stem is round.

Leaf : Leaves are opposite, compound, evenly pinnate, shortly petiolate with 10-14 small oblong leaflets, in unequal pairs.

Floral biology

Flower : Peduncled flowers, solitary in leaf axils, pseudoaxilary, white or yellow; sepals 5, imbricate; petals 5, spreading, imbricate, disc annular, 10 lobed; stamens 10 or 5; ovary sessile 5 to 12 celled, ovules 1 to 5 in each cell (Fig. 2).

Fig. 1. Plant

Fig. 2. Flower

Flowering and pollination : Flowering is rapid, with the first flowers appearing within three weeks and the first fruit within five to six weeks of germination. Flowering may continue for several months until the first frosts in autumn/early winter, when plants die.

Boydston (1990) reported that flowering occurred within 3-4 weeks of emergence when temperatures were consistently above 68°F (20°C), regardless of planting date. *Tribulus terrestris* flowers March through October in Arizona, although primarily from July to August (Parker, 1972). Once the plant begins to flower, it is continuous throughout the plant's life (Reddi *et al.*, 1981). *Tribulus terrestris* flowers are cross pollinated by insects (foragers include: Coleoptera, Diptera, Hymenoptera, Lepidoptera, Thysanoptera) (California Department of Food and Agriculture, EncycloWeedia, 2002, Washington State Noxious Weed Control Board, 2001) along with being self-pollinated, which occurs at the end of each flower's receptive period (within one day) (Reddi *et al.,* 1981). Self-pollination is accomplished when the petals begin to close and push the stamen inward toward the stigma, the longer anthers making direct contact; the potential of this system is 100% seed set (Reddi *et al.,* 1981). Fruits mature in approximately 2 weeks, and subsequently split apart into segments soon afterward (Holm *et al.,* 1991). Plants continue to reproduce and produce fruit until the cool season begins. Boydston (1990) reported during trials in Washington, fruit/burr production stopped in October when average temperatures were under 68°F (20°C)

Fruit : Schizocarp; woody burrs, gray to yellow-tan, hairy, to approximately 0.4-0.7 in. (1-1.8 cm) in diameter, more or less flattened, lobed; separates into 5 (occasionally 4) wedge-shaped, indehiscent nutlets (cocci), each with 2 stout dorsal spines 0.15-0.27 in. (4-7 mm) long and spreading, and several prickles (Fig. 3). Boydston (1990) reported that seeds planted in May, June, July, and August subsequently produced 5600, 5200, 3600, and 200 burrs/plant, respectively.

Fig. 3. Fruit

Seed biology

Seed : Each fruit/burr usually contains 5 nutlets (cocci), each nutlet (coccus) can contain 2-5 seeds (Boydston, 1990). The nutlets or "seeds" are hard and bear two to three sharp spines, 10 mm long and 4 to 6 mm broad point-to point. It is a trailing and spreading herb, densely covered with minute hair (Fig. 4).

More than 1000 seeds can be produced by a single plant and seed can remain viable for many years.

Fig. 4. Seed

Propagation : By seeds carried in cocci (bits of fruit)

Caltrop is able to regenerate from tap root material, if frosted off or if shoot growth is removed by cultivation, chipping or herbicide damage.

Seed dispersal : Each fruit section (coccus) has 2 sharp divergent spines and several other spines and warty protuberances enabling the *Tribulus terrestris* fruits to easily attach to animals and humans and to stick onto vehicle tires (cars, farm, airplane), subsequently facilitating long distance dispersion and spread (Ernst and Tolsma, 1988; Ridley, 1930; Squires, 1979; Whitson 1992).

Holm *et al.* (1991) points out that due to the architecture of the schizocarp/fruit, the large and small spines are arranged at different angles with at least one of the spines always pointing upward no matter how the fruits/burrs fall from the plant, and can easily imbed into feet, hooves, or tires. After getting caught or imbedded into the hooves, feet, and wool of livestock and other animals, the fruits/burrs are subsequently broken off as the animals try to rid themselves of the irritation (Ridley, 1930). Furthermore, they can stick to the shoes and clothing of people, and the fur and feathers of animals (California Department of Food and Agriculture, EncycloWeedia, 2002).

Life cycle : Caltrop is annual plant which germinates after rainfall in late spring and summer when the maximum air temperature reaches 24°–27°C. Flowers may be produced within 3 weeks, fruits/burrs within 6 weeks.

Seed longevity : Buried *Tribulus terrestris* seed can remain viable for several years (California Department of Food and Agriculture, EncycloWeedia, 2002), staying dormant in the soil for 4-5 years (Whitson, 1992).

Seed dormancy and Germination : Caltrop and yellow vine are annual plants which germinate after rainfall in late spring and summer when the maximum air temperature reaches 24°–27°C. There is a succession of germinations throughout the summer following each rainfall event.

Generally, *Tribulus terrestris* has a considerable seed dormancy lasting over fall and winter months (Washington State Noxious Weed Control Board, 2001) with some seeds staying dormant for longer periods of time. Its seedlings emerge in the early spring through summer, often in flushes following increased soil moisture (California Department of Food and Agriculture, EncycloWedia, 2002). It germinates after the start of the monsoon rains, on any type of barren soil, in southern Arizona (Parker, 1972).

Ernst and Tolsma (1988) reported that dormancy of seeds was very high in both fresh seeds and 3-6 year old seeds. Squires (1979) reported that freshly harvested seeds have a germination rate of 10% and dry stored 6 month old seeds have a germination rate of 84%.

Pathak (1970) reported that *Tribulus terrestris* germination is inhibited by low temperatures, low light intensities and wet soil.

Most newly matured *Tribulus terrestris* seeds are dormant and require an after ripening period of approximately 6-12 months. Imbibition of water differed in dormant versus non-dormant seeds: measurements taken at one hour and periodically through one day of imbibition showed dormant seeds had approximately half of the water content of non-dormant seeds (Ernst and Tolsma, 1988). The largest seed within a nutlet (coccus) is usually the first to germinate; the remaining seeds may germinate or remain dormant depending on moisture availability (California Department of Food and Agriculture, EncycloWeedia, 2002); this large seed is usually positioned near the basal end of the burr (coccus) (University of California, 1998).

The optimum temperature range for germination to occur is 81-95°F (27-35°C). In Australia, germination occurs when the maximum air temperature is approximately 75-81°F (24-27°C) (Squires, 1979). During trials, *Tribulus terrestris* emergence was initiated when average soil temperatures reached 59°F (15°C) for at least 2 weeks and approached 68°F (20°C) (Boydston, 1990). After initial emergence occurred, emergence was multipeaked with no relationship existing with temperature for the remainder of the summer (Boydston, 1990). In subsequent years, emergence occurred and peaked at similar times regardless of the age of the seed, indicating environmental stimuli influencing emergence (Boydston, 1990).

Tribulus terrestris seedlings emerge during early spring through summer, often in flushes following increased soil moisture (California Department of Food and Agriculture, EncycloWeedia, 2002, Mirsa, 1962 in Holm *et al.,* 1991). During field observations in Botswana, *Tribulus terrestris* was observed to germinate and emerge following a rainshower having more than 0.04 in. (10 mm) of precipitation (Ernst and Tolsma, 1988). Maximum germination occurred after a series of heavy rains, facilitating a 35% germination rate, with continued germination of seeds lasting for another 4 months (Ernst and Tolsma, 1988).

On sandy soils, seedlings emerge from depths to approximately 5 cm (less on heavy soils) (California Department of Food and Agriculture, EncycloWeedia, 2002). Squires (1979) reported that seeds buried more than 4 in. (10 cm) deep in sandy soils, can successfully emerge.

Germination was irregular in *Tribulus terrestris* seeds, whether seeds remained in the fruit (cocci) or were isolated (Ernst and Tolsma, 1988). In greenhouse trials, the highest germination rate was achieved by isolated seeds with a maximum of 66.9%, and a mean of 37.3±25.1% (ranging from 68/59°F (20/15°C) up to 95/86°F (35/30°C), in 5°C increments, alternating day/night temperatures). Seeds that remained in the fruits rarely germinated synchronously (Ernst and Tolsma, 1988). Ernst and Tolsma (1988) add that because of this pattern, direct competition for water and nutrients is avoided.

Effects of crop production : *Tribulus terrestris* roots can develop nitrogen-fixing nodules. Athar and Mahmood (1985) observed that plants having root nodules had more lush green healthy growth and greater dry weight versus stunted growth in plants without nodules.

Control measures : Ensure good agronomic practices such as appropriate sowing time and depth, adequate fertility and moisture at sowing.

- Use high sowing rates to achieve full ground cover quickly.
- Use narrow row spaces to provide full ground cover much quicker than wide row spaces.
- Direct drill to minimize disturbance.
- Band nitrogen in crops. It will be more directly available to the crop than to the weeds.

Uses : Tribulus terrestris has long been used as a tonic and aphrodisiac in Unani system of medicine. It has been used in India and Pakistan as a treatment for impotence and as a stimulant to enhance sexual drive and performance (Brown *et al.,* 2001). *T. terrestris* has diuretic and uricosuric effects.

REFERENCES

Athar, M. and A. Mahmood. (1985). Observations on nitrogen fixation by *Tribulus terrestris* Linn. under natural habitat. *Geobios,* 12(1): 44-46.

Boydston, R.A. (1990). Time of emergence and seed production of longspine sandbur (*Cenchrus longispinus*) and puncturevine (*Tribulus terrestris*). *Weed Science* 38(1):16-21.

Brown, G.A., M.D. Vukovich and E.R Martini. (2001). Effects of androstenedione-herbal supplementation on serum sex hormone concentrations in 30- to 59-year-old men. *Int. J. Vitam. Nutr. Res.,* 71(5): 293-301.

California Department of Food and Agriculture, *EncycloWeedia.* (2002). Tribulus. Ed by: Healy, E.A., S. Enloe, J.M. DiTomaso, B. Roberson, N. Dechoretz, S. Schoenig, P. Akers, L. Butler, and J. Garvin. Non-Cropland Weed group, UC Extension Service, Weed Science Program, Department of Vegetable Crops, The University of California. Davis, CA. 95616.

Ernst, W.H. and D.J. Tolsma. (1988). Dormancy and germination of semi-arid annual plan species, *Tragus berteronianus* and *Tribulus terrestris. Flora,* 181(3/4):243-251.

Holm, L.G., D.L. Plunknett, J.V. Pancho and J.P. Herberger. (1991). The world's worst weeds. Distribution and biology. *Krieger Publishing Company,* Malabar, Florida, 609.

Mirsa, D. (1962). *Tribulus terrestris* weed in arid zone farming. *Indian Journal of Agronomy* 7(2):136-141.

Parker, K.F. (1972). An illustrated guide to Arizona weeds. The University of Arizona Press, Tucson, AZ. 338 pp. http://www.uapress.arizona.edu/online.bks/weeds/

Pathak, P.S. (1970). Contributions to the ecology of *Tribulus terrestris* Linn. II. Habitat studies. *Agra University Journal of Research Science,* 19(2):149-166.

Reddi, C.S., E.U. Reddi and N.S. Reddi. (1981). Breeding structure and pollination ecology of *Tribulus terrestris. Proceedings of the Indian National Science Academy,* Part B 47(2):185-193.

Ridley, H.N. (1930). The dispersal of plants throughout the world. L. Reeve and Co., Ltd., Lloyds Bank Buildings, Ashford, Kent. 744.

Squires, V.R. (1979). The biology of Australian weeds. 1. *Tribulus terrestris* L. Journal of the Australian Institute of Agricultural Science, 45(2):75-82.

University of California (1998). The grower's weed identification handbook. Cooperative Extension University of California, Division of Agriculture and Natural Resources, Publication 40: 311.

Washington State Noxious Weed Control Board. (2001). Puncturevine (*Tribulus terrestris* L.). Website: http://www.wa.gov/agr/weedboard/weed_info/puncturevine.html

Whitson, T.D., Editor; Burrill, S.A. Dewey, D.W. Cudney, B.E. Nelson, R.D. Lee and R. Parker (1992). Weeds of the West. The Western Society of Weed Science in cooperation with the Western United States Land Grant Universities Cooperative Extension Services and the University of Wyoming. 630.

Website

http://pi.cdfa.ca.gov/weedinfo/TRIBULUS2.htm

INDIAN BORAGE (*Trichodesma indicum*)

Common name : Indian borage

Botanical name : Trichodesma indicum (L.) R. Br.

Tamil names : Kazhuthai Thumbai, Kavizh Thumbai.

Hindi name : Chhota Kalpa

Family : Boraginaceae

Origin and distribution : It is found throughout India and Pakistan.

Fig. 1. Plant

Botany

Habitat : Common dryland weed found in roadsides and stony dry wastelands, upto 1,500 m.

Habit : Diffuse or erect annual hispid herb (Fig. 1).

Stem : Stem is round and is covered with simple and tubercle-based hairs.

Leaf : Leaves mostly sessile lanceolate or cordate - lanceolate, tuberculate on the upper surface.

Floral biology

Flower : Flowers pale blue, pink or white, axillary, solitary, becoming racemose due to gradual reduction of floral leaves; calyx deeply 5 partite, lobes hastate or cordate at the base, greyvillous, enlarged in the fruit; corolla 5 lobes ovate, abruptly apicultate and recurved each with two glandular spots on the throat, as long as the tube; stamens 5, very closely connivent, densely woolly on the back, tips hair like, spirally twisted; ovary superior, four celled, syncarpous, style terminal, filiform, stigma small (Fig. 2).

Flowering and fruiting : T. indicum flower throughout the year but profusely after monsoon rain. Most of the flowers open in the evening at 17.00 – 18.00 h and few in the morning at 6.00 – 7.00 h. Mode of pollination is known as 'Buzz pollination'.

Fruit : Fruit ellipsoid with central quadrangular carpophores, nutlets smooth pitted on the inner surface (Fig. 3).

Fig. 2. Flower

Fig. 3. Fruit

Seed biology

Seed (Nutlet) : Nutlets are about 5 mm long, and rough on the inner surface.

Propagation : By nutlets.

Seed dispersal : Through winds.

Uses : The plant is acrid, bitter in taste. In herbal medicine jargon, it is thermogenic, emollient, alexeteric, anodyne, anti-inflammatory, carminative, constipating, diuretic, depurative, ophthalmic, febrifuge and pectoral.

This herb is also used in arthralgia, inflammations, dyspepsia, diarrhoea, dysentery, strangury, skin diseases and dysmenorrhoea.

COAT BUTTONS (*Tridax procumbens*)

Common name : Coat buttons

Botanical name : Tridax procumbens L.

Tamil name : Thata poondu

Hindi name : Ghamra

Family : Compositeae (Asteraceae)

Origin and distribution : It is native to the Tropical Americas, but it has been introduced to Tropical, Subtropical, and mild Temperate regions worldwide.

Botany

Habitat : Roadsides and waste places in all dry districts.

Habit : Procumbent, hispid, weak, straggling perennial herb (Fig. 1).

Stem : Stem is round and covered with sparsely distributed hispid hairs.

Leaf : Leaves opposite, ovate or lanceolate, deeply serrate.

Floral biology

Flower : Heads solitary, long peduncled, 2-3 times as long as the stem, heterogamous, rayed; ray flowers female, fertile, two lipped, outer lip large, trifid, inner small, two lobed; disc flowers bisexual, fertile, tubular, limb elongate, pentafid; involucre of bracts campanulate, few seriate, outer broad, herbaceous, inner scarious; receptacle flat or convex; palaea membranous; corolla of female flowers ligulate, of bisexual flowers tubular; syngenesious anther with short acute auricles at the base; style armas of bisexual flowers hairy, tips subulate (Fig. 2).

Fig. 1. Plant

Fig. 2. Flower

Fruit : Achens are brown, oblong, and silky; pappus of short feathery bristles, shining (Fig. 3).

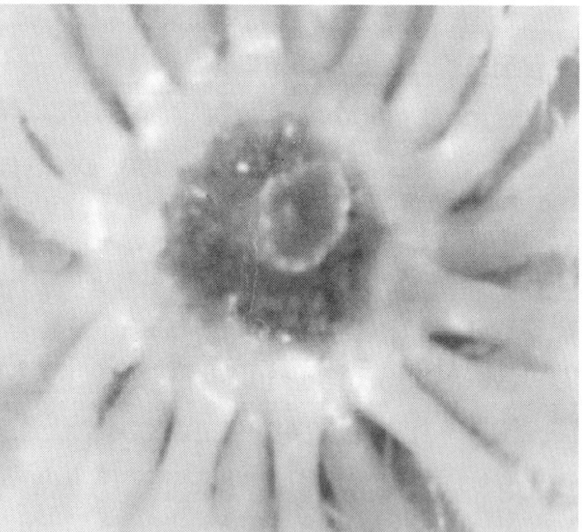

Fig. 3. Fruit

Seed biology

Seed : Non-endospermous..The length of achene- proper is 2.5-2.7 mm. The length of pappus is about 5 mm. Achene is brown in colour, turbinate or oblong, silky, and densely pilose. Pappus is composed of several plumose setae (feather - like). As the pappus in expanded form is voluminous, it helps the seed to remain floating for a longer period (Fig. 4).

Fig. 4. Seed (Achens)

Propagation : By achenes.

Seed dispersal : Achens are dispersed by winds.

Seed dormancy and germination : The seeds require light for germination. The seeds buried in the soil showed no germination or low germination presumably due to the absence of exposure to light.

Germination response was greater at the warmer fluctuating temperatures (30/20 and 35/25°C) than at the colder temperatures (25/15°C). Light stimulated germination; however, some seeds still germinated in the dark. The seeds were moderately tolerant of salt and water stress. At the soil surface, coat buttons emergence was 76%, but this declined rapidly with increasing soil depths. The seed has not emerged from depths exceeding 3 cm. Seedling emergence and seedling dry matter were greatly reduced with the addition of crop residue to the soil surface at rates equivalent to 4 to 6 t ha^{-1} (Chauhan and Johnson, 2008)

Effects on crop production : Coat buttons are among the most troublesome Asteraceae weed species of rainfed rice.

Uses : The leaf juice has antiseptic, insecticidal and parasiticidal properties. It is a good remedy against conjunctives and used to check hemorrhage from cuts, bruises and wounds insect repellent. It is also used as a bio adsorbent for chromium.

Past work done on Seed Science and Technology

Seed position while sowing and germination

Wajid and shaukat (1993) reported that initial germination varied with the position of seeds in the germination medium and essentially completed by 13th day. Generally, the percentage germination in all the positions was low except for horizontally lying position with attached pappus on the soil surface. The diaspore having attached pappus and horizontally placed started germinating on 4th day of imbibition, whereas in case of detached pappus condition, ther was no germination.

The former position showed the maximum germination compared to any other landing position. The standing diaspore with detached pappus uppermost and half buried in the soil showed less germination compared to the seeds with attached pappus in the same position. In the latter, germination started earlier i.e., on 4th day while in the former case germination started on 9th day of imbitition. The seeds which were buried in soil showed low percentage germination (Fig. 5).

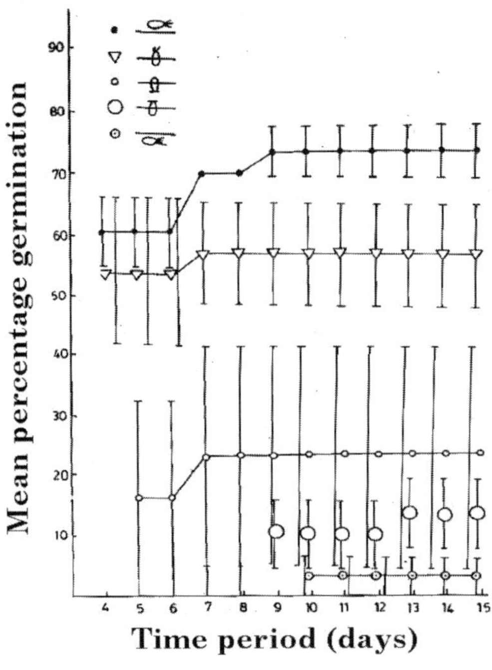

Fig. 5. The progress of germination of *Tridax procumbens* when diaspore was variously placed in or on the soil. Wherever standard errors are very small, they are not shown.

REFERENCES

Chauhan, B.S and D.E. Johnson. (2008). Germination ecology of two troublesome asteraceae species of rainfed rice: siam weed (*Chromolaena odorata*) and coat buttons (*Tridax procumbens*). *Weed Sci.*, 56(4):567-573.

Wajid, A. and S.S. Shaukat. (1993). Effects of seed dispersal unit and the position of seed relative to substrate on germination of three composites, *Pak. J. Bot.*, 25(2): 118-126.

COCKLE BUR (*Xanthium strumarium*)

Common name : Bur-weed

Botanical name : Xanthium strumarium L.

Tamil name : Marul Oomathan.

Hindi name : Chota dhatura

Family : Compositae (Asteraceae)

Origin and distribution : The native is probably North America and it is currently distributed worldwide, although is most common in the temperate zone (Love and Dansereau, 1959).

Botany

Habitat : Waste places and rubbish heaps.

Habit : An annual herb (Fig. 1).

Fig. 1. Plant

Stem : Stem is rather stout and rough with scabrid hairs.

Leaf : Leaves are simple, alternate and ex stipulate. Petiole is 1-3 inches long; blade is broadly triangular ovate up to 3 inches long, somewhat 3-lobed and irregularly toothed; leaf base is somewhat cordate and tapering.

Floral biology

Flower : Flowers are monoecious. Male florets are many in globose heads at the top of the inflorescence. Involucre is short with a few bracts; receptacle is cylindric and paleae are hyaline enclosing the florets. Corolla is tubular and five lobed. Anthers are five and distinct with inflexed tips; filaments are united into a tube. Female florets are in the axils of upper leaves and at the tops of branches; usually two of them are together enclosed in a two horned utricle formed by the union of bracts and covered with hooked spine-like structures. Corolla is absent; style arms are two and long (Fig. 2).

Fig. 2. Flower

Flowering and pollination : Seed is produced through self and cross fertilization and reportedly, apomixis (United States Department of Agriculture, Forest Service: Iverson, 2002; Uchytil, 1992), Although Weaver and Lechowicz (1983) stated that there is no apomixis occurring in *Xanthium strumarium.*

It flowers in late summer when day length shortens enough to stimulate flowering (Holm *et al.,* 1991) with seeds maturing in early autumn (Holm *et al.*, 1991, Pitcher, 1989). Flowering can be induced by only 1-3 photoperiodic (light/dark) cycles early in the plant's life cycle (Holm *et al.*, 1991, Kaul, 1971 in Weaver and Lechowicz, 1983).

It is wind-pollinated and selfcompatible, and is mostly self-pollinated (Love and Dansereau). Outcrossing rates in natural populations were found to range between 0-12 per cent (Moran and Marshal, 1978 in Pitcher, 1989; Weaver and Lechowicz, 1983).

Fruit : Cylindric to ovoid, two-chambered bur; 0.4-1.4 in. (1-3.5 cm) long; glandular, covered with hooked prickles, with two larger, longer incurved prickles/ beaks projecting from the apex of the bur (Fig. 3). Brownish to reddish when mature. *Xanthium strumarium* plants grown on open sites can produce 500-5400 burrs per plant (Weaver and Lechowicz, 1983 in Pitcher, 1989).

Fig. 3. Fruit

Seed biology

Seed : One achene per floret, two per bur; 0.4-0.6 in. (1-1.5 cm) long; dark-brown, oblong and flattened with pointed apex, no pappus (Fig. 4). The seeds encased within a burr are dimorphic, each having differing germination requirements (Esashi *et al.,* 1983). *Xanthium strumarium* is a prolific seed producer (Manitoba Agriculture and Food, 2001) creating large seed banks on a site (Elkhorn Slough National Estuarine Research Reserve, 2000).

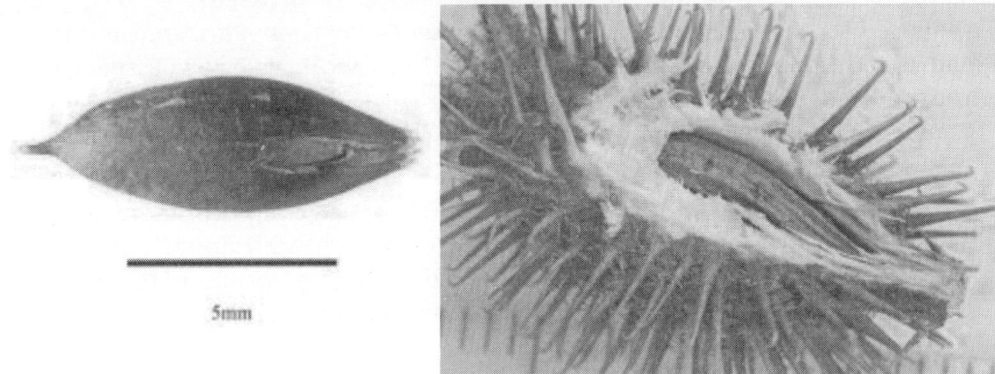

Fig. 4. Seed

Propagation : By the hooked fruit.

Seed dispersal : Burrs and thus, its seeds, are spread by water, humans, and other animals (Pitcher, 1989); floodwaters, and also rivers and streams, can carry burrs for long distances and across large areas (Ridley, 1930). Kaul (1961 in Pitcher, 1989) reported that the burrs are buoyant, floating for up to 30 days.

Burrs are easily dispersed by hooking onto, and then subsequently carried, by man and animals (Manitoba Agriculture and Food, 2001, Ridley, 1930).

Seed dormancy and germination : Seed germination occurs within the fruit/burr (Uchytil, 1992). Seeds germinate from early spring through summer (Uva *et al.*, 1997) when sufficient heat occurs (Lee, 1996) and moisture (Kaul, 1968 in Pitcher, 1989) is available.

Two types of seed are produced within each bur of *Xanthium strumarium* (termed: somatic polymorphism) (Redosevich and Holt, 1984 in Pitcher, 1989), each differing in size and germination requirements (Weaver and Lechowicz, 1983). The smaller seed is often pushed upwards towards the beak of the fruit, and has an innate dormancy (Weaver and Lechowicz, 1983). The lower seed has a shorter dormancy than the upper seed, requiring overwintering or several months of after ripening, then germinating first, usually in the spring following its maturity (Redosevich and Holt, 1984 in Pitcher, 1989; Weaver and Lechowicz, 1983). The upper seed will remain dormant for 1 to several years (Uva *et al.*, 1997). Also, Uchytil (1992) states that the two seeds occasionally will germinate synchronously.

The dormancy in the upper seed may be caused by seed coats impermeability to oxygen (Davis, 1930 in Weaver and Lechowicz, 1983) and germination inhibitors (Porter and Wareing, 1974). The presence of oxygen degrades the inhibitors, although each type degrades at a different rate causing the rupturing of the seed coats to facilitate germination (Redosevich and Holt, 1984 in Pitcher, 1989). This strategy insures germination of seed at different times in the event that conditions are not suitable (Redosevich and Holt, 1984 in Pitcher, 1989).

During trials, the dimorphic *Xanthium strumarium* seeds occurring in each burr appeared to have differing light requirements for their germination; the upper seed had a red-light requirement for germination, while the lower seed was dark germinating (Esashi *et al.*, 1983).

Germination of the upper seeds can be induced using several methods, such as presoaking the seeds in water at 73°F (23°C) under dark conditions for up to 8 days (Esashi *et al.*, 1976) or adding ethylene to the seed's environment (Satoh and Esashi, 1980 in Weaver and Lechowicz, 1983), although both upper and lower seeds reportedly germinate under conditions of microbial decay, pricking of the seed coat, acid scarification and exposure to high temperatures (122°F (50°C) for 10 days) (Kaul, 1965 in Weaver and Lechowicz, 1983).

Seed longevity and seed storage : *Xanthium strumarium* seeds are viable for several years (Elkhorn Slough National Estuarine Research Reserve, 2000; Pitcher, 1989), although viability declines during that time (Wapshere, 1974). Viability of seed distinctly declined after 18 months of storage, regardless of storage conditions (Kaul, 1965 in Weaver and Lechowicz, 1983; Lee 1996).

During a study of buried seeds from the Southern United States, viability declined from a 99% germination percentage to 66% after 6 months of burial, and then to 18% for seeds buried for 30 months; burial depths ranging from 3-9 inches (8-38 cm) didn't affect germination percentages (Egley and Chandler, 1978 in Weaver and Lechowicz, 1983). In contrast, after 30 months of dry storage at 39°F (4°C) seed viability remained at 100% with germination increasing from 61% to 92% (Weaver and Lechowicz, 1983).

Effects on crop production : The leaf leachates of *Xanthium* caused maximum reduction in seedling growth (shoot and root length) of wheat and barley, while fruit leachate were deleterious to maize, rice (expect dry weight) cotton and sunflower. Shoot leachates did not affect the shoot length, dry weight and seedling emergence of wheat, barley and maize. Soil beneath the *X. strumarium* plant also did not affect the seedling growth and seedling emergence of all crops (Tanveer *et al.*, 2008).

Control measures : Hand labor, mechanical removal prior to flowering is recommended for control (Pitcher, 1989). According to him, burning is an effective method of destroying *Xanthium strumarium* seeds, although prescribed fire is a seldom used method.

Rice (1992) lists 2,4-D, MCPA, MSMA, cacodyllic acid, ametryn, amitrole, atrazine, bentazon (Prompt), bromoxynil, dicamba, dichlorprop, diuron, metribuzin, oxyfluorfen, tebuthiuron, triclopyr and clopyralid (Confront) for cocklebur control.

REFERENCES

Davis, W. (1930). The development of dormancy in seeds of cocklebur (*Xanthium*). American Journal of Botany 17:77-87.

Esashi, Y., N. Ishihara, R. Kuraishi and H. Kodama. (1983). Light actions in the germination of cocklebur seeds. *Journal of Experimental Botany,* 34(144): 903- 914.

Esashi, Y., K. Kotaki and Y. Ohhara. (1976). Induction of cocklebur seed germination by anaerobiosis: A question about the "inhibitor hypothesis" of seed dormancy. *Planta,* 129 : 109-112.

Elkhorn Slough National Estuarine Research Reserve. (2000). Weed Control by Species. Elkhorn Slough Foundation, California.

Egley, G.H. and J.M. Chandler. (1978). Germination and viability of weed seeds after 2.5 years in a 50-year buried seed study. *Weed Science* 26(3): 230-239.

Holm, L.G., D.L. Plunknett, J.V. Pancho and J.P. Herberger. (1991). The world's worst weeds. Distribution and biology. *Krieger Publishing Company,* Malabar, Florida, 609.

Kaul, V. (1965). Physiological-ecology of *Xanthium strumarium* L. II. Physiology of seeds in relation to its distribution. *Journal of the Indian Botanical Society* 44: 365-380.

Kaul, V. (1968). Physiological-ecology of *Xanthium strumarium* L.V. Water relations. *Tropical Ecology* 9: 88-102.

Kaul, V. (1971). Physiological-ecology of *Xanthium strumarium* L. IV. Effect of climatic

factors on growth and distribution. New Phytologist 70:799-812.

Lee, J.M. (1996). Common cocklebur *Xanthium strumarium*. Iowa State University, The ISU Weed Biology Library, Agronomy 517: *Weed Biology and Ecology*.

Love, D and P. Dansereau. (1959). Biosystematic studies on *Xanthium*: taxonomic appraisal and ecological status. *Canadian Journal of Botany,* 37:173-208.

Manitoba Agriculture and Food. (2001). Weeds, insects, and diseases: pest management – Weeds – Cocklebur.

Moran, G.F. and D. R. Marshal. (1978). Allozyme uniformity within and variation between races of the colonizing species *Xanthium strumarium* L. (Noogoora burr). *Australian Journal of Biological Science* 31:283-291.

Pitcher, D. (1989). Element stewardship abstract for *Xanthium strumarium,* Rough cocklebur. The Nature Conservancy, 1815 North Lynn St., Arlington, Virginia.

Porter, N.G. and P.F. Wareing. (1974). The role of the oxygen permeability of the seed coat in the dormancy of seed in *Xanthium pensylvanicum* Wallr. *Journal of Experimental Botany* 25:583-594.

Ridley, H.N. (1930). The dispersal of plants throughout the world. L. Reeve and Co., Ltd., Lloyds Bank Buildings, Ashford, Kent. 744 pp.

Redosevich, S.R. and J.S. Holt. (1984). *Weed Ecology*. John Wiley & Sons, New York.

Rice, Jr., R.P. (1992). Nursery and landscape weed control manual. *Thompson Publications*, Fresno, California. 290.

Sotah, S., and Y. Esashi. (1980). D-Amino-acid-stimulated ethylene production in seed tissues. Planta 149:64-68.

Tanveer, A., M. Tahir, M.A. Nadeem, M. Younis, A. Aziz and M. Yaseen. (2008). Allelopathic effects of *Xanthium strumarium* L. on seed germination and seedling growth of crops. *Allelopathy J.* 21:317-328.

Uchytil, R. (1992). *Xanthium strumarium*. In: U.S. Department of Agriculture, Forest Service, Rocky Mountain Research Station, Fire Sciences Laboratory.

Uva, R.H., J.C. Neal, and J.M. DiTomaso (1997). Weeds of the Northeast. Comstock Publishing Associates, Cornell University Press.

Wapshere, A.J. (1974). The regions of infestation of wool by Noogoora bur (*Xanthium strumarium*), their climates and the biological control of the weed. *Australian Journal of Agricultural Research* 25:775-781.

Weaver, S.E., and M.J. Lechowicz. (1983). The biology of Canadian weeds. 56. *Xanthium strumarium* L. Canadian Journal of Plant Science 63:211-225.

United States Department of Agriculture, Forest Service: Iverson, L. (2002). *Xanthium strumarium* var. *canadensis*. Illinois Plant Information Network; compiled by Ketzner, D., and J. Karnes. Illinois Natural History Survey, 607 E. Peabody Dr.,Champaign, Illinois 61820. Website:http://www.fs.fed.us/ne/Delaware/ilpin/3194.co

Websites

http://www.agron.iastate.edu/~weeds/WeedBiolLibrary/u4cockle.html

http://www.fs.fed.us/database/feis/plants/forb/xanstr/all.html

http://www.gov.nb.ca/agriculture/crops/weeds/fab33s00.html

http://www.elkhornslough.org/plants/restoration.html then click on 'The Weed Control by Species Guide'

http://tncweeds.ucdavis.edu/esadocs/documnts/xantstr.html.

PART - II
SEED BIOLOGY OF GARDENLAND WEEDS

C. Karthika[1] and V. Manonmani[2]

1. Ph.D., scholar, 2. Associate professor,
Department of Seed Science and Technology,
Tamil Nadu Agricultural University,
Coimbatore – 641 003,
E-mail: karthi.aug18@gmail.com / mano_2004@yahoo.com

INDIAN NETTLE (*Acalypha indica* L.)

Common name : Indian acalypha, Indian nettle, three-seeded mercury.

Botanical name : Acalypha indica Linn.

Tamil name : Kuppai meni, Kuppivaeni

Family : Euphorbiaceae

Origin and distribution : Occurs in Nigeria and South Africa. It is also widespread in the Indian Ocean islands and occurs furthermore in India, South-East Asia and Oceania. It was introduced into the warmer parts of the New World.

Botany

Habitat : Distributed profusely in wastelands and pastures particularly soon after rains. In cultivated areas grown with crops requiring frequent inter cultivation (Fig. 1).

Fig. 1. Plant

Habit : Erect, annual, pubescent herb growing to a height of 20 to 60cm.

Stem : Round or angular and densely shortly pubescent.

Leaf : Long-petioled, petiole longer than the blade, ovate or rhombic-ovate, crenate, serrate, acute or obtuse, base cuneate entire.

Floral biology

Flower: Minute, green, unisexual, monoecious, in axillary erect spikes; bracteate, bracts cuneiform, many nerved, dentate, folded, alternate; male flowers few and terminal; calyx lobes 4, membranous, valvate; petals; ostamens many on a convex receptacle; female flowers basal: calyx 3-4 lobed, lobes imbricate, petals 0; ovary superior, hispid, syncarpous, 3-celled, ovules solitary in each cell, style filiform, often very long (Fig. 2).

Flowering : *Acalypha indica* flowers throughout the year in regions without a pronounced dry season.

Fruit : Capsules covered by the bract, often one seeded.

Seed biology

Seed : Seeds ovoid, acute, smooth (Fig. 3).

Fig. 2. Flower

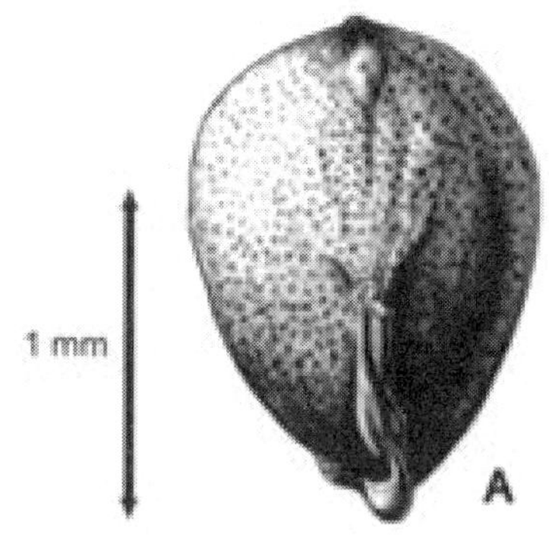

Fig. 3. Seed

Propagation : By seed

Effect on crop : It is a host of several plant viruses, e.g. pumpkin yellow vein mosaic virus, okra yellow vein mosaic virus, leaf curl virus and tobacco virus 10, often resulting in very severe and recurrent outbreaks in crops. Roots may suffer from nematode infestation (*Meloidogyne* spp.). In India, *Acalypha indica* is a host of the spiralling whitefly (*Aleurodicus dispersus*).

Uses : It is an important medicinal plant for its expectorant properties. It also has significant antibacterial and antifungal activities, both against human and plant pathogens. In East Africa, the leaf sap is used as eye drops to treat eye infections. Leaf powder is applied to maggot-infested wounds. In Comoros, a leaf decoction is used as a massage cream to treat pain of the joints. In the Seychelles and Réunion, a root infusion or decoction is taken to treat asthma, and also to clean the liver and kidneys. In the Seychelles, a root decoction is also taken to treat intestinal worms and stomachache. A poultice made from the whole plant is applied to treat headache. The leaf sap is taken as an emetic and the root decoction as a laxative.

GOAT WEED (*Ageratum conyzoides*)

Common name : Billy goat weed, goat weed, tropical whiteweed, Geratum

Botanical name : Ageratum conyzoides

Tamil name : Pumppillu, appakkoti, nechati

Hindi name : Jangli pudina

Family : Compositae (Asteraceae).

Origin and distribution : Origin is tropical America. Also in the subtropics and extending into temperate areas from latitude 30°N to 30's. Widespread in SE Asia. Present in Java prior to 1860.

Botany

Habitat : A common weed in garden lands, pastures and other damp and shady places; generally associated with sugarcane crop (Fig. 1).

Fig. 1. Plant

Habit : Erect softly hairy, annual herb, growing to a height of 70 cm.

Stem : Branched, terete.

Leaf : Leaves opposite, petiolate, ovate, crenate, serrate.

Floral biology

Flower : Heads small, homogamous, in dense corymbose; involucre companulate, bracts 2-3 seriate, striate, sub-equal; receptacle flat, naked or with caducous scales; corolla tubular, equal limb 5-cleft, regular; blue or white; anthers; appendaged, base obtuse; style arms elongate, obtuse (Fig. 2).

Fig. 2. Flower

Flowering and pollination : *A. conyzoides* flowers from July to March in India. The occurrence of male sterility has been documented for both *Ageratum conyzoides* and *Ageratum houstonianum*. The usually male fertile *Ageratum conyzoides* is self-pollinated and cross-pollinated by insects. *Ageratum houstonianum* has been reported as self-incompatible.

Seed biology

Fruit/Seed : The fruit is an achene and it is 5-angled; glabrous or the angles slightly hispid; pappus paleaceous, scaly, 5-awned, serrate below (Fig. 3).

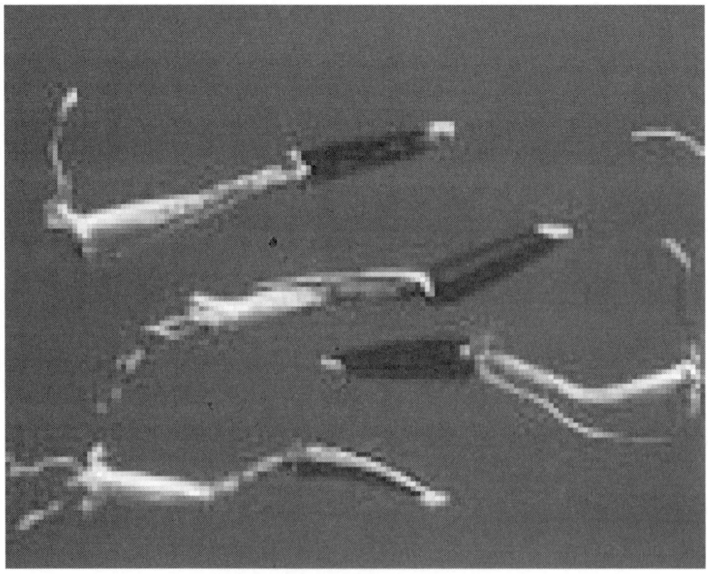

Fig. 3. Fruit/Seed

Propagation : By seeds

Seed dispersal : Seeds are mainly spread by wind and water (Holm *et al.*, 1977 in Pier, 2008). It may produce 40 000 or more seeds per plant.

Life cycle : Hardy annual. Quick flowering and short lived, as little as two months. *A. conyzoides* has the potential to produce many seeds (94,772 seeds per plant) and to shed seeds over extended times (5 to 8 months), as well as its extraordinary physiological plasticity, has enhanced its persistence in arable fields (Ekeleme *et al.*, 2005). Invasiveness is enhanced by its ability to emerge in abundance (1000 plants/m^2) (Anon., 2000, in Ekeleme *et al.*, 2005).

Seed dormancy and germination : No dormancy. The optimum germination temperature ranges from 20 to 25°C (Sauerborn and Koch, 1988 in Ming, 1999). There is no marked dormancy and germination is promoted by light and inhibited by burial. Seed of *Ageratum conyzoides* can germinate at comparatively low temperatures (10-20°C), which explains its occurrence at higher altitudes, whereas maximum temperature for germination is around 30°C. For seed of *Ageratum houstonianum*, the minimum temperature for germination is reported as 20°C and the maximum temperature 35°C.

Seed storage : Seeds are positively photoblastic, and viability is often lost within 12 months (Marks and Nwachuku, 1986; Ladeira *et al.*, 1987).

Effects on crop production : Ageratum conyzoides is a weed in many annual and perennial crops and has been reported as host of many crop diseases (Ekeleme *et al.*, 2005). Weeds interfere with growth and production of crops and therefore exert significant ecological and economic impacts (Singh and Saha, 2001, Batish *et al.*, 2009). For example, in Asia rice yield Asia has been negatively associated with *A. conyzoides* density (Roder *et al.*, 1998). Some studies have demonstrated allelopathy in the weed, however, shoot competition for light appears to be a major mode of interference in crops (Ekeleme *et al.*, 2005). Volatile components of *A. conyzoides* that contribute to phytoinhibition/allelopathy include precocenes and their derivatives monoterpenes and sesquiterpenes. There is much evidence that *A. conyzoides* inhibits germination and growth of other plants through chemicals produced by its root and shoot systems. Fresh leaves and volatile oils of *A. conyzoides* inhibit seedling growth of various crops including peanut, redroot amaranth, cucumber and ryegrass (Kong *et al.,* 2002). Studies have shown that shoot extracts of *A. conyzoides* inhibit germination of *Amaranthus caudatus, Digitaria sanguinalis* and lettuce (*Lactuca sativa*). Extracts of *A. conyzoides* inhibit germination of wheat and rice (*Oryza sativa*) seeds (Jha and Dhakal, 1990 in Ming 1999).

The phenolics present in leaf extracts and residues of *A. conyzoides* negatively interfere with the growth and development of wheat (Singh *et al.*, 2003 in Batish *et al.*, 2009a). Root and shoot length and biomass accumulation of rice are significantly reduced by 18 to 30% when grown in the rhizosphere soil of *A. conyzoides* (Batish *et al.*, 2009a). Leaf debris of *A. conyzoides* deleteriously affects the early growth of rice.

Lastly, *A. conyzoides* causes reduction in chickpea (*Cicer arietinum*) growth and nodulation and (Batish *et al.*, 2004). Phytoinhibition/allelopathy increases when plants are grown in nutrient-deficient conditions and decreases under fungal infection or aphid feeding (Kong *et al.*, 2002).

Plant pathogen transmission: A. conyzoides is a weed host and disease reservoir of the *Tomato yellow leaf curl Tanzania virus* near tomato farms in Tanzania (Kashina *et al.*, 2003).

Control : It can be readily controlled when young by hand pulling or hoeing. Seedlings and young stages readily controlled by 2, 4-D and MCPA.

Uses : Extracts and metabolites from this plant have been found to possess pharmacological and insecticidal activities (Okunade, 2002). In Central Africa it is used to treat pneumonia, but the most common use is to cure wounds and burns (Durodola, 1977, in Ming 1999). Traditional communities in India use this species as a bacteriocide, antidysenteric, and antilithic (Borthakur and Baruah, 1987, in Ming, 1999).

Past work done on Seed Science and Technology

Light and KNO_3 on *Ageratum conyzoides* L. seed germination at constant and alternating temperature

At constant temperature, light promoted the percentage and the speed of germination and KNO_3 reduced the germination percentage and speed of *A. conyzoides* seeds in darkness. At alternating temperatures, KNO_3 with light increased the germination percentage and speed of *A. conyzoides* seeds (Ikeda *et al.*, 2008).

Table 1. Effect of parthenin on germination and early growth of *A. conyzoides*

Germination (µM)	Germination (%)	Plumule length (cm)	Radicle length (cm)	Dry weight (mg/seedling)
0 (Control)	90.0 ± 2.5	3.7 ± 0.31	0.97 ± 0.04	1.90 ± 0.36
50	72.4 ± 2.8*	2.6 ± 0.42*	0.72 ± 0.12*	1.14 ± 0.21*
100	59.4 ± 1.5**	1.9 ± 0.24**	0.57 ± 0.08*	0.76 ± 0.17*
200	10.4 ± 0.5**	1.1 ± 0.20**	0.12 ±0.05**	0.30 ± 0.09**
400	–	–	–	–

* and ** mean significant at $P < 0.05$ and 0.01, respectively.

Singh *et al.* (2002) studied the effect of parthenin - a sesquiterpene lactone from *Parthenium hysterophorus* on the germination, growth, and some associated physiological changes in *Ageratum conyzoides*. The study reveals that germination and growth of *A. conyzoides* was severely reduced by parthenin in the concentration range of 50, 100, and 200 µM, while at 400 µM a complete inhibition of germination was observed. The radicle length of *A. conyzoides* was reduced more than the plumule length (Table 1). Further, the content of the photosynthetic pigment chlorophyll was appreciably reduced in the leaves of test plants spray treated with parthenin at 200 µM concentration. However, this inhibitory effect declined with the passage of time and at 10 days after treatment only a 25% reduction in chlorophyll content was observed compared to 76% on the first day after spray. Likewise, the cellular respiration measured through 2, 3, 5-triphenyl tetrazolium chloride was found to be less in the treated leaves, and the inhibitory response also declined with time. The content of proteins and carbohydrates decreased with the passage of time after parthenin treatment. The specific activities of the enzymes protease and α - and β-amylase were also adversely affected in the parthenin treated leaves. In the case of protease, the activity decreased with passage of time while those of the amylases increased. Thus, we conclude that parthenin affects the germination and growth of *A. conyzoides* by altering the contents of some macromolecules and the specific activities of some enzymes.

Parthenin, an allelochemical of *Parthenium hysterophorus,* inhibited germination of *A. conyzoides* but not wheat (Batish *et al.,* 1997).

REFERENCES

Anonymous. (2000). Upland Rice Ecosystem: Farmers' practice of using salt for weed control in upland rice. Las Banos, Philippines: International Rice Research Program Rep. Pp. 56–59.

Batish, D.R., R.K. Kohli, H.P. Singh and D.B. Saxena. (1997). Studies on herbicidal activity of parthenin, a constituent of *Parthenium hysterophorus* towards billgoat weed (*Ageratum conyzoides*). *Curr. Sci.,* 73:369–371.

Batish, D.R., S. Kohli, H. P. Singh and R.K. Kolhi. (2009). Nature of interference potential of leaf debris of *Ageratum conyzoides. Plant Growth Regul.,* 57:137–144.

Batish, D.R., S. Kaur, H.P. Singh and R.K. Kohli. (2004). Herbicidal activity of volatile oils from *Eucalyptus citriodora.* Proceedings of the 2nd European Allelopathy Symposium Allelopathy from Understanding to Application.

Batish, Daizy R., Kaur, Shalinder, Singh, Harminder Pal, Kohli, Ravinder Kumar. (2009a). Role of root-mediated interactions in phytotoxic interference of *Ageratum conyzoides* with rice (*Oryza sativa*) Flora (Jena), 204 (5): 388-395.

Borthakur, N. and A.K.S. Baruah. (1987). Search for precocenes in *Ageratum conyzoides* Linn. of North-East India. *J. Indian Chem. Soc.,* 64:580–581.

Durodola, J.J. (1977). Antibacterial property of crude extracts from herbal wound healing remedy — *Ageratum conyzoides. Planta Med.,* 32:388–390.

Ekeleme, F.F. Forcella, D.W. Archer and I. Okezie Akobundu. (2005). Seedling emergence model for tropic ageratum (*Ageratum conyzoides*). *Weed Science,* 53: 55–61.

Holm, L.G., Pluckett, D.L., Pancho, J.V. and Herberger, J.P. (1977). The world's worst weeds. Univ. Press of Hawaii, Hionolulu. pp 212-216.

Ikeda, F.S., R. Carmona, D. Mitja and R.M. Guimaraes. (2008). Luz e KNO_3 na germinação de sementes de *Ageratum conyzoides* L. sob temperaturas constantes e alternadas. Revista Brasileira de Sementes, v., 30(2): 93-199.

Jha, S. and M. Dhakal. (1990). Allelopathic effects of various extracts of some herbs on rice and wheat. *J. Inst. Agr. Anim. Sci.,* 11: 121-123.

Kashina, B.D., R.B. Mabagala and M.A. Mpunami. (2003). First report of *Ageratum conyzoides* L. and *Sida acuta* Burm F. as new weed hosts of *Tomato yellow leaf curl Tanzania virus*. *Plant Protect. Sci.,* 39: 18–22.

Kong, C., F. Hu and X. Xu. (2002). Allelopathic potential and chemical constituents of volatiles from *Ageratum conyzoides* under stress. *Journal of Chemical Ecology, Vol. 28, No. 6.*

Ladeira, A.M, L.B.P. Zaidan and R.C.L. Figueiredo-Ribeiro. (1987). *Ageratum conyzoides* L. (Compositae): Germinação, floração e ocorrência de derivados fenólicos em diferentes estádios de desenvolvimento. *Hoehnea,* 15:53–62.

Marks, M.K. and A.C. Nwachuku. (1986). Seed-bank characteristics in a group of tropical weed. *Weed Res.,* 26(3):151–157.

Ming, L.C. (1996). Produção de biomassa e teor de óleo essencial em função de fases de desenvolvimento, calagem e adubações mineral e orgânica em *Ageratum conyzoides* L. PhD Thesis, UNESP, Jaboticabal.

Ming, L.C. (1999). *Ageratum conyzoides*: A tropical source of medicinal and agricultural products. **In:** Prosperspectives on new crops and new uses, Janick, J. and V.A. Alexandria (Eds.). ASHS Press, Alexandria, VA., USA., pp: 469-473.

Okunade, A.L. (2002). *Ageratum conyzoides* L. (Asteraceae). *Fitoterapia,* 73:1–16.

Pacific Island Ecosystems at Risk (Pier). *Ageratum conyzoides.* (2008). Summary: Available from: http://www.hear.org/Pier/species/ageratum_conyzoides.htm [Accessed 16 October 2009]

Roder, Keoboulapha, Phengchanh, Prot and Matias. (1998). Effect of residue management and fallow length on weeds and rice yield. *Weed Research,* 38(3): 167-174.

Sauerborn, J. and W. Koch. (1988). Untersuchungen zur Keimungsbiologie von sechs tropischen Segetalaten. *Weed Res.,* 28(1): 47–52.

Singh, H.P, D.R. Batish and R.K. Kohli. (2003). Phytotoxic interference of *Ageratum conyzoides* with wheat (*Triticum aestivum*). *Journal of Agronomy and Crop Science,* 189: 341-346.

Singh, H.P., D.R. Batish and R.K. Kohli. (2002). Allelopathic effect of two volatile monoterpenes against bill goat weed (*Ageratum conyzoides* L.). *Crop Prot.,* 21:347–350.

Singh, S.K. and G.P. Saha. (2001). Effect of weed management practices on performance of prominent wet-season crops. *Indian J. Agron.,* 46: 489–495.

SPINY AMARANTH (*Amaranthus spinosus*)

Common name : Spiny amaranth, Prickly amaranth, Thorny amaranth

Botanical name : Amaranthus spinosus

Tamil name : Mullukkeerai

Hindi name : Kanta nutiya

Family : Amaranthaceae

Origin and distribution : Tropical America. It is mainly tropical and subtropical in distribution, but also extends into the temperate zone from latitude 30°N to 30"s.

Botany

Habitat : Cultivated beds, open waste place, gregarious and abundant along sand bars and margins of streams.

Habit : Erect monoecious herb, glabrous, annual, bushes with branches growing to a height of 100 to 130 cm (Fig. 1).

Fig. 1. Plant

Stem : Round and stout and green or reddish.

Leaf : Long petioled, ovate or oblong, leaf axils with spines

Floral biology

Inflorescence : Inflorescence consisting of dense clusters, lower ones axillary, higher ones often collected in an axillary and terminal spike which is often branched in its lower part; axillary clusters usually armed with (1–)2(–3) very sharp spines up to 2 cm long (Fig. 2)

Fig. 2. Inflorescence

Flower : Flowers unisexual, solitary in the axil of a bract, subtended by 2 bracteoles; bracts and bracteoles scarious, mucronate from a broad base, shorter or as long as the perianth; male flowers usually arranged in a terminal spike above the base of the inflorescence, green; tepals 5 or in male flowers often 3, free, subequal, ovate-oblong to oblong-spatulate, up to 2.5 mm long, very convex, membranous, with transparent margins and green or purple median band; male flowers with 5 stamens about as long as tepals; female flowers with superior, oblong ovary, 1-celled, styles 2–3, ultimately recurved (Fig 3a, b).

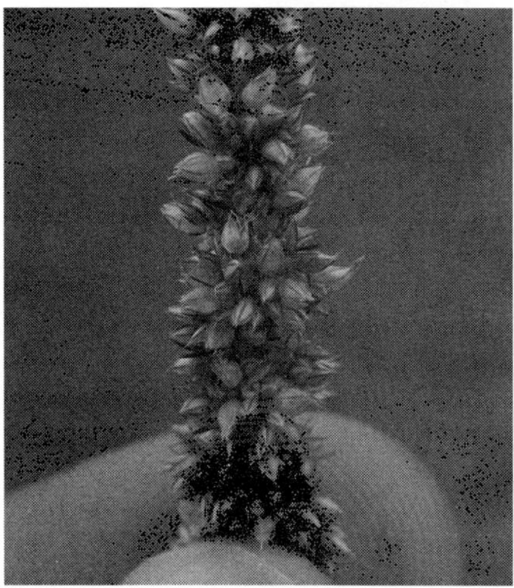

Fig. 3a. Flower

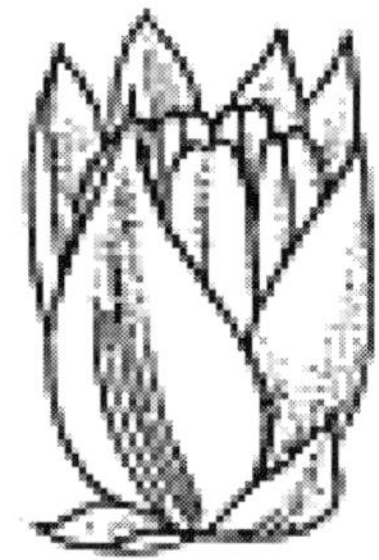

Fig. 3b. Male flower with bracteoles

Flowering and pollination: The flowers are monoecious (individual flowers are either male or female, but both sexes can be found on the same plant) and are pollinated by wind and self. The plant is self-fertile. A light intensity of less than 30% completely suppresses flowering. Flowering is earliest and most abundant in areas with daylength of 11-12h.

Fruit : Fruit an oblong capsule with persisting styles, circumscissile a little below the middle or indehiscent, 1-seeded (Fig. 4).

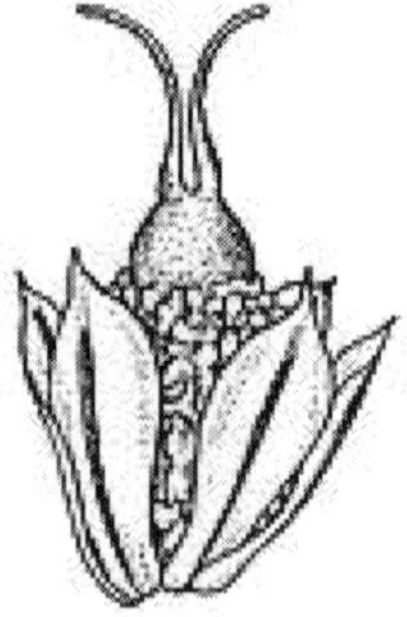

Fig. 4. Fruit

Seed biology

Seed : Seed about 1 mm in diameter, shiny black or brownish-black with thin margin. Seedling with epigeal germination; cotyledons leafy, glabrous, apex rounded to slightly acute; hypocotyl up to 12 mm long, epicotyl absent (Fig. 5)

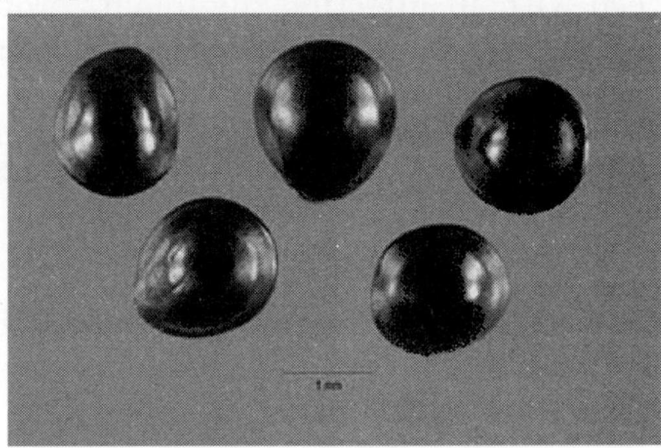

Fig. 5. Seed

Propagation : By seeds. Cuttings of growing plants root easily.

Seed dispersal : Seeds are dispersed by water. It produces 235, 00 seeds per plant. The weight of 1000 seeds is 140–250 mg.

Life cycle : It is in leaf stage in April. It flowers from July to September and the seeds ripen from August to October. Seed matures about one month after flowering.

Seed dormancy and germination : The seeds will usually germinate in 10-15 days. Freshly collected seeds may germinate at temperatures as high as 40°C, with a germination rate of up to 95%. Exposure to light increases seed dormancy. Ingestion by sheep enhances germination.

Seed storage : Have long viability. After storage, however, temperature requirements are lower. Seeds stored for one month at room temperature have almost 100% germination, and after 5 months they have approximately 90% germination. When they are stored for one year at 20°C, the germination rate will drop to about 50%, but storage at lower temperatures gives a higher rate.

Effect on crop : Shrefler *et al.* (1994) reported that it can reduce yield by up to 40% on corn and soybean if not properly controlled. Spiny amaranth is a very noxious weed in many parts of the world. It is, for instance, troublesome in maize, cassava and groundnut in Ghana, in cotton in Mozambique, and in sugar cane in South Africa. *Amaranthus spinosus* is a host plant for, among others, tobacco mosaic virus, groundnut rosette virus, cucumber mosaic virus and root-knot nematodes (*Meloidogyne* spp.), which attack some commercial crops. When the world's worst weeds are ranked on the basis of the number of pests hosted, *Amaranthus spinosus* is placed number 6, hosting 15 pests that may affect crops.

Control : Mechanical control: The preferred method for is mechanical removal, i.e. grubbing out.

Chemical control: Can be controlled by 2, 4-D and metolachlor in various formulations.

Biological control: As a weed in tomato in India, spiny amaranth has been successfully controlled by the application of geraniol, which completely blocked the germination of the weed without affecting the tomato crop. An ethanolic extract of seeds of *Coffea Arabica* L. (with 1, 3, 7-trimethylxanthine as active ingredient) at a concentration of 1.2 g l^{-1}, completely inhibited germination of spiny amaranth in a crop of black gram (*Vigna mungo* (L.) Hepper) without negative effects to this pulse crop. Some insects attacking *Amaranthus spinosus* have been recorded from Mexico: the pyralid *Herpetogramma bipunctalis* and the curculionid *Conotrachelus seniculus*. These might be useful for biological control.

Uses : In tropical Africa and elsewhere, *Amaranthus spinosus* leaves and young plants are collected for home consumption as a cooked, steamed or fried vegetable, especially during periods of drought.

It is also used as forage and said to increase the yield of milk in cattle.

In Uganda, the ash of burnt *Amaranthus spinosus* plants is used as a tenderizer in cooking tough vegetables such as cowpea leaves and pigeon peas. The ash is also used as a vegetable salt and in southern Africa it is used as a snuff, alone or with tobacco. The root is known as an effective diuretic. In South-East Asia, a decoction of the root is used to treat gonorrhoea and is also applied as an emmenagogue and antipyretic.

In many countries, including those in Africa, the bruised leaves are considered a good emollient and applied externally in cases of eczema, burns, wounds, boils, earache and haemorroids. In mainland South-East Asia, it is also used as a sudorific, febrifuge, antidote to snake poison, galactagogue, and to treat menorrhagia. Some tribes in India apply *Amaranthus spinosus* to induce abortion.

Past work done on Seed Science and Technology

1. Effect of salts on germination

Mali and Telave (2011) showed that untreated seeds gave 94% germination. The seeds treated with the different salts showed a decrease in percent germination. The seeds treated with 2% NaCl and KCl remained ungerminated. Comparatively, the sulphates of magnesium and sodium were less toxic. With $MgSO_4$ and Na_2SO_4, the percent germination at 2% concentration was 46 and 24% respectively (Table 1).

Table 1. Effect of salts on germination of *Amaranthus spinosus* L.

Salt concentration %	Percentage germination			
	NaCl	KCl	$MgSO_4$	Na_2SO_4
0	94	94	94	94
0.5	64	58	88	92

1.0	12	08	66	46
1.5	00	02	48	16
2.0	00	00	46	24

Allelopathic effects

Rizvi *et al.* (1987) observed that seeds of *A. spinosus* treated with 1,000ppm of 1, 3, 7-T showed a marked decline in amylase activity during germination. The extent of inhibition varied with the duration of germination. It showed nearly 7.6% inhibition after 27h of incubation and about 29.5% after 63h (Fig. 6).

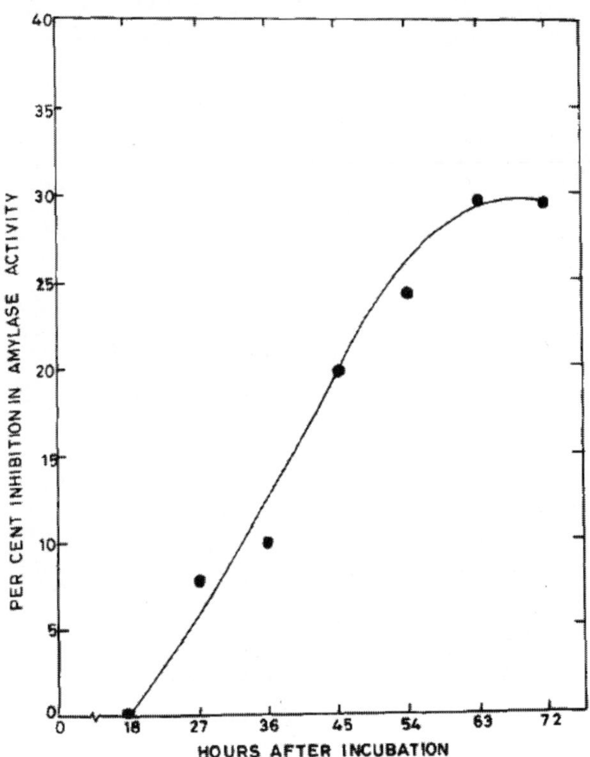

Fig. 6. *Amaranthus spinosus:* Effect of 1000 ppm 1,3,7-trimethylxanthine on amylase activity in seedlings.

The observed inhibition in amylase activity could be due to: (i) inhibited catalytic property of the enzyme, (ii) inhibited synthesis of the enzyme (iii) both. The Km value (0.45%) as derived from double reciprocal plot (Fig. 7) was unaffected, indicating no effect on the catalytic property of the enzyme. *In vitro* treatment of the enzyme extracted from untreated germinating seeds had no effect at any of the concentration (300, 600, 900, 1000 and 1500 ppm) used (Table 2). Variation in the incubation time had no effect either (Table 3).

Table 2. *Amaranthus spinosus:* Amylase activity in cell free extracts obtained from untreated germinating seeds when incubated for different periods in the presence of 1,000ppm of 1,3,7- trimethylxanthine

Incubation period (minutes)	Enzyme activity (mg maltose h^{-1} g^{-1} f.w.)
0	251.2 ± 1.4
10	250.8 ± 1.8
20	254.5 ± 1.3
30	248.3 ± 2.2
40	250.4 ± 2.1
50	252.3 ± 1.9
60	249.5 ± 2.0

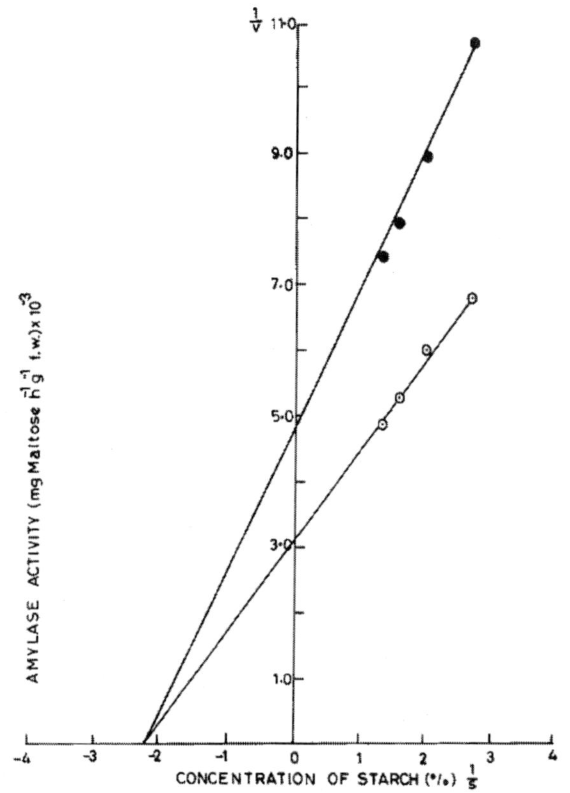

Fig. 7. *Amaranthus spinosus:* Lineweaver-Burk plot of amylase activity in cell free extracts obtained from germinating seeds treated with 1,3,7-trimethylxanthine (e) and in that obtained from untreated germinating seeds (O).

Table 3. *Amaranthus spinosus:* Amylase activity in germinating seeds treated with 1,3,7-trimethylxanthine (1,000 ppm) in the presence of different concentrations of gibberellic acid

Concentration of gibberellic acid (μM)	Per cent inhibition
0	30.0 ± 2.0
2	3.12 ± 1.7
5	30.6 ± 1.6
10	31.3 ± 2.0
15	29.8 ± 1.9
20	30.2 ± 1.5

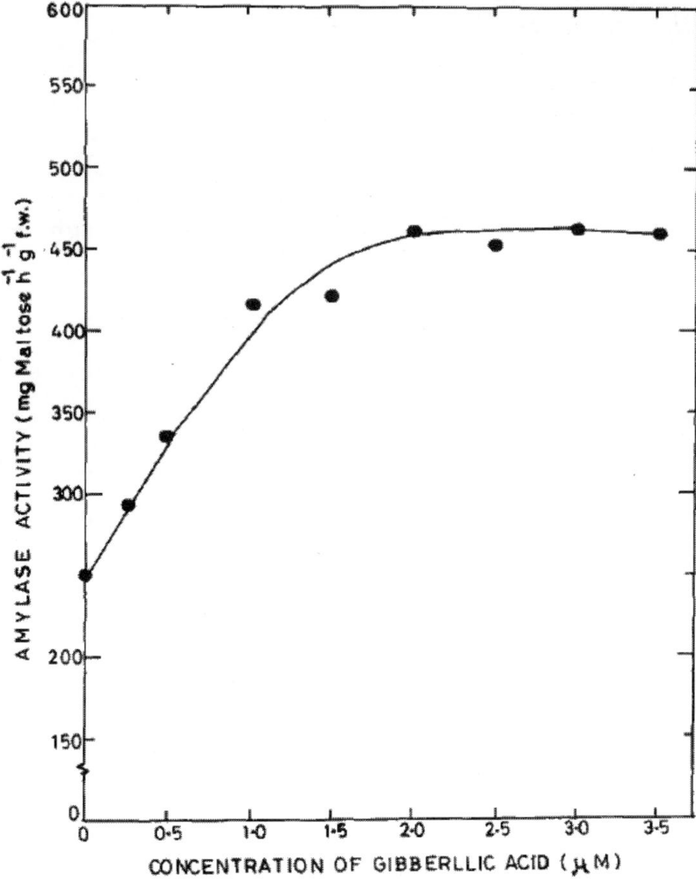

Fig. 8. *Amranthus spinosus*: Effect of GA_3 on amylase activity in seedlings

It is clear from the above results that 1,3,7-T possibly affected the synthesis of amylase. Thus, one of the possible causes of reduced synthesis might be a reduced production of gibberellic acid (GA). While all concentrations (0.25, 0.5, 1.0. 1.5, 3.0, 3.5 #m) of GA_3 (in the absence of 1,3,7-T) caused an appropriate increase in amylase activity (Fig. 8), it could not overcome the inhibitory effect of 1,3,7-T, even at concentrations much higher than the saturating one (Table 3). When 1,3,7-T (1000 ppm) treated seeds were simultaneously treated with various concentrations (2.0, 5.0, 10.0, 15.0, 20.0 # m) of GA_3, amylase activity decreased by nearly 30% at all the concentrations (Table 3).

Macharia and Peffley (1995) stated that the percentage germination of amaranth seeds moistened with Heshiko root exudates was significantly lower than the control seeds and those moistened with root exudates from the other genotypes (Table 4). The highest germination percentage was obtained when seeds were irrigated with exudates from 81215. Germination of amaranth seeds moistened with root exudates from Heshiko was 44-53% less than germination of controls or when moistened with exudates from other genotypes, providing evidence of allelopathy against *Amaranthus spinosus.*

Table. 4. Percentage germination of *Amaranthus spinosus* seeds irrigated with *Allium* root exudates

	Germination	
Treatment	**Experiment 1**	**Experiment 2**
Heshiko	15.5*	27.0*
Ishikura	77.5	78.0
NMYG	76.5	67.0
81215	85.5**	86.0**
8273	64.0	66.0
Control	68.5	71.0
l.s.d	4.3	3.3

[d] Mean separation by Fisher's l.s.d. = 5%; * inhibition of germination in relation to control; ** stimulation of germination in relation to control

REFERENCES

Macharia, C. and E.B. Peffley. (1995). Suppression of *Amaranthus spinosus* and *Kochia scoparia:* evidence of competition or allelopathy in *Allium fistulosum. Crop Protection,* 14(2): 155-158.

Mali, B.S. and A.B. Telave. (2011). Effect of salts on gemination of *Amaranthus spinosus. Bioinfolet,* 8 (3): 312-313.

Rizvi, S.J.H., V. Rizvi and D. Mukerjee. (1987). 1,3,7-Trimethylxanthine, an allelochemical from seeds of *Coffea arabica:* Some aspects of its mode of action as a natural herbicide. *Plant and Soil,* (98): 81-91.

Shrefler, J.W, J.A. Dusky, D.G. Shilling, B.J. Brecke and C.A. Sanchez. (1994). Effects of phosphorus fertility on competition between lettuce (*Lactuca sativa*) and spiny amaranth (*Amaranthus spinosus*). *Weed Sci.,* 42: 556–560.

PIG WEED *(Amaranthus viridis* L.)

Common name : Green amaranthus, pig weed

Botanical name : Amaranthus viridis L.

Tamil name : Kuppai keerai

Hindi name : Jungali chaulayi

Family : Amaranthaceae

Origin and distribution : It is possibly of Asian origin, but now a cosmopolitan weed in the tropical and subtropical regions of the World, also penetrating far into temperate regions (e.g. in Europe, North America, Asia and Australia). In tropical Africa, it is also a widespread and common weed. It is occasionally cultivated in Nigeria, Gabon and DR Congo.

Botany

Habitat : Common wastelands weed and along bunds in cultivated lands.

Habit : The plant is a perennial with long prostrate or ascending branches and a woody main stem growing to a height of 30-60cm (Fig. 1).

Fig. 1. Plant

Stem : Stem is striate, round and softly woolly tomentose.

Leaf : Leaves are simple, alternate, shortly stalked, sub-orbicular or obovate with an obtuse tip. The base is tapering. The upper surface of the leaf is pubescent, while the lower one is densely hairy. The leaves on the side branches are much reduced.

Floral biology

Inflorescence and flower : A dense spike, simple or branched at the top of the stem, and in the leaf axils, 2.5 to 12cm long and 2 to 5mm wide (Fig 2a).

Flower is very small, sessile, greenish white, in dense axillary short spikes, which are solitary or clusterd. Bract is one and bracteoles are 2, membranous and apiculate. Perianth consists of 5 free oblong sepals, wooly on the back. Stamens are 5, united at the base into a cup with interposed staminodes. Ovary is superior one-celled, with one pendulous ovule (Fig. 2b).

Fig. 2a. Inflorescence

Fig. 2b. Flower

Flowering, pollination and fruit set : The flowers are monoecious (individual flowers are either male or female, but both sexes can be found on the same plant) and are pollinated by wind, self. The plant is self-fertile.

Fruit : Fruit is an ovoid bladder-like structure (utricle) (Fig. 3).

Fig. 3. Fruit

Seed biology

Seed : Seed is lenticular, smooth and shining black (Fig. 4).

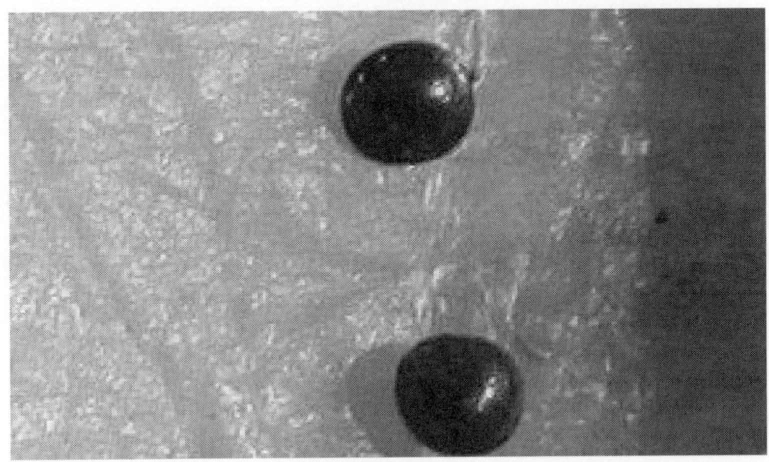

Fig. 4. Seed

Propagation : By seeds. Cuttings of growing plants root easily.

Seed dispersal : The plant can produce 7000 seeds which are dispersed by the water or wind.

Life cycle : It is in leaf stage in April. It is in flowering from July to September and the seeds ripen from August to October.

Seed dormancy and germination : Primary dormancy in *Amaranthus* plays a fundamental role in extending germination over a longer period, so that the probability of seedling survival is maximized. Germination is usually rapid and good, if the soil is warm. A drop in temperature overnight aids germination. A temperature of 30°C was the optimum for germination. Germination of slender amaranth seed at 21 days was similar with 35/25, 35/20, 30/25, and 30/20°C alternating temperature regimes. As temperatures in alternating regimes increased, time to onset of germination decreased and rate of germination increased. Slender amaranth germination was greater with acidic than with basic pH conditions. Germination declined with increasing water stress and was completely inhibited at water potentials below -0.6 MPa. Slender amaranth emergence was greatest at depths of 0.5 to 2 cm, but some seeds emerged from as deep as 6 cm.

Seed storage and burial : The seeds of weedy amaranths have remarkably long viability; some have germinated after 40 years.

Effect on crop : It is considered as one of the most harmful weeds which may lead to heavy losses in crop yield. Its weedy status is more important than its culinary value. In Bangladesh, an yield reduction up to 80% was observed. Green and redroot pigweed are two of the most important weed species in Ontario (Weaver, 1984). Previous studied have determined that pigweed emerging several weeks after crop emergence causes crop yield losses of less than 5%; this usually is tolerable to farmers and does not reduce profit (Dieleman *et al.*, 1995; Knezevic *et al.*, 1994).

Control : Cultivation favours *A. viridis* emergence and is a useful means of killing seedlings before sowing. Biological control agents like *Hypolixus trunculatus* and *Hymenia recurvalis* are used. Butachlor applied pre-emergence @ 2.0 to 2.5 kg ha^{-1} gives excellent control. Post-emergence application of 2, 4-D @ 500 g ha^{-1} also controls this weed.

Uses : It is a astringent and vermifuge. A decoction of the entire plant is used to stop dysentery and inflammation. The root juice is used to treat inflammation during urination. It is also taken to treat constipation. Yellow and green dyes can be obtained from the whole plant. *A. viridis* is used as a vegetable.

Past work done on Seed Science and Technology

1. Allelopathic effect

Supriya Vaid *et al.* (2011) observed that the germination of the weed was significantly inhibited even at very low concentrations (0.1 and 0.5 µL) of limonene where about 4 and 11% inhibition were observed respectively (Table 1). Not only germination, but the seeding growth of the test weed in terms of radicle length, seedling length and seedling dry weight were appreciably reduced in response to limonene. At the concentration of 0.7 µL, the seedling length was reduced by about 25% while seedling dry weight was reduced by about 9%.

Table 1: Effect of limonene on percent germination and radical length of *A. viridis*

Concentration (µL)	Percent germination	Radicle length (cm)
Control	100 ± 1.25^{a}	3.51 ± 0.86^{a}
0.1	94.46 ± 1.48^{a}	3.36 ± 0.09 $(95.73)^{b}$
0.5	88.66 ± 2.35^{b}	3.19 ± 0.08 $(90.88)^{b}$
0.7	86.36 ± 1.93^{b}	2.97 ± 1.16 $(84.62)^{c}$
1	44.31 ± 5.77^{c}	1.39 ± 0.43 $(39.60)^{d}$
2	19.32 ± 1.92^{d}	0.54 ± 0.26 $(15.38)^{e}$
5	18.18 ± 1.92^{d}	0.34 ± 0.14 $(9.69)^{e}$
7	0^{e}	–

Different superscripts in a column represent significant difference atp <0.05

Similarly, there was a significant reduction in the chlorophyll content and respiratory activity of the test weed which were reduced by over 50% upon treatment with 0.7 µL limonene (Table 2).

Table 2. Effect of limonene on chlorophyll content and percent respiration of *A. viridis*

Concentration (μL)	Chlorophyll content (μg/mg)	Percent respiration
Control	15.5 ± 0.07[a]	100 ± 1.0.[a]
0.1	12.34 ± 0.47 (79.61)[b]	72.34 ± 0.48[b]
0.5	9.85 ± 0.36 (63.55)[c]	53.78 ± 0.06[c]
0.7	7.32 ± 0.32 (47.23)[d]	45.60 ± 1.62[d]
1	3.47 ± 0.42 (22.39)[e]	36.06 ± 1.24[e]
2	2.41 ± 0.05 (15.55)[f]	28.97 ± 0.41[f]
5	0.68 ± 0.08 (4.4)[g]	13.38 ± 4.26[g]
7	–	–

Different superscripts in a column represent significant difference atp <0.05

On the basis of this study, it is concluded that limonene has a weed suppressing potential and therefore, can be used for future weed management programmes either directly or by serving as a lead molecule.

Batish *et al.* (2002) undertook a study to explore the effect of parthenin (chemically a sesquiterpene lactone from *Parthenium hysterophorus* L. on *Amaranthus viridis*. In a laboratory bioassay, germination, seedling growth and dry weight accumulation of the weeds were significantly reduced by treating with parthenin (Table 3). The chlorophyll content and the respiratory activity were also significantly affected in the treated plants. The study concluded that phytotoxicity of parthenin could be useful as a natural herbicide for future weed management programmes.

Table 3. Percent inhibition of seed germination, seedling length and chlorophyll content of *A. viridis* in response to parthenin[1].

Concentration (μM)	Germination	Seedling length	Chlorophyll content
50	22.4*	29.3*	21.7*
100	42.6*	52.4*	43.4*
300	72.1**	69.7**	67.2**
500	100**		

[1]Data presented as percent inhibition with respect to the control;
* and ** represent significant difference at P<0.01 and 0.05, respectively.

Muhammad Uzair *et al.* (2009) studied the effect of chemical industry waste (organic with heavy metals) on seed bank of A. *viridis*. Highest number of seedlings (40 ± 11) was recorded in 20 kg ash treatment with *D. Muricata*, while lowest number of seedlings appeared in 20 kg sludge treatment. Higher amount of sludge reduced the seed germination of this species while higher amount of ash promoted the germination and this effect was signigicantly different (Table 4). Though maximum height was recorded in control, while maximum fresh weight and dry weight were recorded in 20kg sludge however, no significant differences were recorded between the treatments.

Table 4. Effect of Industrial sludge and ash on the mean number of seedling, height, fresh weight and dry weight of *Digera muricata* (L.) Mart.

Treatments	Number of seedlings	Height (cm)	Fresh weight (g)	Dry weight (g)
Control	ab	a	a	a
	22±3	28.36±1.84	641.71±184.09	116.37±30.66
10kg sludge	ab	a	a	a
	26±1	26.67±0.73	1217.96±591.45	179.73±68.46
20kg sludge	b	a	a	a
	8±4	21.16±10.60	1766.66±1257.42	255±156.60
10kg ash	b	a	a	a
	19±5	24.97±1.12	870.69±192.00	175.37±101.59
20kg ash	a	a	a	a
	40±11	27.94±1.46	703.92±96.48	129.64±18.22
LSD	18.95	15.68	2145.25	319.01

The maximum number of seedlings (9 ± 4) was found in 10 kg sludge while fresh weight and dry weight (630 ± 429.22 and 148.33 ± 117.34 respectively) were maximum in 20 kg sludge treatment. The lowest number of seedlings was observed in both 20 kg sludge and 20 kg ash treatment. It shows that high amount of sludge and ash reduces the seed germination while 20 kg sludge treatment showed better response (Table 5).

Table 5. Effect of Industrial sludge and ash on the mean number of seedling, height, fresh weight and dry weight of *Amaranthus viridis* L.

Treatments	Number of seedlings	Height (cm)	Fresh weight (g)	Dry weight (g)
Control	a	a	a	a
	8±3	20.74±10.37	140.57±87.11	43.91±26
10kg sludge	a	a	a	a
	9±4	19.80±8.66	496.66±400.84	129.5±110.80
20kg sludge	a	a	a	a
	3±1	21.14±11.13	630±429.22	148.33±117.34
10kg ash	a	a	a	a
	5±1	25.82±1.12	630±429.22	127.5±36.99
20kg ash	a	a	a	a
	3±1	19.47±9.76	348.33±209.37	103±52.65
LSD	12.83	29.30	1014.61	276.96

Germination requirements

Oryokot *et al.* (1997) constructed a mathematical model of seed germination for green and redroot pigweed based on temperature and water potential (moisture) and expressing cumulative germination in terms of thermal time (degree days). Empirical observations indicated green pigweed germinated at a lower base temperature than redroot pigweed but the germination rate of redroot pigweed is much faster as mean temperature increases. Moisture limitation delayed seed germination until 23.8°C (green pigweed) or 27.9 (redroot pigweed); thereafter, germination was independent of water potential as mean temperatures approached germination optima. The germination model, based on a cumulative normal distribution function, accounted for 80 to 95% of the variation in seed germination and accurately predicted that redroot pigweed would have a faster germination rate than green pigweed. However, the model predicted that redroot pigweed would germinate before green pigweed (in thermal time) and was generally less accurate during the early period of seed germination. The model also predicted that moisture limitation would increase, rather than delay, seed germination. Also tested the effect of mean temperature on shoot elongation (emergence) and described the relationship by a linear model. Base temperatures for shoot elongation were higher than for seed germination. Shoot elongation began at 15.6 and 14.4°C for green and redroot pigweed, respectively; they increased linearly with temperature until the optimum of 27.9°C was reached. Elongation was dependent on completion of the rate-limiting step of radicle emergence and was sensitive to temperature but not moisture; hence, elongation was sensitive to a much smaller temperature range.

Gutterman *et al.* (1992) investigated the effects of temperature, light and oxygen on the germination of *Amaranthus caudatus* (var. *viridis*) seeds. At the highest temperature studied (35°C), seed germination was found to be most rapid and highest in both the dark and the light. Germination was inhibited by continuous white light. The lower the temperature, the greater the inhibitory effect of light. The far-red and blue spectral regions were the most effective in light inhibition. The light effect was also dependent on the percentage of oxygen in the atmosphere around the seeds; it was less marked in O_2-enriched atmospheres. Light/ oxygen dependency resulted from the action of the seed coat and endosperm. Scarified seeds were much less sensitive to light and hypoxia than non-scarified seeds.

REFERENCES

Batish, D.R., H.P. Singh, D.B. Saxena and R.K. Kohli. (2002). Weed suppressing ability of parthenin – a sesquiterpene lactone from *Parthenium hysterophorus*. *New Zealand Plant Protection,* 55:218-221.

Dieleman A., A.S. Hamill, S.F. Weise and C.J. Swanton. (1995). Empirical models of pigweed (*Amaranthus* spp.) interference in soybean (*Glycine max*). *Weed Sci.* 43:612–618.

Gutterman, Y., F. Corbineau and D. Come. (1992). Inter-related effects of temperature, light and oxygen on *Amaranthus caudatus* L. seed germination. *Weed Res.*, 32(2):111–117.

Knezevic S.Z., S.F. Weise, C.J. Swanton. (1994). Interference of redroot pigweed (*Amaranthus retroflexus* L.) in corn (*Zea mays* L.). *Weed Sci.,* 42: 568–573.

Muhammad Uzair, Moinuddin Ahmed and Kanwal Nazim. (2009). Effect of industrial waste on seed bank and growth of wild plants in Dhabeji area, Karachi, Pakistan. *Pak. J. Bot.,* 41(4): 1659-1665.

Oryokot, J.O.E., S.D. Murphy, A.G. Thomas and C.J. Swanton. (1997). Temperature- and moisture-dependent models of seed germination and shoot elongation in green and redroot pigweed (*Amaranthus powellii, A. retroflexus*). *Weed Sci.,* 45(4): 488-496.

Supriya Vaid, Daizy R. Batish, H.P. Singh and R.K. Kohli. (2011). Phytotoxicity of limonene against *Amaranthus viridis.* L. *The Bioscan An International Quarterly Journal of Luife Science.* 6(1): 163-165.

Weaver, S. E. (1984). Differential growth and competitive ability of *Amaranthus retroflexus, A. powellii* and *A. hybridus. Can. J. Plant Sci.* 64:715-724.

GANGES PRIMROSE (*Asystacia gangetica* (L.))

Common name : Ganges Primrose, Chinese violet, Creeping foxglove

Botanical name : Asystasia gangetica (L.) T. Anderson

Tamil name : Parchorri, chorri, mekampokki, medday keerai

Family : Acanthaceae

Origin and distribution : Asystasia gangetica is native in tropical Africa, Arabia and tropical Asia, but has been introduced in many other tropical regions, where it often naturalized. It occurs throughout tropical Africa.

Botany

Habitat : Common straggler in bushes and hedges, often grow as weed in cultivated fields.

Fig. 1. Plant

Habit : Straggling or climbing perennial herb (Fig. 1).

Stem : Branched, tall, internodes long.

Leaf : Opposite, ovate, acute, rounded or subcordate, narrowed at base; bracts very small.

Floral biology

Flower : Inflorescence a terminal raceme up to 25 cm long, with flowers directed to one side. Flowers white, blue purple or yellow; sepals linear, lanceolate, pubescent at the back; corolla pubescent, tube yellow, limb pink or pale purple; stamens 4, didynamous; ovary 2-celled; style linear (Fig. 2).

Fig. 2. Flower

Flowering and pollination : Year around flowering. The flowers attract bees and butterflies, especially Fiery Skipper (*Hylephila phyleus*) butterflies. Pollination by insects or birds.

Fruit : Fruit a club-shaped capsule 2–3 cm long, pubescent and glandular, usually 4-seeded (Fig. 3).

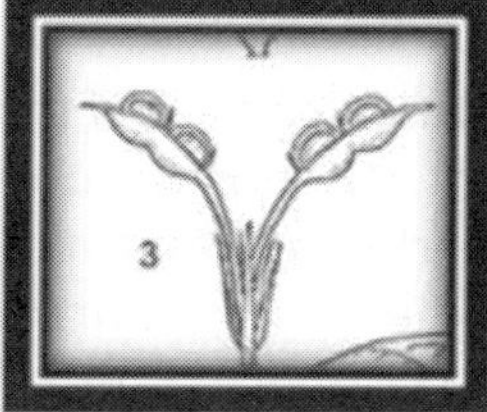

Fig. 3. Fruit

Seed biology

Seed : Seeds ovoid, flattened, 4–5 mm long, grey to brown, with crenate margins, tuberculate, supported by retinacula (Fig. 4).

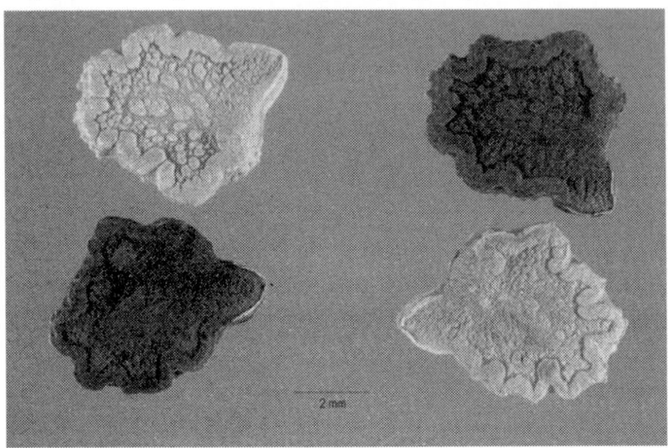

Fig. 4. Seed

Propagation : Asystasia gangetica can be propagated by seed and stem cuttings with 1–3 nodes.

Seed dispersal : The seeds are thrown as far as 6 m by an explosive opening mechanism of the fruits, triggered by hot afternoons and by man (Smith, 1985).

Life cycle : Single-node cuttings buried in soil produce flowers and fruits within 6 weeks. The period from seedling emergence to seed dispersal can be as short as 8 weeks in open areas, but it can take 2 weeks longer in partially shaded areas. It takes one month from flower development to seed dispersal.

Seed dormancy and germination : Freshly harvested seed does not germinate well under ordinary conditions. This seed normally requires a storage period of several months at room temperature before it germinates. After a period of approximately 135 days in ordinary storage, it germinates readily at room temperature (Akamine, 1947).

Effect on crop : A. gangetica subsp. *micrantha* is a known troublesome weed in tropical crops such as coffee, oil-palm and rubber plantations. It competes effectively for soil nutrients, reducing productivity and increasing management costs. It could also become and agricultural weed in Australia. It could potentially affect crops such as soybeans, vegetables, cut flowers and oil-tea tree. It can smother all vegetation in the herbaceous layer (Pier, 2006) reducing the availability of habitat for native plants and animals, hence reducing biodiversity.

Control : Seedlings can be removed by hand and bagged for disposal ensuring no rhizomes are left as these can develop into new infestations. The weed is also controlled by the herbicide mixtures paraquat + 2,4-D dimethylamine @ 0.25 + 0.5 kg ai/ha, MCPA + diuron + amitrole + paraquat @ 0372 + 0.48 + 0.95 + 0.25 kg ai/ha, MSMA + 2,4-D dimethylamine + $NaClO_3$ @ 1.0 + 0.5 + + 5.0kg ai/ha and 2,4-D isoocryl ester + ioxynil octanoate @ 0.01 + 0,08 kg ai/ha.

Uses : It is locally used as a pot herb and leafy vegetable. It sometimes promoted as a cover plant in orchards because it checks erosion and prevents infestation by noxious weeds, and because it attracts bees to the orchard. Because of its ability to grow under shade and its high nutritive value, it is used as forage. In Africa, an infusion of the plant is used to ease pain during child birth and the sap is applied to sores, wounds and piles and in embrocations to treat stiff neck and enlarged spleen in children. Powdered roots are considered analgesic and used in treating stomach-ache and snake bites.

REFERENCES

Akamine, E. K. (1947). Germination of *Asystasia gangetica* L. seed with special reference to the effect of age on the temperature requirement for germination. *Plant Physiology*, 22: 603-607.

Smith, C. W. (1985). Impact of Alien Plants on Hawai'i's Native Biota. In: Stone, Charles P. and Scott, J. Michael, (Eds.) Hawai'i's terrestrial ecosystems: Preservation and Management. Cooperative National Park Resources Studies Unit, University of Hawaii, Manoa.

Pacific Island Ecosystems at Risk (Pier). *Asystasia gangetica.* (2006). Summary: Available from: http://www.hear.org/pier/species/asystasia_gangetica.htm

RED HOGWEED (*Boerhaavia diffusa* L.)

Common name : Spreading hogweed, red hogweed, tar vine, red spiderling

Botanical name : Boerhaavia diffusa Linn. Syn. *B. repens; B. repens* var. *diffusa*

Tamil name : Mukkurattai, Mukkarattai keerai

Hindi name : Snathikari

Family : Nyctaginaceae

Origin and distribution : Boerhaavia diffusa is also indigenous to India; it is found throughout the warmer parts of the country up to an altitude of 2000 m in the Himalayan region (Chopra, 1969). The plant is also cultivated to some extent in West Bengal (Anon., 1988).

Botany

Habitat : A common weed of pastures and waste land.

Habit : A perennial diffuse herb; stem profusely branching, angular, reaching a length of 5 feet (Fig. 1).

Stem : Prostrate, decumbent or ascending, 4-10cm long, rather slender, divaricately branched

Leaf : Opposite or sub-opposite, two of a node unequal, broadly ovate or suborbicular, obtuse to rounded or subcordate at the base.

Floral biology

Flower : In penduncululate, glomerulate clusters arranged in slender, long stalked, axillary or terminal corymbs (Fig. 2).

Fig. 1. Plant

Fig. 2. Flower

Flowering, pollination and fruit set : It is flowering and fruiting throughout the year, when sufficient water is available. The seeds germinate before the onset of the monsoon. The plant grows profusely in the rainy season, and mature seeds are formed in October– November. Self-pollinated when not crosspollinated. The flowers offer nectar for pollinators, and during observations, *Boerhavia diffusa* was visited by small, medium, and large bees, and small and medium flies; ants were also observed visiting *Boerhavia diffusa* during a observation.

Fruit : Obovoid or sub-ellipsoid, rounded above, slightly cuneate, below, broadly and bluntly 5-ribbed, very glandular throughout (Fig. 3).

Fig. 3. Fruit

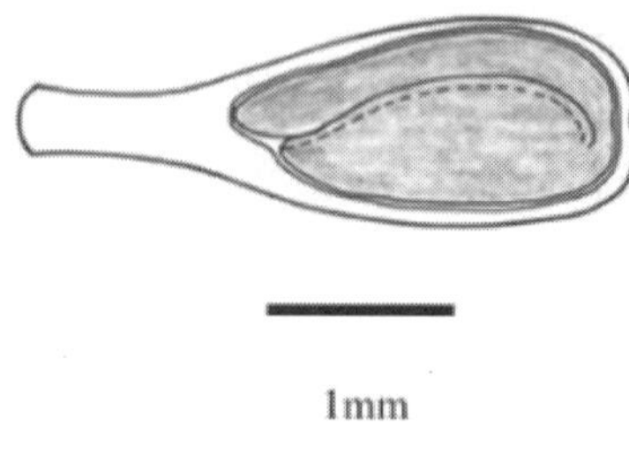

Fig. 4. Seed

Seed biology

Seed : Seed obovoid, pale brown (Fig. 4).

Propagation : It is propagated by seed. *Boerhavia diffusa* is successfully propagated by *in vitro* induction of adventitious roots on stem explants, leaf or shoot tip cultures.

Seed dispersal : The mucous coat of the anthocarp shows a distinct sticky swelling when ripe, with which it clings to mammals and birds which helps in its dispersal from one place to another. It has been observed, while doing fieldwork, that entire branches of the plant and inflorescence get stuck and tangled around legs and shoes, being carried some distance.

Seed dormancy and germination : The optimum temperature for seed germination is around 25°C. According to Spellenberg (2000), *Boerhavia diffusa* seeds have dormancy.

Seed storage and burial : The seed viability is poor and has very low germination percentage.

Effect on crop : Uncontrolled weed infestation in monocultures reduced the maize grain by an average 54% and the cassava-root yields by 70% for the 2 years.

Control measure : Foliar applications of triclopyr amine provide good control of *Boerhavia* (Motooka *et al.* 2002 in Pier 2002).

Uses : According to Ayurveda, Punarnava is bitter, cooling, astringent to bowels, useful in biliousness, blood impurities, leucorrhoea, anaemia, inflammations, heart diseases, asthma, alternatives etc. The leaves are useful in dyspepsia, tumours, spleen enlargement, and abdominal pains. According to Unani system of medicine, the leaves are appetizer, alexiteric, useful in opthalmia, in joint pains. Seeds are tonic expectorant, carminative, useful in lumbago, scabies. The seeds are considered as promising blood purifier. In many parts of India, different parts of Punarnava are used as folk medicine.

Past work done on Seed Science and technology

Satsangi *et al.* (2010) determine the allelopathic effect of wheat straw on *Boerhavia diffusa* L. Straw treatments linearly decreased the germination and the length of roots and shoots of the test weeds relative to controls. At the higher concentrations of wheat straw leachates, seed germination was affected in *Boerhavia diffusa* under burned and unburned straw treatments, whereas, at lower concentrations (25%) of the burned straw treatment, it remained similar to the water control. Seedling growth was less affected under lower concentrations. Thus, straw management practices should be considered when developing an integrated weed management program (Table. 1).

Table.1 Effect of wheat straw leachate on the seed germination (%), root length (cm) and shoot length (cm) *Boerhavia diffusa* (*in vitro*). Values are the mean ± SD

Leachate concentration (%)	Leachate (unburned straw)			Leachate (burned straw)		
	Germination	Root length	Shoot length	Germination	Root length	Shoot length
0 (Water control)	62.5 ± 2.9	2.0 ± 0.24	4.0 ± 0.30	62.5 ± 2.7	2.0 ± 0.27	4.0 ± 0.19
25	50 ± 2.0	1.8 ± 0.25	3.5 ± 0.26	51.25 ± 2.7	2.0 ± 0.22	4.0 ± 0.23
50	40 ± 2.2	1.5 ± 0.3	3.5 ± 0.26	45 ± 2.3	2.0 ± 0.34	3.8 ± 0.34
100	37.5 ± 2.8	1.5 ± 0.27	2.8 ± 0.19	47.5 ± 2.1	1.5 ± 0.27	2.0 ± 0.27
CD at 5%	3.2	0.22	0.34	4.6	0.76	0.63

CD = critical difference; level of significance ($P<0.05$)

REFERENCES

Anonymous. (1988). The Wealth of India: Raw Materials. CSIR Publication, New Delhi, India, p. 174.

Chopra, G.L. (1969). Angiosperms. Systematics and Life Cycle. S. Nagin & Co., Jalandhar, Punjab, India. pp. 361–365.

Motooka, Philip/Ching, Lincoln/Nagai, Guy. (2002). Herbicidal Weed Control Methods for Pasture and Natural Areas of Hawaii. Cooperative Extension Service, College of Tropical Agriculture and Human Resources, University of Hawai'i. CTAHR free publication WC-8.

Pacific Island Ecosystems at Risk (Pier). (2002). *Boerhavia coccinea* Mill., Nyctaginaceae. Pacific Island Ecosystems at Risk, Website: http://www.hear.org/pier/bococ.htm

Satsangi, G. P., Garima Bartariya and Anjali Saxena. (2010). Allelopathic response of wheat (*Triticum aestivum* L.) straw on germination and growth of selected weed species. *Seventeenth Australasian Weeds Conference*. 147-149.

Spellenberg, R. (2000). Blooming 'behavior' in five species of *Boerhavia* (Nyctaginaceae). SIDA Contributions to Botany 19(2):311-323.

RUNNING GRASS (*Brachiaria reptans* (L.))

Common name : Running grass, para grass

Botanical name : Brachiaria reptans (L.)

Hindi name : Para ghas

Family : Poaceae

Origin and distribution : Originated in Africa and this weed has reached tropics of the New and Old World like in the middle east, Indian and south east Asian subcontinents, China, Philippines, Indonesia, Australia and Pacific Islands.

Botany

Habitat : Garden land weed

Habit : Tufted annuals (Grass) (Fig. 1)

Stem : Culms tufted to 30cm.

Leaf : Blades oblong to 3.5 x 1 cm. Racemes open to 20cm long.

Floral biology

Inflorescence : Spikelets oblong, usually paired, solitary at ends (Fig. 2).

Flowering, pollination and fruit set : The flowering and fruiting time is July – September.

Seed biology

Seed : Caryopsis oblong, apex beaked (Fig. 3).

Propagation : Seeds

Seed dispersal : Seed is dispersed by birds and bats in Fiji (Ghazanfar *et al.*, 2001)

Fig. 1. Plant

Fig. 2. Inflorescence

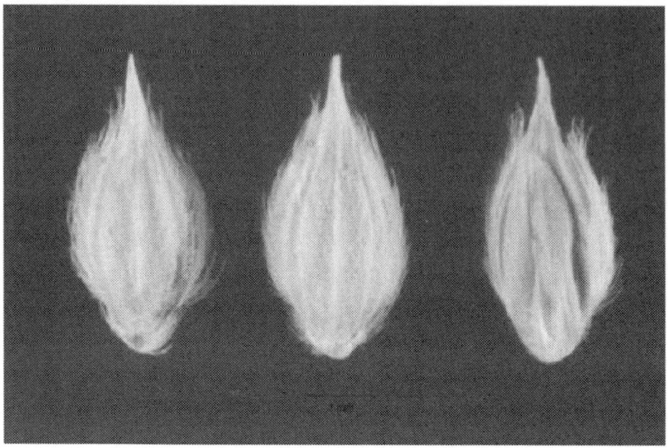

Fig. 3. Seed

Seed dormancy and germination : Permeability barrier to oxygen uptake is the main cause of seed dormancy.

Seed storage : Low seed viability is probably one of the chief limitations to the wide use of *Brachiaria* spp.

Effect on crop : It is a weed of minor agricultural importance. Can survive through the dry season as well as reproduce from seed and can provide strong competition for upland crops.

Control measure : Problematic weeds, which require continuous hoeing and weeding to reduce the competition amongst the desired species. Pre emergence application of atrazine at 2.2 kg/ha in sorghum; linuron at 2.2 kg/ha in peanuts; nitrofen at 3.5 kg/ha; alachlor at 2.4 kg/ha; butachlor at 1.5 kg /ha; anilophos at 400 g/ha; pretilachlor at 1.0 kg/ha.

Uses : Plant is a source of fodder (Datta and Banerjee, 1978) and it is nutritious. Grains are edible. The leaf sap is used for anaemia. This weed grass is also used for erosion control.

Past work done in Seed Science and Technology

Rewatt Jantawinuragz (1988) applied extracts from mature papaya (*Carica papaya* L.) seeds and roots at 5 months old plants to seeds and seedlings of test species. The extract from papaya seeds were found to promote germination of *Pennisetum polystachyon* (L.) Schult seeds. The extract from roots promoted cucumber seed germination but inhibited germination of *Abutilon hirtum* Sweet seeds. The extract from seeds promoted the growth of *A. hirtum* and *Brachiaria reptans* (L.) Gard. et Hubb. but inhibited the growth of *Bidens pilosa* L. var. *minor* (Bl.) Sherff., *Phaseolus lathyroides* L. and *Trianthema portulacastrum* L. while the extract from root promoted the growth of *Echinochloa crusgalli* (L.) Beauv. but inhibited the growth of *Mimosa pigra, P. polystachyon*, and *Leptochloa chinensis*.

Elsafori *et al.* (2011) analysed the soil seed bank for the number of live and dead seeds at three soil depths (0 to 5, 5 to 10 and 10 to 15 cm). The analysis revealed the following: The seed bank density was higher in the upper soil depths (0 to 5 and 5 to 10 cm) as compared to the lower ones (10 to 15 cm). It was found that the seed density had decreased with increasing depth. The live seed density ranged from 1015 to 5371 seeds/m^2, whereas that of the dead seeds ranged from 3215 to 6957 seeds/m^2 (Fig. 4). The dominant plant species to which the live seeds belonged were grasses including *Brachiaria* spp.

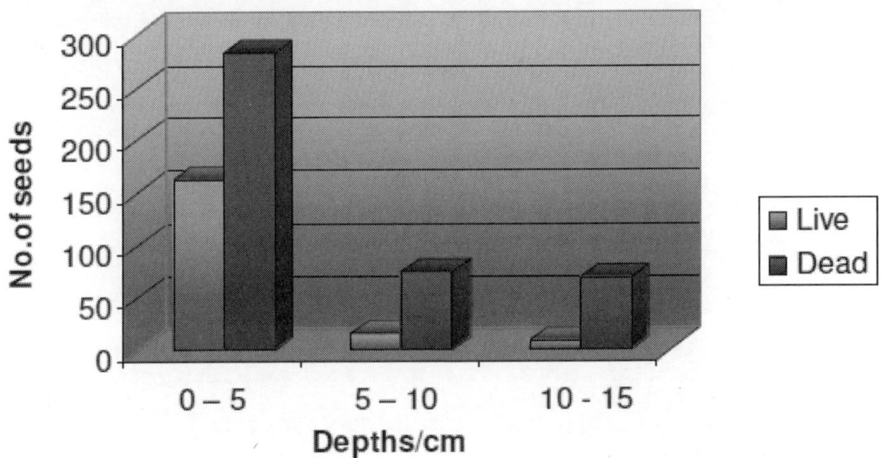

Fig. 4. Histogram showing the number of live and dead seeds at each soil depth for Site A.

REFERENCES

Ghazanfar, S. A., G. Keppel and S. Khan. (2001). Coastal vegetation of small islands near Viti Levu and Ovalan, Fiji. *New Zealand Journal of Botany* 39: 587-600.

Rewatt Jantawinuragz. (1988). Effect of 'Khaek Dam' papaya extracts on germination and growth of some plant species. *Weeds.*

Dutta, S. C. and A. K. Banerjee. (1978). Useful weeds of West Bengal rice fields. *Econ. Bot.,* 32: 297-310.

Elsafori, A.K., A.N. Guma and M.A. El Nour. (2011). Soil seed banks of a rangeland area White Nile State, Sudan. *J. of Horticulture and Forestry,* 3(6), 178-185.

SILVER SPIKED COCKS COMB : (*Celosia argentea* L.)

Common name : Silver spiked cocks comb

Botanical name : Celosia argentea L.

Tamil name : Pannai keerai

Family : Amaranthaceae

Origin and distribution : Probably originally from tropical Africa, but spreading early throughout tropical Asia and Malesia; many cultivars are widely grown in tropical and subtropical areas.

Botany

Habitat : Distributed in all plains, mainly a dry land weed up to 1200 m, associated with ragi and groundnut. Typical red soil weed.

Fig. 1. Plant

Habit : Erect glabrous herb, branched annual, growing upto 1m(Fig. 1).

Stem : Glabrous, somewhat angled and striate.

Leaf : Leaves alternate, linear or linear – lanceolate.

Floral biology

Flower : Spikes long peduncled, dense (Fig. 2); Flowers white or pink, shining; perianth acute; stamens 5, united below into a cup, anthers 2 celled; ovary 1 celled, ovoid; style filiform (Fig. 3).

Fig. 2. Inflorescence

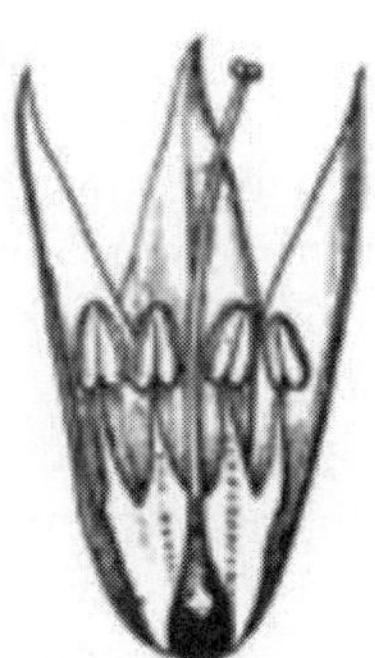

Fig. 3. Flower

Flowering and pollination : Flowering in August to October. Pollination is by wind and insects, especially bees and flies, which visit the flowers regularly.

Fruit : Utricle circumsessile, membranous (Fig. 4).

Seed biology

Seed : Seeds lenticular, 1–1.5 mm long, black and shining, shallowly reticulate (Fig. 5).

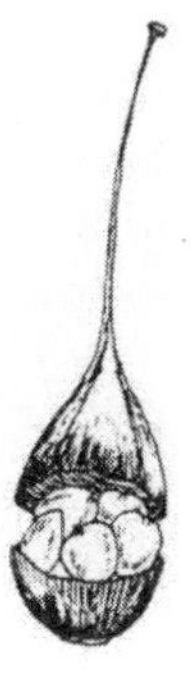

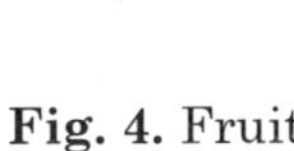

Fig. 4. Fruit

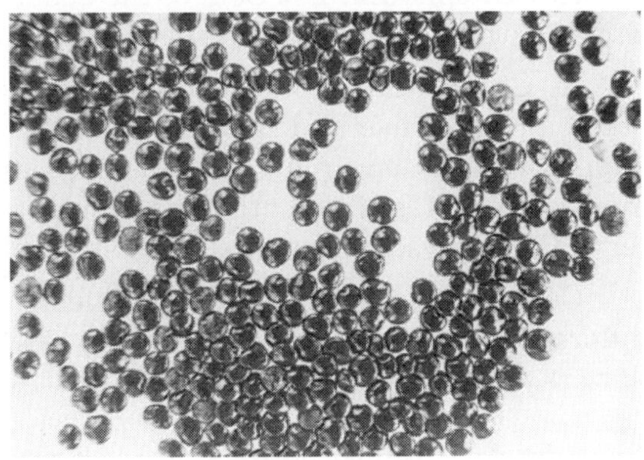

Fig. 5. Seed

Propagation : It is propagated by seed

Seed dispersal : Seed dispersal by wind and water.

Life cycle : The seedling emerges 5–7 days after sowing. Early vegetative growth is slow, but flowering may occur already 6–7 weeks after sowing. Seed maturity starts from the basal part of the inflorescence and gradually moves up to the tip. Seeds are mature in 10–20 weeks from sowing and shatter when the inflorescence is dry.

Seed dormancy and germination : In some cases, very low fluence rates are required to induce germination as in *Celosia* (Dixit and Amritphale, 1996). In other cases, increasing the irradiance level (up to 150 mol. sec^{-1}. m^{-2}) impacts the time required to satisfy photodormancy. Seeds of *Celosia argenata* posses dormancy which could be overcome by GA (100 ppm). They have also reported that nitrogenous substances like ammonium nitrate accelerated the germination promoting activity of gibberellic acid. Almost all seeds are waiting in a dormant state for some outside stimulus to break their dormancy, some just need sufficiently high ambient humidity, others need scarification, vernalization or to be passed through the intestines of an animal. *Celosia argentea* seeds will usually germinate in 10-15 days. Normally, will only germinate with light so surface sow. Sow *Celosia argentea* seeds about 1mm deep in a well drained seed sowing mix at about 22°C.

Seed storage and burial : They remain dormant on the soil surface until the start of the next rainy season.

Effect on crop : It is found in many crops such as groundnut, finger millet and maize.

Control measure : Hand weeding and pre emergence herbicide like Butachlor is used to control.

Uses : Celosia is primarily used as a leafy vegetable. The whole plant is used as an antidote for snake bites and the roots to treat colic, gonorrhoea and eczema.

In Ethiopia and DR Congo, the seeds are used as medicine for diarrhoea, and in Ethiopia the flowers to treat dysentery and muscle troubles.

In China and Japan, seed extracts have traditionally been used as a therapeutic drug for eye and hepatic diseases. In India, the leaves mixed with honey are applied to inflated areas or abscesses, and the seeds are widely used for the treatment of diabetes. In South-East Asia, the flowers are used as medicine for dysentery, haemoptysis and menstruation problems.

Celosia can also be used as a livestock feed. Forms with fasciated, yellow to red inflorescences are widely grown as a bedding plant in gardens and also used as cut flowers. Also used in Africa to help control growth of the parasitic Striga plant.

Past work done on Seed Science and Technology

1. Seed production

Ravindra *et al.* (2008) reported that *C. argentea* produces 2000 to 3000 seed $plant^{-1}$ which add to soil seed bank and germination of *C. argentea* seed started at 3 DAS and continued up to 8 DAS. Branching started at 10 DAS and flowering commenced at 20 DAS, the plant continued to flower until 40 DAS. Some plants started to mature as early as 44 DAS and continued up to 80 DAS. At maturity *C. argentea* produced 12.97 ± 0.83 seed heads with 52.67 ± 3.52 seeds seed $head^{-1}$, which gave 690.1 ± 70.9 seeds $plant^{-1}$ and a 1,000 seed weight of 6.67 ± 0.32 g (Table

1). It can be concluded from that controlling the weed *C. argentea* before it attains 30 d would be useful in reducing crop weed competition in the field condition as the rate of growth and DM accumulation was greatest during this period.

Table 1. Plant characteristics of *Celosia argentea* at maturity

Parameter	Mean ± SE	Range
Seed heads plant^{-1}	13.0 ± 0.83	5 – 20
Seed seed head^{-1}	52.7 ± 3.52	24 – 95
Seeds plant^{-1}	690.1 ± 70.87	200 – 1672
1000 seed weight (g)	6.7 ± 0.32	4 – 10

2. Seed dormancy and germination

The optimum temperature for germination of *Celosia* was between 20 and 30°C. The germination was inhibited with osmotic potentials increasing from -0.4 to -1.0 MPa. Seed germination was not altered much when pH of the moisture supply varied between 4.0 and 9.0. *Celosia* seed germination was slightly inhibited in dark.

3. Germination improvement

Germination percentage was increased upto 100% after 7 days of priming treatment when compared to control (78%) (Fig. 6).

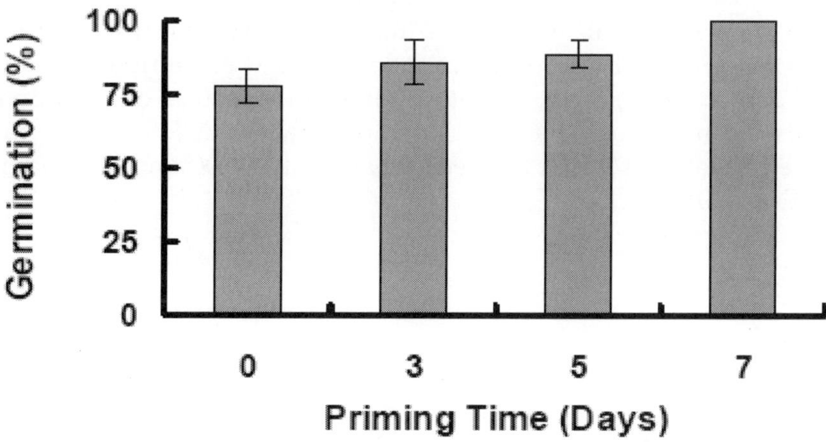

Fig. 6. Effect of artificial priming on germination percentage of *Celosia argentea* seeds

Radicle length was significantly increased after 7 day of priming. It increased upto 3.2 cm in 7 day treatment compared to 2.51 cm (control) while the radicle lengths of 3 and 5 day treated germinating seeds were 2.79 and 3.1 cm, respectively (Fig. 7).

The EC of seed steep water remained constant (14.52 mS/cm) after 24 h observations, even after 7 day treatment.

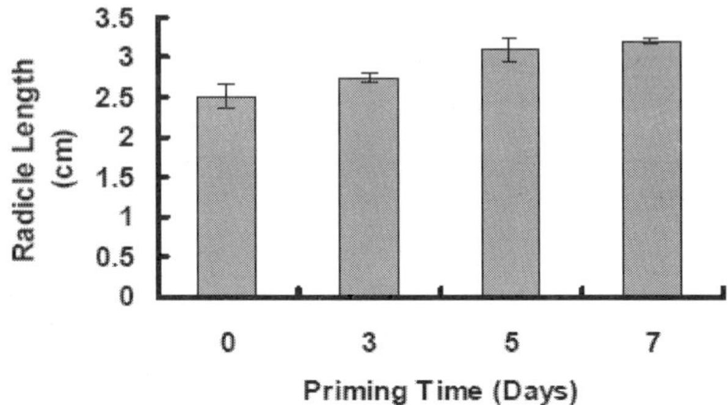

Fig. 7. Effect of artificial priming on the radicle length of *Celosia argentea*

4. Allelopathic effect

The effects of different concentrations of *Senna alata* crude water extracts on the germination, growth and flowering of *Celosia argentea*. All concentrations of this extract induced a consistent decrease in absolute percentage and rate of germination, and inhibited radical elongation in *Celosia argentea*. Earliness in flowering and overall increase in height was obtained from seedlings sprayed with 75% C and 100% C crude water extracts.

There was significant decrease ($p<0.05$) in the percentage and rate of germination of *C.argentea* in 50 and 100% brewery effluent. The highest percentage (35%) of germination occurred at 3h in 100% effluent below the control level (45%) at 2h (Fig. 8a-c and 9a –c). One hundred percent textile brewery effluent was completely toxic to *C. argentea* (Fig. 8, 9). Paint effluent inhibited seed germination with increasing presoaking period. Germination was very low in *C. argentea* (5-15%) and toxicity resulted beyond 11/2 and 2 h seed treatment in 100 and 50 5 paint effluent, respectively (Fig. 8, 9).

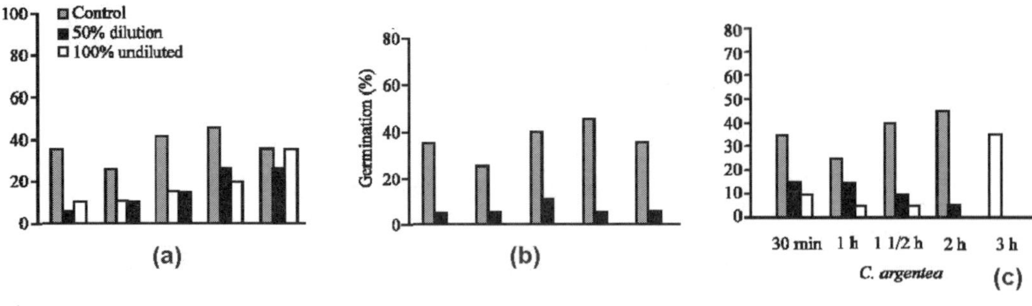

Fig. 8. Percentage germination of *Celosia argentea* treated with brewery, textile and paint effluent, (a) brewery (b) textile and (c) paint

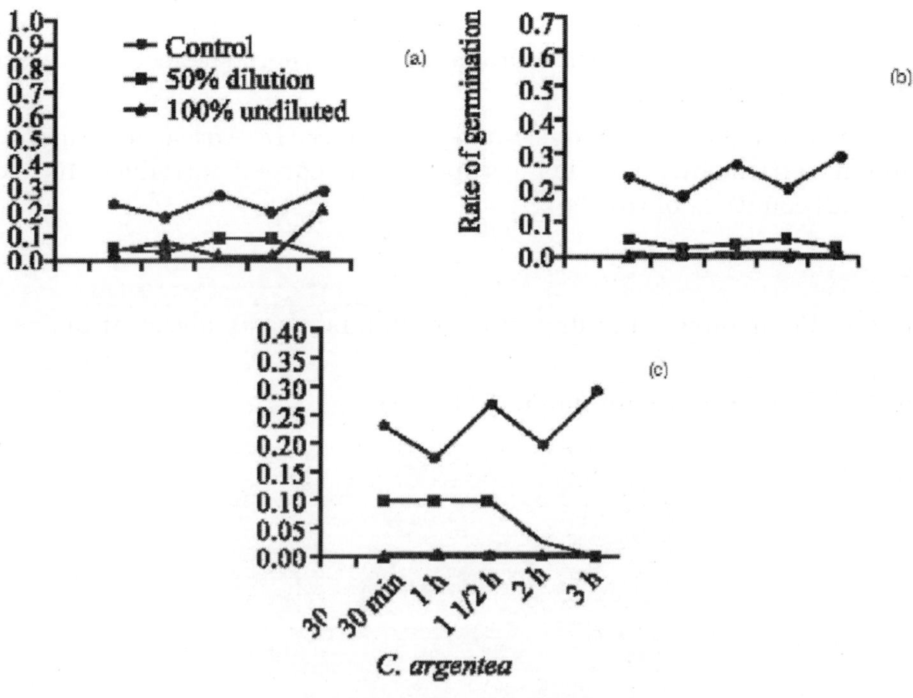

Duration of pretreatment

Fig. 9. Rate of germination of *Celosia argentea* treated with brewery, textile and paint effluent, (a) brewery (b) textile and (c) paint

REFERENCES

Dixit, S. and D. Amritphale. (1996). Very low fluence response in the induction and inhibition of seed germination in *Celosia argentea*. *Seed Sci. Res.* 6:43-48.

Ravindra, G.M., S. Shankarappa, G.K. Girijesh and V.N. Hooradahalli. (2008). Weed biology and growth analysis of *Celosia argentea* L., a weed associated with groundnut and finger millet crops in southern India. *Communications in Biometry and Crop Science,* 3 (2) 80–87.

WILD SPIDER FLOWER (*Cleome gynandra*)

Common name: Wild spider flower, African spider-flower, cat whiskers, bastard-mustard

Botanical name: Cleome gynandra L. (Syn. *Gynandropsis pentaphylla* DC)

Tamil name: Taivelai, nalvelai, velai, acakanta

Hindi name: Jakhiya, safed hulhul, parhar, safed bagro

Family : Capparaceae

Origin and distribution : The origin of Cleome gynandra is not known. There are claims that it has a southern Asian origin, but others suggest that it originates from Africa or Central America.

It occurs throughout the tropics and subtropics. In Africa, it is mainly found near human settlements, possibly escapes from earlier introductions. It occurs probably in all countries of tropical Africa.

Botany

Habitat : Common weed in dry and garden land, waste places at plains and low elevations.

Habit : Tall, pubescent annual herb (Fig. 1).

Fig. 1. Plant

*Root :*It has a long tap root, with a few secondary roots with root hairs.

Stem : Striate covered with hairs

Leaf : Leaves 5 foliate, long petioled, leaflets broad – obvoate, acute or obtuse, entire or serrulate.

Floral biology

Flower : Flowers pale purplish in corymbs, hypogynous, bisexual, bracteate, bracteolate; sepals 4, free; petals 4, clawed; stamens 6, adnate below to the gynophore; ovary superior, 1 – celled, on elongated gynophore, ovules many on parietal placentae. While the continued development of axillary branches and terminal flowers extends the flowering period to 6-12 months, fruit development and maturation (Chweya and Mnzava, 1997) (Fig. 2).

Fig. 2. Flower

Flowering, pollination and fruit set : Cleome gynandra is both self- and cross-pollinated. In Venezuela, plant populations with either male or female flowers were found. In studies carried out in Zimbabwe in 2001, it was observed that some flowers first develop stamens, others first the pistil. A population has been found with a few plants that show male sterility, producing short anthers that do not shed pollen. Flower from July to August, and the seeds ripen from August to September.

Fruit : Capsule cylinderical, nearly glabrous striate; valves 2, spreading from the placentae (Fig. 3).

Seed biology

Seed : Seeds kidney – shaped, black, glabrous (Fig. 4). The thousand seed weight varied from 0.8 g to 0.92 g (Schippers, 2002).

Propagation : By seeds

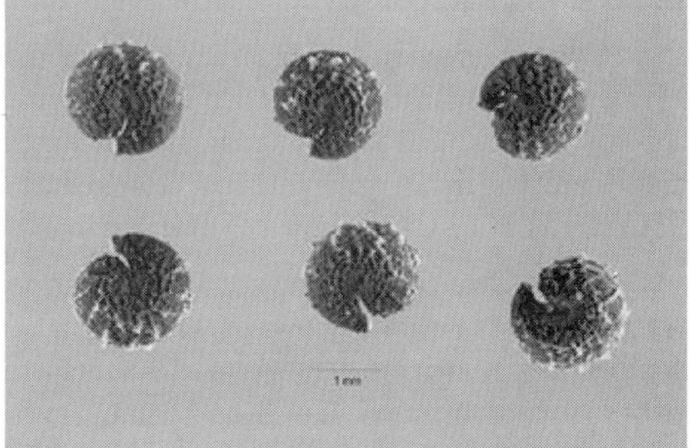

Fig. 3. Fruit

Fig. 4. Seed

*Seed dispersal :*Natural dispersal mechanisms: It appears to have no specialised means of dispersal. Seeds drop to the ground close to or beneath the parent plant. The wind may subsequently move seed short distances. No information found regarding avian dispersers.

Human dispersal mechanisms: Human mediated dispersal is probable in contaminated machinery, produce, and soil or stock feed.

Life cycle : After sowing the first seedlings emerge after 6–8 days. However, germination is erratic, occurring over an extended period. Dormancy is much reduced some 6 months after the seeds have dried, and after 12 months most seeds germinate readily. Plants tend to start flowering 4–6 weeks after germination and lasting for 2 months, usually when 60–90 cm tall; stress can trigger flowering even at the seedling stages. During the reproductive stage, which may take up to three months under favourable conditions, vegetative growth declines and leaves become senescent quickly. The fruits dry after a few weeks and dehisce to release dry seeds.

Seed dormancy and germination : Seeds have no dormancy and germinate readily after shedding (Jansen, 2004). The relatively large seed is likely to be persistent in the seedbank under favourable conditions. Scarified seed germinated best at 32°C, and decreased at lower temperatures. At 14°C, there was almost no germination. Seeds germinated equally well in continuous light or darkness, with a decrease in germination under conditions of decreased light intensity or duration. Emergence was the same on the soil surface and at 1cm depth (Menon and Kulkarni, 1987).

Seed storage and burial : Yepes (1978) planted freshly harvested seed at monthly intervals, for 13 consecutive months. It was determined that seeds have a rest period (latency) that extends to the 5th month after collection. Active germination starts 6 months after harvest, and increases to 88% in 3 months. Highest germination occurs after 12 months of storage.

Effect on crop : In south east Asia, the principal crops affected are rice and tobacco, with impacts also on tomatoes, soybeans, maize, sugarcane, sweet potato, taro, cotton, and ground nut. The direct effects are reduced growth and yield of crop plants through competition for light, water and nutrients.

Control measure : Hand weeding and pre-emergence application of metolachlor (1 kg ha^{-1}) and alachlor (1 kg ha^{-1}) in soybeans and tomatoes; CGA 362 + ametryn (3 kg ha^{-1}) in sugarcane; atrazine (1 kg ha^{-1}) and metolachlor (1 kg ha^{-1}) in maize; pendimethalin in groundnut (Reddy *et al.,* 2007; Gana *et al.*, 2006, Praveen and Murthy, 2005; Chandrika 2004; Reddy *et al.,* 2003; Hongyuan *et al.,* 1989)

Uses: The tender leaves, young shoots and occasionally flowers are eaten boiled as pot herb, relish, stew or side dish. In several communities, boiled spider plant leaves are traditionally given to mothers before and after delivery of a child, and in other situations where blood has been lost, e.g. to warriors.. The glands on the stems and leaves have insect repellent properties; cabbage and related crops intercropped with spider plant suffer less from diamond back moth larvae. Similarly, in french bean intercropped with spider plant, the beans are less affected by flower thrips and are therefore of better quality for export.

The seeds are used to feed birds. The seed contains an edible polyunsaturated oil, which is extracted by simple pressing and does not need refining. The seed cake can be used as animal food. In Kenya, *Cleome gynandra* was shown to exhibit repellent and acaricidal properties to larvae, nymphs and adults of the ticks (*Rhipicephalus appendiculatus* and *Amblyomma variegatum*) indicating a potential for tick control.

Past work done on Seed Science and Technology

1. Fruit position

According to K'Opondo (2011), pod position had effects on thousand seed weight, germination rate (time to reach 50% germination, T50) and percentage germination. As regards TSW, while bottom fruits in GP and PP produced heavier seeds than middle and top fruits; for GP, PG and PP, the middle fruits produced heavier seeds than top fruits. Seeds from bottom and middle fruits in GP and PP germinated faster than those from top fruits. Bottom fruits in GP and middle fruits in GP and PP respectively, produced seeds that showed higher germination than those from top fruits (Table. 1). The results indicated that often bottom and medium fruits produce heavier seeds compared to top fruits; also that heavier seeds germinate faster and result into higher germination percentage.

Table 1. Figures followed by different letters along the rows, for Gmax are significantly ($P \leq 0.05$) different, according to Student's t-test. Data are means of four replicates of 50 seeds for the four spiderplant morphotypes.

Morphotype	Maximum percentage germination ± S.E. (Gmax)		
	Bottom fruits	Middle fruits	Top fruits
GG	98.0 ± 0.82a	96.0 ± 1.41ab	95.0 ± 2.38ab
GP	93.5 ± 2.06a	92.5 ± 2.75ac	77.0 ± 2.38bd
PG	85.5 ± 1.26a	88.0 ± 1.63ab	79.5 ± 5.12ab
PP	92.5 ± 3.10a	96.0 ± 1.15ab	84.0 ± 1.83ac

2. Seed drying

K'Opondo (2011) suggested that sun drying achieved a significantly ($P \leq 0.05$) higher percentage germination (Gmax) than shade drying with morphotypes GP, PG and PP (Fig. 5). Significantly higher germination occurred in morphotype GG with sun drying, while PG recorded significantly ($P \leq 0.05$) lower germination than the rest (Fig. 5).

3. Seed dormancy and germination

Cleome gynandra exhibited negative photosensitivity on exposure to continuous white light at 20°C during germination, but photoinhibition was greatly reduced at alternating 20/30°C or at constant 30°C (Ochuodho and Modi, 2005).

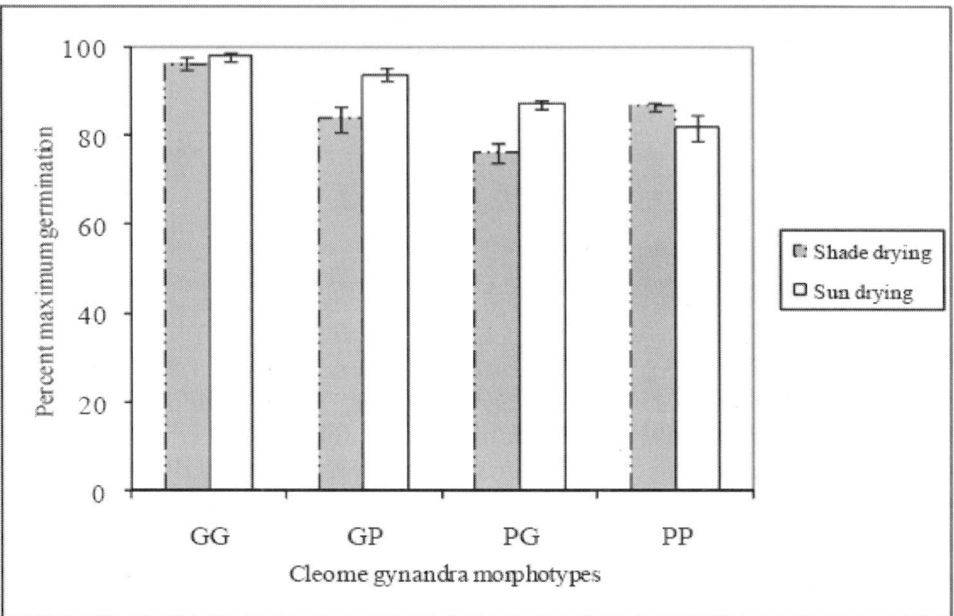

Fig. 5. Influence of natural seed drying method on percent maximum germination (G_{max}) with four spiderplant morphotypes from western Kenya. The vertical line indicates S.E. for comparisons of significant differences between natural drying methods.

Ochuodho and Modi (2007) reported that seeds from the Kenya and Agricultural Research Council (ARC) showed a germination of 57% and 41% at alternating 20/30°C in darkness and in continuous light, respectively, after four days of incubation. Germination at 20°C was 43% in the dark, but it decreased to 2% in continuous light (Figure 6A). The seeds from Kenya Seed Company (KSC) showed a similar germination pattern as the ARC seeds under the same conditions (Figure 6B). At alternating 20/30°C, the KSC seeds had a germination of 90% in darkness and 74% in light, after four days. Seed germination at 20°C was 82% in the dark, but the seeds failed to germinate in continuous white light. Mature (black) seeds from green pods attained more than 90% germination within 96 h of incubation at alternating 20/30°C in the presence of light or in darkness. The germination of seeds from yellow pods was intermediate, reaching about 70% in both light and darkness, after four days (Fig. 7). Germination reduced significantly ($P < 0.001$) when seeds were exposed to more than 12 h in a day compared to less exposure periods (Fig. 8). The seeds produced at 21/16°C showed 76% germination; those produced at 27/22°C had a germination of 93% while those produced at 33/28°C obtained a germination of 84% at the germination temperature of 20/30°C after 4 days of incubation in darkness. The seeds showed better germination at 20°C in darkness compared to alternating 20/ 30°C in continuous light. It is concluded that seed germination of *C. gynandra* is negatively photosensitive and germinates better and faster in darkness.

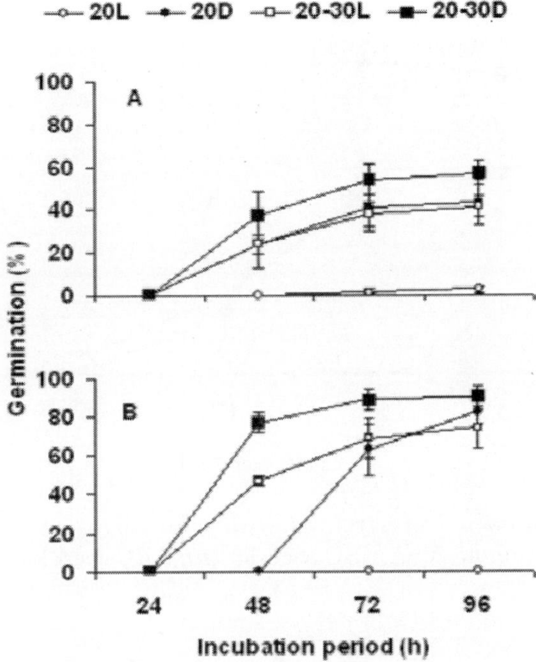

Fig. 6. Germination of the original Cleoma seeds from ARC (A) and KSC (B) at 20°C or alternating 20/30 (16/8h) in continuous white light or darkness. 20°C with light (o), 20°C in darkness (•), 20/30°C with light (□) and 20/30°C in darkness (■). Error bars represent SD, n=6.

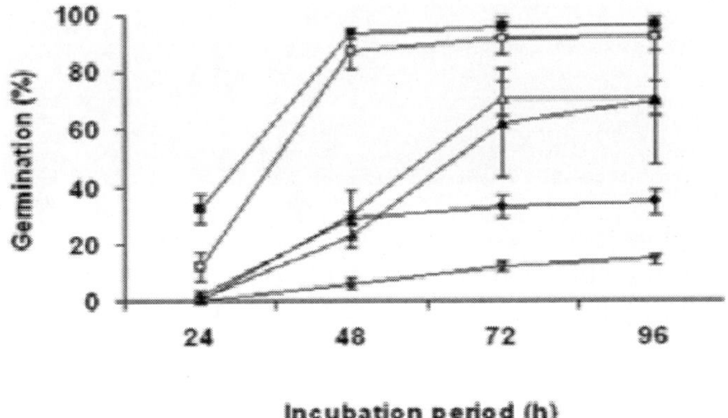

Fig. 7. Germination of black seeds of Cleome grown in the field and harvested at different stages of pod maturity. The seeds were germinated at alternating 20/30°C (16 h night/8h day) in continuous white light or darkness. Seeds from green pods incubated in continuous white light (□) or darkness (■), yellow pods is incubated in continuous white light (Δ) or in darkness (▲) and brown pods incubated in continuous white light (o) or in darkness (•), Error bars represent SD, n = 4.

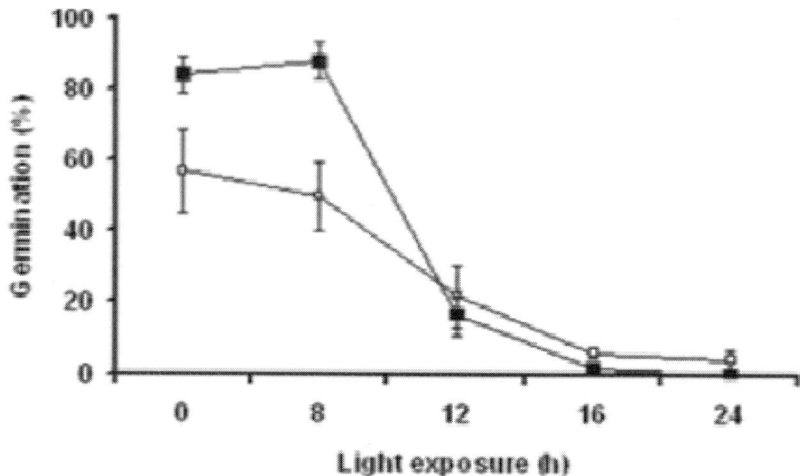

Fig. 8. Seeds of *C. gynandra* were treated with white light at varying exposure periods during their germination at 20°C. KSC seed lot (■); ARC seed lot (□). Error bars represent SD, n = 4.

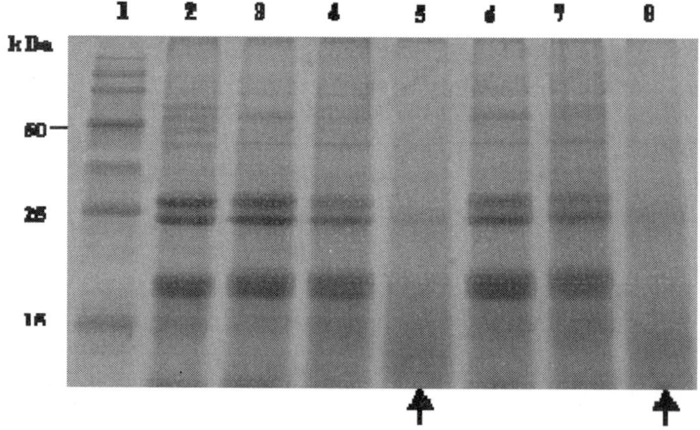

Fig. 9. Analysis of the proteins of KSC seeds with 10% SDS-PAGE gel. Molecular weight marker (Lane 1); dry seeds (Lane 2); seeds germinated at alternating 20/30°C in continuous white light for 24h (Lane 3), for 48h (Lane 4) and 72h (Lane 5); seeds germinated at alternating 20/30°C in darkness for 24h (Lane 6), for 48 h (Lane 7) and 72h (Lane 8). Arrows indicate the lanes where the protein expression diminished.

Ochuodho *et al.* (2008) examined changes in protein expression during germination of *Cleome gynandra* in the presence or absence of light and at constant or alternating temperatures. Protein bands of KSC seeds incubated at 20/30°C in darkness and in the presence of white light had diminished by 72h of germination (Fig. 9). However, the analysis of band intensity using the computer package Analysis showed that the proteins of seeds germinated in darkness diminished faster (24h) than the bands of seeds incubated in the presence of light (48h) (Fig. 10). By 72h of germination, the intensity of the bands in both light and dark-germinated seeds was less than half the original values (Figure 10). *Cleome* seeds incubated at 20°C in continuous light failed to germinate or had low germination. The seeds did not show any protein accumulation change on SDS PAGE analysis, but showed protein bands similar to those of the dry ungerminated seeds (Fig. 11). However, the high molecular weight proteins of 28 kDa began diminishing by the fourth day (96h) in the seeds that were germinated in darkness.

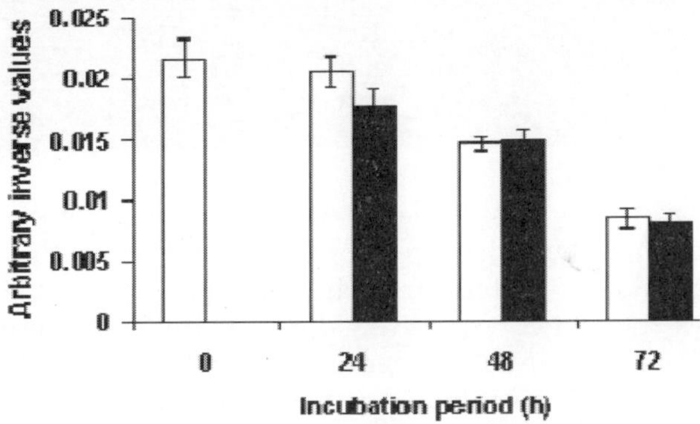

Fig. 10. Analysis of band intensity of the gel in Fig. 9 using computer soft image system (Analysis). Seeds were germinated at alternating 20/30°C with light (□) or in darkness (■). Seed proteins were extracted at intervals and analysed by SDS-PAGE. Error bars represent SD, n =30.

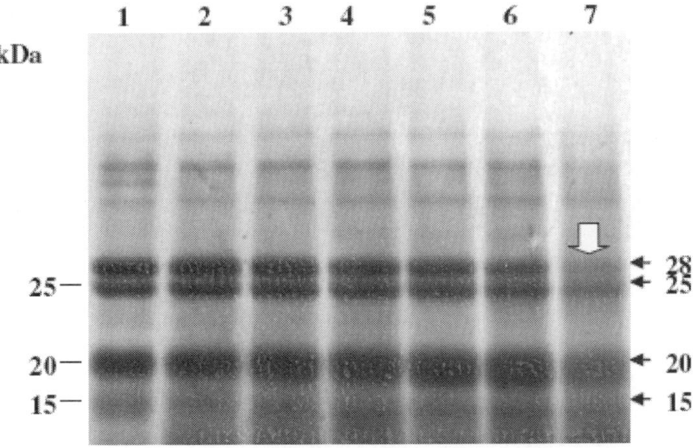

Fig. 11. Seeds from KSC germinated at 20°C in continuous light and in darkness. Lane 1, dry seed; Lanes 2, 3 and 4, seeds incubated with light for 48, 72 and 96h, respectively; Lanes 5, 6 and 7, seeds incubated in darkness for 48, 72 and 96h, respectively. Open arrow indicates the band that diminished and closed arrows indicate the proteins observed.

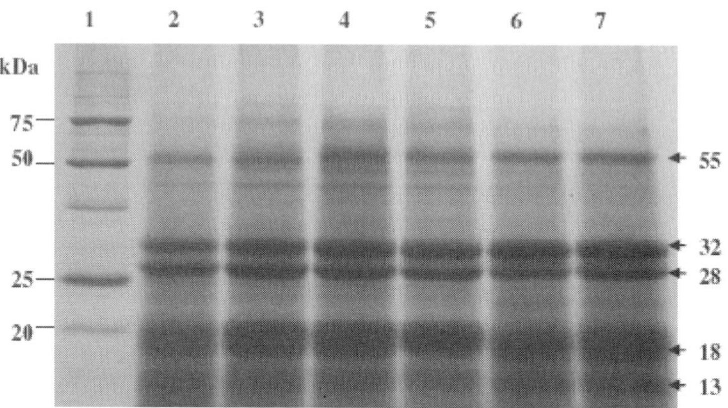

Fig. 12. Seeds were grown in glasshouses at day/night temperatures of 21/16, 27/22 and 33/28°C, harvested when the pods were brown and analysed with 10% SDS-PAGE. Lane 1, molecular weight marker. Proteins from dry seeds produced at 21/16, 27/22 and 33/28°C are shown in Lanes 2, 3 and 4, respectively. Proteins from seeds produced at 21/16, 27/22 and 33/28°C, and incubated at 20°C in continuous white light for 7 d are shown in Lanes 5, 6 and 7, respectively. Arrows indicate the sizes of proteins that were differentially expressed due to germination treatment.

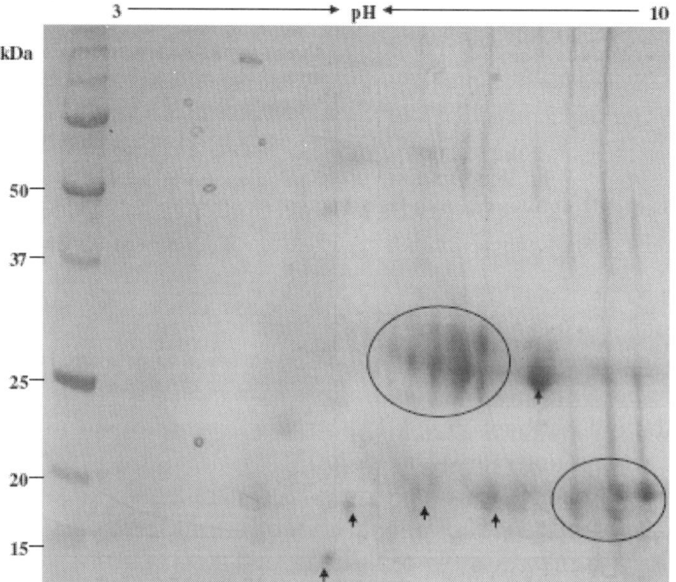

Fig. 13. Two-dimensional polyacrylamide gel showing polypeptides isolated from dry seeds of *C. gynandra* produced in glasshouse at day/night temperatures of 27/22°C. Prominent spots are encircled and marked with arrows. These spots correspond with the bands observed in Figure 11.

Photoinhibition of seed germination for up to 7 days neither changed the number nor band intensity of the storage proteins expressed (Fig. 12). The protein profile for dry seeds harvested from the glasshouses was similar to that of the seeds incubated at 20°C in continuous light for 7 d, although there were slight indications of diminished content of the heavy bands between 40 - 50 kDa in lanes 6 and 7. The 2-D gels of dry seeds (Fig. 13) and photoinhibited seeds (Fig. 14) showed the same polypeptides but they differed slightly in content.

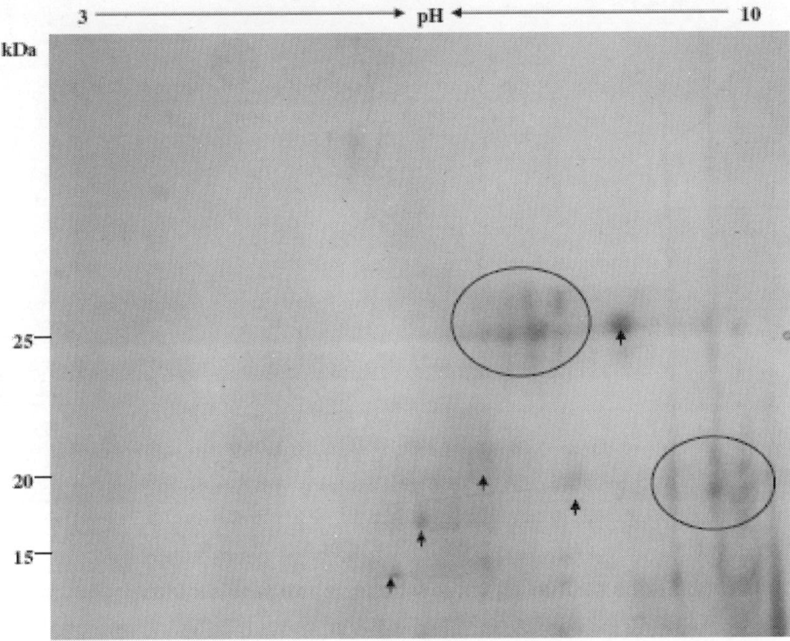

Fig. 14. Two-dimensional polyacrylamide gel showing polypeptides isolated from seeds of *C. gynandra* incubated on moist filter papers at 20°C in continuous white light for 7 d. The seeds were produced in a glasshouse at day/night temperatures of 27/22°C. The prominent polypeptides correspond with those shown in Figure 12.

In conclusion, these results indicate that during the germination of the seeds of *C. gynandra* seed storage proteins are mobilised. Neither the protein profiles nor the protein content changed during photoinhibition. The content of the four proteins of Mr 28-32 and 18-20 kDa consistently observed in mature seeds of *C. gynandra* did not change during photoinhibition. It can be stated that seed storage proteins are not utilised during photoinhibition in white light.

Accumulation of seed proteins

In a recent review, Shutov *et al.* (2003) emphasised that during formation and deposition, storage proteins are protected against premature breakdown, whereas they become completely degradable during initiation of germination.

The 12 S globulins are synthesised and accumulate during seed development and maturation stages (Laroche *et al.*, 1984; Bewley and Black, 1994).

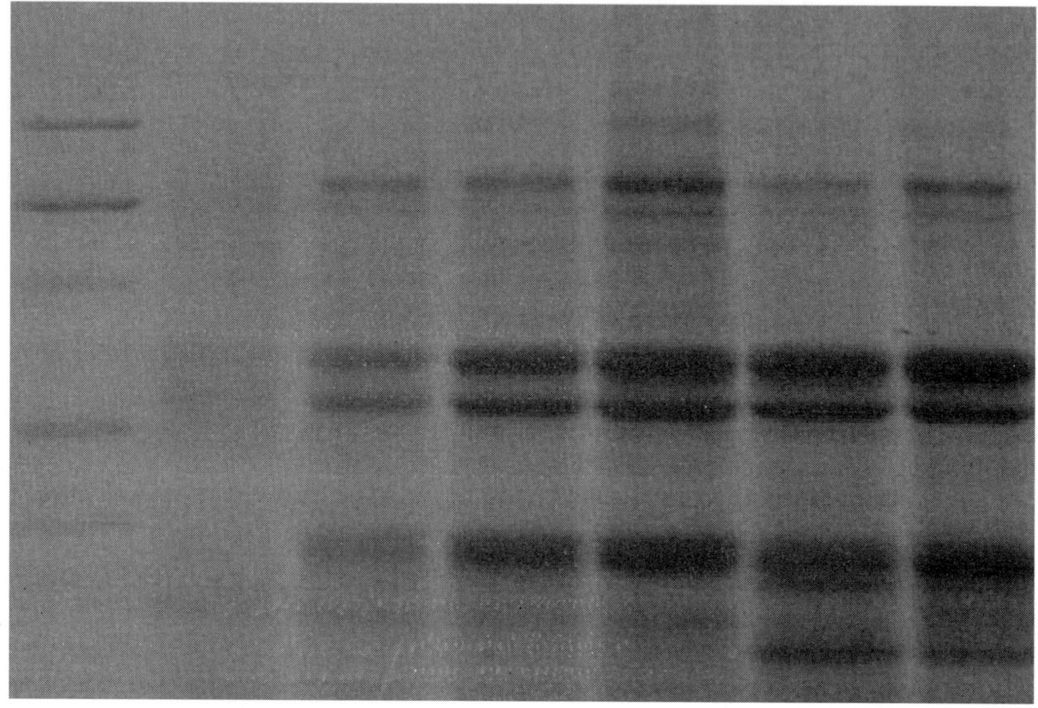

Fig. 15. One-dimensional gel electrophoresis of proteins isolated from *Cleome gynandra* seeds during development. Proteins were loaded at 10 Ag per lane and proteins were detected with standard Coomassie blue stain after electrophoresis. Storage proteins are identified by arrows on the right. Lane 1, molecular weight marker in kilodaltons; 2, 3, 4, green, brown and black seeds selected from green pods; 5, seeds from green pods; 6, black seeds from yellow pods; 7, black seeds from brown pods. Arrows indicate the prominent proteins.

Ochuodo *et al.* (2006) characterized the seed proteins of *Cleome gynandra*. The content of seed proteins increased as the seeds of *Cleome* matured as shown by sequential increase in the number of bands (Fig. 15). Green seeds from green pods showed the least number of protein bands. Brown seeds selected from the green pods accumulated more seed proteins compared to the green seeds, but less than the black seeds. Black seeds selected from green pods and those obtained from yellow and brown pods had four protein bands with molecular weights of 55.19 kDa, 30.06 kDa, 26.63 kDa and 17.76 kDa. The seeds from yellow and brown pods further accumulated a low molecular weight protein band of about 13.31 kDa. The intensity of band expression increased significantly as the seeds matured (Fig. 16). The green seeds from green pods showed the lowest expression whereas the black seeds from yellow and brown pods showed the highest band intensity.

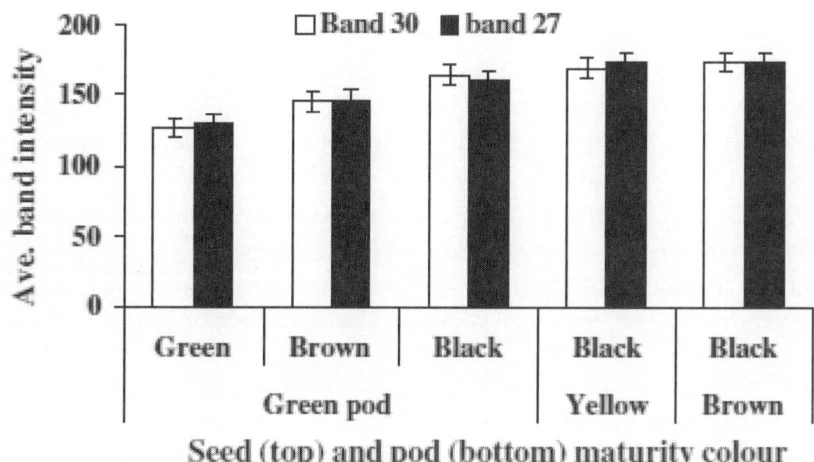

Fig. 16. Average accumulation intensity of bands of Mr 30 and 27 kDa was determined with Quantity One computer package. One-dimensional gel electrophoresis of proteins isolated from Cleome gynandra seeds during development. Proteins were loaded at 10 Ag per lane and proteins were detected with standard Coomassie blue stain after electrophoresis (Fig. 15). The error bars represent SE, n =3.

REFERENCES

Bewley, J.D. and M. Black. (1994). Protein synthesis during germination and its dependence on mRNA synthesis (pages 178– 188) and Environmental control of germination (pages 257–268). SEEDS: Physiology of Development and Germination, 2nd edn Plenum Press, New York.

Chandrika, V. (2004). Integrated weed management in groundnut (*Arachis hypogaea* L.) during rabi season. *Legume Research* 27(4): 243-248.

Chweya, J.A. and N.A. Mnzava. (1997). Cat's whiskers. *Cleome gynandra* L. Promoting the conservation and use of underutilized and neglected crops. Institute of Plant Genetics and Crop Plant Research, Gatersleben/International Plant Genetic Resources Institute, Rome, Italy.

Gana, A.K., A.A. Ndarubu and L.D. Busari. (2006). Efficacy of CGA 362 and ametryn with pre-emergence herbicides on weed control in sugarcane. *Sugar Technology* 8(1): 88-90.

Hongyuan, T., W. Xuee and Y. Zi. (1989). Distribution and infestation of major weeds in *Cleome gynandra* L. *Rev. Comalfi* 5(1):49-53.

Jansen, P.C.M. (2004). *Cleome viscosa* L. Internet Record from Protabase. Grubben, G.J.H. & Denton, O.A. (Eds). PROTA (Plant Resources of Tropical Africa / Ressources végétales de l'Afrique tropicale), Wageningen, Netherlands.

K' Opondo, F.B.O. (2011). Influence of drying method and fruit position on the mother plant on seed quality of spiderplant (*Cleome gynandra* L.) morphotypes from western Kenya. *Advances in Applied Science Research*, 2 (3): 74-83.

Laroche, M., L. Aspart, M. Delseny and P. Penon. (1984). Characterization of radish (*Raphanus sativus*) storage protein. *Plant Physiology,* 74, 487– 493.

Menon, A. and A.R. Kulkarni. (1987). Ecological studies in *Cleome viscosa* L. seed and seed germination. *Indian Botanical Reporter* 6 (1): 1-7.

Ochuodho, J.O. and A.T. Modi. (2005). Temperature and light requirements for seed germination of *Cleome gynandra. S. Afr. J. Plant Soil,* 22: 49-54.

Ochuodho, J.O. and A.T. Modi. (2007). Light-induced transient dormancy in *Cleome gynandra* L. seeds. *African Journal of Agricultural Research,* 2 (11), 587-591.

Ochuodho, J.O., A.T. Modi and M. Beukes. (2006). Accumulation of seed storage proteins in *Cleome gynandra* L. and *Brassica kaber* L. *South African Journal of Botany,* 72: 238 – 244.

Ochuodho, J.O., A.T. Modi and M. Beukes. (2008). Photoinhibition influences protein utilisation during seed germination in *Cleome gynandra* L. *African Journal of Biotechnology,* 7 (10), 1468-1472.

Praveen, V.L. and B.B. Murthy. (2005). Relative efficiency of herbicides in maize and cowpea intercropping system for green fodder. *Indian Journal of Weed Science,* 37(1): 123- 125.

Reddy, M.M., A.M. Vilath and L.J. Rao. (2007). Integrated weed management in pigeonpea (*Cajanus cajan*) and soybean (*Glycine max*) intercropping system on vertisol under rainfed conditions. *Indian Journal of Agricultural Sciences,* 77(3): 177-178.

Reddy, S.M., G.P. Reddy and B.S. Reddy. (2003). Weed management in soybean, *Glycine max* L. *Journal of Oilseeds Research,* 20(2): 292-294.

Schippers, R.R. (2002). African indigenous vegetables, an overview of the cultivated species 2002. Revised edition on CD-ROM. National Resources International Limited, Aylesford, United Kingdom.

Shutov, A.D., H. Baumlein, F.R. Blattner, K. Muntz. (2003). Storage and mobilisation as antagonistic functional constraints on seed storage globulin evolution. *Journal of Experimental Botany,* 54, 1645– 1654.

Yepes, J.H. (1978). Estudio de lamaleza platanito (*Cleome gynandra* L.) Study of the weed cotton fields in China and their control. *Acta Agriculturae Shanghai,* 5(3): 51-58.

STICKY SPIDER-FLOWER (*Cleome viscosa* L.)

Common name : Sticky spider-flower,

Botanical name : Cleome viscosa L.

Tamil name : Nai kadugu

Family : Capparaceae

Origin and distribution : Cleome viscosa occurs in northern tropical Africa, from Cape Verde and Senegal to Egypt, Ethiopia and Zanzibar; It also occurs in Madagascar and other Indian Ocean islands. Outside Africa, it is widespread in peninsular Arabia, tropical Asia, Australia and tropical and subtropical America (http://database.prota.org).

Botany

Habitat : Cultivated beds. Sandy and freely draining soils in open woodland scrub and on scree slopes in dry areas (http://www.pfaf.org/).

Fig. 1. Plant

Habit : Erect annual, softly pubescent, viscous herb with stalked glands on all parts of the plant (Fig. 1).

Leaf : Leaves 3-5 foliate, leaflets ovate or obovate, equaling or shorter than the petiole

Stem : Round, striate and densely covered with glandular hairs.

Floral biology

Flower : The flowers are hermaphrodite. Flowers yellow, in axillary racemes; calyx of 4 spreading sepals; corolla of 4 imbricate petals, reflexed; stamens 12-20, inserted on the disc; ovary densely glandular, sessile (Fig. 2).

Flowering and fruiting : Flowering May to September: fruiting August to November.

Fruit : Capsule thinly glandular, pubescent, striate, narrowed to the tip (Fig. 3).

Fig. 2. Flower

Seed biology

Seed : Seed 1.3 mm across, 1 mm thick, reddish brown, cleft narrow and granular. The seed weight of *Cleome viscosa* varied from 0.0145 to 0.0421 g (range = 0.0277 g), and mean seed mass of this population was 0.0247 g. (Fig. 4).

Fig. 3. Fruit

Fig. 4. Seed

Propagation : Reproducing by seeds.

Seed dispersal : No specialised means of dispersal. Seeds drop to the ground close to, or beneath the parent plant. The wind may subsequently move seed short distances. Human mediated dispersal is probable in contaminated machinery, produce and soil or stock feed.

Seed dormancy and germination : Seeds have no dormancy and germinate readily after shedding (Jansen, 2004). Seeds germinated equally well in continuous light or darkness, with a decrease in germination under conditions of decreased light intensity or duration. Emergence was the same on the soil surface and at 1cm depth (Menon and Kulkarni, 1987). Higher germination percent was observed when the seeds were irradiated with red light.

Effect on crop : Occasionally a noxious weed. *C. viscosa* is likely to have negligible economic impacts in intensive cropping areas and nurseries, and is unlikely to be a threat to pastoral agriculture.

Uses : The leaves are diaphoretic, rubefacient and vesicant. They are used as an external application to wounds and ulcers. The juice of the leaves has been used to relieve earache.

The seeds are anthelmintic, carminative, rubefacient and vesicant.

Control measure : Easily controlled by hand or hoe and by cultivation. Monolinuron, trifluralin, chlorbromuron, atrazine, prometryne, terbutryne, metribuzin, diuron and oxadiazon are successful, but alachlor and fluorodifen are inefficient.

Past work done on Seed Science and Technology

Seed size and germination

Aziz and Shaukat (2010) stated that large-sized category gave significantly greater germination percentage than those placed in medium-size class followed by small-size seeds (Fig. 5). Root: shoot ratio was significantly higher ($p<0.01$) for seedlings from large-sized seeds followed by medium and small seeds. Higher amounts of dry weights (g) of both root and shoot were observed in the plants emerged from large-sized seeds in all three species (Table 1). The highest amount of reproductive output in large-seeded individuals followed by medium and small seeded individuals was observed.

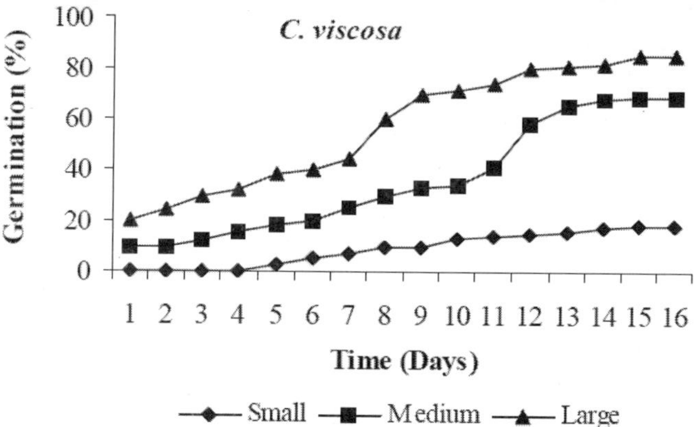

Fig. 5. Cumulative germination percentages over time of *Cleome viscosa* seeds belonging to three size categories.

Table 1. The effect of seed size on the reproductive structures of *Cleome viscosa*. (mean ± std. error)

Species	Seed size	No. of fruits/plant	No. of seeds/plant
C. viscosa	Large	22 ± 1.20 ***	660.2 ± 2.45***
	Medium	18.0 ± 0.32 ***	542.3 ± 1.63***
	Small	11.0 ± 0.75 n.s.	315.6 ± 1.55**

Note: Level of significance; * = p<0.01, ** = p<0.01, *** = p<0.001; n.s. = Non-significant

REFERENCES

Aziz, S. and S.S. Shaukat. (2010). Effect of seed mass variations on the germination and survival of three desert annuals. *Pak. J. Bot.*, 42(4): 2813-2825.

Jansen, P.C.M. (2004). *Cleome viscosa* L. Internet Record from Protabase. Grubben, G.J.H. & Denton, O.A. (Editors). PROTA (Plant Resources of Tropical Africa / Ressources végétales de l'Afrique tropicale), Wageningen, Netherlands. < http://database.prota.org/search.htm> (28 April 2008).

Menon, A. and A.R. Kulkarni. (1987). Ecological studies in *Cleome viscosa* L. seed and seed germination. *Indian Botanical Reporter*, 6 (1): 1-7.

BUTTERFLY PEA (*Clitoria ternata* L.)

Common name : Butterfly pea

Botanical name : Clitoria ternatea L.

Tamil name : Kuvalai

Hindi name : Aparijita

Family : Fabaceae

Origin and distribution : Native to *Africa*. Now, widespread throughout humid and sub-humid lowlands of Asia, the caribbean, central and south America, and more recently in semi-arid (600–800 mm) tropical Australia.

Botany

Habitat : Hot low altitudes, usually beside rivers and lake shores, riverine and coastal associations, in wooded and open grassland, and in disturbed places.

Habit : Annual vine to 6m (Fig. 1).

Stem : It is a vigorous, strongly persistent, herbaceous perennial legume; stems fine twining, sparsely pubescent, suberect at base, 0.5-3 m long.

Leaf : Leaves odd-pinnate, 9 x 6 cm; leaflets 2 or 3 pairs, opp, ovate, 2.5-4 x 1.5-2.5 cm; stipules persistent.

Floral biology

Flower : Flowers 3 cm across, resupinate, single or in clusters; bracteoles orbicular, persistent. Corolla exserted, deep blue. Stamen 9 + 1 (Fig. 2).

Fig. 1. Plant

Flowering, pollination and fruit : Flowers can develop in 4-6 weeks after sowing and continue to flower while temperature and moisture are adequate. Flowering can occur throughout the year given sufficient soil moisture and frost-free conditions. Predominantly self-fertile, but with some out-crossing.

Fruit : Pod oblong, 9 x 0.8 cm, compressed, subcontinuous, apically beaked (Fig. 3).

Seed biology

Seed : Seeds 10-15, reniform (Fig. 4).

Fig. 2. Flower

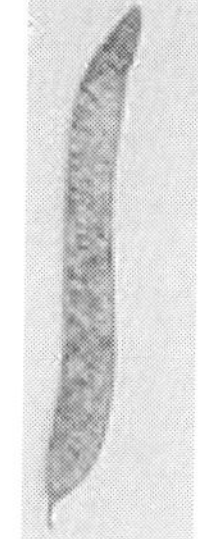

Fig. 3. Fruit

Fig. 4. Seed

Propagation : It is propagated by seed and cuttings.

Seed dispersal : The seeds are sticky in texture and are dispersed when they adhere to animals.

Seed dormancy and germination : Dormancy can be broken by treatment with concentrated sulphuric acid of sp. gr. 1.8 for 20 minutes (Prodonoff, 1968). Soaking of seeds for 12h in water, followed by 12h in a freezer at 15°C, then defrosted for planting improved the germination.

Seed storage : Clitoria ternatea seeds of forage species that have been stored at ILRI under conditions of 5% moisture content and 8°C for 5 years without substantial loss of viability.

Effect on crop : Not much effect on crop. It fixes nitrogen fixation in the soil.

Control measure : Hand pull isolated plants and small infestations, making sure that all roots and stem fragments are removed.

Uses : In traditional Ayurvedic medicine, it has been used for centuries as a memory enhancer, nootropic, antistress, anxiolytic, antidepressant, anticonvulsant, tranquilizing and sedative agent. Widely planted as an ornamental on fencerows. Increases soil fertility to improve yields of subsequent crops (maize, sorghum, wheat) when grown as green manure or ley pasture. Also used for cut-and-carry and conserved as hay. Hay suitable for goats in Sudan. Used as a revegetation species on coal mines in central Queensland, Australia.

Past work done on Seed Science and Technology

Effect of sodium cyanide on germination

Presoaking seeds of *Clitoria ternatea* in suitable concentrations of aqueous sodium cyanide solution was found to increase the percentage of imbibition and germination and favourably effected the growth of the seedlings. Usually values of germination shown by *Clitoria ternatea* and *Rhyncosia minima* under ordinary condition were not appreciable (Chatterji, 1966; Baxi, 1965).

Table 1. Effect of Sodium cyanide solution on growth of the seedlings at room temperature

Concentration of sodium cyanide in ppm	Total length of the seedlings in centimetre Clitoria ternatea Time in days		
	2 nd day	4 th day	6 th day
100	8.5	12.7	14.7
250	5.4	7.6	9.4
500	4.6	5.9	7.2
1000	3.2	3.8	5.2
Control	5.4	8.4	10.5

Prakash Mullick and Chatterji (1967) studied the effect of aqueous sodimn cyanide on germination of *C. ternatea* and *R. minima* seeds. The suitable concentration of this chemical accelerates imbibition, germination, and promotes the early growth of the seedlings. The best result is obtained by soaking the seeds for 24 hours in a solution of 100 ppm of sodium cyanide. As concentrations are increased, the inhibitory effect becomes more and more pronounced leading to the eventual suppression of normal germination and growth of the seeds (Table. 1). The mean germination time of the seeds, treated with effective and critical concentrations of the chemical, is decreased. Light and temperature conditions do not influence the chemical effect on the seeds.

REFERENCES

Baxi, D. (1965). Contributions to the comparative study of physiological and anatomical aspects of arid zone plants II Morphological and Eco-physiological studies on the germination of certain leguminous seeds, Ph. D. Thesis, University of Jodhpur.

Chetterji, U. N. (1966). Germination experiment with the seeds of *Clitoria ternatea*. *Proc. Ind. Sci. Cong.* III (IV), 348-349.

Prakash Mullick and U. N. Chatterji. (1967). Effect of sodium cyanide on germination of two leguminous seeds. Plant Systematics and Evolution. Volume 114, No. 1 (1967), 88-91.

Prodonoff, E. T. (1968). Seed testing in Queensland. (Department of Primary Industries: Brisbane).

DAY FLOWER (*Commelina benghalensis* L.)

Common name : Day flower, dew flower

Botanical name : Commelina benghalensis L.

Tamil name : Kannan keerai

Hindi name : Kanchara, Kaua-kaini, Kanuraka.

Family : Commelinaceae

Origin and distribution : Origin is old World tropics and it is widespread in the tropics and subtropics, but also in temperate areas.

Botany

Habitat : It is a weed of cultivated lands, field borders, wet pastures, gardens, roadsides and waste places. It can withstand flooding and waterlogged conditions. In rice and other lowland crops, it is subaquatic and grows readily on banks of irrigation ditches.

Habit : Annual either perennial herb. Glabrous or pubescent, herb, growing upto 60cm (Fig. 1).

Stem : Branches many, some of them growing underground and bearing reduced leaves and spathes with apetalous closed flowers and capsules

Leaf : Leaves ovate, oblong or suborbicular, obtuse or acute, base contracted into a petiole.

Floral biology

Flower : Pubescent or villous; sheaths pubescent or villous, often bearded or ciliate, diffuse, short or long; spathes funnel – shaped; 1 –3 together, auricled on one side, pubescent or hirsute, upper branch of cymes 2 – 3 flowered, lower 1-2 flowered or suppressed; Flowers blue, underground flowers white, sepals 3, membranous, inner two connate; petals 3 two long clawed, third sometimes wanting; stamens 3, perfect, 2 – 3 imperfect, filaments filiform, spirally coiled; ovary 3 – celled, free, two cells 2 – ovuled, third cell one – ovuled (Figs. 2; 3).

Fig. 1. Plant

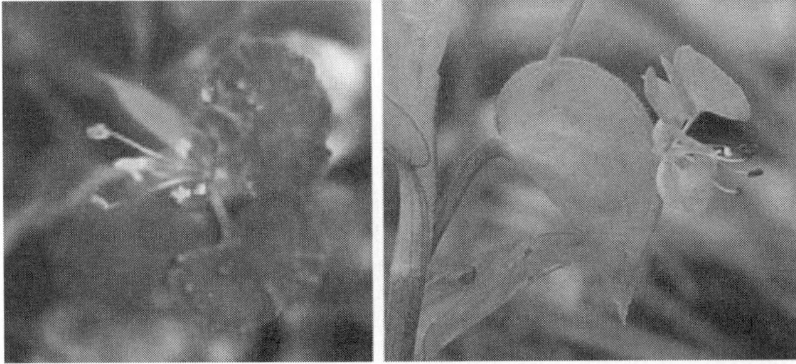

Fig. 2. Flower

Fig . 3. Underground flowers

Flowering, pollination and fruit set : Flowering July – October, chasmocleistogamy. Aerial flowers are chasmogamous and self fertilizing, producing one large seed and 4 small. Subterranean flowers are cleistogamous (self fertilizing and do not open), producing one large seed and two small.

Fruit : Capsule membranous, shining, dorsal valve striolate, 5 – seeded (Fig. 4).

Seed biology

Seed : Seeds closely pitted (Fig. 5).

Fig. 4. Opened spathe with fruits

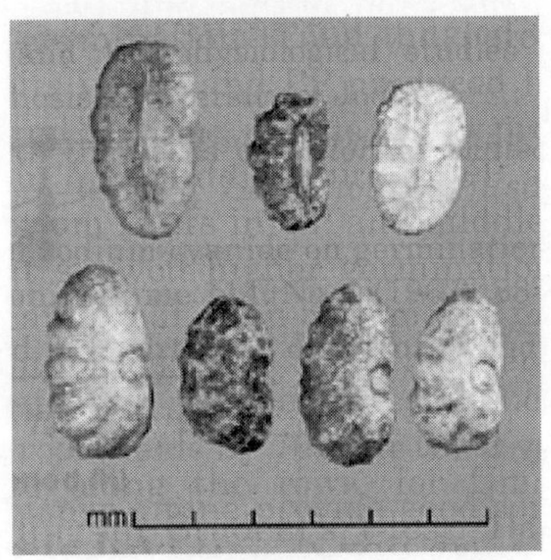

Fig. 5. Seed

Propagation : It can propagate easily by layering (very resistant to dryness) but also by seeds. A plant can produce 1,600 seeds.

Seed dispersal : Through farm equipment and products. Subterranean seeds of amphicarpic plants typically have limited dispersal and longer than aerial seeds (Kaul *et al.,* 2000).

Life cycle : This difference in lifecycle can be associated with a difference in ploidy levels, with tropical *C. benghalensis* being hexaploid and temperate being diploid. Tropical hexaploid plants rarely have subterranean flowers. It can produce seeds within 40-45 days of emergence and has multiple generations per year. Subterranean flowers develop about 6 weeks after emergence; aerial flowers develop about 8-10 weeks after emergence. Fruits are produced within 3 days after flowering, with viable seeds within 25 days after flowering. There are four categories of seeds, large and small aerial and large and small subterranean. Small aerial seeds account for 73-79% of all seeds found.

Seed dormancy and germination : Small aerial seeds have a stronger dormancy than large. Clipping the seed coat or exposing the seeds to temperatures in excess of 90°C for 2 h removed dormancy for all seeds. The optimal temperature for germination of aerial seeds is 18-25°C and 21-28°C for subterranean seeds. The optimal depth for emergence is 2 inches, with larger seeds emerging from depths up to 6 inches.

Effect on crop : Benghal day flower is most problematic in peanuts, soybeans and corn. It is listed as one of the World's worst weeds, affecting 25 crops in 29 countries (Webster *et al.,* 2005).

Control : Cultural control: It is particularly difficult to control by cultivation, partly because broken pieces of above and below ground stems readily take root. With hand weeding, it is necessary to uproot all the plant from the soil to ensure effectiveness.

Biological control: Very few natural enemies have been reported to attack *C. benghalensis* and, of this, only two agromyzid leaf miners from the Americas appear promising (*Amauromyza* sp. and *Liriomyza* commelinae from Diptera family).

Chemical control: Butachlor has been reported to provide good pre-emergence control of *C. benghalensis,* for post-emergence control, propanil @ 2 kg ha^{-1} or 2-4-D with propanil can increase effectivness or almix (Chlorimuron+metsulfuron) @ 4g ha^{-1}. Single treatment with contact herbicides destroys the foliage, but does not prevent regeneration. Application of 2, 4 - D (1.4-2.8kg ha^{-1}), metribuzin (2.5-3 kg ha^{-1}), and glyphosate (1.8-2kg ha^{-1}) resulted in almost complete kill and the treated plots were free of the weed for upto 4 months.

Uses : In Africa and India, the leaves and stems of *Commelina benghalensis* are chopped and cooked as vegetables and used as feed for livestock. Although a weed in crop, dayflower is a useful forage plant in Indonesia and Philippines. Leaves are used as human food, as medicine for infected wounds, for treating sterility and as mucilage for treating burns, sore eyes and sore throats.

In Philippines and India, the whole plants are considered demulcent, laxative and astringent, and applied for strangury. In Taiwan, the aerial parts are used as a maturative poultice.

Past work done on Seed Science and Technology

1. Seed production and dispersal

Benghal dayflower is amphicarpic, possessing the ability to produce two types of fruits (aerial and subterranean) with different roles (Maheshwari and Singh, 1934; Maheshwari and Maheshwari, 1955). Amphicarpy is characterized by two reproductive strategies: (1) one that is successful in a predictable environment in which the plant will have significant vegetative growth prior to abundant fruit production and (2) another in which an unpredictable environment dictates that reproduction occur early in the growing season without regard to maximizing fruit production (Zeide, 1978). Aerial seeds of benghal dayflower are produced by chasmogamous (typical, open pollinated) flowers as part of the reproductive strategy in a predictable environment; while subterranean seeds, formed from cleistogamous (closed, self pollinated) flowers, ensure reproduction in an unpredictable environment.

According to Webster and Grey (2008) both the aerial seeds (produced by chasmogamous flowers) and subterranean seeds (produced by cleistogamous flowers) were related to irrigation volume in a linear manner (Fig. 6) and duration of drought stress in an inverse linear manner (Fig. 7). Seed production occurred at the lowest moisture level and was similar for both aerial and subterranean fruits (Fig. 6). However, the rate of aerial seed production was 8.6 times greater than subterranean seeds (t 5 4.14df), indicating that aerial seed production was more sensitive to relative moisture status than was production of subterranean seeds.

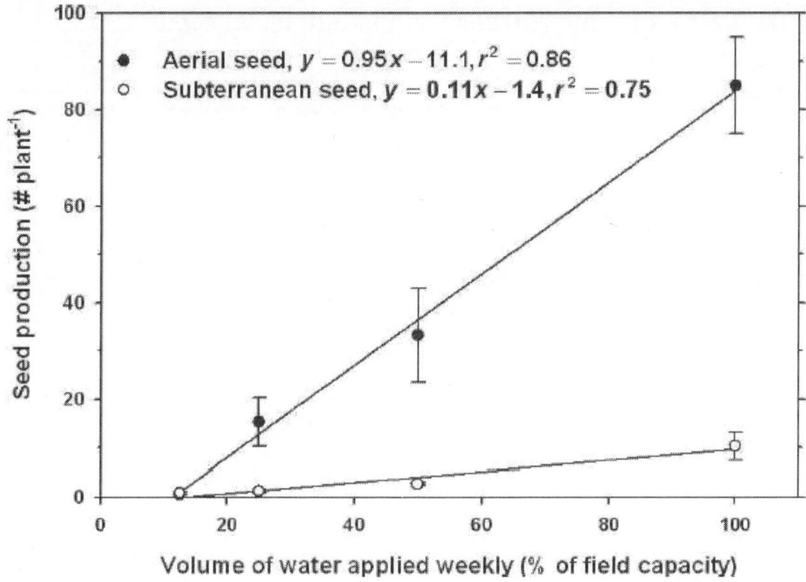

Fig. 6. The linear relationship between benghal dayflower seed production and volume of water applied weekly for aerial and subterranean seeds over the 9- wk study. A t-test indicated that rates of seed production as a function of irrigation volume for aerial and subterranean seeds were different (t 5 4.14df). The interval around the treatment means represent the standard error.

As duration of drought stress increased, rate of aerial seed production was reduced more (t 5 4.66df) than subterranean seed production (Fig. 7). Subterranean seed production ranged from 1 to 10.3 seeds plant^{-1} at all durations of drought stress, while 49 and 56 d of drought stress prevented aerial seed production (Fig. 7). Production of subterranean seeds under all moisture levels ensures survival of the population in an unpredictable environment and contributes to the weedy nature of this exotic species.

Subterranean seeds of amphicarpic plants typically have limited dispersal and are larger than aerial seeds (Kaul *et al.,* 2000). Previous studies determined that subterranean seeds accounted for less than 4% of the total number of seeds produced (Kim, 1998; Walker and Evenson, 1985). Aerial flower production under field conditions was initiated at the 17- to 19-leaf stage of benghal dayflower, while subterranean flower initiation occurred at the 7- to 10-leaf stage (Kaul *et al.*, 2002). The short duration of the current study and the relative earliness for subterranean seed production compared to aerial seeds may account for the greater proportion (12%) of the subterranean seeds in this study at the highest irrigation volume compared to previous reports. In addition to producing seeds earlier in the growing season, seedlings produced by the larger subterranean seed of amphicarpic plants are more vigorous and have a greater competitive ability than those produced by aerial seeds (Kaul *et al.* 2000). Benghal dayflower produces two types of subterranean seeds, large (8.81 mg seed^{-1}) and small (3.51 mg seed^{-1}) (Santos *et al.*, 2001), which are 1.8 to 4.6 times larger than small aerial seeds.

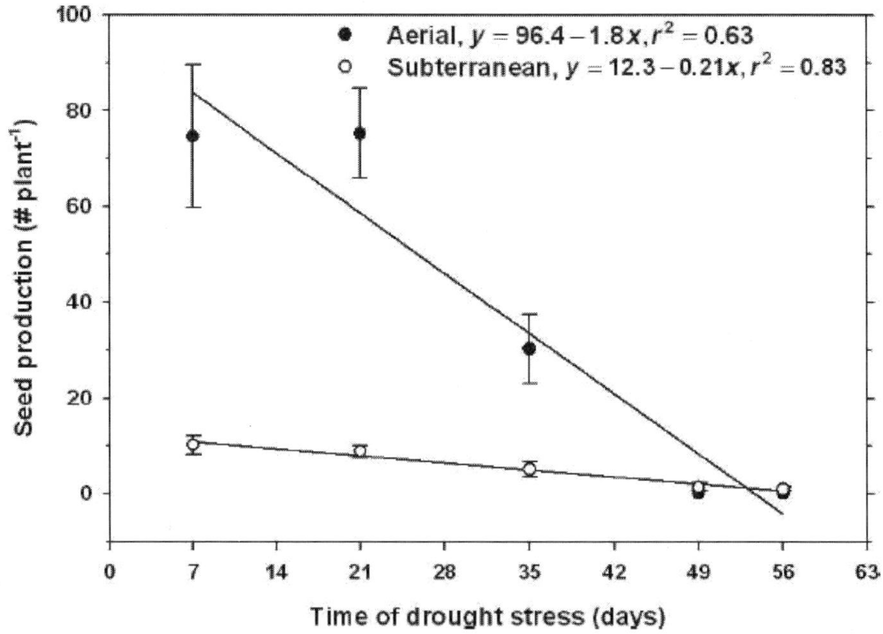

Fig. 7. The inverse linear relationship benghal dayflower seed production and volume of water applied weekly for aerial and subterranean seeds over the 9-wk study. A t-test indicated that rates of seed production as a function of duration of drought stress for aerial and subterranean seeds were different (t 5 4.66df). The interval around the treatment means represent the standard error.

Aerial flowers of amphicarpic plants will produce numerous light-weight seed that are readily dispersed from the parent plant and allows for the possibility of cross-pollination, which encourages adaptability and plasticity within the population (Cheplick, 1994; Kaul *et al.,* 2000). Benghal dayflower produces two types of aerial seed, with each flower containing two to four small (1.9 mg seed^{-1}) seed and one large (3.65 mg seed21) seed (Faden, 1993, 2000; Santos *et al.,* 2001). Birds have been shown to consume aerial seeds, however, it is not known if these seeds can survive the digestive tract (Carter *et al.,* 2006). Aside from human-mediated seed movement (i.e., farm machinery, as contaminants of harvested crops), there are no other obvious means of dispersal for these seed. Gibson and Tomlinson (2002) suggested that the lower competitive ability of smaller seeds in a heterocarpic species may offset greater dispersal ability to new sites compared to larger seeds, which may also be the case with benghal dayflower. Some of the variability in the current study may be associated with the size (small vs. large) and type of seed (aerial vs. subterranean) that generated the seedling. Seedlings in the nondrought study developed from aerial seeds, both large and small. In the drought stress studies, cotyledon-stage seedlings were collected from a grower's field, therefore we were unable to determine the nature of the seed that produced each seedling.

Seed viability and impact on dispersal

According to Mandeep Riar *et al.* (2010), benghal dayflower seeds lose viability at a faster rate in soil in a warmer climate than in milder climates (Fig. 8). benghal dayflower seed banks in North Carolina will have longer persistence than those in Georgia. All types of Benghal dayflower seeds remain viable at 45 and 50°C temperatures found in uncovered hay bales for at least 14 days. The higher hay bale temperature of 65°C kills the seeds much faster than lower temperatures (Fig. 9). Rumen digestion for up to 48 h enhances germination of large aerial seeds (Fig.10A). Digestion duration of 12 and 24 h also increases the seed germination. Germination of small aerial seeds is also enhanced by digestion of any duration compared to the control (Fig.10B). The process of digestion in rumen generally lasts for 12–48 hours. Thus, dispersal of viable benghal dayflower seeds via ruminants is very likely (Fig. 11).

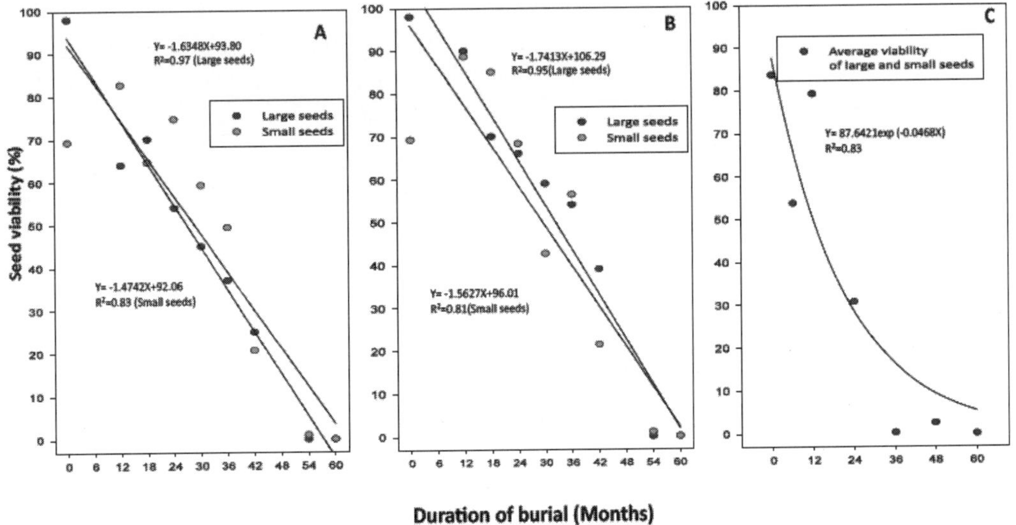

Fig. 8. Percent of viable benghal dayflower seeds in (A) North Carolina start year 2004 (B) NC start year 2005 (C) GA start year 2005.

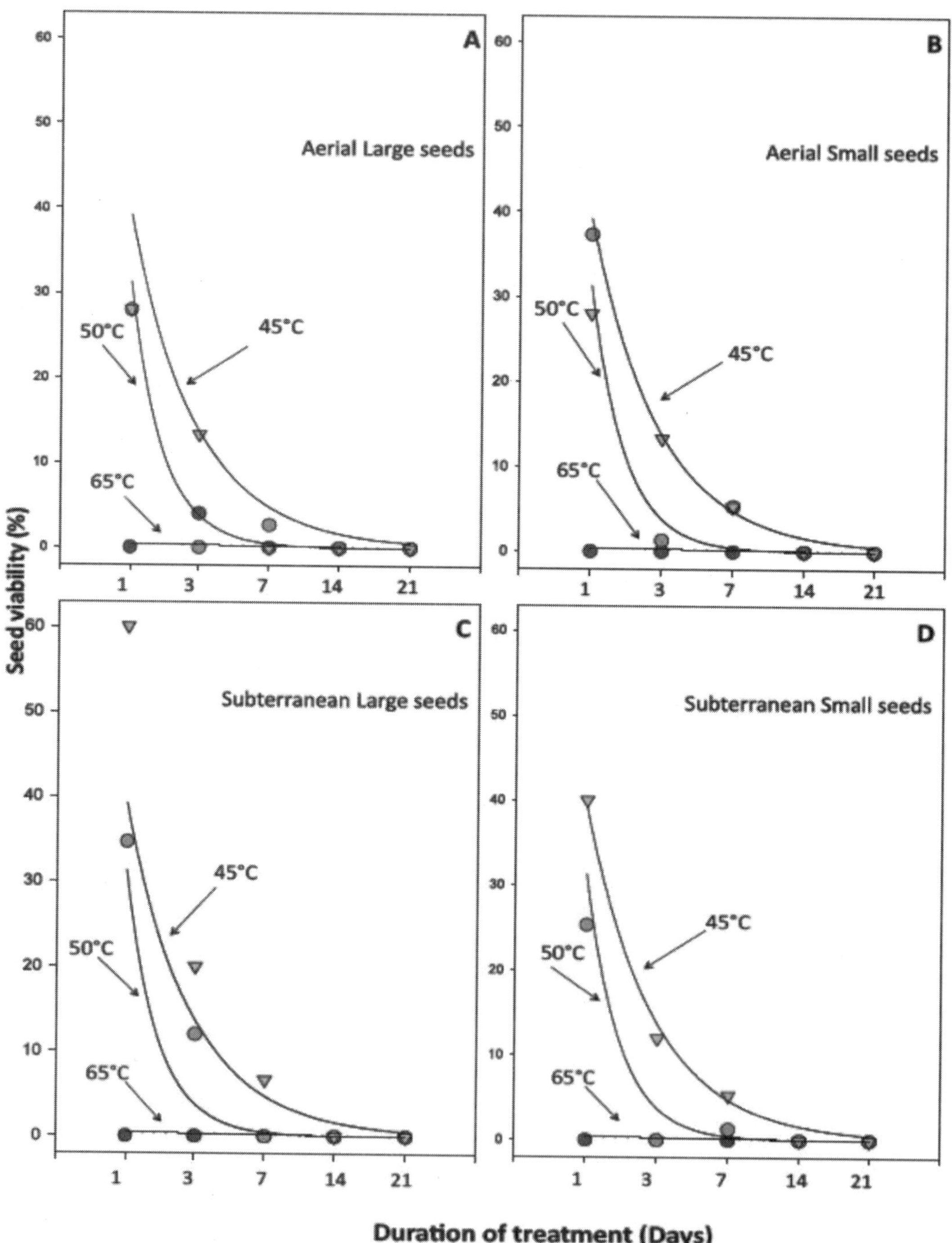

Fig. 9. Viability of (A) aerial large (B) aerial small (C) subterranean large and (D) subterranean small seeds after exposure to a range of simulated hay bale temperatures.

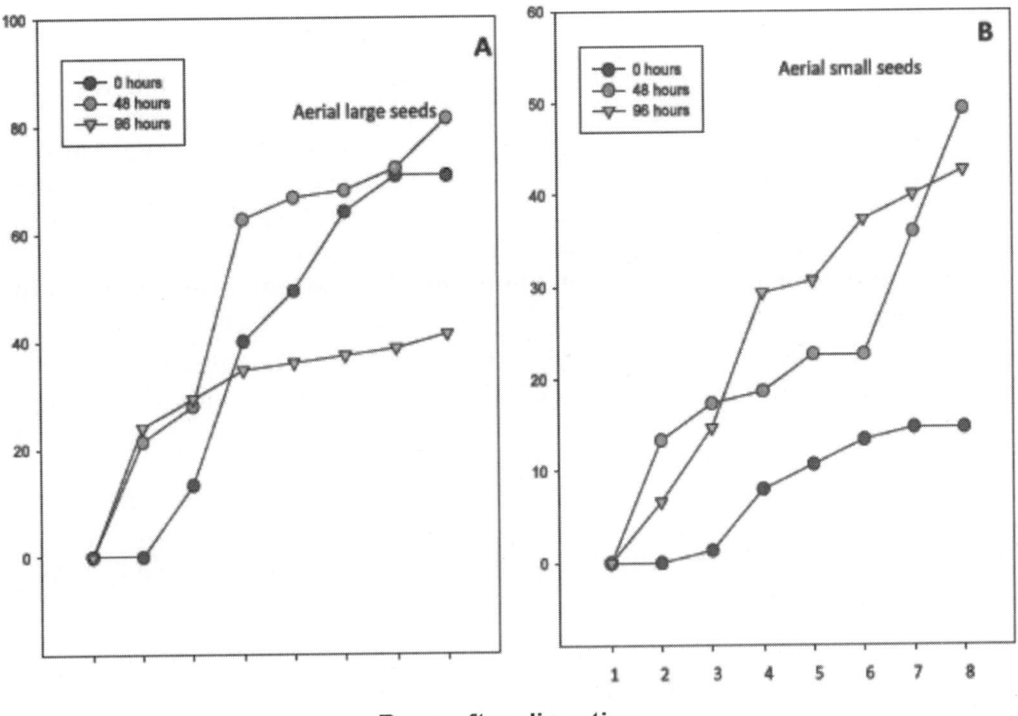

Fig. 10. Germination of (A) large and (B) small aerial benghal dayflower seeds after rumen digestion

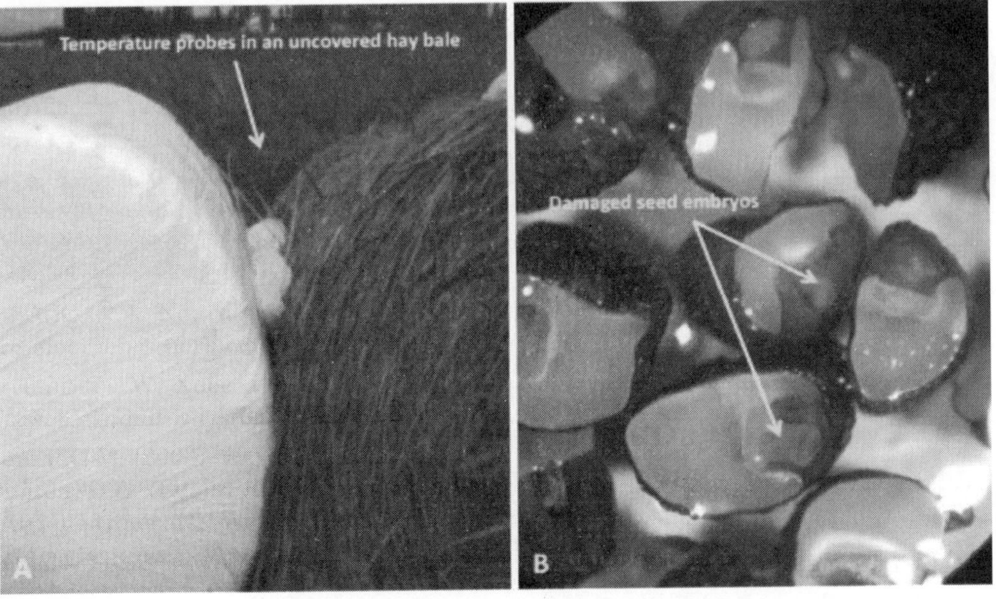

Fig. 11. (A) Temperature measurements in hay bales (B) Damaged Benghal dayflower seed embryos at 65°C after 3 days.

2. Germination improvement

Sabila and Mercy Hellen (2008) stated that the aerial seed had a higher germination per cent for the temperature treatment. The highest germination per cent was 90% at 30°C. The 20°C treatment had a cumulative germination of only 30% indicating that emergence could still be sustained even at these low temperatures. Germination was 45 and 15% for the 10 and 20mM salt concentration, respectively. No germination occurred with the 40 mM salt concentration. The highest germination was at 0 and 1 cm plant depth. No germination occurred in the 12 cm plant depth.

Goddard *et al.* (2009) studied the potential dispersal of benghal dayflower seeds by mourning doves in southern Georgia, U.S.A. The gut contents (both crop and gizzard) of mourning doves harvested in the autumn months were investigated to determine if mourning doves fed on benghal dayflower and whether seeds can survive conditions in the bird gut. Research indicated that mourning doves fed selectively on benghal dayflower with some harvested birds containing hundreds of benghal dayflower seeds and capsules in their guts. Further, some seeds recovered remained highly viable.

Table 1. Number of benghal dayflower seeds recovered separately from crop and gizzard in doves harvested in 2006 from Berrien County.

Bird number	Organ	Seed recovered
Berrien.01	Crop	562
	Gizzard	123
Berrien.02	Crop	149
	Gizzard	122
Berrien.03	Crop	310
	Gizzard	104
Berrian.04[a]		115
Berrien.05	Crop	3
	Gizzard	42
Berrien.06	Crop	271
	Gizzard	123
Berrien.07	Crop	not recovered
	Gizzard	105

a. Crop and gizzard combined.

For the initial seed viability test, data indicated a small percentage of seeds from doves taken in 2004 tested strongly positive with TZ, indicating that at least some ingested seeds were viable with staining similar to controls (Fig. 12a–c).

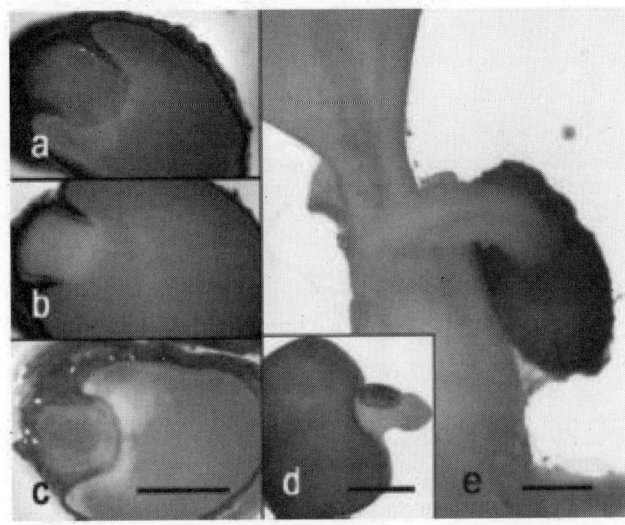

Fig. 12. Seed viability and germination of benghal dayflower. (a–c) Strong positive results of viability testing with tetrazolium (TZ). (a) Control seed that was imbibed unboiled and stained in 1% TZ demonstrates a red-stained embryo. (b) Control seed that was imbibed then boiled to kill cells of the embryo, then stained in 1% TZ demonstrates no staining of the embryo. (c) Seed extracted from gut contents, stained with 1% TZ showing a strongly positive red-stained embryo from bird 2004.01. (d–e) Germinating seeds of benghal dayflower. (d) The embryotega is lifted as the embryo emerges while the seed coat remains intact. (e) Germinated seedling of benghal dayflower. The still intact seed/seed coat is connected to the vertically oriented seedling by the taenia or cotyledonary stalk (bars 5 0.5 mm).

Seeds of *Commelina* species germinate by rupturing through the micropyle region and lifting the embryotega, a callus-like covering over the micropyle that functions essentially as an operculum (Fig. 12d). Even after imbibition, the seed coat never ruptures, an attribute apparently related to its strength, but the germinating seedling emerges by pushing the embryotega aside (Fig. 12d). Most of the embryo is extruded and develops outside the seed coat connected by a taenia, or cotyledonary stalk, to the scutellum inside the seed (Fig. 12e). Benghal dayflower seeds have a distinctive shape with a clearly visible embryotega and linear hilum (Fig. 13a). The seed surface has a relief pattern with large and smaller reticulations (Fig. 13b). There is a thin papery outer layer of the seed coat that covers the entire seed including the embryotega, which likely developed from the epidermis postanthesis. In unimbibed seeds, this layer is intact with no visible cracks or breaks when viewed by SEM (Fig. 13b and 13c). After imbibition, small cracks can first be seen over the embryotega and linear hilum regions (Fig. 13d and 13e, respectively).

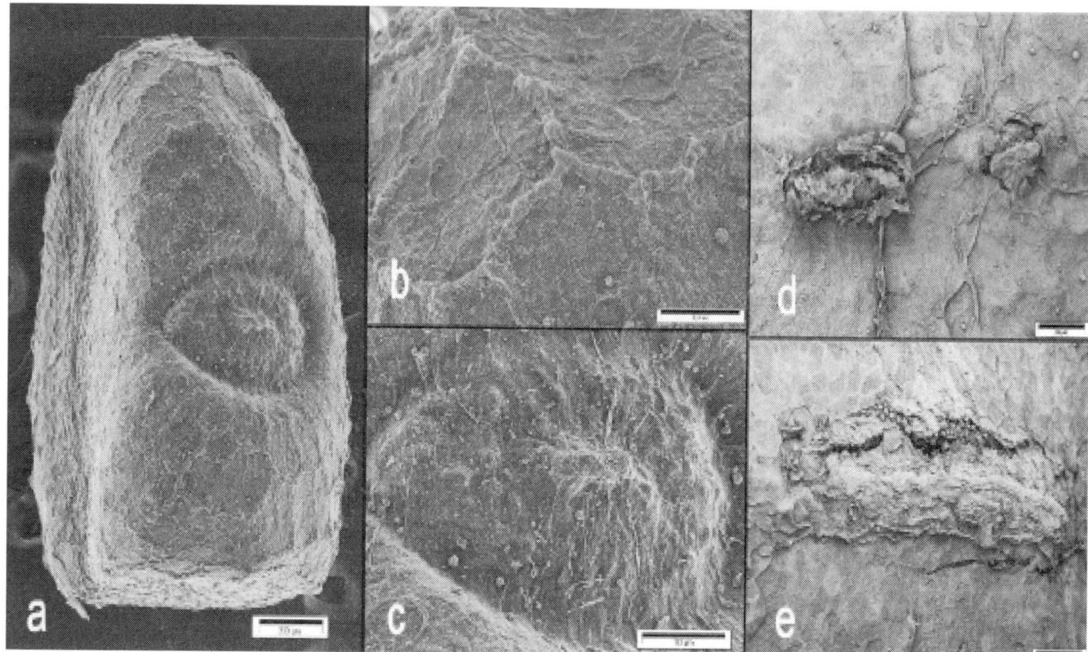

Fig. 13. Surface morphology of wild, control seeds that had been imbibed or not. (a) Intact unimbibed aerial control seed collected from benghal dayflower plants in the field. Distinctive features include the elliptical embryotega at the center and the linear hilum apparent on the flattened surface at left in the image (bar 5 200 mm). (b) A distinctive feature of benghal dayflower seeds is the raised surface over the seed coat in a reticulate pattern. Both large and smaller reticulations are clearly evident (bar 5 100 mm). (c) Higher magnification image of the unimbibed control seed embryotega in Figure 13a. The embryotega is covered, with no cracks or breaks, by a layer of the seed coat that is continuous over the entire seed (bar 5 100 mm). (d) In an imbibed control seed, cracks are shown in the outer surface layer of the seed coat over the embryotega (bar 5 50 mm). (e) The linear hilum region of a control seed that had been imbibed also demonstrates extensive cracking of the surface layer of the seed coat (bar 5 100 mm).

In crop extracts, many capsules were found still containing benghal dayflower seeds (Fig. 14a). Seeds extracted from bird crops were not identical to control seeds as perceived by light microscopy, but had extensive cracks in the surface layer of the seed coat, particularly around the embryotega when viewed by SEM (Fig. 14b; compare with Fig. 13a). Even seeds still attached to dehisced capsules demonstrated surface abrasion in the crop (Fig. 14c). In many seeds extracted from the crop, the surface layer of the seed coat had already been abraded away, or chemically removed, revealing a subtending layer of cells with interconnected thick walls in a honeycomb pattern (Fig. 14d). This layer of cells and the subtending wall appear to form a resilient barrier of the seed coat, which is discussed further with regard to the morphology of seeds extracted from the gizzard.

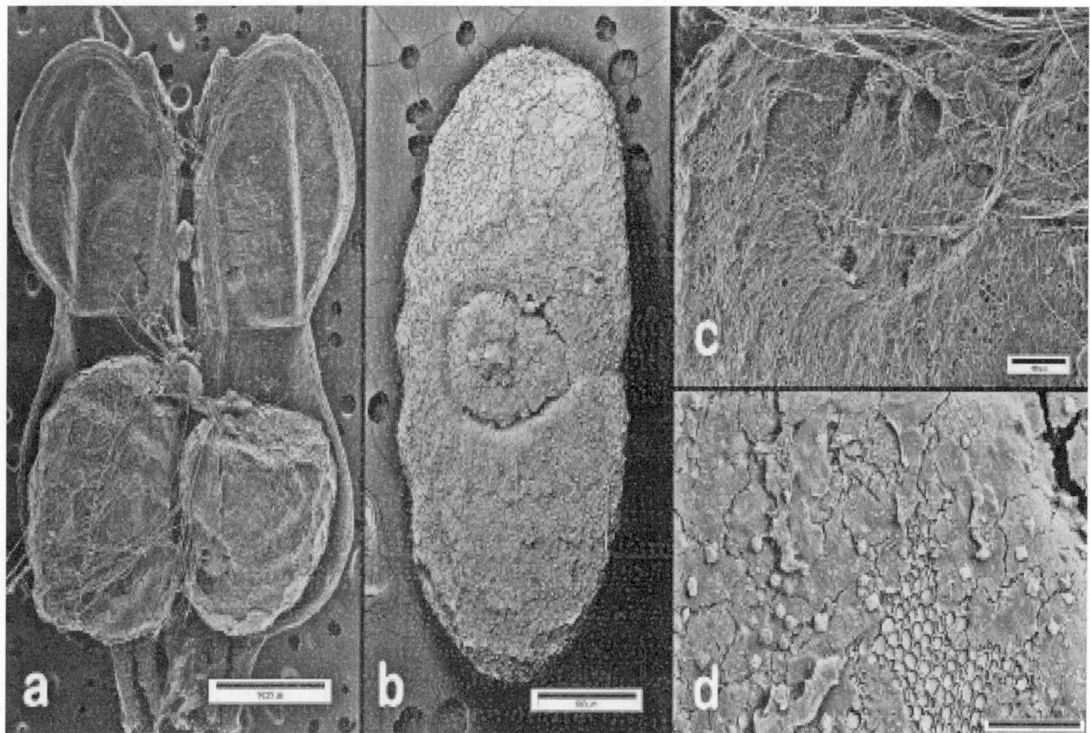

Fig. 14. Morphology of seeds extracted from dove crops. (a) Image of an intact dehisced capsule with two seeds in place extracted from a dove crop (bar 5 1 mm). (b) Backscatter electron image of a whole seed extracted from dove crop. The outer surface layer of the seed coat is extensively cracked, particularly around the edge of the embryotega (bar 5 0.5 mm). (c) High-magnification image of one of the seeds present in Figure 14a, from dove crop. Surface of the seed (shown near the embryotega) is already cracked and abraded revealing underlying cells in a seed that has not yet been mechanically dislodged from the fruit (bar 5 100 mm). (d) Higher magnification backscatter-electron scanning electron micrograph of the seed surface from a bird crop showing parts of the outer layer of the seed coat already fully removed revealing a tightly packed layer of cells whose walls form a honeycomb pattern (bar 5 100 mm).

Some seeds present in the gizzard had been stripped of surface layers of the seed coat, revealing the embryotega completely (Fig. 15a). Other seeds extracted from the gizzard were encrusted with debris (Fig. 15b and 15c) but with their papillate embryotega revealed (Fig. 15c). Large sections of the outer layer of the seed coat were removed in the gizzard revealing the honeycombed network of cells seen in seeds from the bird crop (Fig. 15d). The surface reticulations noted in intact seeds (Fig. 13b) are often not apparent in seeds extracted from the gizzard or are present only as dislodged fragments of seed coat surface material among other debris in the gizzard (Fig. 15d and 15e). These fragments appear to be exclusively a part of the outer layer that is removed during digestion.

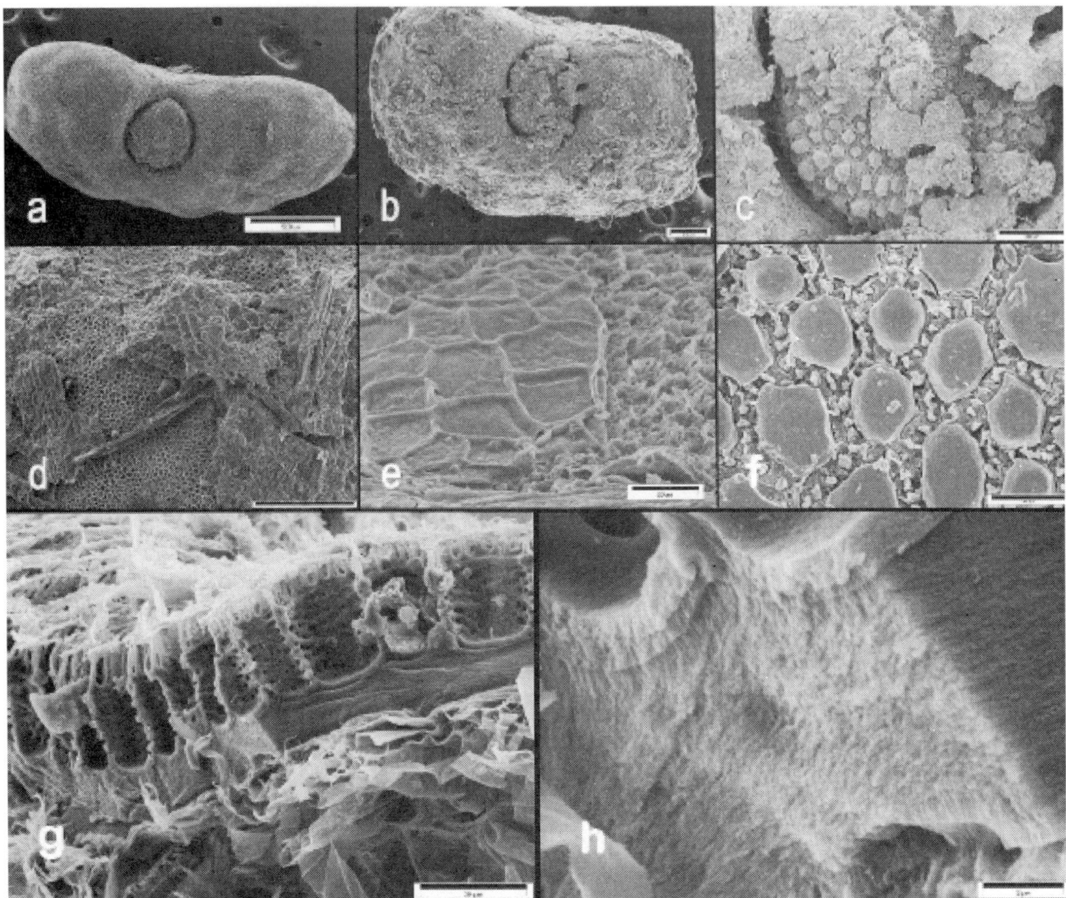

Fig. 15. Morphology of benghal dayflower seeds extracted from dove gizzard. (a) Whole seed with the surface covering fully removed and exposing the embryotega covering the micropyle (bar 5 0.5 mm). (b) Common appearance of seeds in the gizzard demonstrates debris covering the surface with most of the seed coat covering abraded, revealing underlying layers of the seed coat (bar 5 200 mm). (c) Higher magnification scanning electron micrograph (SEM) of embryotega with papillate cells (bar 5 100 mm). (d) SEM image of the seed surface with debris or seed coat layers partially occluding the continuous layer of cells forming the underlying surface of the seed coat (bar 5 200 mm). (e) SEM image of a portion of the seed surface with a remnant of the outer layer of the seed coat with minor reticulations. Minor reticulations are larger than the underlying cells (bar 5 20 mm). (f) SEM image of the layer of cells underlying the seed coat removed in the gizzard. Lateral walls have projecting spires that give them a punctate appearance when seen in surface view. Lateral walls are closely appressed between cells and provide no identifiable intercellular spaces between cells. Cells are covered with a lid-like structure (bar 5 10 mm). (g–h) Lateral view of a seed coat cell layer (similar to surface view of the same layer in Figure 15f) in a seed fractured with a razor blade prior to processing for SEM. (g) Surface cells have no intercellular spaces and trabeculate lateral walls, open at the surface. All surface cells are appressed to a continuous tangential wall layer forming part of the seed coat (bar 5 20 mm). (h) High magnification image of the tangential wall at the base of the seed coat (seen in Fig. 15g; bar 52 mm).

The underlying cell layer of the seed coat has complex lateral cell walls that form a honeycomb pattern and have little or no intercellular spaces (Fig. 15f). In one seed that was forcibly fractured with a razor blade during preparation, the outer honeycomb-patterned cell layer of the seed coat was revealed in cross-section (Fig. 15g). In the fractured seed coat layer, the cell lumen is exposed, revealing thick, almost trabeculate walls with a subtending continuous tangentially oriented wall layer (Fig. 15g). This subtending wall layer is microfibrillar in structure and forms a barrier beneath the honeycomb-patterned cells upon which each cell in the layer is continuous (Fig. 15h). The benghal dayflower seed coat appears adapted for strength and rigidity. Although the honeycombed network of cells in this layer appears to provide structural reinforcement, this layer of cells is sometimes scraped away in the gizzard (Fig. 16a and 16b) revealing only the subtending continuous tangential wall layer (Fig. 16b). The honeycomb pattern of the abraded cells is still apparent on the surface that is revealed (Fig. 16b). Few seeds were found that were cracked or in pieces relative to the number of intact seeds.

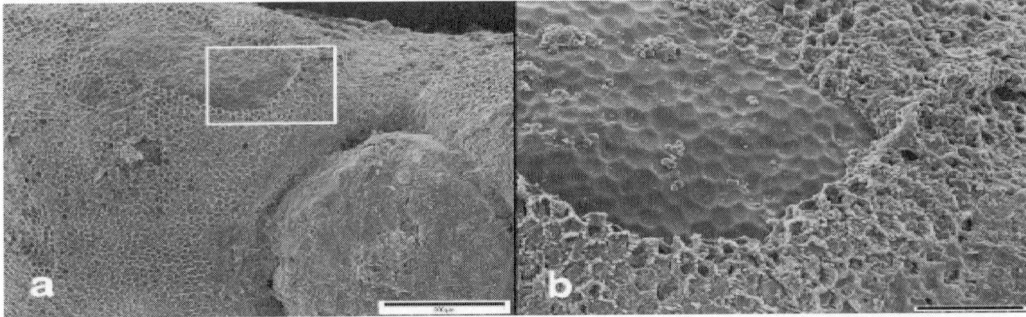

Fig. 16. Scanning electron micrograph of the surface of a seed extracted from dove gizzard. (a) Parts of the trabeculate cell layer have been abraded and removed exposing the continuous tangential cell layer (bar 5 200 mm). (b) Detail of the portion of Figure 6a outlined by the white box. Seed coat is not cracked even with the surface layer of cells abraded away. Tangential wall that remains shows clearly the honeycombed pattern of the cell walls that were removed (bar 5 50 mm).

The benghal dayflower seed coat appears at least two layered with a rigid thick inner layer and a thin outer covering. From the morphological data acquired in this study and the germination data presented subsequently, it appears that maintaining the integrity of the outer covering of the seed coat is necessary to preserve seed dormancy. The inner layer of the seed coat appears to be structural in nature and may protect the seed from harsh mechanical perturbation in the bird gizzard. The question remains whether seeds that pass through dove digestive tracts remain intact and viable, germinating after passing, a test not accomplished in this study.

Germination rates in seeds taken from bird crops were similar to controls over the first 4 wk of germination and enhanced over control treatments during the latter 16 wk of a 20-wk germination study. Ultimately, seeds extracted from dove crops had 92% germination as compared to 80% for control seeds. Seeds extracted from dove gizzards had 45% germination, about half that of controls (Table 2).

Benghal dayflower seeds have a structurally reinforced seed coat that probably aids in survival of mechanical damage through bird intestinal tracts. Benghal dayflower seeds exposed to 1.0 M HCl treatment for 2 h had little loss in viability, successfully germinating after such treatment (Table 3). When evaluating mechanisms for the eradication of benghal dayflower from agricultural crops, consideration needs to be given to the large number of mourning doves and other bird species that visit cropland and potentially aid in its dispersal.

Table 2. Summary results of seed germination in seeds extracted from bird crop and gizzard in birds Berrien.01 and Berrien.06 harvested in 2006, and between PPMa-treated control vs. no PPM treatment.

	4 wk	12 wk	20 wk
		% seed germination	
Organ of seed recovery			
Crop	49	87	93
Gizzard	47	45	45
F-value	3.11	76.93	79.88
P-value	0.099	< 0.0001	< 0.0001
Control treatment			
PPM	50	79	80
No PPM	24	37	51
F-value	15.23	89.62	47.74
P-value	0.0169	0.0007	0.0023

a Abbreviation: PPM plant preservative plant

Table 3. Summary results of percentage of seed germination at 4, 12, or 20 wk after planting, in different acid treatments on wild seed. Analysis of variance was performed using PROC Mixed in SAS, with variances partitioned into random effects of trial and replication. Transformed treatment means were separated using Fisher's Protected LSD0.05.

Treatment	4 wk		12 wk		20 wk	
			% seed germination[a]			
Light control	54	a	79	ab	92	a
Dark control	12	d	30	c	87	b
0.1 M HCl for 1 h	41	bc	84	a	93	a
0.1 M HCl for 2 h	37	c	75	ab	93	a
0.1 M HCl for 4 h	49	ab	79	ab	92	a
1 M HCl for 1 h	40	c	70	b	84	b
1 M HCl for 2 h	41	bc	70	b	82	b
12 M HCl for 1 h	0	e	0	d	0	c

Means with different letters within a column are significantly different

3. Seed dormancy and germination

Kim *et al.* (1990) reported that concentrated sulphuric acid and NaOCl treatments were more effective than dry heat and hot water treatment in breaking the dormancy of *C. benghalensis* aerial seeds. Treatments increased germination of both large and small dormant aerial seeds due to its effect on the seed coat integrity. A SEM revealed that changes as a result of scarification occurred in the hilum region of the seed and in the seed coat surface. The large aerial seeds were affected by all treatments more than the small aerial seeds. Difference in germinability between the two seed types was related to the difference in their seed vigour.

REFERENCES

Cheplick, G.P. (1994). Life history evolution in amphicarpic plants. *Plant Species Biol.* 9:119–131.

Carter, J.R., R.H. Goddard, T.M. Webster, J.T. Flanders, A.S. Culpepper and T.L. Grey. (2006). Do mourning doves disperse seed of tropical spiderwort? Page 85 in C. Butts, ed. American Peanut Research and Education Society Proceedings. Savannah, GA: American Peanut Research and Education Society.

Faden, R. B. (1993). The misconstrued and rare species of Commelina (Commelinaceae) in the eastern United States. Ann. Missouri Bot. Gard. 80:208–218.

Faden, R. B. (2000). Commelina. Pages 192–197 in N. R. Morin, ed. Flora of North America, Vol. 22. New York: Oxford University Press.

Gibson, J.P. and A.D. Tomlinson (2002). Genetic diversity and mating system comparisons between ray and disc achene seed pools of the heterocarpic species *Heterotheca subaxillaris* (Asteraceae). *Int. J. Plant Sci.* 163:1025–1034.

Goddard, R.H., Theodore M. Webster, Richard Carter and Timothy L. Grey. (2009). Resistance of benghal dayflower (*Commelina benghalensis*) seeds to harsh environments and the implications for dispersal by mourning doves (*Zenaida macroura*) in Georgia, U.S.A. *Weed Sci.,* 57:603–612.

Kaul, V., A.K. Koul and M.C. Sharma. (2000). The underground flower. *Curr. Sci.* India 78: 39-44.

Kaul, V., N. Sharma, and A.K. Koul. (2002). Reproductive effort and sex allocation strategy in *Commelina benghalensis* L., a common monsoon weed. *Bot. J. Linnean Soc.*, 140:403–413.

Kim, S.Y., S.K. De Datta and B.L. Mercado. (1990). The effect of chemical and heat treatments on germination of *Commelina benghalensis* L. aerial seeds. *Weed Res.,* 30:109–116.

Kim, S. Y. (1998). Growth and development of *Commelina benghalensis* L. from four seed types. Kor. J. Weed Sci. 18:42–47.

Maheshwari, P. and J.K. Maheshwari. (1955). Floral dimorphism in *Commelina forskalaei* Vahl. and *C. benghalensis* L. *Phytomorphology* 5:413–422.

Maheshwari, P. and B. Singh. (1934). A preliminary note on the morphology of the aerial and underground flowers of *Commelina benghalensis* Linn. *Curr. Sci.,* India 3:158–160.

Mandeep Riar, Janet F. Spears, Joseph C. Burns, Theodore M. Webster, Danesha S. Carley and Thomas Ruay. (2010). Benghal dayflower (*Commelina benghalensis* L.) seed viability and impact on dispersal. North Carolina State University, Raleigh, NC.

Sabila and Mercy Hellen. (2008). Environmental factors affecting benghal dayflower (*Commelina benghalensis*) seed germination. University of Georgia Theses and Dissertations.

Santos, I.C., F.A. Ferreira, G.V. Miranda and L.D.T. Santos. (2001). Germination of aerial and underground seeds of *Commelina benghalensis*. *Planta Daninha*, 19:163–170.

Walker, S. R. and J. P. Evenson. (1985). Biology of *Commelina benghalensis* L. in south-eastern Queensland. 1. Growth, development and seed production. *Weed Res.* 25:239–244.

Webster, T.M. (2005). Weed survey—southern states: Broad leaf crops subsection. *Proceedings of the Southern Weed Science Society*, 58:291–306.

Webster, T.M. and T.L. Grey. (2008). Growth and reproduction of benghal dayflower (*Commelina benghalensis*) in response to drought stress. *Weed Sci.*, 56:561–566.

Zeide, B. (1978). Reproductive behavior of plants in time. *Am. Nat.,* 112:636–639.

FIELD BINDWEED (*Convolvulus arvensis* L.)

Common name : Field bindweed, morning glory flower, lesser bind weed

Botanical name : Convolvulus arvensis L.

Tamil name : Bhoomi chakra poondu

Hindi name : Hiranpug

Family : Convolvulaceae

Origin and distribution : Native to Europe and Asia. It is found wild throughout Europe, Siberia, China, Persia, India, north America, where it has been introduced and in Chile.

Botany

Habitat : Habitats include lawns, gardens, fields, clay banks, areas along roadsides and railroads (including ballast), vacant lots, and miscellaneous waste areas. This plant occurs primarily in disturbed areas. Common in red soils associated with horsegram.

Habit : It is a climbing or creeping herbaceous perennial plant growing to 0.5–2 m high (Fig. 1).

Stem : Trailing, twining, herbaceous, glabrous to pubescent, appearing 4-5 angled because of twisting of stem.

Leaf : The leaves are spirally arranged, linear to arrow head-shaped, 2–5 cm long and alternate, with a 1–3 cm petiole.

Floral biology

Flower : Solitary flowers or in cymes of 2-5, axillary, on peduncles (1-10 cm long), 1 –3 flowered; Flowers white to pale pink, funnel shaped approximately 1.9 – 2.5 cm across and subtended by small bracts, bracteate, bracts narrow; sepals 5, subequal, obtuse, glabrous, petals 5, fused, broadly funnel – shaped, limb plicate; stamens 5, included, filaments unequal; ovary superior, bicarpellary, syncarpous, 2 – celled; style filiform, stigma 2 (Fig. 2).

Fig. 1. Plant

Fig. 2. Flower

Flowering, pollination and fruit set : It flowers from May to September and insect-pollinated. It is self-incompatible, requiring pollinators. It also produces significant number of seeds (600/plant), which become viable 10-15 days after pollination. Thirty days after pollination, the seed matures. Flowering begins in late June and continues into the fall as long as conditions are favourable. An individual flower persists for one day only, beginning expansion and nectar production in early morning (Weaver and Riley, 1982). Observations in Lawrence, Kansas, determined that flowers were fully open by 8:00 a.m. during late June and early July. Nectar is produced at the base of the tube of fused petals, and attracts various pollinators including Halictid bees, honeybees, bumblebees, butterflies and moths. Halictid foraging is highest from 8:30-11:30 a.m (Waddington, 1976). Compared to other weed species found in Canada, field bindweed is intermediate in pollinator visitation (Mulligan and Kevan, 1973).

Fruit : Fruit are light brown, rounded and 1/8 in. (0.3 cm) wide. Each fruit contains 2 seeds (Fig. 3).

Seed biology

Seed : The dark seeds are 3-angled and oblong, but tapering somewhat at the ends. Each seed usually has 2 flat sides and 1 convex side; it is about 1/8" long (Fig. 4).

Propagation : Seeds, creeping roots and rhizomes.

Seed dispersal : Seeds fall near the parent plant (Rolston, 1978), but can be transported by water (Weaver and Riley, 1982) or birds (Proctor, 1968). Field bindweed seeds remain viable in the stomachs of migrating Killdeer for up to 144 hours, and can pass through animals with little or no damage (Callihan *et al.,* 1990). Quail may retain the seed for 24 hours, ducks, 5 geese, 19 lesser yellowlegs, 6 green jays, 13 ravens, 8 mocking birds, 2 and starlings 10 (Callihan *et al.,* 1990).

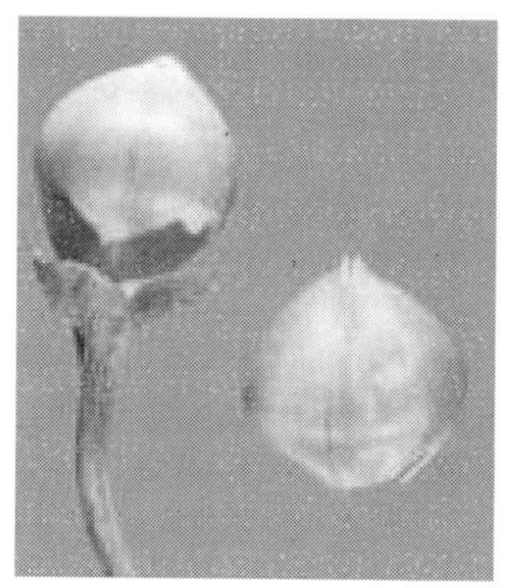

Fig. 3. Fruit

Fig. 4. Seed

Life cycle : Perennial

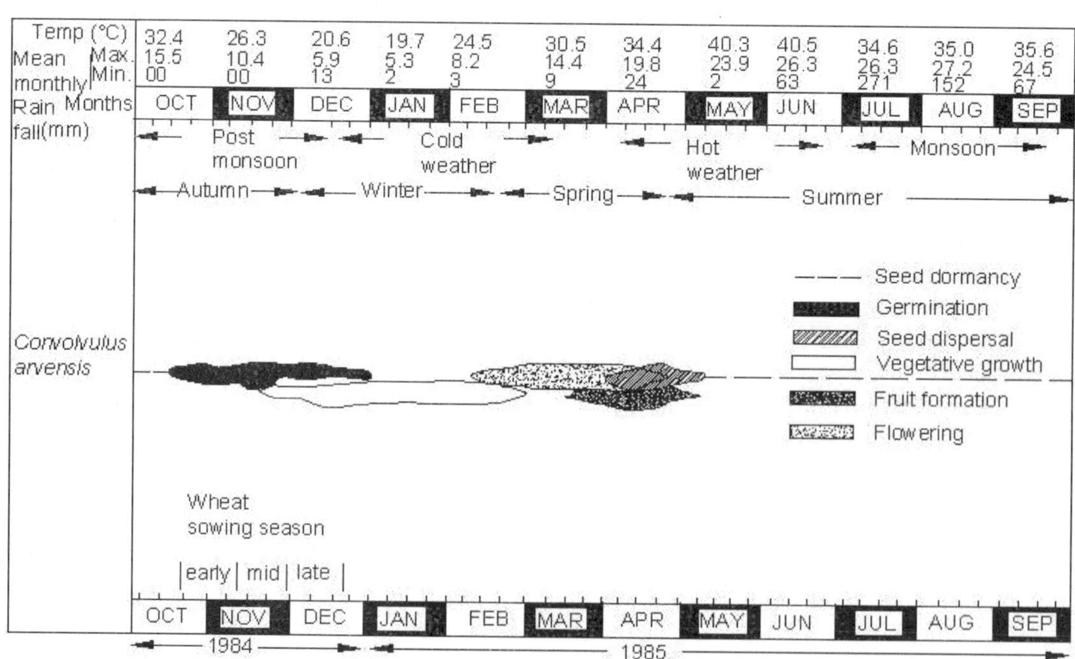

Fig. 5. Calendar of various phenological events of life-cycles of *C. arvensis* growing in irrigated wheat fields in villages around New University Campus. Lahore along with climatic data.

Seed dormancy and germination mechanism : Hard seededness is apparent in field bindweed (*Convolvulus arvensis* L.), (Cardina and Sparrow, 1997; Egley *et al.,* 1986; Riggio Bevilacqua *et al.,* 1987).

Field bindweed is extremely persistent as seed. The moisture content drops to 10-13% and they grow hard, dark, and their coats become impermeable (Sripleng and Smith, 1960; Brown and Porter, 1942). The seed coat is tough and can remain viable in the soil for years (Wiese and Phillips, 1976). Impermeable seeds must wait at least until the next season to germinate, and they may lie dormant in the soil for many years (Brown and Porter, 1942; Bellue *et al.,* 1959).

In order for a seed with a mature coat to germinate, its coat must become permeable to water. Seed coats can become permeable if exposed to adequate water, moist air, or fluctuating soil temperatures in the top 8cm of soil (Brown and Porter, 1942; Swan, 1980; Weaver and Riley, 1982). Under laboratory conditions optimal temperatures for germination of field bindweed ranged from 20-35°C, but seeds may also germinate at temperature as low as 10°C and as high as 40°C (Brown and Porter, 1942). Germination of fresh seed is not stimulated by light (Weaver and Riley, 1982). Permeability can be achieved artificially by treatments with concentrated sulphuric acid for 45-60 minutes (Brown and Porter, 1942; Rolston, 1978), ethyl alcohol for 20 hours (Rolston, 1978) or by mechanical scarification (Weaver and Riley, 1982).

Fresh seeds from Iowa, Kansas, Nebraska, S. Dakota, and California collected over five years differed in percent germination rate (5-47%), impermeability (31-97%), and viability (47-94%), depending on the year and collection site (Brown and Porter, 1942). Seeds sown at two-week intervals had various germination rates 60-75% (spring), 44-57% (early summer), and 1-10% (midsummer). Seedlings that germinated in August and September reached a soil depth of 25-50 cm by November, but they were unlikely to overwinter because their root systems were insufficiently established (Swan, 1980)

Some seeds germinate in autumn, most seedlings emerge in late spring and rapidly start to develop its vertical root system in the upper 30 cm of the soil. At the nodes and along the stems, field bindweed is able to produce a new plant from each bud. Once established, this weed is able to spread up to 3 m each year radially.

Seed storage and burial : Brown and Porter (1942) found that 65% of a carefully selected set of plump 55-year old dry-stored seeds were still germinable, and those buried deeply in the soil can persist nearly as long. Most herbicides are not absorbed by dormant seeds and are therefore ineffective at controlling them (Mulligan and Findlay, 1970; Swan, 1980). Over 95% of seeds possess this quality and field bindweed seeds are viable in soil for decades. It also retains its longevity in dung and compost, and also in silage. Seeds pass the digestive tracts of migrating birds unharmed that are eaten by birds and can remain viable in the soil for decades. It is difficult to eradicate because the seeds remain viable in soil for up to 20 years.

Effect on crop : Like other weeds, field bindweed takes nutrients and water that would otherwise be available to desirable species (Swan, 1980; Frazier, 1943). It can reduce the available soil moisture in the top 60cm of soil to below the wilting point for many species (Weaver and Riley, 1982). Overall, crop yields can be reduced to 50-60% where infestations are dense (Callihan *et al.*, 1990). Wiese *et al.,* (1996) found a significant increase in available soil water following increased control of field bindweed in an agricultural setting. Finally, exudates from field bindweed roots decrease germination of some crop seeds (Swan, 1980).

Control measure : Manual control: Discing, tilling or hand pulling,

Chemical control: Application of 2,4-D or glyphosate (Roundup) before seeds set controls the weed very effectively.

Uses : The root of field bindweed and also a resin made from the root, has agents that increase the flow of bile and its discharge from the body. It is also urine-inducing, laxative and strongly purgative.

Past work done on Seed Science and Technology

1. Germination requirements

Archana Kumari *et al.* (2010) recorded maximum germination of *C. arvensis* (41%) at 20°C which decreased with any increase or decrease from optimum temperature. *C. arvensis* had maximum germination with distilled water compared with salt solution. *C. arvensis* germination was 20%, at 200 mM NaCl conc. Osmotic potential of -0.8 MPa reduced the germination of *C. arvensis* to zero. Light was not pre-requisite for the germination of this weed species. Optimum depth for the germination of *C. arvensis* was 1.0 cm where corresponding germination was 40%. Reduction in germination and growth was recorded with increase and decrease from the optimum depth. Tolerance towards flooding was significant as *C. arvensis* tolerated 20 days of flooding.

2. Allelopathic effect

Mohamed Abdel Aziz Balah (2011) reported that germination and seedling length of *C. arvensis* inhibited under different concentration of *J. montana* (aerial parts) aqueous extracts (Table 1). Percentage germination and seedling length inhibition increased as the concentration of water extract increased. The highest applied concentration of 10 g DW 100 ml^{-1} completely inhibited germination of *C. arvensis.* The LD_{50} values of aqueous extracts were 3.75 g DW 100 ml^{-1} for shoot length, 2.8 g DW 100ml^{-1} for root length and 3.0 g DW 100ml^{-1} for germination in *C. arvensis.* Finally, at the highest concentration of 10 g DW 100ml^{-1}, aqueous extracts of *J. montana* decreased *T. aestivum* germination by 65.52% and root length by 85.16% and shoot length by 64.59% when compared to the untreated controls. Dose response curve of *J. montana* aqueous extracts showed that LD50 values in *T. aestivum* were 5.25, 3.22, and 6.5 g DW 100 $ml^{-1}$for shoot length, root length, and germination, respectively (Table 1).

Table 1. Allelopathic potential of *Jasonia montana aerial* parts aqueous extracts on germination and seedling length of the study species

Concentration (g dry weight/100 ml water)	Shoot length (cm)	Root length (cm)	Germination %
Wheat (*T. aestivum* L.)			
Control	15.53 ± 0.95	20.67 ± 1.53	86.7 ± 0.58
1.25	16.83 ± 2.02	12.33 ± 3.21	76.7 ± 0.58
2.50	15.33 ± 1.04	7.67 ± 1.53	80.0 ± 1.0
5.00	9.17 ± 0.58	3.67 ± 0.58	46.7 ± 0.58
10.00	5.50 ± 0.50	3.07 ± 0.90	36.7 ± 0.58
LSD (0.05)	4.374	7.814	33.21
Bindweed (*C. arvensis*)			
Control	7.67 ± 1.53	5.23 ± 1.17	76.7 ± 0.58
1.25	5.50 ± 0.50	2.50 ± 0.50	50.0 ± 0.0
2.50	5.00 ± 1.15	2.00 ± 0.58	26.7 ± 0.58
5.00	5.00 ± 0.58	1.20 ± 0.06	16.7 ± 0.58
10.00	0.00 ± 0.0	0.00 ± 0.0	0.00 ± 0.0
LSD (0.05)	2.481	2.779	4.865

3. Seed storage and burial

Leishman *et al.*, (2000) buried seed bags containing fungicide-treated and non-fungicide-treated seeds of *Convolvulus arvensis* L. with persistent seed banks under control and simulated climate change (winter warming plus supplemented summer rain) conditions, and monitored seed survival over 1 to 2 years. The only significant factor in the percentage of intact seeds was time ($F_{1,30}$ = 7·43, P = 0·01) where the percentage of intact seeds decreased with time (Fig. 6).

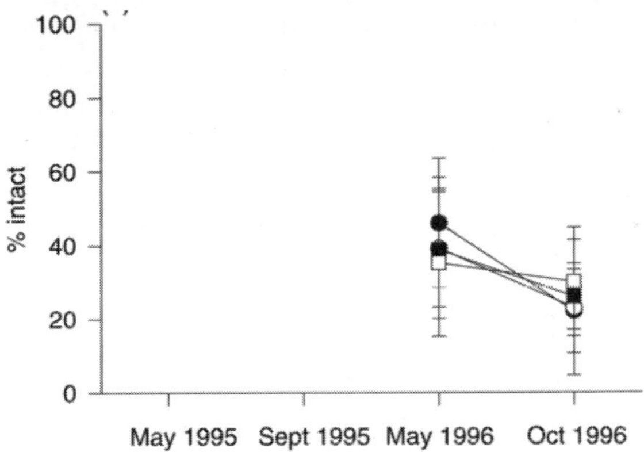

Fig. 6. The percentage of intact (viable) seeds remaining after burial in seed bags for *Convolvulus arvensis*. The symbols represent the two climate treatments: □ supplemented summer rainfall/winter warming; ○, control. Solid symbols represent fungicide-treated seeds, while open symbols represent non-fungicide-treated seeds.

However, both fungicide treatment ($F_{1,30}$ = 10·37, P = 0·003) and time ($F_{1,30}$ = 12·70, P = 0·001) affected the percentage germination of seeds, with higher percentage germination for fungicide treated seeds and increased germination over time (Fig. 7).

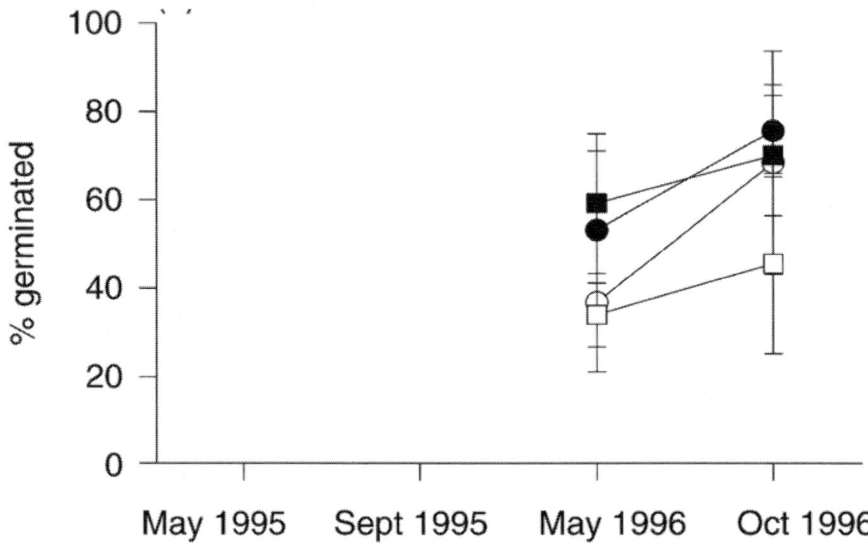

Fig. 7. The percentage of germinated seeds after burial in seed bags for *Convolvulus arvensis*. The symbols represent the two climate treatments: □, supplemented summer rainfall/winter warming; ○, control. Solid symbols represent fungicide-treated seeds, while open symbols represent non-fungicide-treated seeds.

REFERENCES

Archana Kumari, Kuldeep Singh, Anil Yadav and Samunder Singh. (2010). Factors affecting seed germination of *Convolvulus arvensis* and *Lathyrus aphaca*. *Indian J. Weed Sci.,* 42 (3 & 4): 203-211.

Bellue, M.K., E.A. Britton and T.C. Fuller. (1959). Weed seed handbook. State of California, Department of Agriculture, Sacramento, CA.

Brown, E.O. and R.H. Porter. (1942). The viability and germination of seeds of *Convolvulus arvensis* L. and other perennial weeds. Agricultural Exp. Station, Iowa State College, Research Bulletin, # 294.

Callihan, R.H., C.V. Eberlein, J.P. McCaffrey and D.C. Thill. (1990). Field bindweed: Biology and management. University of Idaho, Cooperative Extension System, College of Agriculture Bulletin, #719.

Cardina, J. and D.H. Sparrow. (1997). Temporal changes in velvetleaf (*Abutilon theophrasti*) seed dormancy. *Weed Sci.,* 45: 61–66.

Egley, G.H., R.N. Paul and A.R. Lax. (1986). Seed coat imposed dormancy: histochemistry of the region controlling onset of water entry into *Sida spinosa* L. *Physiol. Plant,* 67: 320–327.

Frazier, J.C. (1943). Food reserve depletion and synthesis in field bindweed, *Convolvulus arvensis* L., as related to 7-day and 14-day intervals of cultivation. *Plant Physiology,* 18(3): 315-323.

Leishman, M.R., G.J. Masters, I.P. Clarke and V.K. Brown. (2000). Seed bank dynamics: the role of fungal pathogens and climate change. *Functional Ecology,* 14: 293–299.

Mohamed Abdel Aziz Balah. (2011). Allelopathic characteristics of *Jasonia montana* for weeds control. *Australian Journal of Basic and Applied Sciences,* 5(11): 98-106.

Mulligan, G.A. and J.N. Findlay. (1970). Reproductive systems and colonization in Canadian weeds. *Canadian Journal of Botany,* 48(2): 859-860.

Mulligan, G.A. and P.G. Kevan. (1973). Color, brightness and other floral characteristics attracting insects to the blossoms of some Canadian weeds. *Canadian Journal of Botany,* 51:1939-1952.

Proctor, V.W. (1968). Long-distance dispersal of seeds by retention in digestive tract of birds. Science 160:321-322.

Riggio Bevilacqua, L.R., F. Fossati and G. Dondero. (1987). "Callose" in the impermeable seed coat of *Sesbania punicea.* Ann. Bot., 59: 335–341.

Rolston, M.P. (1978). Water-impermeable seed dormancy. Botanical Review, 44:365-396.

Sripleng, A. and F.H. Smith. (1960). Anatomy of the seed of *Convolvulus arvensis. American Journal of Botany,* 47(5): 386-392.

Swan, D.G. (1980). Field bindweed, *Convolvulus arvensis* L. Washington State University, College of Agriculture Research Center, Bulletin #0888.

Waddington, K.D. (1976). Foraging patterns of Halictid bees at flowers of *Convolvulus arvensis.* Psyche, 83:112-119.

Weaver, S.A. and W.R. Riley. (1982). The biology of Canadian weeds. 53. *Convolvulus arvensis* L. *Canadian Journal of Plant Science,* 62:461-472.

Wiese, A. F. and W. M. Phillips. (1976). Field bindweed. *Weeds Today,* 7: 22-23.

Wiese, A.F., C.D. Salisbury, B.W. Bean, M.G. Schoenhals and S. Amosson. (1996). Economic evaluation of field bindweed (*Convolvulus arvensis*) control in a winter wheat-fallow rotation. *Weed Science,* 44: 622-628.

RATTLEPOD (*Crotalaria pallida*)

Common name : Rattlepod, salts rattlebox, smooth crotalaria, smooth rattlebox, smooth rattlepod, streaked rattlepod, striped crotalaria

Botanical name : Crotalaria pallida Aiton (Syn : *C. striata* DC.)

Hindi name : San

Family : Fabaceae

Origin and distribution : C. pallida is probably a native of tropical Africa, But, its natural distribution is obscured by widespread cultivation and subsequent pan tropical naturalization. In Asia, it is common in India and Sri Lanka and throughout south-earth Asia.

Botany

Habitat : A common weed gregariously growing in waste and cultivated land.

Fig. 1. Plant

Habit : Perennial herbs or subshrubs (Fig. 1).

Stem : Stems ascending or erect, 10-20 cm long, striate, glabrous or appressed short-pubescent.

Leaf : Leaves 3-foliolate, to 8 cm; leaflets obovate, 3.5-6.5 x 2-4cm.

Floral biology

Flower : Flowers 9-15 mm long, 22-75 in terminal, somewhat lax inflorescences, rachis 15-40 cm long (incl. peduncle), bracts linear-triangular, 3-4 mm long, ascending, caducous, bracteoles at base of calyx, narrowly triangular; calyx densely appressed short-pubescent, the tube 3-4 mm long, the lobes 3.5-4.5 mm long, lateral lobes distinct at apex; petals yellow with lines near base, standard 9-12 mm long, 7-9 mm wide, striate, wings oblong-obovate, 6-10 mm long, 2.8-3.8 mm wide, keel petals striate, with nontwisted prolonged beak, margins ciliate; elongated anthers 2.5-3.2 mm long, short ones 0.5-0.8 mm long (Fig. 2).

Fig. 2. Flower

Flowering, pollination and fruit set : Flowering and fruiting throughout the year. This species is partially selfing and bee pollinated. Few insects visited the inflorescences. Legitimate pollinators are two large carpenter bees, *Xylocopa frontalis* and *Xylocopa grisescens*, which together, accounted for more than 90 per cent of the visits (Claudia *et al.,* 2005).

Fruit : Pod oblong, 4 x 0.7 cm, glabrescent, much exceeding calyx (Fig. 3).

Fig. 3. Fruit

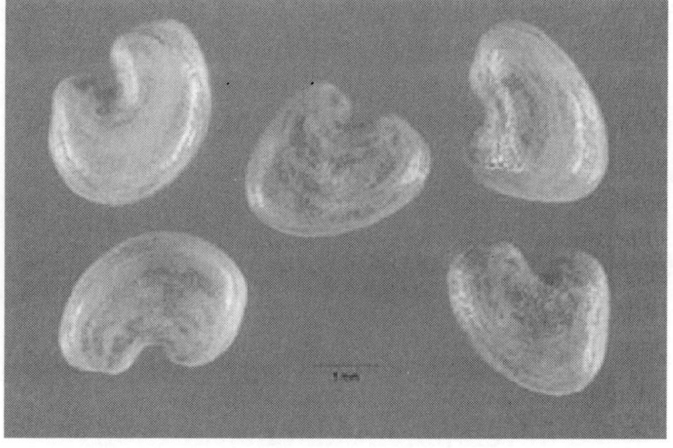

Fig. 4. Seed

Seed biology

Seed : Caryopsis, kidney-shaped (Fig. 4).

Seed dispersal : Seed predation by ants (Moore, 1975). *Crotalaria pallida* is an annual plant with limited dispersal and lacks any mechanism for long-distance seed dispersal. (Heywood, 1991; Linhart and Grant, 1996).

Life cycle : Annuals and shortest vegetative period of 3 months

Propagation : By seed

Seed dormancy and germination: Fresh seed of *C. pallid* has a higher germination rate than store seed. Germination was found to be improved by treatment with hot water.

Effect on crop : Alien weed species rank among the most important threats to conservation of biodiversity, making understanding the extent to which protected natural areas are vulnerable to invasion by weeds pivotal in long-term maintenance and conservation of biodiversity. It is susceptible to the root-knot nematodes *Meloidogyne incognita* and *M. javanica.* It is one of the alternative hosts of *Maruca testulalis,* which is a pest in cowpea. It also acts as a vector for the 'kette' virus affecting cardamom.

Control measure : Biological control: During dry periods, flea-beetles may cause severe defoliation. The bug *Ragnus importunatas* causes leaf fall at the beginning of the rainy season. Other insects causing damage are the beetle *Langitarsus* sp. and the caterpillar *Utetheisa lotrix.*

Chemical control: Pre-emergence atrazine plus fenac, post-emergence diuron plus surfactant, or ametryne, 2,4-D.

Uses : The plant is used as green manure, young seeds as food, roasted ripe seeds as a substitute for coffee, stem fiber as a material for thread. It is used as a ground cover and a green manure crop throughout the humid tropics, through on a limited scale.

It also has some value as a forage crop. However, its use is no longer recommended as the seed occasionally gets mixed into fodder grains, causing poisoning.

In Indo-China a kind of coffee is prepared from roasted seed. In Vietnam, the roots are sometimes chewed with betelnut.

In traditional medicine, *C. pallida* is used to treat urinary problems. A poultice made of the roots is applied to painful swelling of joints, and an extract of the leaves is taken as a vermifuge. In Laos, the plant is used to reduce fever.

Past work done on Seed Science and Technology

1. Seed development and maturation

Yang *et al.* (2004) observed that between 20 to 65 days after flowering (DAF), both fresh and dry weight of seeds increased gradually. The natural desiccation period occurred from 65 to 75 DAF. Moisture content of seeds decreased gradually during this period with significant maturation desiccation at the later stages. Seed germinability gradually reached up to a maximum from 25 to 70 DAF, but decreased somewhat thereafter. The leachate conductivity of seeds gradually decreased to a minimum from 25 to 70 DAF, but then increased at 75 DAF; leachate conductivity of immature seeds increased quickly with prolong of soaking time. During 30 to 60 DAF, seed desiccation tolerance increased and reached a maximum at 65-70 DAF, but decreased at 75 DAF. The lowest seed MC that the mature seeds could tolerate was about 10%. The responses to illumination and temperature on seed germination were independent interactive and dependent on developmental stages.

2. Seed dormancy and germination

Williams *et al.*, (2003) stated that exposure to elevated temperatures between 80 and 100°C significantly increased the germination. The treatments that produced the highest germination overall were exposure to 80°C water for 5 min. and 85°C dry heat for 45 min. No enhancement of germination by heat shock was observed (Fig. 5). *Crotalaria pallida* had a significantly higher percentage of non-dormant seeds. A tetrazolium test of the *Crotalaria pallida* seeds exposed to 120°C for 5 min. showed that none were viable, indicating that exposure to this temperature is lethal to the embryos. Exposure to smoke or nitrate in isolation did not significantly promote germination.

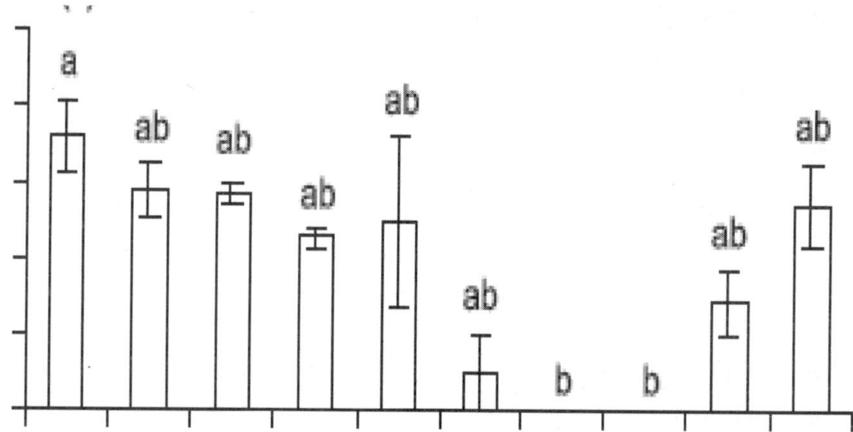

Fig. 5. Mean (±SE) percentage germination of viable seed in *Crotalaria pallida* after exposure to various levels of heat shock, as well as smoke and nitrate. Columns in each graph with the same lower-case letter are not significantly different at $P > 0.05$. All temperature exposures are for 5 min, except the 45 min duration of 85°C 45 min. 80°C/W, exposure to 80°C water for 5 min.

According to Kak *et al.* (2006) in *C. pallida,* soaking seeds in hot water at 60°C for 30 min. enhanced germination. Acid scarification was also found to be effective in breaking the dormancy, but had undesirable effects and resulted in maximum of abnormal seedlings and dead seeds. Subjecting seeds to dry heat and mechanical scarification was found ineffective in removing hardseededness.

Tauro *et al.*, (2007) expressed that sulphuric acid treatment significantly ($P < 0.05$) increased the germination by > 60% and reduced seed hardness to < 15%. Response to potassium nitrate pre-chilling and hot water treatment were insignificant.

REFERENCES

Claudia M. Jacobi, Mauro Ramalho and Maise Silva. (2005). Pollination Biology of the Exotic Rattleweed *Crotalaria retusa* L. (Fabaceae) in NE Brazil. Biotropica. ***37 (3):*** 357–363.

Heywood, J.S. (1991). Spatial analysis of genetic variation in plant populations. *Annual Review of Ecology and Systematics,* 22**:** 335–355.

Kak, A., Devi, C. & Gupta, T. (2006). Response of seed treatment on seed germination in wild crotalaria species. First international conference on indigenous vegetables and legumes. December 12 – 15, ICRISAT Campus, Patancheru, Hyderabad, India. 66.

Linhart, Y.B. and M.C. Grant. (1996). Evolutionary significance of local genetic differentiation in plants. *Annual Review of Ecology and Systematics,* 27: 237–277.

Tauro, T.P. Nezomba, H., F. Mtambanengwe and P. Mapfumo. (2007). Field emergence and establishment of indigenous N_2-fixing legumes for soil fertility restoration. *African Crop Science Conference Proceedings,* 8: 1929-1935.

Williams, P.R., R.A. Congdon, A.C. Grice and P.J. Clarke. (2003). Fire related cues break seed dormancy of six legumes of tropical eucalypt savannas in northeastern Australia. *Austral. Ecology.* 28, 507–514.

Yang, Q.H., S.H. Yin, S.Q. Song and W.H. Ye. (2004). Development of desiccation tolerance and germination physiology of *Crotalaria pallida* ait seeds.

DODDER (*Cuscuta* spp L.)

Common name : Dodder

Botanical name : Cuscuta spp.

Tamil name : Thanga kodi

Hindi name : Amar bel

Family : Cuscutaceae

Origin and distribution : Native to U.S; Worldwide distribution.

Botany

Habitat : Many found in low wet areas and pond margins, others found in prairies, glades, thickets, and cultivated ground.

Habit : Dodder is a parasitic vine with smooth, wiry, twining stems that attach to a host plant with tiny suckers. Only the seedlings have roots (Fig. 1).

Stem : Typically greenish-yellow to orange, thin, herbaceous, twining, glabrous, typically achlorophyllous or with sparse chlorophyll, attaching to host with haustoria.

Leaf : Absent or scalelike and alternate.

Floral biology

Flower : Typically loose to compact cymose globular clusters (but very dense in a couple of species). Dodder flowers are numerous, tiny, whitish to pinkish and form in small clusters along the stems. Flowers sessile or on pedicillate. Corolla lobes acute to obtuse or rounded, spreading to reflexed or erect. Stamens typically with scale like appendages. Appendages often fimbriate. Styles typically 2, sometimes single or united. Ovary globose, 2-locular (Fig. 2).

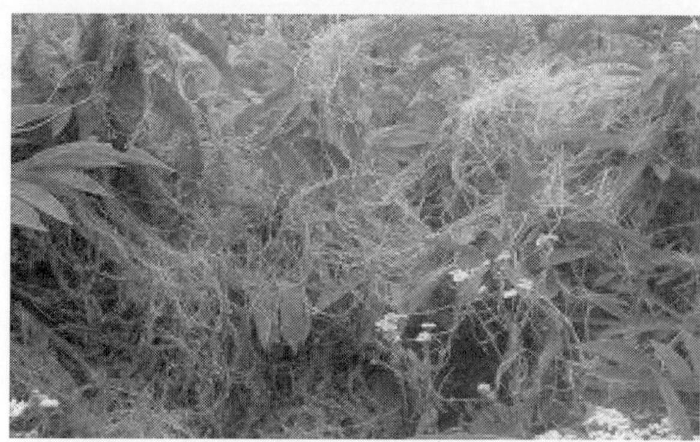

Fig. 1. Plant

Fig. 2. Flower

Flowering : Flowering takes place from May to October, depending on the species and location.

Fruit : Fruit typically a globose circumsissle capsule with 2 seeds per locule (Fig. 3).

Fig. 3. Fruit

Fig. 4. Seed

Seed biology

Seed : The seeds are minute and produced in large quantities. They have a hard coating, and can survive in the soil for 5–10 years or more (Fig. 4).

Seed dispersal : A single plant may produce thousands of seeds. Dodder seeds are often inadvertently harvested, distributed and planted along with host plant seeds. They may also be spread in hay, soil, water, or animal manure

Life cycle : 1 year

Propagation : Dodder reproduces readily from seed. Broken stems will readily attach to new hosts.

Seed dormancy and germination : Dodder seeds sprout at or near the surface of the soil. While dodder germination can occur without a host, it has to reach a green plant quickly; dodder grows toward the smell of nearby plants.

If a plant is not reached within 5 to 10 days of germination, the dodder seedling will die. Before a host plant is reached, the dodder, as other plants, relies on food reserves in the embryo; the cotyledons, though present, are vestigial. Soaking the seeds for 30 minutes in concentrated sulfuric acid will ensure high germination. In *Cuscuta* spp. seeds do not need any stimulation by the host. Germination occurs (permeability of seed testa provided) when temperature and moisture conditions are favorable, even immediately after harvest. Upon germination, the epicotyl (3-6 cm long) searches for a host by circumnutation; it is even able to move for some distance in order to reach a host and to establish contact to it.

Cuscuta seed therefore, may stay dormant for several years upon the testa is rotted; on the other hand, part of the seeds may germinate directly after maturity on the mother plant. Before the seeds of parasites will respond to germination stimulants, they require a period of conditioning subsequent to afterripening. Seed coat permeability may increase during conditioning and/or changes in the levels or activities of endogenous germination promotors or inhibitors may occur. The duration of conditioning is dependent on certain factors like temperature, origin, and age of the seeds. The period of time necessary for conditioning ranges from several days up to several weeks. Prolonged conditioning without subsequent germination induction by stimulants can cause secondary dormancy.

Storage : The seeds have rough coats and vary in size, depending on the species, and may be able to survive over 20 years in the soil.

Effect on crop : After a dodder attaches itself to a plant, it wraps itself around it. If the host contains food beneficial to dodder, the dodder produces haustoria that insert themselves into the vascular system of the host. The original root of the dodder in the soil then dies. The dodder can grow and attach itself to multiple plants. In tropical areas, it can grow more or less continuously, and may reach high into the canopy of shrubs and trees; in temperate regions it is an annual plant and is restricted to relatively low vegetation that can be reached by new seedlings each spring.

Dodder is parasitic on a very wide variety of plants, including a number of agricultural and horticultural crop species, such as alfalfa, lespedeza, flax, clover, potatoes, chrysanthemum, dahlia, hlenium, trumpet vine, ivy and petunias, among others. Dodder ranges in severity based on its species and the species of the host, the time of attack, and whether any viruses are also present in the host plant. By debilitating the host plant, dodder decreases the ability of plants to resist virus diseases, and dodder can also spread plant diseases from one host to another, if it is attached to more than one plant.

Control : Many countries have laws prohibiting import of dodder seed, requiring crop seeds to be free of dodder seed contamination. Before planting, all clothes should be inspected for dodder seed when moving from an infested area to a non-infested crop. When dealing with an infested area, swift action is necessary. Recommendations include planting a non-host crop for several years after the infestation, pulling up host crops immediately, particularly before the dodder produces seed and use of preemergent herbicides like Dacthal in the spring. Examples of non-host crops include grasses and many other monocotyledons. If dodder is found before it chokes a host plant, it may be simply removed from the soil. If choking has begun, the host plant must be pruned significantly lower than the dodder, as dodder is versatile and can grow back, if present from haustoria

Uses : The petroleum ether extract of *C. reflexa* may be useful in treatment of androgen induced alopecia by inhibiting the enzyme 5 alpha-reductase. *C. Chinensis* seeds have long been used for the treatment of osteoporosis in China and some Asian countries. Kaempferol and hyperoside may be the active compounds that have osteogenic effect. *C. Chinensis* is a commonly used traditional Chinese medicine to nourish the liver and kidney. Various dodder species have been used to treat a great variety of different ailments in numerous cultures. Some are thought to take on and enhance the medicinal properties of the plants they parasitize. The dodder that grows on thyme (*Cuscuta epithymum*) has been a favorite in European folk medicine.

Past work done on Seed Science and Technology

Seed dormancy and germination

Hutchison and Ashton (1979) used petroleum gel to block the putative area of the water gap in seeds of *Cuscuta campestris* and concluded that nondormant (made nondormant by treating seeds with conc. H_2SO_4) *C. campestris* seeds take up water throughout the entire seed coat. They observed damaged papillae in the seed coat, and thus concluded that water enters the seed through these structures. Lyshede (1984, 1992) compared electron micrographs of nondormant seeds of *C. pedicellata* and of dormant and nondormant (made nondor-mant by percussion) seeds of *C. campestris* and also concluded that papillae on the seed surface are responsible for water uptake by nondormant seeds of Cuscuta. However, papillae are formed superficially from bulging cells of the epidermal layer (Hamed and Mourad, 1994), which lies above the palisade layer that is normally considered to be the water impermeable layer in water-impermeable seeds.

Extensive studies have been done on seed anatomy and embryology of Cuscuta (Tiagi, 1951; Johri and Tiagi, 1952; Teryokhin and Nikiticheva, 1982; Sampathkumar and Ayyangar, 1982; Hamed and Mourad, 1994). However, most of these studies concentrated on taxonomy. Hamed and Mourad (1994) described seed coat anatomy and surface morphology of the seed coat of *C. gronovii* and *C. chinensis* away from the hilum. Johri and Tiagi (1952) described seed development, including the hilum, of *C. reflexa* but from a taxonomic aspect. There have been no studies focusing on the anatomy or morphology of the water gap of any Cuscuta species.

Cycling of dormancy is a common phenomenon in seeds with PD (Baskin and Baskin, 1985), and there are claims that dormancy cycling can also occur in seeds with PY (Rolston, 1978; Norsworthy and Oliviera, 2007). However, seeds with PY become nondormant by formation of an opening in a specific morpho-anatomical area in the seed coat referred to as the water gap. Once it opens, resealing of this structure would not seem to be possible. To explain the cyclic germinability patterns of physically dormant seeds of some legumes (Taylor, 1981, 2005; Taylor and Revell, 1999; Van Assche *et al.,* 2003) and *Ipomoea lacunosa* (Jayasuriya *et al.,* 2008), a sensitivity cycling model has been proposed in which seeds cycle between sensitive and insensitive stages. In the sensitive stage, seeds respond to dormancy-breaking treatments, and in the insensitive stage, they fail to respond to dormancy-breaking treatments. Sensitivity cycling has been documented in only two of 16 families that produce seeds with PY: Convolvulaceae (Tribe : Ipomoeeae) and Fabaceae (Subfamily : Papilionoideae).

Gaertner (1950) studied seed germination of 16 species of Cuscuta and came to the conclusion that a water impermeable seed coat (along with PD in some species) was responsible for dormancy in all of them. However, germination of freshly collected seeds of *C. epilinum* and *C. epithymum* was not facilitated by scarification with concentrated sulfuric acid, which worked well for the other 14 species. Meulebrouck *et al.* (2008) found that scarified seeds of *C. epithymum* require a period of cold stratification to break physiological dormancy (PD) of the embryo. Thus, they concluded that seeds of this species have combinational dormancy (PY+PD). Germination of *C. europea* seeds was not facilitated by mechanical scarification or by complete removal of the seed coat (Gaertner, 1956), and a significant percentage of these seeds germinated only after they were cold stratified at –88°C and 58°C. This suggests that seeds of *C. europea* also have combinational dormancy. Tingey and Allred (1961) clearly showed that seeds of *C. approximata* have combinational dormancy since scarified seeds required 2 weeks stratification at (58°C) to germinate. PY has been reported for seeds of several Cuscuta species (Drummitt, 1946): *C. campestris* (Hutchison and Ashton, 1980; Lados, 1999; Benvenuti *et al.,* 2005), *C. trifolii* (Lados, 1999), *C. monogyna* and *C. planiflora* (Salimi and Shahraeen, 2000), *C. chinensis* (Marambe *et al.,* 2002), *C. gronovii, C. umbrosa, C. epithymum* and *C. epilinum* (Costea and Tardif, 2006). However, the percentages of hard seeds at dispersal can vary from plant to plant [e.g. *C. campestris* (Hutchison and Ashton 1979) and *C. chinensis* (Marambe *et al.*, 2002)]. Lyshede (1984) claimed that seeds of *C. pedicellata* were nondormant at the time of dispersal; however, she did not present any data to support her conclusion.

Hutchison and Ashton (1979) blocked the hilum area of nondormant seeds (made nondormant by treating with conc. H_2SO_4) of *C. campestris* with petroleum gel and observed no difference in germination between blocked and nontreated seeds. Hence, they concluded that nondormant seeds take up water throughout the entire seed coat. Further, these authors observed damaged papillae on seed coats of nondormant seeds under the scanning electron microscope and suggested that these were sites of water entry into the seeds. They also stated that the impermeable layer in the seed coat of this species is located above the light line. Lyshede (1984) studied nondormant seeds of *C. pedicellata* and compared them with nondormant and dormant seeds of *C. campestris*. She also came to the conclusion that nondormant seeds take up water through the entire seed coat. Further, Lyshede (1984) suggested that swollen papillae in nondormant seeds of *C. pedicellata* and swollen and damaged papillae in those of *C. campestris* are responsible for water uptake. Benvenuti *et al.* (2005) reported cycling of dormancy in initially physically dormant seeds of *C. campestris*. Their interpretation of dormancy cycling in this species seems to be that seeds cycle between PD and nondormancy (PD/$ND) after breaking of PY. Thus, the two annual cycles of dormancy reported for seeds of *C. campestris* would be as follows: PY (fresh seeds in autumn) → ND (1st spring) → PD (2nd autumn) → ND (2nd spring) → PD (3rd autumn).

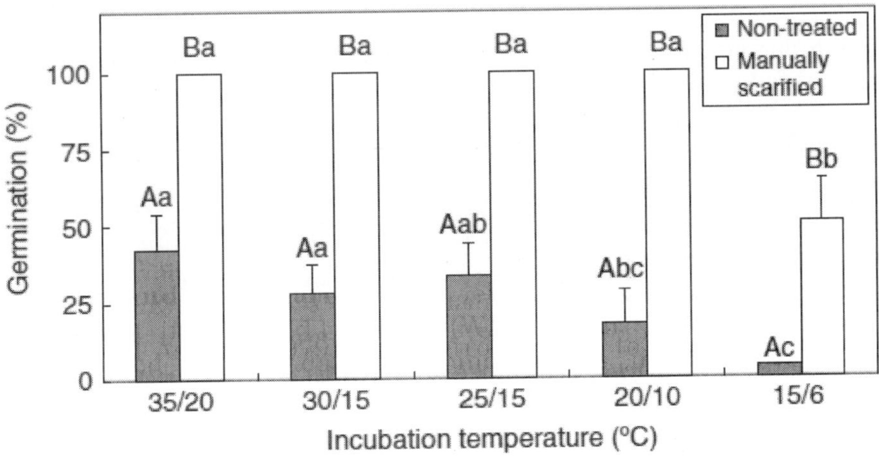

Fig. 5. Germination of manually scarified and of non-treated *C. australis* seeds in light/dark after 30 d incubation at five temperature regimes. Different upper-case letters indicate significant differences between treatments within same temperature regimes and different lower-case letters significant differences between temperature regimes within the same treatment. UT, Non-treated; MS, manually scarified. Bars are +1 s.e.

Jayasuriya *et al.* (2008) reported that the mass of manually scarified seeds increased >175% in 52 h, whereas that of non-treated (intact) seeds increased by only about 40%. Thus, it is concluded that a high percentage (78%) of fresh seeds are physically dormant. Manually scarified seeds germinated to 100% within 30 d in all temperature regimes except 15/6°C in light/dark (Fig. 6) and 15/6°C and 20/10°C in dark. Non-treated seeds germinated >30% at 35/20, 30/15 and 25/15°C and < 25% at 20/10 and 15/ 6°C in both light / dark (Fig. 5).

There is a significant 2nd degree polynomial correlation between swelling (release from PY) and maximum incubation temperature in both light/dark and dark (Fig. 7).

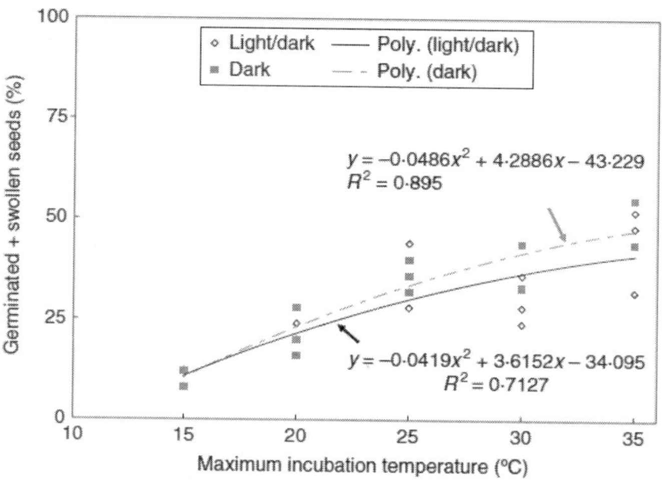

Fig. 6. Correlation between swollen seeds (germinated þ swollen) and maximum incubation temperature in light/dark and in dark.

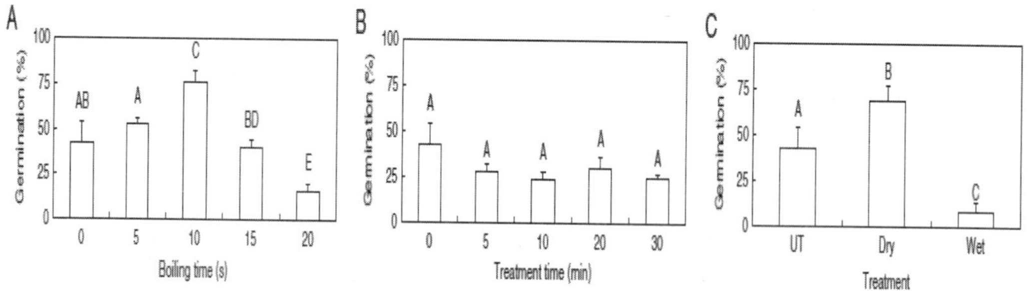

Fig.7. Germination (¼ imbibition) of *C. australis* seeds at 35/20 8°C following (A) boiling, (B) dry-heat treatment and (C) dry and wet storage at ambient laboratory temperature for 2 months. Different upper-case letters indicate significant differences between treatments. UT, Non-treated. Bars are +1 s.e.

Seeds germinated >75% at 35/20 °C (Fig. 7A) and at 25/15° C after dipping in boiling water for 10 s. Germination percentage was lower after dipping seeds in boiling water for 15 and 20 s. Seeds exposed to 90°C dry heat for 5–30 min germinated <30% (Fig. 7B) at 35/20°C. Seeds stored dry at ambient laboratory temperature for 2 months germinated >65%, whereas wet-stored seeds germinated to only 8.9 % (Fig. 7C). Thus, high temperature acts as the dormancy breaking treatment.

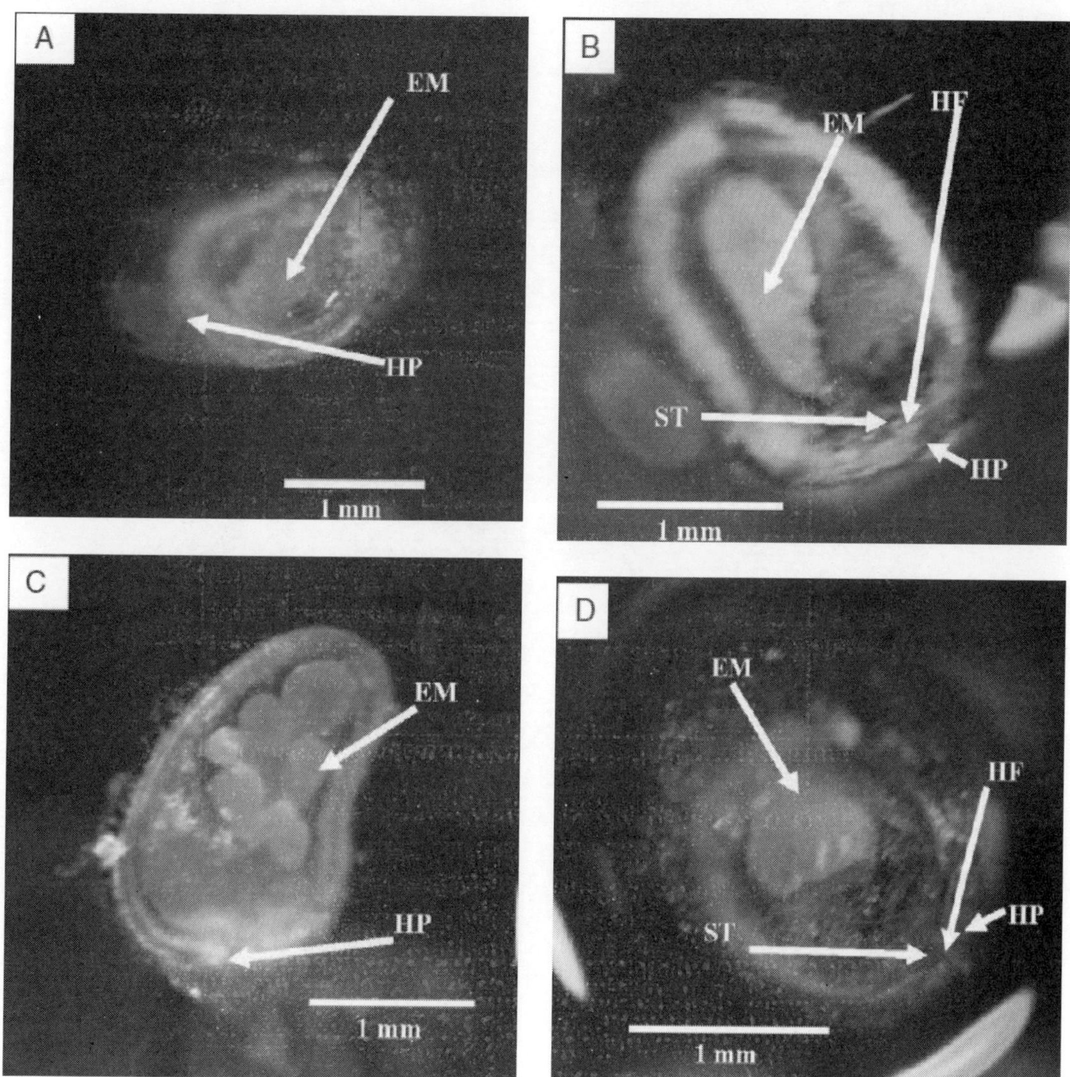

Fig. 8. Photographs of (A) dormant seed, (B) nondormant seed, (C) hilum-painted nondormant seed, each of which was immersed in aniline blue for 1 h, and (D) nondormant seed immersed in aniline blue for 2 h. EM, Embryo; HF, hilar fissure; HP, hilar pad; ST, stain.

No stain had penetrated the seed coat of dormant seeds after 1 h (Fig. 8), whereas stain was observed in tissues under the hilar fissure in nondormant seeds soaked for 1 h in aniline blue (Fig. 8B). No staining could be seen in tissues under the seed coat away from the hilar fissure. After soaking for 2 h (Fig. 8D) in aniline blue, stain was observed throughout the embryo and seed coat of nondormant seeds, whereas in dormant seeds and in hilum-painted nondormant seeds (Fig. 8C) no stain was observed.one dormancy breaking treatments caused the highly closed hilum tissue to open. In the centre of the hilar pad, a linear-shaped fissure can be seen (Fig. 9.E, F).

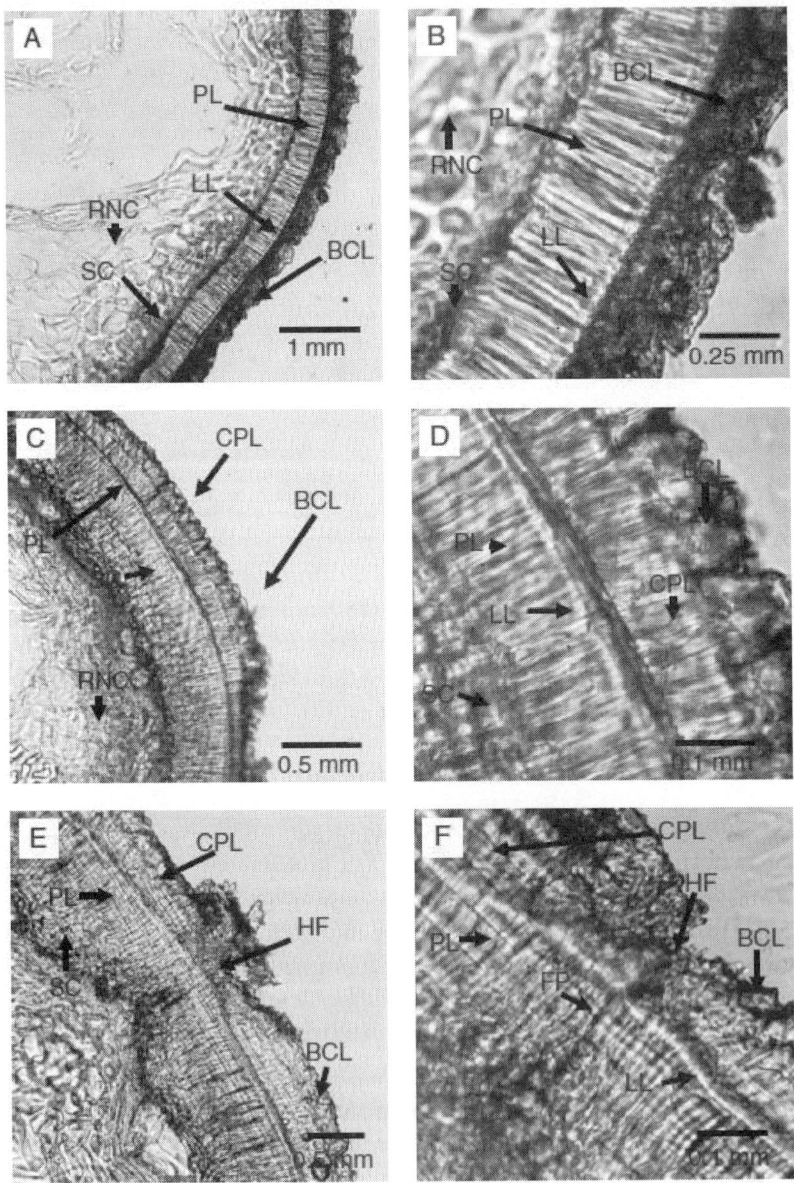

Fig. 9. Cross-sections through seeds of *C. australis* showing (A and B) seed coat away from the hilar pad; (C and D) hilar pad; and (E and F) hilar fissure. BCL, Bulging cell layer; CPL, counter palisade; FP, fissure through palisade layer; HF, hilar fissure; LL, light line; PL, palisad layer; RNC, remnants of nucellar cells; SC, sclereid cells.

Seed stroage

According to Jayasuriya *et al.* (2008), germination and imbibition percentages at 35/20°C of seeds stored at ambient laboratory temperature increased gradually to >70% after 3 months of storage, compared with 42% in non-treated control. However, the germination then declined rapidly from >70% to 42% after 6 months of dry storage. After 2 months of dry storage at ambient laboratory temperature, 65% of the seeds germinated at 35/20°C within 30 d, but after further dry storage at 35/20°C for 1 and 2 weeks germination percentage had decreased to 48 and 32, respectively (Fig. 10A). However, germination of seeds stored wet at 35/20°C for 1 and 2 weeks was >65 % and 72%, respectively. Germination of non-treated seeds at 25/15°C was 33%, and during 2 months dry storage at ambient laboratory temperature it decreased to 20%. Germination did not change significantly after subsequent dry storage at 35/20°C (Fig. 10B), whereas germination of seeds stored wet at 35/20°C for 1 and 2 weeks increased again to 38% and 44%, respectively.

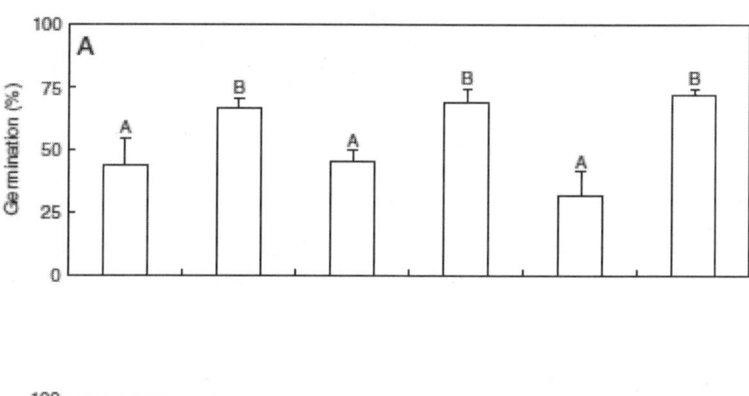

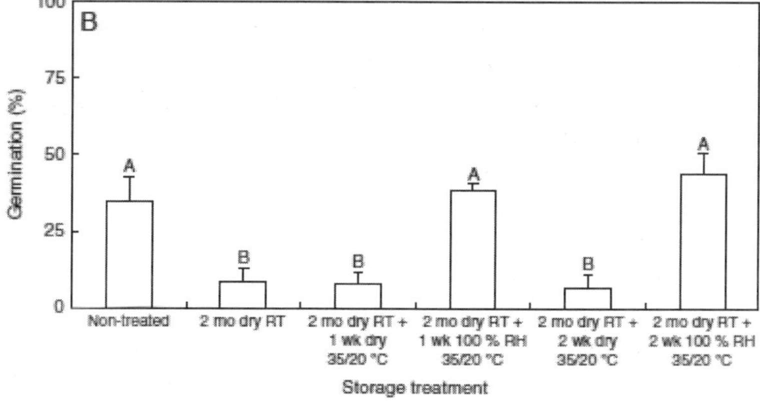

Fig. 10. Effect of wet vs. dry storage at 35/20°C on germination (= imbibition) of sensitive seeds of *C. australis* at (A) 35/20°C and (B) 25/15°C. Seeds were made sensitive by storing them dry at ambient laboratory temperature for 2 months, after which they were stored wet to dry at 35/20°C for 1 and 2 weeks. Different upper-case letters indicate significant different between treatments. RT, Ambient laboratory temperature. Bars are ±1 s.e.

Seed burial

Jayasuriya *et al.* (2008) reported that more than 5% of seeds in both wet and dry/wet treatments had swollen after 1 month of burial. As temperatures increased in the non heated glasshouse in spring, so did the percentage of swollen seeds in both wet and dry/wet treatments. However, swelling percentage of the dry/wet treatment was significantly higher than that of the wet treatment from June to September 2007 (Fig. 11).

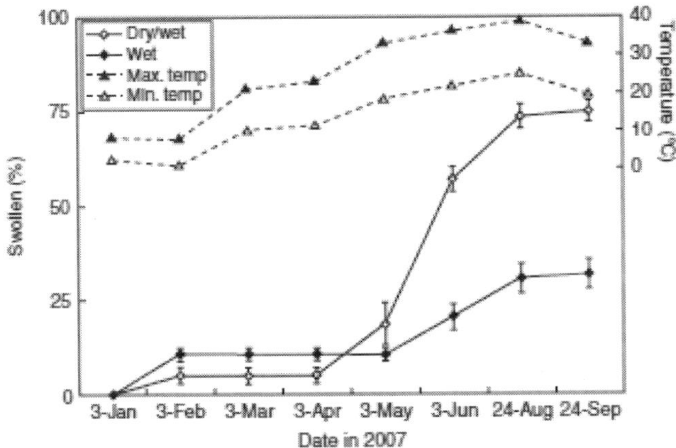

Fig. 11. Swelling of *C. australis* seeds buried on continuously wet and in alternately wet/dry soil in a nonheated glasshouse. Mean daily minimum and maximum air temperatures are shown for each month of the study. Bars are ± 1 s.e.

REFERENCES

Baskin, J.M. and C.C. Baskin. (1985). The annual dormancy cycle in buried weed seeds: a continuum. *BioScience,* 35: 492–498.

Benvenuti, S., G. Dinelli, A. Bonetti and P. Catizone. (2005). Germination ecology, emergence and host detection in *Cuscuta campestris. Weed Research,* 45: 270–278.

Costea, M. and F.J. Tardif. (2006). The biology of Canadian weeds. 133. *Cuscuta campestris* Yuncker, *C. gronovii* Willd. ex Schult., *C. umbrosa* Beyr. ex Hook., *C. epithymum* (L.) L. and *C. epilinum* Weihe. *Canadian Journal of Plant Science,* 86: 293–316.

Drummitt, M. (1946). The germination of dodder seed occurring in lespedeza. *Proceedings of the Association of Official Seed Analysts,* 36: 125–131.

Gaertner, E.E. (1950). Studies of seed germination, seed identification, and host relationships in dodders, *Cuscuta* spp. Memoirs of the Cornell University Agricultural Experiment Station No. 294, 3–56.

Gaertner, E.E. (1956). Dormancy in the seeds of *Cuscuta europea. Ecology* 32: 389.

Hamed, K.A., M.M. Mourad. (1994). Seed exomorphic and anatomical characters of some species of Convolvulaceae. *Egyptian Journal of Botany* 34: 1–16.

Hutchison J.M. and F.M. Ashton. (1979). Effect of desiccation and scarification on the permeability and structure of the seed coat of *Cuscuta campestris. American Journal of Botany* 66: 40–46.

Hutchison, J.M. and F.M. Ashton. (1980). Germination of field dodder (*Cuscuta campestris*). *Weed Science* 28: 330–333.

Jayasuriya, K.M.G.G, J.M. Baskin and C.C. Baskin. (2008). Cycling of sensitivity to physical dormancy-break in seeds of *Ipomoea lacunose* (Convolvulaceae) and ecological significance. *Annals of Botany,* 101: 341–352.

Jayasuriya, K.M.G.G., J.M. Baskin, R.L. Geneve, C.C. Baskin and Ching-Te Chien. (2008). Physical dormancy in seeds of the holoparasitic angiosperm *Cuscuta australis* (convolvulaceae, cuscuteae): dormancy-breaking requirements, anatomy of the water gap and sensitivity cycling. *Annals of Botany,* 102: 39–48.

Johri, M.B. and B. Tiagi. (1952). Floral morphology and seed formation in *Cuscuta reflexa. Phytomorphology,* 2: 162–180.

Lados, M. (1999). Effect of temperature, pH and host plant extract on the germination of *Cuscuta trifolii* and *C. campestris* seeds. *Novenytermeles* 48: 367–376.

Lyshede, O.B. (1984). Seed structure and germination in *Cuscuta pedicellata* with some notes on *C. campestris. Nordic Journal of Botany,* 4: 669–674.

Lyshede, O.B. (1992). Studies on mature seeds of *Cuscuta pedicellata* and *C. campestris* by electron microscopy. *Annals of Botany,* 69: 365–371.

Marambe, B., S. Wijesundara, K. Tennekoon, D. Pindeniya and C. Jayasinghe. (2002). Growth and development of *Cuscuta chinensis* Lam. and its impact on selected crops. *Weed Biology and Management,* 2: 79–83.

Meulebrouck, K., E. Ameloot, J.A. Van Assche, K. Verheyen, M. Hermy and C.C. Baskin. (2008). Germination ecology of the holoparasite *Cuscuta epithymum. Seed Science Research,* 18, 25–34.

Norsworthy, J.K. and M.J. Oliveira. (2007). Role of light quality and temperature on pitted morning glory (*Ipomoea lacunosa*) germination with afterripening. *Weed Science* 55: 111–118.

Rolston, M.P. (1978). Water impermeable seed dormancy. *The Botanical Review,* 44: 365–390.

Salimi, H. and N. Shahraeen. (2000). Study on comparison of seed dormancy and germination of three species of dodder. *Rostaniha,* 1: 33–36.

Sampathkumar, R. and K.R. Ayyangar. (1982). Seed morpho-anatomy of Convolvulaceae. *Journal of the Indian Botanical Society,* 61: 346–354.

Taylor, G.B. (1981). Effect of constant temperature treatments followed by fluctuating temperatures on the softening of hard seeds of *Trifolium subterraneum* L. *Australian Journal of Plant Physiology,* 8: 547–558.

Taylor, G. B. (2005). Hardseededness in Mediterranean annual pasture legumes in Australia: a review. *Australian Journal of Agricultural Research,* 56: 645–661.

Taylor, G.B. and C.K. Revell. (1999). Effect of pod burial, light, and temperature on seed softening in yellow serradella. *Australian Journal of Agricultural Research,* 50: 1203–1209.

Teryokhin, E.S. and Z.I. Nikiticheva. (1982). Biology and evolution of embryo and endosperm in parasitic flowering plants. *Phytomorphology,* 32: 335–339.

Tiagi, D.C. (1951). Contribution to the morphology and embryology of *Cuscuta hyaline* Roth and *C. planiflora* Lenore. *Phytomorphology,* 1: 9–21.

Tingey D.C. and K.R. Allred. (1961). Breaking dormancy in seeds of Cuscuta approximata. Weeds 9: 429–436.

Van Assche, J.A., L.A. Debucquoy and A.F. Rommens. (2003). Seasonal cycles in the germination capacity of buried seeds of some Leguminosae (Fabaceae). *New Phytologist,* 158: 315–323.

BERMUDA GRASS (*Cynodon dactylon*)

Common name : Bermuda grass, devil's grass, couch grass, wire grass, Indian doab

Botanical name : Cynodon dactylon Pers.

Tamil name : Arugampullu

Hindi name : Doop, durva

Family : Graminae (Poaceae)

Origin and distribution : Native to East Africa where it is widely distributed from sea level to 2,160 m altitude. It can now be found throughout the World in temperate and subtropical regions.

Botany

Habitat : It is a highly nutritious fodder grass and is extensively used as lawn grass. In black soils and in dryland. It is a noxious weed and once a field is infested with it, cultivation becomes impossible. It is considered as one of the ten worst weeds in the World and problematic in all tropical and subtropical areas in dry land and garden lands.

Habit : Terrestrial, annual or perennial, tufted herb (Fig. 1).

Stem : Creeping, to 40cm long, glabrous, terete, mat-forming, from rhizomes and stolons.

Leaf : Leaf blades to 8cm long, 4mm broad, flat to slightly keeled, pubescent toward the base. Sheaths slightly keeled, pubescent at the apex and along the apical margins. Ligule a short membrane with a ciliate apex.

Floral biology

Inflorescence - Palmate arrangement of 2-6 spikes. Spikes 8cm long, with many spikelets. Spikelets appressed into 2 rows along the axis.

Fig. 1. Plant

Flowers - Spikelets with 1 perfect floret. Glumes to 2mm long, lanceolate, acute at the apex. Lower glume slightly smaller than the upper. Lemmas 2.8mm long, 3-nerved, acute, pubescent or scabrous on the nerves. Anthers 1.3mm long, tan to yellow. Styles purple (Fig. 2).

Fig. 2. Inflorescence and flower

Flowering and pollination : Flowers throughout the growing season . Wind pollinated, also some insects including bees, harvest pollen.

Fruit : A nut

Seed biology

Seed : The seed heads are on 1-3 inch (3-7 cm) spikes and are themselves about 2 inches long. Seeds are ovoid, about 1.5 mm (1/16 inch) long, and yellow to reddish (Fig. 3).

Fig. 3. Seed

Propagation : Spreading mainly by below-ground rhizomes and above-ground stolons. It can also be spread by some extent from seed, but germination was observed to be very low.

Life cycle : In the spring, when the temperature begins to increase new stolons elongate and aerial shoots sprout (Horowitz, 1972a). The characteristic prostrate growth of bermuda grass lasts for one to several months, early in the season, before flowering culms develop. Most of the lateral growth, produced in concentric circles from the original rhizome, occurs throughout the summer (Horowitz, 1972 a,b; Holm *et al.*, 1977). Sprouts between three days to a week from planting,

Seed dispersal : Seeds are dispersed by water (including irrigation), soil movement, agricultural and landscape machinery, as a commercial seed contaminant, in livestock feed and bedding, human activities and in ship ballast and packing materials.

Seed dormancy and germination : Seed viability of six variants of *Cynodon dactylon* tested from southern Africa ranged from 0-3.5% (Fuls and Bosch, 1990 in Carey, 1995). Anderson (1999) reported that *Cynodon dactylon* seeds germinate when daily temperatures are above 65°F (18°C).

In California, *Cynodon dactylon* emerged when soil temperatures reached 63°F (17°C) at a depth of 5 cm (Keeley and Thullen, 1989 in Carey 1995).

Scarification promoted the germination of viable seeds; seeds treated with sulfuric acid for 10 min had 68% germination after 4 days; untreated seeds had 4.5% germination (Bryan, 1918 in Carey, 1995).

Seeds passing through the digestive tract of cattle and sheep retain viability, and possibly show improved germinability (Anderson, 1999; Holm *et al.*, 1991).

Heady (1954 *in* Carey, 1995) reported that *Cynodon dactylon* seeds observed in domestic sheep dung germinated in large numbers.

Laboratory germination trials testing acid conditions on *Cynodon dactylon* seed showed that germination could occur under pH 4.0, 5.0, 7.0 but not at 3.0 and below (Ryan *et al.*, 1975).

Seed storage and burial : *Cynodon dactylon* seeds can survive and remain viable 3-4 years under field conditions, although hybrids and some *Cynodon* rarely produce viable seeds. Holm *et al.* (1991) noted that *Cynodon dactylon* seeds can survive at least 50 days of submergence in water.

Effect on crop production : Troublesome weed of agriculture worldwide, reported to occur in more than 80 countries and in some 40 different crops such as rice, sugarcane, maize, vineyards and plantation crops. In Bangladesh, it can reduce rice yield by 50%. In addition to competing with native plants for nutrients, *Cynodon dactylon* presents a direct threat to agricultural crops and possibly to natural vegetation by acting as an alternate host to eleven arthropods, twelve nematodes and numerous viruses (Holm *et al.*, 1977; Akobundu, 1987).

Uses : Cynodon dactylon plays an important role in conservation, because it prevents soil erosion. It provides good grazing and is very useful as a lawn grass and is recommended for the protection of waterways. In traditional medicine, it is used for indigestion and the treatment of wounds. According to an old Venda tradition, it is used in the fermentation process to make beer sour.

Control : Cultural control: Seedlings can be destroyed by cultivation, but perennating organs are difficult to eradicate. The plants with creeping stolons on the soil surface can be controlled by frequent shallow cultivation and removing the weed to prevent re-rooting. Rhizomatous ones are more difficult to control. Several dry-season cultivations can be effective, if the rhizomes are brought to the surface, where they are killed by desiccation. This usually requires tractor-drawn implements, especially where deep rhizomes are present.

Chemical control: Use of glyphosate application under non cropped area effectively controls *C. dactylon* in upland rice.

Past work done on Seed Science and Technology

1. Vegetative propagation

Abdessatar Omezine and Fethia Skhiri-Harzallah (2011) reported that the excavation of *C. Dactylon* rhizomatous system shows that the majority of rhizomes are in the top soil. About 85% of rhizomes are located within 20 cm of soil and few rhizomes are able to reach a depth exceeding 30 cm (Fig. 4). The observation reveals that these rhizomes have different colors from white to black, indicating that they have different ages. Top soil is a densely populated area by the young rhizomes and deep layers are populated by rhizomes a little older.

The emergence period of shoot is longer for deep planting than the shallower planting. The first aerial shoot emerged 7 days after planting at the 5 cm deep. Similarly, the aerial shoot emerged 11 days after planting the rhizomes at 10 cm depth. For rhizomes buried at a depth of 20 cm, shoots emerging after one month, the rhizomes buried deeper, beyond 30 cm, did not emerge. The excavation of buried rhizomes showed that some buds had resumed but they are not able to reach the surface, they are rotten and dead halfway. The rhizomes deeply burial significantly reduced shoot growth when compared to a lowly burial. The rhizomes less deeply buried give more biomass than the rhizomes more deeply buried and the rhizomes are able to produce more perennial organs (Fig. 5).

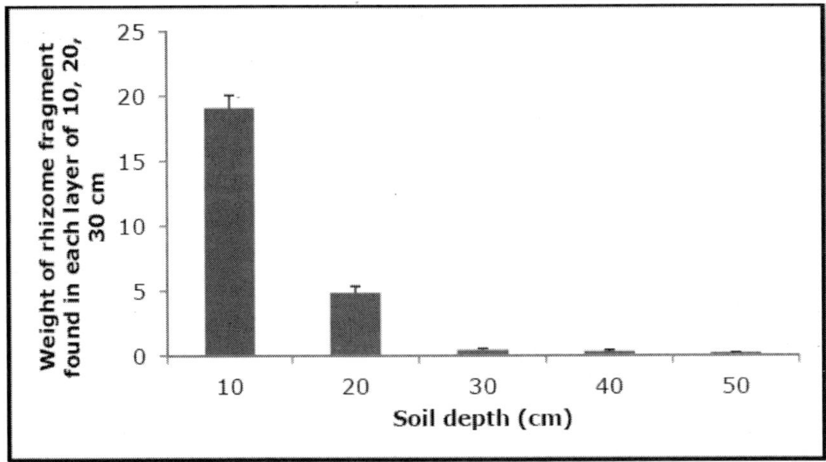

Fig. 4. Spatial distribution of *C. dactylon* rhizomes

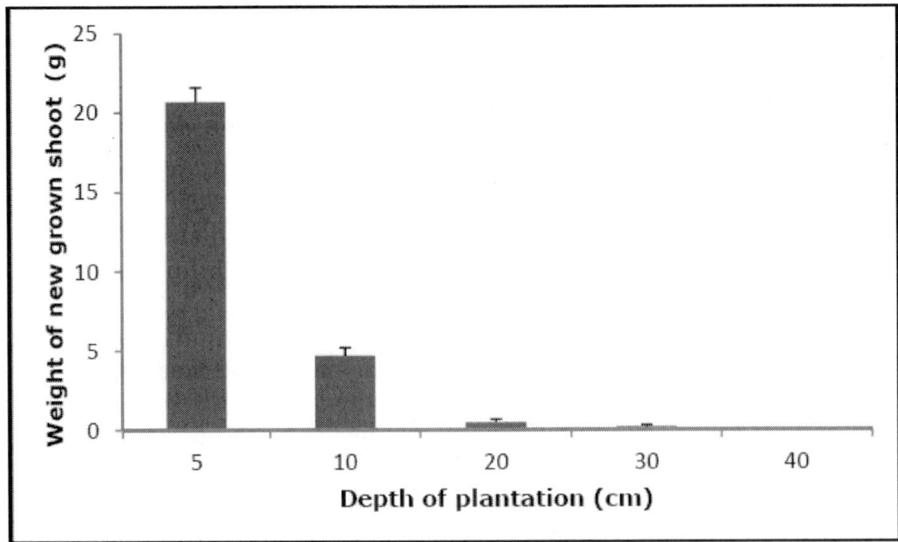

Fig. 5. Effect of burial depth on new growth of shoots

2. Germination improvement

Gokhan Hacisalihoglu (2007) stated that final germination increased 29% (from 70% to 90%) in matriconditioned seeds compared with nonprimed control (Fig. 6A). Furthermore, MC significantly reduced MGT (30%) compared with nonprimed control seeds (Fig. 2D). Both 7-d and 14-d aging treatments decreased germination (Figs. 2B–C). Matriconditioned seeds had 11% (aged 7 d) and 2.5-fold (aged 14 d) higher final GP compared with nontreated control seeds (Fig. 2). There was no significant difference in MGT between matriconditioned and control seeds after 7 or 14 d of aging (Figs. 6E–F).

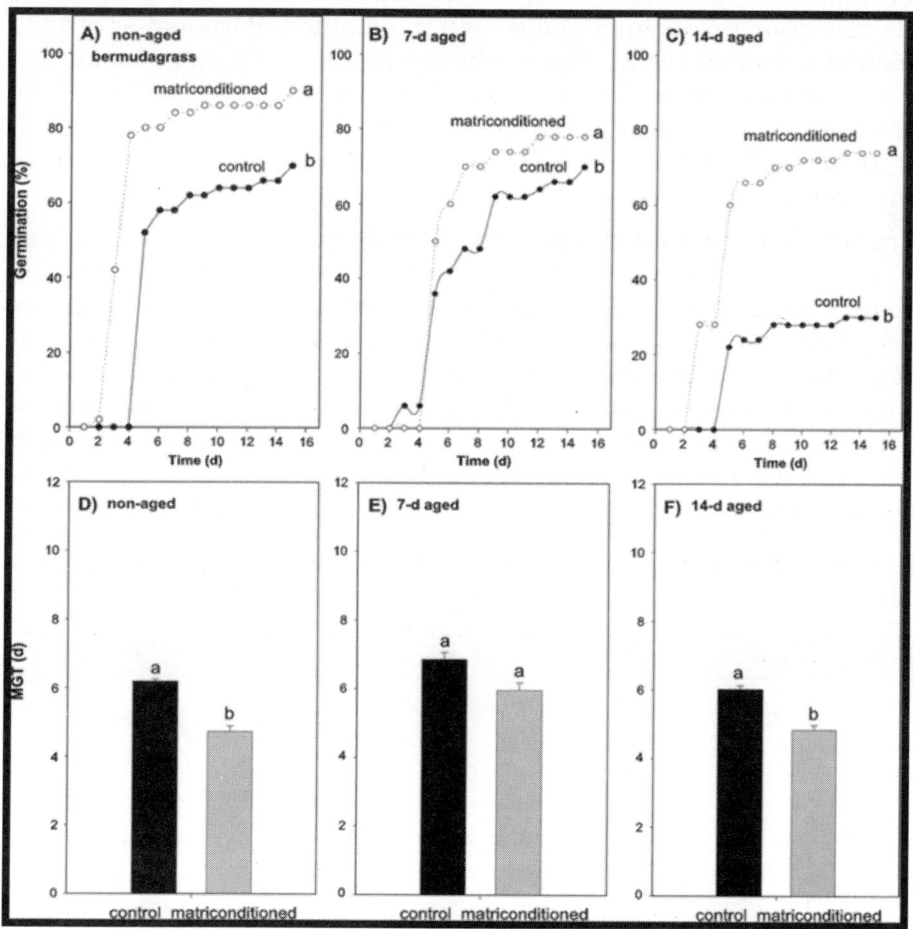

Fig. 6. Germination percentage and mean germination time (MGT) of 'Princess' bermuda grass seeds with three aging treatments. (A) Germination of nonaged seeds, (B) germination of seeds aged 7 d, (C) germination of seeds aged 14 d, (D) MGT of nonaged seeds, (E) MGT of seeds aged 7 d, and (F) MGT of seeds aged 14 d. Accelerated aging was induced by storing seeds for 0, 7, and 14 d at 42°C (107.6°F) and 95% relative humidity. Within each box, lines or bars with the same letters are not significantly different at $P < 0.05$.

3. Germination inhibition

Wheeler and Hill (1957) reported that hulled bermudagrass seed germinates more quickly than unhulled seed.

Ahring and Stritzke (1982) studied the effects of chemical desiccation of bermudagrass [*Cynodon dactylon* (L.) Pers.] before seed harvest on seed yield and quality. Harvest of 57.3% of available seed was obtained with the stripper harvester after chemical desiccation with paraquat (1,1′-dimethyl-4,4′-bipyridinium ion). Only 44.6% of the available seed was harvested by direct combining after paraquat treatment. This was significantly better than the 31.2% seed obtained by harvesting using the conventional swathing and combining method. Paraquat severely reduced germination of unhulled seeds. The inhibitory effects of paraquat application were removed by hulling or by hulling and scarifying the caryopses before germination. Significant improvement in germination of all seed (unhulled, hulled, and hulled-scarified) resulted from use of 0.2% KNO_3 rather than water as a substrate moistening agent.

The results indicate that the percentage of germination decreased significantly with the increase of storage period of seeds in all the parameters in all the experimental species. In *Cynodon dactylon* (L) Pers., the highest germination was in 52.00 in 90 days stored seeds of control,whereas in CW and CS, it was 28.40 and 12.40, respectively in 90 and 120 days stored seeds. Similarly, the lowest germination of 19.20 was observed in 180 days stored control seeds, whereas in CW and CS, it was 7.00 and 4.00, respectively in 180 days stored seeds. Inhibition in germination was high due to fresh crude oil (CO) followed by crude oil contaminated water of oil fields (CW) > Crude oil contaminated soil of oil fields (CS) (Table 1).

Table 1. Effect of Assam petroleum crude oil on the germination of oil resistant herb

Name of the species	Duration	Control	CW	CS	CO
	30 days	37.2 ± 2.48	16.6 ± 2.15 (55.37)***	4.4 ± 1.74 (88.17)***	0 (100)
	60 days	38.0 ± 2.61	27.0 ± 2.28 (28.95)***	9.2 ± 1.72 (75.78)***	0 (100)
	90 days	49.8 ± 3.18	28.8 ± 1.72 (42.17)***	11.4 ± 2.40 (77.11)***	0 (100)
Cynodon dactylon	120 days	52.0 ± 2.97	24.6 ± 2.42 (52.69)***	12.4 ± 2.33 (76.17)***	0 (100)
	150 days	37.4 ± 1.85	14.0 ± 1.40 (62.57)***	8.4 ± 1.35 (77.54)***	0 (100)
	180 days	19.2 ± 1.72	74 ± 2.48 (61.46)***	40 ± 1.41 (79.17)***	0 (100)

CW, CS and CO for crude oil contaminated water, crude oil contaminated soil, fresh crude oil

It is concluded from the above study that crude oil certainly endanger for germination of seeds. Spilled crude oil not only destroies the seed bank through mixing in soil and seeds, but also inhibits in plant growth, lowers the species diversity and degrades the soil environment (Debojit Barua *et al.,* 2011).

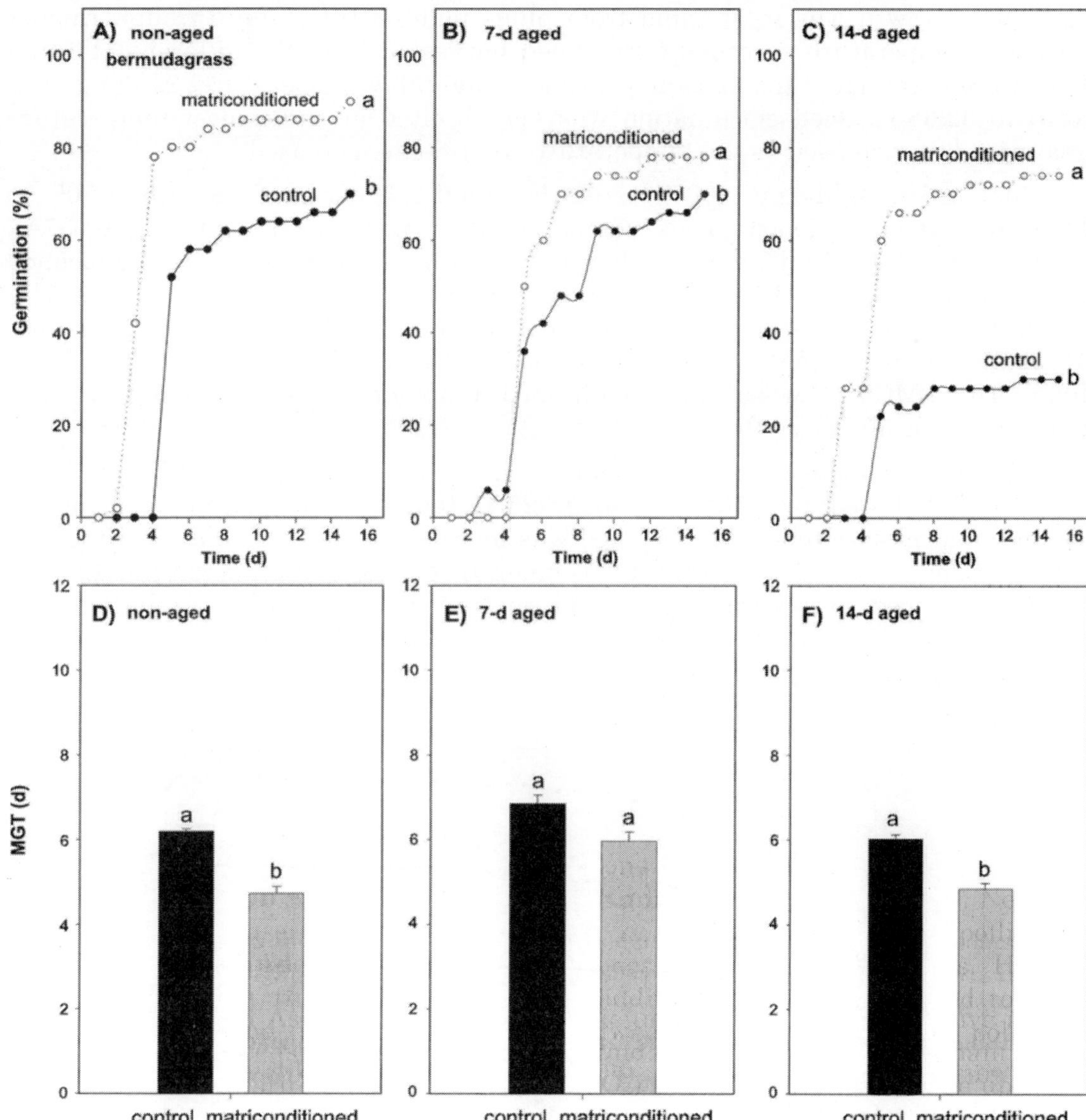

Fig. 7. Germination percentage and mean germination time (MGT) of 'Princess' bermuda grass seeds with three aging treatments. (A) Germination of nonaged seeds, (B) germination of seeds aged 7 d, (C) germination of seeds aged 14 d, (D) MGT of nonaged seeds, (E) MGT of seeds aged 7 d and (F) MGT of seeds aged 14 d. Accelerated aging was induced by storing seeds for 0,7 and 14 d at 42°C (107.6° F) and 95% relative humidity. Within each box, lines or bars with the same letters are not significantly different at $P < 0.05$.

Keeley and Thullen (1989) examined planting date on common bermudagrass establishment for two years at Shafter, CA (35°47′N 119°18′W). Plugs grown in the greenhouse were transplanted monthly from March through October. Emergence began from the March plantings when the soil temperature reached 17°C at the 5 cm depth. Growth was most rapid from plugs planted from May through August when air temperature exceeded 20°C. Seed harvested from the different planting dates were germinated the following spring at night/day temperatures ranging from 10/15 to 32/38°C. Seed germination was very limited at the coldest temperature treatment but increased as the temperature treatments increased.

Gokhan Hacisalihoglu (2007) stated that final GP increased 29% (from 70% to 90%) in matriconditioned seeds compared with nonprimed control (Fig. 7A). Furthermore, MC significantly reduced MGT (30%) compared with nonprimed control seeds (Fig. 7D). Both 7-d and 14-d aging treatments decreased GP (Figs. 7B–C). Matriconditioned seeds had 11% (aged 7 d) and 2.5-fold (aged 14 d) higher final GP compared with nontreated control seeds (Fig. 7). There was no significant difference in MGT between matriconditioned and control seeds after 7 or 14 d of aging (Figs. 7E–F).

Evers and Parsons (2010) reported that the best germination of hulled seed was at 20/30°C followed by 25/35°C and 15/25°C temperature treatments (Fig. 8). Optimum germination of unhulled seed was more specific with the most rapid and total germination at 25/35°C (Fig. 9). Temperature treatments lower than 15/25°C severely reduced germination rate and total germination of both hulled and unhulled seed (Table.2). In the southeastern US, hulled bermudagrass seed should be planted from mid-April through June and unhulled seed from mid-May through June.

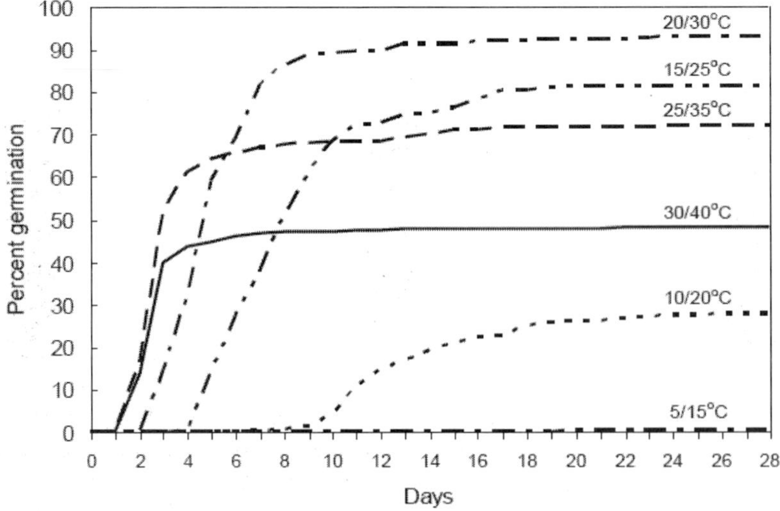

Fig. 8. Germination of hulled common bermuda grass seed over 28 days at six night/day temperatures

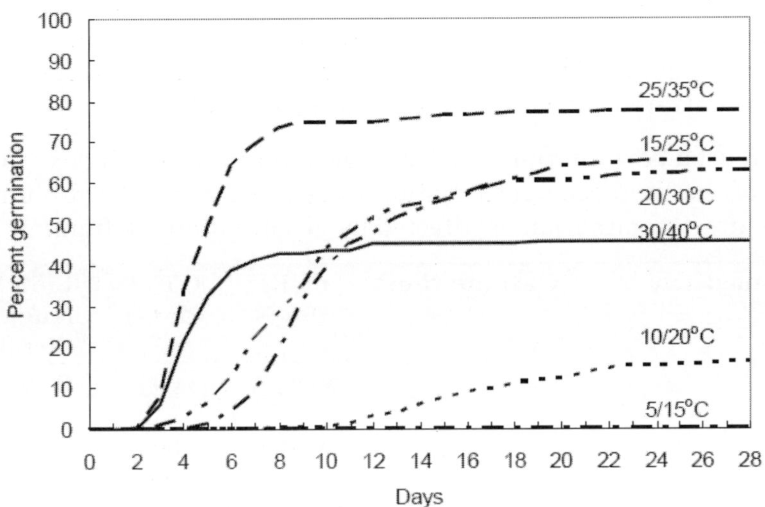

Fig. 9. Germination of unhulled common bermuda grass seed over 28 days at six night/day temperatures

Table 2. Germination rate index (GRI) of hulled and unhulled common bermuda grass seed at different temperatures

Temperature night/day (°C)	Hulled seed	Unhulled seed
	GRI	
5/15	0.02 f A+	0.00 d A
10/20	2.19 e A	1.03 d B
15/25	11.02 d A	6.69 c B
30/30	19.26 b A	7.65 c B
25/35	23.93 a A	16.26 a B
30/40	17.17 cA	9.86 b B
LSD (0.05)	1.41	1.98

+ Values in a column followed by the same lower-case letter are not significantly different and values in a row followed by the same upper-case letter are not significantly different using Fisher's Protected LSD at the 0.05 level.

4. Allelopathic effects

An inhibition on seed germination and seedling growth of *Cynodon dactylon* (Al-Naib and Rice, 1971) was attributed to the allelopathic effects of sycamore (*Platanus occidentalis* L.).

Al-Humaid and Warrag (1998) reported that the final germination percentage and the germination rate, as judged by corrected germination rate index (CGRI) and to a lesser extent by the time to 50% of the final germination (GT_{50}), were significantly reduced by the extracts in comparison with the distilled water control.

This reduction increased with increasing extract concentration, until germination was totally inhibited with the most concentrated extract. Both radicle and plumule lengths were also significantly retarded. The radicles, which did not elongate at all in the most concentrated extract, were more affected (Table 3).

Table 3. Final seed germination, corrected germination rage index (CGRI), days to 50% of the final seed germination (GT_{50}) and radicle and plumula length 5 days after germination as affected by germination medium.

Germination medium	Germination* (%)	CGRI (day^{-1})	$GT_{50}$$^{+}$ (days)	Radicle length (mm)	Plumule length (mm)
Distilled water	64.8**	86.97 A	2.00 A	23.6 A	8.6 A
Extract concentration (g l^{-1})					
10	51.3 B	79.95 B	2.00 A	4.5 B	8.2 A
20	37.5 C	79.07 B	2.00 A	2.9 B	8.3 A
30	13.3 D	70.21 C	2.75 AB	1.9 D	6.9 AB
40	11.8 D	61.12 D	3.00 AB	1.2 E	5.7 B
50	4.3 E	57.22 E	3.50 B	0.4 F	2.4 C
60	0.0 F	–	–	0.0 G	1.6 C

*Percent germination data were arcsin transformed for analysis.
** Means in the same column followed by the same letter were not significantly different at the 5% level using Duncan's multiple range test (DMRT)
+Days to 50% germination data were square-root transformed (X+0.5)1/2 for analysis.

Table 4. Final seed germination and corrected germination rate index (CGRI) in distilled water control, leaf aqueous extracts and polythene glycol (PEG) solutions.

	pH						
	6.34	5.91	5.75	5.64	5.60	5.48	4.97
	Osmotic potential (-mPa)						
Growth medium	0.003	0.043	0.096	0.141	0.188	0.230	0.275
Germination (%)*							
Distilled water	66.3**	–	–	–	–	–	–
Leaf extract	–	54.8 B	35.3 C	17.0 C	13.8 C	3.3 C	0.0 C
PEG solution	–	63.3 A	57.0 B	56.8 B	54.5 B	52.5 B	41.8 B
CGRI (days^{-1})							
Distilled water	87.0 A	–	–	–	–	–	–
Leaf extract	–	80.0 B	79.0 B	70.2 B	63.1 B	57.2 B	–
PEG solution	–	87.3 A	88.4 A	87.8 A	86.8 A	87.2 A	86.3 A

*Percent germination data were arcsin transformed for analysis
** Means in the same column followed by the same letter were not significantly different at the 5% level using DMRT.

Polyethylene glycol 2000 solutions, with the same pH and osmotic potentials as the extracts, resulted in a significantly higher germination percentage, CGRI and radicle growth rate than their corresponding extracts. A similar trend was exhibited by the plumules at the three highest concentrations only (Table 3, 4).

REFERENCES

Abdessatar Omezine and Fethia Skhiri-Harzallah. (2011). Resumption and growth of *Cynodon dactylon* rhizome fragments. *Pak. J. Weed Sci. Res.* 17(30: 215-227.

Ahring and Stritzke. (1982). Effects of paraquat on harvesting and quality of bermudagrass (*Cynodon dactylon*) seed. *Weed Sci.*, 30(1): 73-76.

Akobundu, I. (1987). Weed science in the tropics; principles and practices. John Wiley and Sons Ltd., NY. 522 pp.

Al-Humaid, A.I. and M.O.A. Warrag. (1998). Allelopathic effects of mesquite (*Prosopis juliflora*) foliage on seed germination and seedling growth of bermudagrass (*Cynodon dactylon*). *Journal of Arid Environments,* 38: 237–243.

Al-Naib, F.A. and E.L. Rice. (1971). Allelopathic effects of *Platanus occidentalis. Bulletin of Torrey Botanical Club*, 98: 75–82.

Anderson, W.P. (1999). Perennial weeds: characteristics and identification of selected herbaceous species. Iowa State University Press, Ames, Iowa. 228 pp.

Bryan W.E. (1918). Hastening the germination of bermuda grass seed by the sulfuric acid treatment. *Journal of the American Society of Agronomy*, 10: 279-281.

Carey, J.H. (1995). *Cynodon dactylon. In*: Fischer, W.C., Compiler. The Fire Effects Information System [Data base]. U.S. Department of Agriculture, Forest Service, Intermountain Research Station, Intermountain Fire Sciences Laboratory, Missoula, MT. Magnetic tape reels; 9 track; 1600 bpi, ASCII with Common LISP present.

Debojit Barua, Jity Buragohain and Sarada Kanta Sarma. (2011). Impact on assam petroleum crude oil on the germination of four crude oil resistant species. *Asian J. of Plant Sci. and Res.,* 1(3):68-76.

Evers G.W. and M.J. Parsons. (2010). Temperature influence on seeded bermudagrass germination. *The Texas Journal of Agriculture and Natural Resource* 22:87-93.

Fuls, E.R. and O.J.H. Bosch. (1990). Environmental stress resistance and propagation studies of six *Cynodon dactylon* strains to assess reclamation suitability. Landscape and Urban Planning 19:281-289.

Gokhan Hacisalihoglu. (2007). Germination characteristics of three warm-season turfgrasses subjected to matriconditioning and aging. *Hort.Technology*, 17(4): 480-485.

Heady, H.F. (1954). Viable seed recovered from fecal pellets of sheep and deer. *Journal of Range Management,* 7:259-261.

Holm, L.G., D.L. Plunknett, J.V. Pancho, and J.P. Herberger. (1991). The world's worst weeds. Distribution and biology. Krieger Publishing Company, Malabar, Florida. 609 pp.

Holm, L., D.L. Plucknett, J.V.R. Pancho and J.P. Herberger. (1977). A geographical atlas of world weeds. New York, John Wiley and Sons.

Horowitz, M. (1972a). Development of *Cynodon dactylon* (L.) Pers. *Weed Res.* 12:207–220.

Horowitz, M. (1972b). Spatial growth of *Cynodon dactylon* (L.) Pers. *Weed Res.* 12:373–383.

Keeley, P.E. and R.J. Thullen. (1989). Influence of planting date on growth of bermudagrass (*Cynodon dactylon*). *Weed Sci.,* 37(4):531-537.

Ryan, J., S. Miyamoto and J.I. Stroehlein. (1975). Effect of acidity on germination of some grasses and alfalfa. *Journal of Range Management* 28(2):154-155.

Wheeler, W.A. and D.D. Hill. (1957). Southern grasses. p. 544-561. *In* grassland seeds. D. Van Nostrand Company, Inc. Princeton, New Jersey.

NUTGRASS (*Cyperus rotundus*)

Common name : Coco grass, nutgrass, common nutsedge, purple nutsedge.

Botanical name : Cyperus rotundus

Tamil name : Korai, Korai kilangu,

Hindi name : Motha, Mutha

Family : Cyperaceae

Origin and distribution : C. rotundus is a species of sedge, native to Africa, southern and central Europe (north to France and Austria) and southern Asia. It is one of the most invasive weeds known, having spread out to a worldwide distribution in tropical and temperate regions.

Botany

Habitat : A very troublesome perennial weed, occurring in dry cultivated lands, gardens and in semi irrigated land. It is an obnoxious weed with a very high propagation potential. This is one of the world's top ten worst weeds.

Fig. 1. Plant

Habit : Stoloniferous, stolons not bulbiferous; stems trigonous, up to 1 m high; tubers not zoned, perennial (Fig. 1).

Stem : Stem sparsely tufted.

Leaf : Leaves shorter or longer than stem, narrow, numerous; bracts usually 3, up to 60 cm long

Floral biology

Flower : The inflorescence consists of a stalk with 2-4 green bracts underneath a cluster of 3-8 stems (varying in length, up to 10 cm long), spikelets spreading, linear to lanceolate, up to 2.5 cm long; glumes imbricate, plicate, pale-brown or chestnut-brown, sometimes with a reddish tinge; nut trigonous, broadly obovoid, grayish-black (Fig. 2a,b,c,d,e).

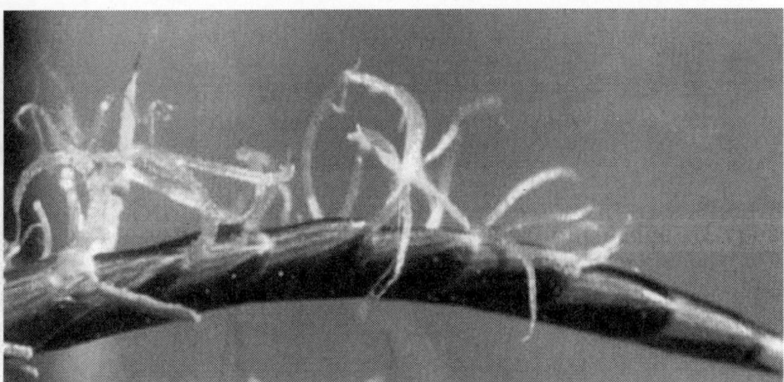

Fig. 2a. Inflorescence **Fig. 2b.** Flower

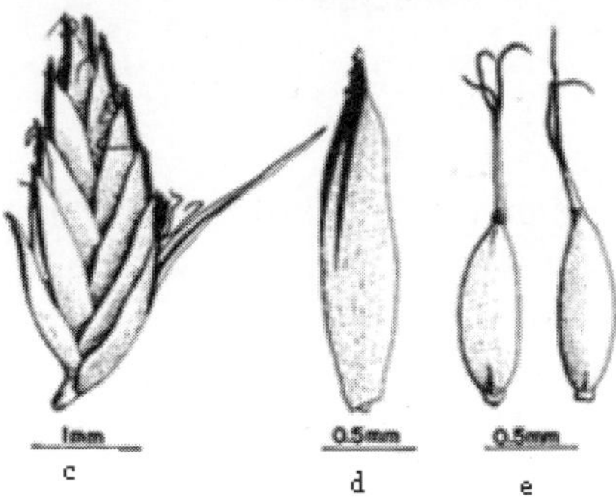

Fig. 2 c. Spikelet, d. glume lateral view, e. the two kinds of pistils, with 3 and 2 stigmas

Fruit : Fruit is a broadly obovoid, trigonous, seed-like nut, grayish black in colour.

Seed biology

Seed : Purple and yellow nutsedges have been documented to produce achenes (an achene is a small dry fruit with one seed) from aerial inflorescence (Fig. 3).

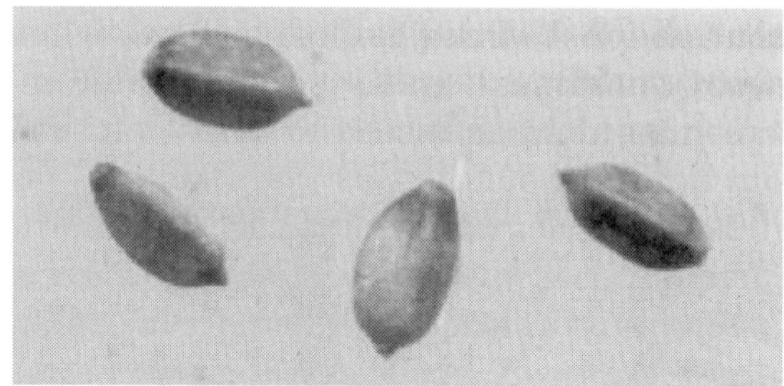

Fig. 3. Seed

Vegetative propagule : Purple nutsedge produces a complex underground system of basal bulbs, rhizomes, and tubers (Stoller and Sweet, 1987); rhizomes are initially white and fleshy with scale leaves, later becoming woody or wiry tubers are recognized as the primary dispersal unit (Fig. 4).

Fig. 4. Tuber

Tuber storage and longevity : Tuber dormancy may last for at least 7 years, cultivation stimulates the growth of *C. rotundus.* Both purple and yellow nutsedge tuber longevity is a function of tuber depth in the soil; tuber survival time increases as its depth in the soil increases. Yellow nutsedge tubers had a half-life of 4 and 6 months at 10 and 20 cm, respectively, in a non- crop environment. Desiccation and temperature extremes can kill both yellow and purple nutsedge tubers. Tubers desiccate quickly when detached from the rhizome-root system under dry conditions and can survive flooded soils for at least 200 days. Tuber longevity is variable and depends on environmental conditions. In most cases, tubers survive about 3-4 years, but under certain conditions they can remain viable for up to 10 years or more.

Dormancy and germination : The tubers contain quiescent buds and function like the seeds of annuals, acting as the primary dispersal units. Tubers are produced on rhizomes and lie dormant in the soil for extended periods. They germinate under the appropriate environmental conditions to produce plants that perpetuate the infestations. Tubers require a chilling period to break dormancy and produce a primary basal bulb 1-2 cm (0.39 - 0.78 inch) beneath the soil surface. Temperature has a marked effect on the germination of tubers. Sprouting has been reported at temperatures of 13-43°C but the range varies with biotype. Desiccation kills tubers, but the duration and temperature of the drying period affects this process. The critical moisture level for tuber germination seems to be in the region of 11.5-15. Tubers can remain dormant for long periods and can become dormant after sprouting. Tuber dormancy often increases with age.

Propagation : Cyperus rotundus mainly reproduces by rhizomes and tubers.

Seed dispersal : Most seeds of *C. rotundus* are not viable. Dispersal by seed is generally regarded as being unimportant. Translocation of machinery/equipment. Both tubers and seeds might be transported by surface water movement. Seeds also disperse with wind.

Dispersal of *C. rotundus* occurs when tubers are moved by tillage equipment or other farm machinery. Flood water may also carry the tubers. Contaminated soil in nursery stock is an avoidable, but common method of dispersal. Under natural conditions, a population of *C. rotundus* extends its boundary by a few meters in a year.

Life cycle : In purple nutsedge, tuberization can begin within 17 days after shoot emergence (Hammerton, 1974 in Stoller and Sweet 1987), but dormant tubers are not found until 6 to 8 weeks after sprouting (Hauser, 1962a in Stoller and Sweet 1987); this is followed by chain formation 10 weeks after shoot emergence (Hauser, 1962b in Bangarwa *et al.*, 2008). The typical life cycle of *C. rotundus* starts with growth of the apical bud of a tuber. As the tuber shoot extends, it swells to form a basal bulb (sometimes called a corm), usually near the soil surface, from which an aerial shoot and roots are produced. Meanwhile, the new shoots grow and produce flowers within 3-8 weeks of emergence, if stimulated by short photoperiods of 6-8 hours. As many as 600 plants have been produced in a single year from a single tuber.

Effect on crop : It is considered the worst weed based on its occurrence in 52 crops in 92 countries and its capacity to cause substantial yield losses. Examples of crop loss caused by *C. rotundus* include 35-89% in vegetables, 30% in cotton and 75% in sugarcane. It may cause 23–89% reduction in yield of associated crops. Reduction in yield of upland rice was as much as 38% in the presence of *Cyperus* (Okafor and DeDatta, 1976).

Even the growth of tree crops can be reduced, for example, mulberries in Japan, citrus in Israel and coffee in Kenya. Much of this can be attributed to the capacity of *C. rotundus* to remove nutrients from the soil and store them in its tubers, making them unavailable to crops.

Adding nitrogen to a crop can actually improve the competitiveness of *C. rotundus*, causing even greater crop loss than where no fertilizer is added. *C. rotundus* undoubtedly competes with crops for water, but also for light when it grows tall enough.

There is evidence that extracts from *C. rotundus* suppress the growth of plants but it is difficult, under field conditions, to separate the effects of allelopathy (the roots releasing substances harmful to other plants) from competition.

Uses : C. rotundus is an important medicine in India and China and noted for its use by pharmaceutical companies to produce diuretics, anthelminthics and treatments for coughs, bronchial asthma and fever. It makes a poor fodder but has value in binding together soil. Useful in checking soil erosion. The tuber yields an essential oil (0.5-0.9%), which is used in perfumery and incense.

Past work done on Seed Science and Technology

1. Tuber dormancy and germination

Purple nutsedge tuber sprouting involves two steps – budbreak and elongation of the emerged bud, and both steps may be thermally regulated. Budbreak occurred readily for most field-collected tubers when held at a relatively constant temperature (84% at 22 to 24%). However, only about 10% of the emerged buds elongated >1 cm when tubers were held for 3 weeks at a relatively constant temperature. For those tubers that received daily fluctuating temperatures of 22.5 and 27.5°C for 12h each, 98% of the tubers had budbreak and 85% of the buds elongated >1 cm. Thus, daily alternating temperature appears to stimulate budbreak and elongation of the emerged buds.

According to Wen-Hao Sun and Nishimoto (1997), a single 30 minute temperature pulse from 20-35°C caused 80% bud break, while tubers without the thermal pulse had only 25% bud break (Fig. 5).

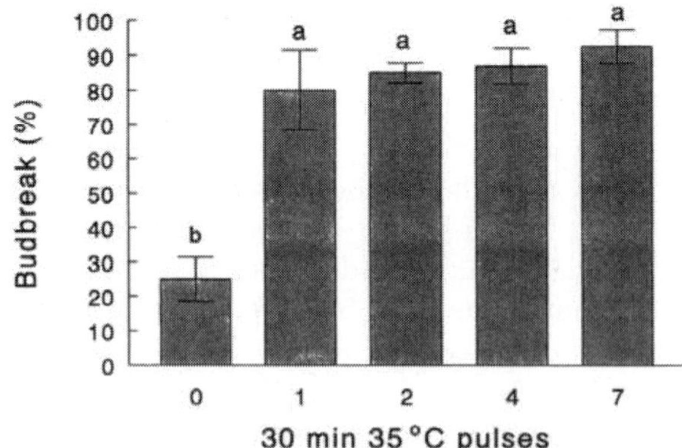

Fig. 5. Budbreak of purple nutsedge tubers in response to increasing numbers of thermal pulses (35°C for 30 min). All tubers were kept at constant 20°C between pulses. Horizontal bars represent ± SE. Bars with different letters differ significantly by Duncan's multiple range test, P = 0.05.

Increasing the duration of the single 35°C temperature pulse from 30 minutes to 12 hours did not increase budbreak. Even a 3- minutes 35°C pulse caused 63% bud break (Fig.6).

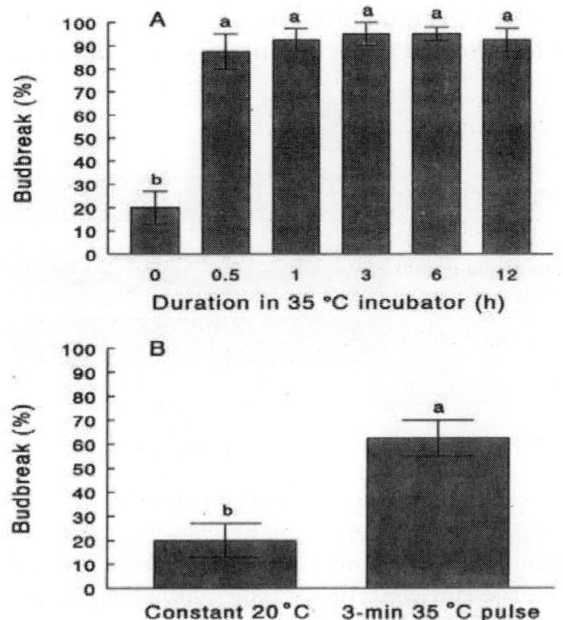

Fig. 6. Effect of duration at 35°C in a single thermal fluctuation (A) and a 30 min 35°C pulse (B) on budbreak of purple nutsedge tubers. The 3-min 35°C pulse was achieved by incubation in a 55°C incubator for 3 min, when tuber surface temperature reached 35°C. After the treatment, tubers were returned to constant 20°C. The value for sprouting at constant 20°C is at 0 h at 35°C period. Horizontal bars represent + SE. Bars with different letters differ significantly by Duncan's multiple range test, P = 0.05.

A single gradual shift (0.02°C per) from 20 to 35°C was an effective as a rapid shift (0.5°C per minute) in stimulating budbreak (Fig. 7).

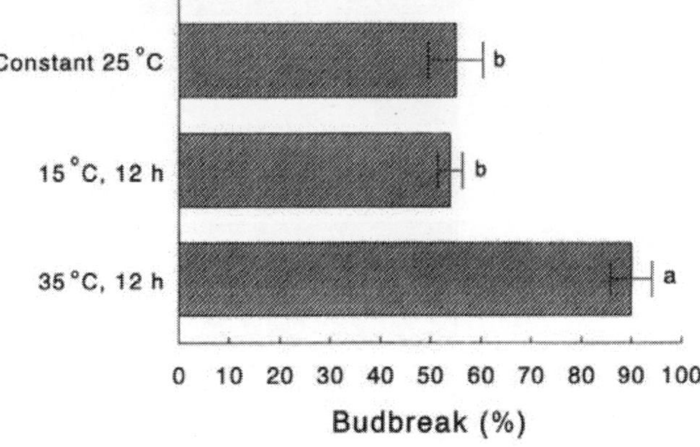

Fig. 7. Effect of rate of temperature change from 20 to 35°C on budbreak of purple nutsedge tubers. The rate of temperature change was 0.5 and 0.02°C per min, for the rapid and gradual shift, respectively. Horizontal bars represent ± SE. Bars with different letters differ significantly by Duncan's multiple range test, P = 0.05.

Nonsprouting tubers preincubated at 25°C were stimulated to cause budbreak by exposure at 35°C for 12 hours, but not at 15°C for 12 hours (Fig. 8).

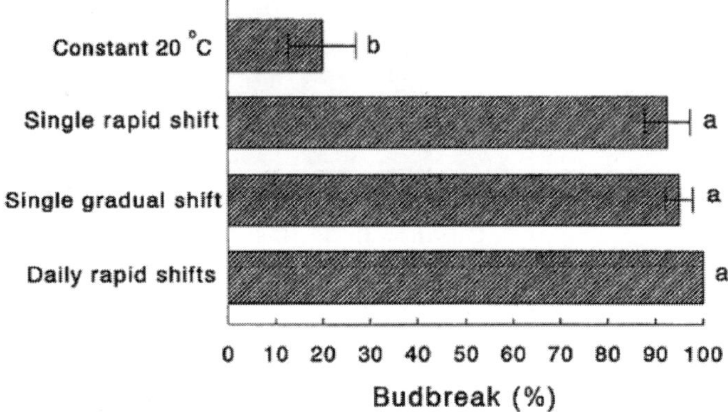

Fig. 8. Effect of increase versus decrease in temperature on budbreak of purple nutsedge tubers 5d after treatment. In this experiment, constant 25°C was used instead of constant 20°C for preincubation and selection of dormant tubers and for the base temperature in both temperature shifts. Horizontal bars represent ± SE. Bars with different letters differ significantly by Duncan's multiple range test, P = 0.05.

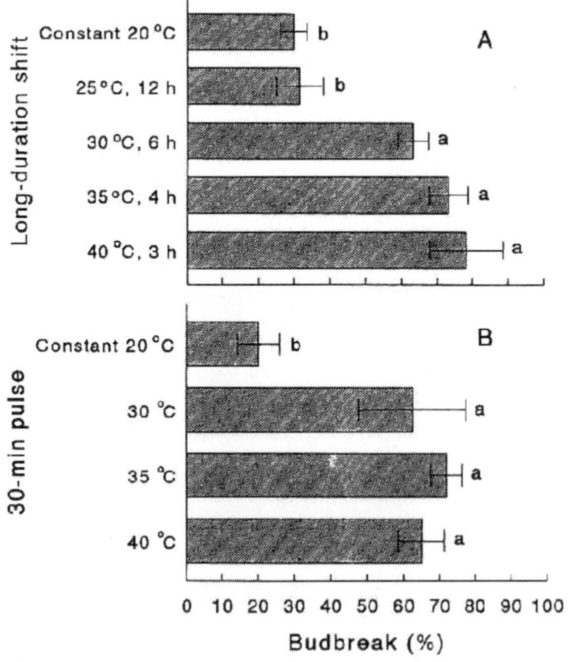

Fig. 9. Effect of magnitude of a single temperature shift on budbreak of purple nutsedge tubers. Each long-duration treatment (A) was equivalent in heat units (°C >20°C x time (min) = 3600°C/min). In the 30-min pulse treatments (B), tubers were exposed to the high temperature for 30 min. Horizontal bars represent ± SE. Bars with different letters differ significantly by Duncan's multiple range test, P = 0.05.

A single temperature pulse of at least 10°C differential (from 20 to 30, 35, or 40°C) stimulated budbreak, while a 5°C differential (from 20 to 25°C) did not; all pulse had the same heat unit (°C > 20°C x time (minutes)= 3600°C/minute) (Fig. 9). These results suggest that budbreak of purple nutsdge tubers responds to high temperature as a trigger action.

Miles *et al.* (1996) stated that tuber sprouting was more rapid and complete with alternating temperatures than with constant temperatures. Increasing temperature fluctuation from 0 to 6°C for 12 h daily linearly increased total tuber sprouting. As little as 30 min exposure to high temperature per day provided nearly the same level of sprouting as a 12 h alternating temperature cycle.

2. Seed storage and burial

Elsafori *et al.* (2011) stated that the seed bank density was higher in the upper soil depths (0 to 5 and 5 to 10 cm) as compared to the lower ones (10 to 15 cm). It was found that the seed density had decreased with increasing depth. The live seed density ranged from 1015 to 5371 seeds/m^{-2}, whereas that of the dead seeds ranged from 3215 to 6957 seeds m^{-2}.

REFERENCES

Bangarwa, S.K., Norsworthy, J.K. Jha, P., Malik, M. (2008). Purple nutsedge (*Cyperus rotundus*) management in an organic production system, *Weed Science 56* (4): 606-613.

Elsafori, A.K., A.N. Gumaa and M.A. El Nour. (2011). Soil seed banks of a rangeland area white Nile State, Sudan. *Journal of Horticulture and Forestry,* 3(6), pp. 178-185.

Hammerton, J.L. (1975). Experiments with *Cyperus rotundus* L. III. Seasonal variations in growth. *Weed Research,* 15(5): 339-348.

Hauser, E.W. (1962a). Development of purple nutsedge under field conditions. *Weeds,* 10(4): 315-321.

Hauser, E.W. (1962b). The establishment of nutsedge from space planted tubers. *Weeds,* 10(4): 209-212.

Miles, J.E., R.K. Nishimoto and Osamu Kawabata. (1996). Diurnally alternating temperatures stimulate sprouting of purple nutsedge (*Cyperus rotundus*) Tubers. *Weed Science,* 44(1): 122-125.

Okafor, L.I. and S.K. De Datta. (1976). Competition between upland rice and purple nutsedge for nitrogen, moisture and light. *Weed Sci.,* 24: 43–46.

Stoller, E.W. and R.D. Sweet. (1987). Biology and life cycle of purple and yellow nut sedges (*Cyperus rotundus* and *C. esculentus*). *Weed Technology,* 1: 66-73.

Wen – Hao Sun and Roy K. Nishimito. (1997). Dormancy release of purple nutsedge tuber buds by a single thermal pulse. *J. Amer. Soc.* 122 (3): 306-309.

YELLOW NUTSEDGE (*Cyperus esculentus* L)

Common name : Yellow nutsedge

Botanical name : Cyperus esculentus L.

Tamil name : Korai

Hindi name : Bara-nagar-motha

Family : Cyperaceae

Origin and distribution : It is listed as being native to the North America and distributed in all temperate to tropical regions, Worldwide.

Botany

Habitat : Cyperus esculentus preferred habitat in typically disturbed areas, low ground and cultivated fields (especially irrigated), wet fields, irrigation canals/ ditches, orchards, turf (California Department of Food and Agriculture, EncycloWeedia 2002; Holm *et al.,* 1991).

In Arizona, *Cyperus esculentus* occurs between 100-8200 ft. (30-2500 m) in elevation (Parker, 1972).

In other areas, *Cyperus esculentus* typically inhabits warm areas, it seemingly is advancing and enlarging its range into cooler climates of the temperate zones; it can be found both at the equator and in Alaska (Holm *et al.,* 1991).

Habit : Perennial sedge (Fig. 1).

Stem : Stout, .erect, triangular (trigonous) in cross-section below the inflorescence, solid/pithy, glabrous, smooth, 6-20 in. (1.5-5 decimeter) long and up to 3 ft. (9 dm), 0.2-0.4 in. (4- 9 mm) wide; stem base slightly triangular, some what swollen basally.

Leaf : Leaves grass-like, basal, erect; approximately equal to or longer than culms, up to 3 ft. (9 dm) tall; 3-ranked. 1 or more leaf-like bracts forming an involucres subtending the inflorescence, with longest bract surpassing the inflorescence.

Floral biology

Inflorescence and flower : Inflorescence terminal, umbellate, umbel simple to compound, loose, with 1-10 narrow, unequal rays 0-15 cm long; and with 2-6 subtending leaf-like, unequal bracts forming an involucre with longest involucral bract much exceeding the umbel, often wider than basal leaves. Spikelets arranged in spikes at distal end of ray, spikelets numerous, divaricate, yellowish to golden brown; spikes loose, 2-4 cm long, 1.5-2.5 mm wide (Fig. 2).

Fig. 1. Plant

Fig. 2. Inflorescence

Flowering and pollination : The flowers are hermaphrodite (have both male and female organs) and are pollinated by wind.

Seed biology

Seed : Achenes light brown to greyish, obovoid, triangular in cross-section (trigonous); about 1-1.5 mm long; impressed-puncticulate (Fig. 3).

Propagation : By seeds. There was a linear relationship between seed weight and germination percentage (Jthullen, 1979; Ma *et al.*, 2011). Yellow nutsedge has potential to reproduce by seed, but probability of purple nutsedge reproducing by seed is low (Thullen and Keeley, 1979).

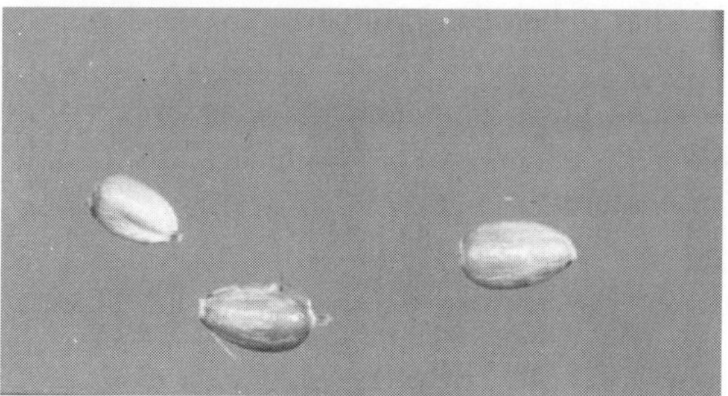

Fig. 3. Seed

Seed production : Thullen and Keeley (1979) stated that yellow nutsedge (*Cyperus esculentus* L.) produced an average of 1227 and 6685 flowers per inflorescence from eight California fields sampled during 1974, 1975 and 1976. Up to 17% of the flowers produced seeds and up to 78% of the seed germinated. Purple nutsedge (*C. rotundus* L.) averaged between 186 and 1354 flowers per inflorescence from three California fields sampled in 1974 and 1976. Only 43 purple nutsedge seed were collected and none germinated. Yellow nutsedge produced viable seed about 2 weeks after anthesis.

Seed dispersal : Seeds disperse with agricultural/nursery activities, soil movement through water, especially flooding (Anderson, 1999; California Department of Food and Agriculture, EncycloWeedia 2002).

Seed dormancy and germination : At maturity, some *Cyperus esculentus* seeds are dormant, while others have a semidormant to non-dormant status (Lapham and Drennan, 1990 *in* Mero-Macias, 1996; Mulligan and Junkins, 1976).

Within the first 6 months after sowing, proportions of dormant seed were reduced rapidly at all sowing depths (Lapham and Drennan, 1990 in Mero-Macias, 1996).

When *Cyperus esculentus* seeds were germinated under laboratory conditions, germination occurred at highest percentages when temperatures were between 20-35°C; seed germination is inhibited at temperatures below 10°C (Mero-Macias, 1996; Mulligan and Junkins, 1976). Maximum germination rates occur with a fluctuation of day/night temperatures of 38/32°C under light conditions (Mero-Macias, 1996).

With alternating temperatures of 35/20°C, seeds germinated at a percentage of 59% in continuous light conditions, and 46% in complete dark conditions; and for alternating temperatures of 30/20°C, seeds germinated at a percentage of 36% germination in continuous light conditions and 1.5% in complete dark conditions (Lapham and Drennan, 1990 *in* Mero-Macias, 1996). Seed did not germinate under dark conditions except when held at optimal temperatures (Durfee, 1960 *in* Mulligan and Junkins, 1976).

Seed storage : Seed stored for 20 years recorded 17% germination percentage (Bellue, 1946 *in* Mulligan and Junkins, 1976).

Effect on crop : For each 100 *Cyperus esculentus* plants/m^2, corn crops can potentially experience a yield loss of 8% (Mero-Macias, 1996). When growing in cotton for a full season, cotton yields were reduced 34% (Keeley and Thullen, 1975 *in* Mulligan and Junkins, 1976).

Cyperus esculentus has reported allelopathic effects, with its dried plant matter and tuber residues inhibiting the growth and development of neighboring plants (Mero-Macias, 1996; Mulligan and Junkins, 1976), improving *Cyperus esculentus* ability to compete. When occurring in high densities, *Cyperus esculentus* is noted to have intraspecific impacts, such as reduced flowering (Mero-Macias, 1996).

Control measure : Tumbleson and Kommedahl (1961 *in* Holm *et al.*, 1991) reported that during tillage operations, tubers rolled to the top of the soil dried, reducing their germinability by 80% within 2 days. *Cyperus esculentus* can be controlled in non-crop areas with 2, 4-D, dicamba, picloram; however, these herbicides cannot be used in wet areas. Chlorimuron, nicosulfuron, primisulfuron, and other sulfonylurea herbicides are preferable (Colorado Natural Areas Program, 2000).

Uses : The tubers are edible, with a slightly sweet, nutty flavour, compared to more bitter-tasting tuber of the related *Cyperus rotundus* (purple nutsedge). Since the tubers contain 20-36% oil, *C. esculentus* has been suggested as potential oil crop for the production of biodiesel. It is regarded as a digestive tonic, having a heating and drying effect on the digestive system and alleviating flatulence. They also promote urine production and mensuration. The tubers are said to be aphrodisiac, carminative, diuretic, emmenagogue, stimulant and tonic. In Ayurvedic medicine, they are used in the treatment of flatulence, indigestion, colic, diarrhoea, dysentery, debility and excessive thirst. The oil extracted from yellow nut-sedge can be used as food oil as well as for industrial and medicinal purposes. Nut-sedge tuber has high nutritive value. It contains 3 to 15% protein, 15 to 20% sugar, 20 to 25% starch, 4 to 14% cellulose and trace amounts of natural resin.

Past work done in Seed Science and Technology

1. Germination improvement

1.1. Effect of temperature

According to Thullen and Keeley (1979), there was a linear relationship between seed weight and germination percentage. A day/night temperature of 38-32°C was best for seed germination, especially when accompanied by a light period.

Yan Shen *et al.* (2011) studied the effect of temperature on germination of yellow nut-sedge seeds and observed less germination at higher temperatures (65°C). The maximum germination rate was observed at 35°C and significantly greater than other treatments. However, the germination percent was not significantly different among all treatments at 27, 45 and 55°C. The growth of radicals was significantly affected by temperature and radical length was highest at 35°C. The radical length was generally higher in the light at all temperatures. At 35°C, they took one more day. The highest germination percentage was 56% at 35°C (Fig. 4).

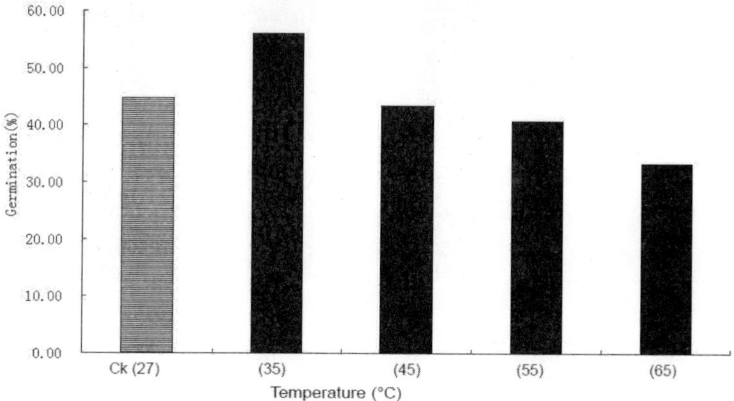

Fig. 4. Germination percentages of yellow nut-sedge seeds under different temperature.

They also reported that in yellow nut-sedge, the seeds germination was obviously enhanced by IAA 10 and IAA 20, whilst it was reduced by the other concentration. Germination of control (CK) was 23.33% and that of seeds treated by IAA 10 and IAA 20 was 96.67 and 100% respectively, but only 20, 63.33 and 6.67% for IAA 5, IAA 15 and IAA 25. Thus, IAA 20 was the best concentration for seeds germinating (Fig. 5).

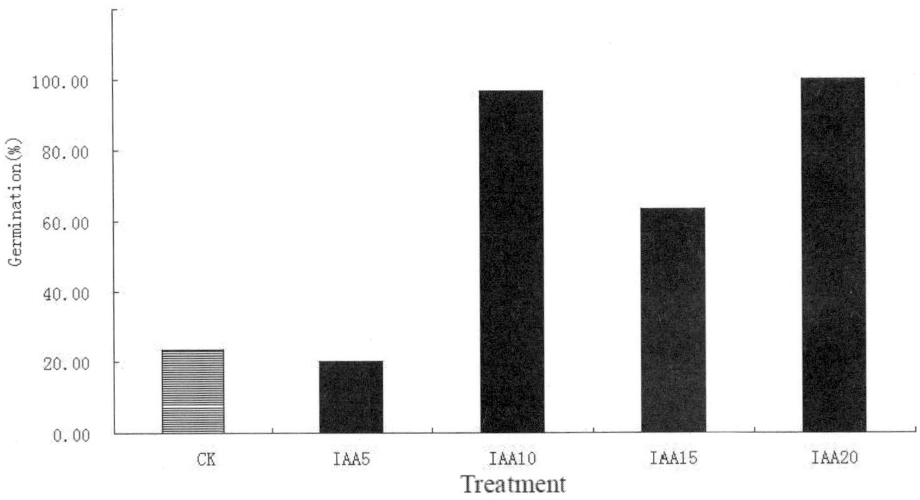

Fig. 5. Germination of yellow nut-sedge seeds from different plant growth regulators treatments.

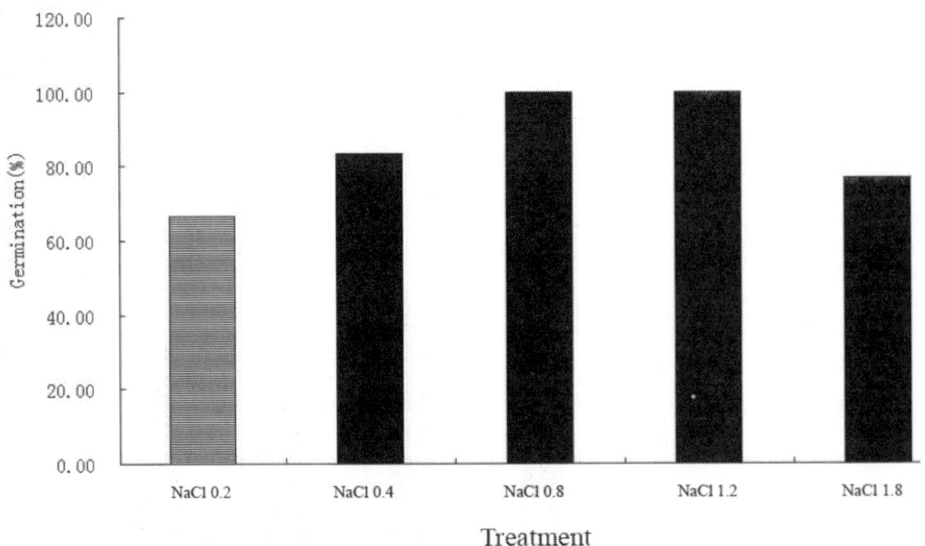

Fig. 6. Germination of yellow nut-sedge seeds under different plant growth regulators treatments

NaCl treatment enhanced the seeds germination obviously. Much higher than 23.33% for control (CK), 66.67, 83.33, 100, and 76.67% seeds germinated for NaCl 0.2, 0.4, NaCl 0.8, 1.2 and 1.8, respectively. The best treatment was NaCl 0.8 and NaCl 1.2, in which all the seeds germinated (Fig. 6). NAA had relatively less effect on the seed germination than IAA and NaCl. NAA had different effect, similarly to the IAA effects above. NAA5 and NAA 10 reduced the seed germination. In NAA5 and NAA 10 had 10.00 and 3.33% seed germination, less than 23.33% for control (CK) whilst in NAA 25 and NAA 50, 70.00 and 56.67% seed germinated. In NAA 35, the seed germination was 26.67%, which was similar to control (CK) treatments (Fig. 7).

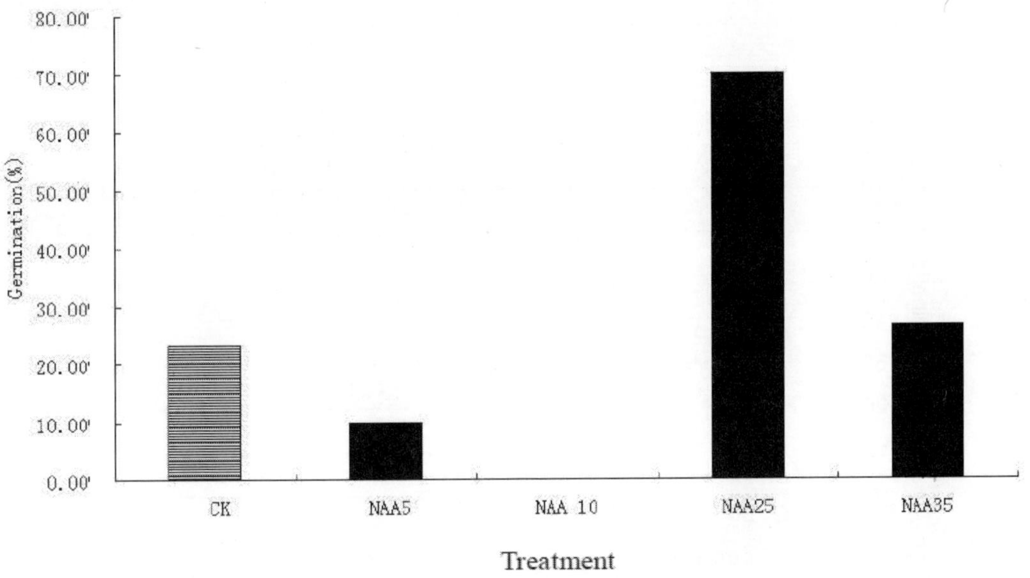

Fig. 7. Germination of yellow nut-sedge seeds under different plant growth regulators treatments.

2. Effect of temperature

A lower temperature of 15°C was equally inhibitory to that of 45°C in reducing the germination of yellow nutsedge. Yellow nutsedge had significantly higher germination at 20 to 40°C (Singh and Singh, 2009).

3. Effect of water stress

Singh and Singh (2009) expressed that yellow nutsedge was more tolerant to water stress than other species as germination was similar to no stress at -0.2 MPa and 4% tubers were able to germinate at -0.8 MPa which was not observed in any other test species (Table 1).

Table 1. Effect of osmotic potential and duration on per cent germination of yellow nutsedge seed*

Duration (DAT)	Control	-0.1 Mpa	-0.2 Mpa	-0.4 Mpa	-0.8 Mpa	-1.2 MPa
7	33	26	29	11	0	0
14	38	36	41	14	4	0
21	43	41	44	14	4	0

*Means followed by same letter within species do not significantly differ using Student-Newman-Keuls (P=0.5). DAT=Days after transplanting.

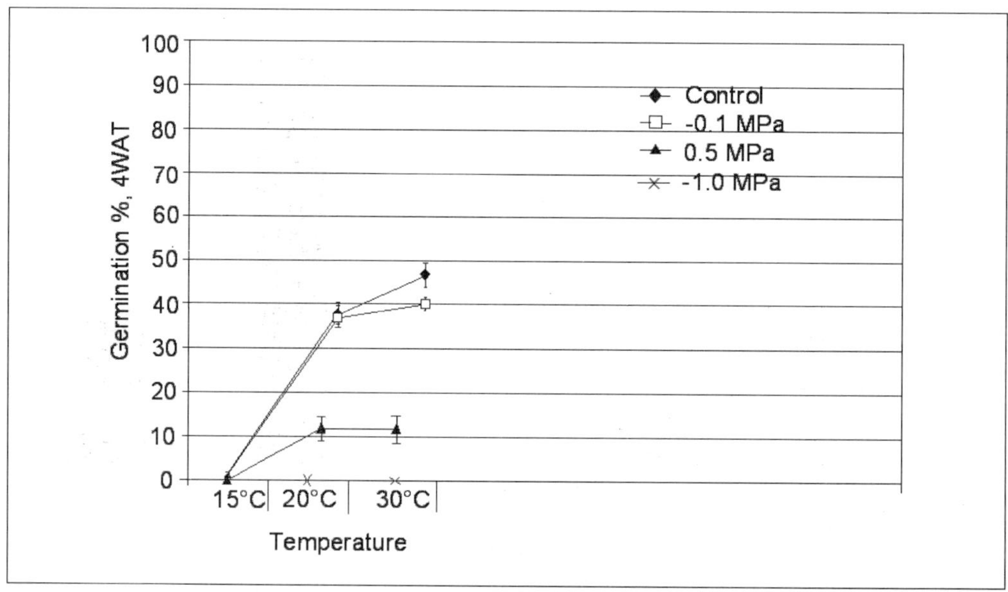

Fig. 8. Effect of temperature and water potential on germination of yellow nutsedge. Bars indicate standard error of means (Sem).

Delayed emergence of yellow nutsedge was observed under water stress conditions in pots buried in fields in California and Arizona (Wilen *et al.,* 1996). Tubers subjected to the dry treatment needed a higher number of accumulated degree-day units before emergence occurred and had fewer emerged shoots as compared to the wet treatment.

They also expressed that increasing the temperature from 15 to 20 and 30° C increased germination of weed species. Germination was reduced drastically at -0.1 and -0.5 osmotic potential compared to control treatments. There was no germination of weed species at -1.0 Mpa osmotic potential. Yellow nutsedge germination increased with increase in temperature from 15 to 20° C and increase in water stress (-0.5 Mpa) had similar effect on germination at 20 or 30 C (Fig. 8).

REFERENCES

Anderson, W.P. (1999). Perennial weeds: characteristics and identification of selected herbaceous species. Iowa State University Press, Ames, Iowa, p. 228.

Bellue, M.K. (1946). Weed seed handbook - Series V. Bulletin of the Department of Agriculture, California. 35:87-94.

California Department of Food and Agriculture, EncycloWeedia. (2002). *Cyperus*. Healy, E. A., S. Enloe, J. M. DiTomaso, B. Roberson, N. Dechoretz, S. Schoenig, P. Akers, L. Butler and J. Garvin (eds.). Non-Cropland Weed group, UC Extension Service, Weed Science Program, Department of Vegetable Crops, The University of California. Davis, CA. 95616.

Colorado Natural Areas Program. (2000). Yellow nutsedge, *Cyperus esculentus* L. *In*: Creating an integrated weed management plan: A handbook for owners and manager of lands with natural values. Colorado Natural Areas Program, Colorado State Parks, Colorado Department of Natural Resources, Division of Plant Industry, Colorado Department of Agriculture. Denver, Colorado. 349 pp.

Durfee, J.W. (1960). Life history and the control of northern nutgrass *Cyperus esculentus* L. Masters thesis. University of Massachusetts, Amherset, Massachusetts. 68 pp.

Holm, L.G., D.L. Plunknett, J.V. Pancho and J.P. Herberger. (1991). The world's worst weeds. Distribution and biology. Krieger Publishing Company, Malabar, Florida. 609 pp.

Jthullen R. (1979). Seed production and Germination in Yellow nut-sedge (Cyperus esculentus).

Keeley, P.E. and R.J. Thullen. (1975). Influences of yellow nutsedge competition on furrow irrigated cotton. *Weed Sci.,* 23:171-175.

Lapham, J. and D.S. Drennan. (1990). The fate of yellow nutsedge (*Cyperus esculentus*) seed and seedlings in soil. *Weed Sci.,* 38: 125-128.

Ma, X.Y., L.H. Zhang, H.B. Shao, G. Xu, F. Zhang, F.T. Ni, M. Brestic. (2011). Jerusalem artichoke (*Helianthus tuberosus*), a medicinal salt-resistant plant has high adaptability and multiple-use values. *J. Med. Plant Res.*, 5: 1275-1282.

Mero-Macias, H. (1996). The weedy nut sedges (*Cyperus* spp.) outline. Outline for: Agronomy 517; Weed Biology and Ecology, Spring Semester, 1996.

Mulligan, G. A, B. E. Junkins (1976). The biology of Canadian weeds. 17. *Cyperus esculentus* L. *Canadian J. Plant Sci.*, 56(2): 339-350.

Parker, K.F. (1972). An illustrated guide to Arizona weeds. The University of Arizona Press, Tucson, AZ. 338 pp.

Singh, S. and M. Singh. (2009). Effect of temperature, light and pH on germination of twelve weed species. *Indian Journal of Weed Science,* 41(3&4): 113-126.

Thullen, R.J. and P.E. Keeley. (1979). Seed production and germination in *Cyperus esculentus* and *C. rotundus. Weed Sci.,* 27(5): 502-505.

Tumbleson, M.E. and T. Kommedahl. (1961). Reproductive potential of yellow nutsedge. *Weeds,* 9:646-653.

Yan Shen, Wei-Jun Chen, Xin-Tao Lei, Hong-Bo Shao, Min-MinTang and Yan Li. (2011). Effects of plant growth regulators and temperature on seed germination of yellow nut-sedge (*Cyperus esculentus* L.). *J. of Medicinal Plants Res.*, 5(31): 6759-6765.

CROWFOOT GRASS (*Dactyloctenium aegyptium*)

Common name : Crowfoot grass, beach wiregrass, coast button grass, comb fringe grass, duck grass, durban crowfoot, Egyptian fingergrass, Egyptian grass, finger comb grass, four-finger grass

Botanical name : Dactyloctenium aegyptium

Tamil name : Kakakal pul

Hindi name : Makra

Family : Poaceae

Origin and distribution : Crowfoot grass is native of old World tropics, but naturalized Worldwide (Holm *et al.,* 1979; Simpson, 1990).

Botany

Habitat : The genus *Dactyloctenium* is widespread and member species occur mostly in dry, sandy soils ranging in habitat from meso- to xerophytic (Holm *et al.*, 1979; Watson and Dallwitz, 1992). A common weed of open ground and waste places in the tropics and subtropics of the World (Wiggins and Porter, 1971). It prefers light, dry soils in Java and grows mainly on sandy areas in Sudan. Despite its preference for a habitat with light soils and low moisture, it has been reported to be an important weed in many countries in the humid tropics (Holm *et al.,* 1977).

Habit : Annual grass (Fig. 1).

Fig. 1. Plant

Stem : Culms 10-62.5 cm high.

Leaf : Leaves flat, 5 to 20 cm long, glabrous

Floral biology

Flower : Inflorescence of 2 to 5 spikes in terminal umbel, dark olive-grey, digitately radiating; the rachis projecting in a point beyond the spikelets (Fig. 2 a,b).

Flowering : The grass flowers in January – April.

a b

Fig. 2. a: Inflorescence; b: flower

Seed biology

Seed : Caryopsis is broadly obovate to triangular, 1 mm long, transversely rugose. Crowfootgrass seed is 0.7 to 1.0 mm in diameter (Fig. 3).

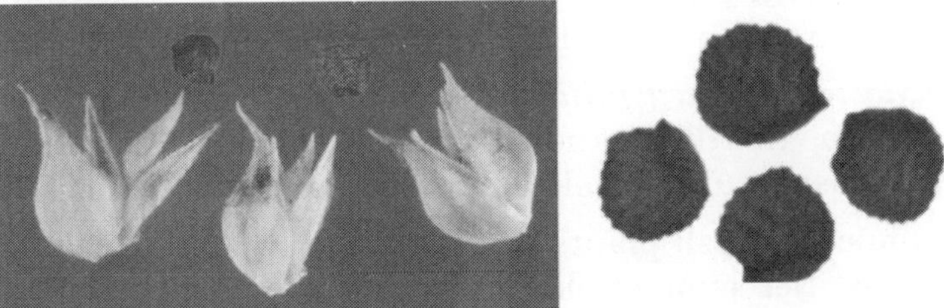

Fig. 3. Seed

Seed dispersal : Dispersed by ants and possibly internally by animals, although this is inconclusive; intact seeds survived digestion, but there were high numbers of non-germinating seeds both before and after digestion (Gardiner *et al.,* 1993). Wind/gravity is also likely to disperse the seed. Human mediated dispersal likely in contaminated machinery, produce and soil or stock feed.

Propagation : Reproduces mainly by seeds, but has creeping or spreading stems which root at the lower nodes (Holm *et al.,* 1977).

Seed dormancy and germination : The seeds exhibit primary dormancy due to hard seed coat (Jenik and Longman, 1965; Qadir and Shahzadi, 1969). Gupta (1973) found that 5 months after-ripening was required to break seed dormancy, while Popay (1974) observed that fresh seed was quite dormant and that 22 months storage of dry seed at room temperature was needed to raise germination to 20%.

Pricking the seed coat with a needle raised germination at room temperature (in Kenya) to 60-84% (Popay, 1974). Low temperatures (5°C) prolonged dormancy (Gupta, 1973).

Seed storage and burial : The seed retains viability for long periods in storage and it is likely to re-establish from the seed bank following disturbance or cultivation.

Effect on crop : According to Holm *et al.* (1977), *D. aegyptium*as is a serious weed of cotton in Thailand, and a principal weed in many crops in different countries. It is considered a weed in peanut production regions of the southeast United States, infesting 25 and 75% of the southeastern and Virginia-Carolina production regions, respectively (Bridges *et al.*, 1994a; 1994b).

Control measure : Pre-emergence herbicides like oxyfluorfen, pendimethalin, metolachlor, atrazine and simazine are effective in various crops (Murthy *et al.*, 2007; Ferrell *et al.*, 2006). Good biological control with fungal pathogens (*Drechslera gigantea, Exserohilium longirostratum* and *E. rostratum*) in the USA.

The response of crowfootgrass seed to fluctuating temperatures and shallow burial depths may indicate adaptation by crowfootgrass to disturbed bare ground (Adu *et al.*, 1994; Gupta, 1973; Nishimoto and McCarty, 1997; Sharma and Chivinge, 1982), where the greatest diurnal fluctuations would be expected. Shallow cultivation for weed control could potentially stimulate germination by placing crowfootgrass seed at the optimum depth for germination. The aforementioned attributes may also contribute to poor control later in the season by soil-applied herbicides as they degrade in the soil.

Uses : Dactyloctenium aegyptium is still a traditional food plant used as a famine food in Africa, this little-known grain has potential to improve nutrition, boost food security, foster rural development and support sustainable landcare.

In Manipur, juice of fresh plants is prescribed in fevers. Decoction of the plant is given in small pox. Decoction of plant, in fresh or in dry state, used internally for dysentery and acute hemoptysis. In Africa, decoction of seeds used to relieve pains in the region of the kidney; stems and leaves applied externally for treatment of ulcers.

In traditional Thai medicine, it is used for diuretic, anti-inflammatory and anti-pyretic effects. In India, juice of fresh plant used for fevers. Used externally for wounds and ulcers.

Past work done on Seed Science and Technology

Seed dormancy and germination

In previous experiments in India and the United States, crowfootgrass required either after-ripening treatments or a 5 months or more period of storage for after-ripening (Gupta, 1973; Taylorson and Brown, 1977). Crowfootgrass seed was 93 ±

4% viable by tetrazolium chloride tests (Peters, 2000) before each study was conducted. Crowfootgrass seed required a considerable period of storage or after-ripening before dormancy was relieved (Gupta, 1973). Consequently, seed was stored at room temperature (28°C) for a period of 6 months before experimentation was begun. Germination before and after storage was 9 ± 4% and 91 ± 4%, respectively, in a 20 and 30°C regime. It was also necessary to remove the pericarp before germination experiments because greater germination was observed without it in place (Gupta, 1973). Gupta (1973) and Kumar *et al.* (1971) reported an optimal constant temperature of 30°C for crowfootgrass seed germination.

Light was required for germination in several studies (Okusanya and Sonaike, 1991; Taylorson and Brown, 1977), but was not found to be a requirement by Kumar *et al.* (1971). Most research has focused on relieving the dormancy of crowfootgrass; however, several studies have examined depth of emergence and effects of substrate pH, salinity, and dry matter production (Adu *et al.*, 1994; Gupta, 1973; Kumar *et al.*, 1971; Okusanya and Sonaike 1991; Sharma and Chivinge, 1982).

Asif Saeed and Waheed Sabir (1993) reported that freshly harvested seeds of this wild grass showed a complete dormant behaviour when placed under favorable conditions for germination. Alternate temperature of 35°/23°±3.5°C for 8/16 hours significantly increased the germination rate and had dormancy breaking action on nearly half of both the intact and mechanically rubbed seeds when compared with lower stratified or higher temperature or with controlled experimental seeds at room temperature. However, mechanically rubbing the seeds alone had little effect. The pH 5.3 treated seeds also showed significant increase in germination rate when compared with the controlled experimental seeds; while the other pH levels had adverse effects. Pre-soaking the seeds in water for 10 days also improved the germination.

Taylarson and Brown (1977) studied the initial dormancy in mature seeds of 12 out of 14 species of grass weeds and reported that dormancy was significantly overcome by pretreatment of the dry seeds at 50°C in stoppered vials. The time required for maximum effect varied depending upon species and lots within species. The procedure, termed accelerated after-ripening, can produce seeds in various stages of dormancy simply by varying the duration of the pretreatment. One of the first stages that becomes apparent is an increased response to light.

Okusanya and Sonaike (1991) determined the germination response of *Dactyloctenium aegyptium* (L.) Beauv from two localities in Nigeria to the effects of some environmental factors. The two populations (Otta and University) germinated equally well in alkaline medium, but in acidic medium the Otta population germinated significantly better than the University of Lagos population. They germinated to the same extent in light and dark and equally well in humic soil. The Otta population had significantly better germination in the red earth and sand than the University population, even though germination in both populations in sand was significantly poorer than in humic soil and red earth. Under the various soil moisture contents tested, they both showed an increase in germination as soil moisture content decreased. The populations exhibited the usual pattern of minimum, optimum and maximum germination in response to the constant temperatures tested. The University population had a salinity tolerance range of 20% sea water with the percentage germination decreasing with increase in salinity, while the Otta population had a range of 100% with maximum germination in up to 50% sea water.

Effect of moisture stress

Dactyloctenium aegyptium was able to germinate at -0.6 MPa water potential (Hoveland and Buchanan, 1973).

Allelopathic effect

Sajjad Hussain *et al.* (2007) studied the allelopathic effect of senna (*Cassia angustifolia*) on germination and growth of *Dactyloctenium aegyptium*. Extract application of senna reduced germination to 54.55% followed by that of soil incorporation and mulcing. The shoot length was highest in extract application of senna, followed by mulching and soil incorporation. Mulching and extract application of senna both significantly affected the shoot length (Table 1).

Table 1. Germination, shoot and root growth parameters of *D. aegypticum* as affected by various senna treatments.

Treatment	Germination % over control	Shoot length (mm)	Root length (mm)
Control	100	13.67	29.83
Soil incorp	72.73	18.00	25.67
Mulching	63.64	32.50	78.83
Extract	54.55	56.00	51.83
LSD 0.05	-	9.43	37.11

Seed storage and burial

Burke *et al.* (2003) conducted laboratory and greenhouse studies to determine the effect of temperature, pH, water stress, and planting depth on crowfootgrass germination. When treated with constant temperature, crowfootgrass germinated over a range of 15 to 40°C, with the optimum germination occurring at 30°C (42%) (Fig. 4). Onset, rate, and total germination (94%) were greatest in an alternating 20 and 35°C temperature regime (Fig. 5). Germination decreased as pH increased, with greatest germination occurring at pH 4 and 5 (Fig. 6). Germination was reduced when seed was subjected to water stress and no germination occurred below -0.8 mPa (Fig. 7). Emergence was similar when seeds were placed on the soil surface or buried at depths of 0.5 or 1 cm. Germination decreased with burial depth, and no seed emerged from 10 cm (Fig. 8).

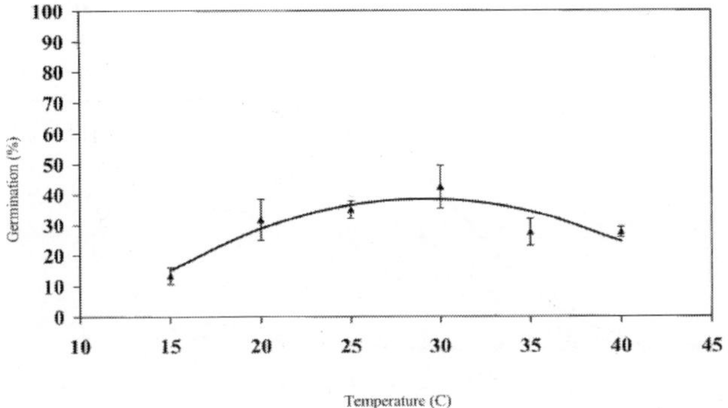

Fig. 4. Influence of constant temperature on cumulative germination of crowfootgrass seed at 8 d as described by the equation $y = -61.3 + 6.9\,(\text{temp}) - 0.12\,(\text{temp})^2$, $R^2 = 0.81$. Vertical bars represent standard errors of the mean.

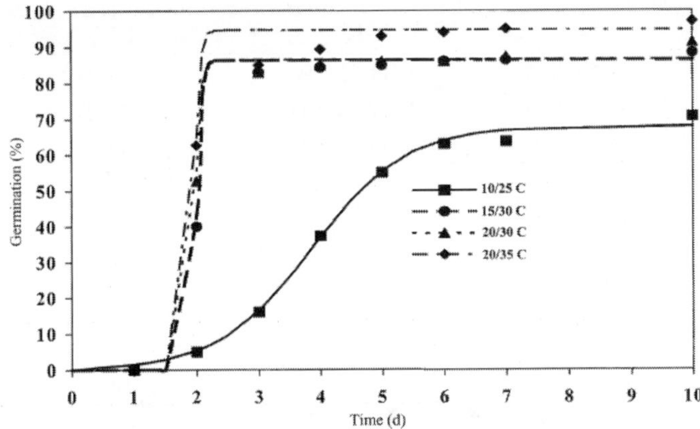

Fig. 5. Influence of four temperature regimes on crowfootgrass germination, modeled using equation y 5 $M\,(1 + \exp[-\,K(t - L)])^{-1}$ with estimated parameters (and standard errors) as follows:

Temperature	M	K	L	R^2
10/18	67.8 (2.9)	1.32 (0.37)	3.9 (0.2)	0.83
15/30	86.3 (2.9)	20.40 (0.24)	2.0 (0.1)	0.81
20/30	86.6 (3.2)	19.65 (0.38)	2.0 (0.1)	0.77
20/35	94.7 (2.3)	19.51 (0.27)	2.0 (0.1)	0.88

$y = M\,(1 + \exp[-\,K(t - L)])^{-1}$ [2] where y is the cumulative percentage germination at time t,

M is the asymptote or theoretical maximum for y, L is the time scale constant or lag to onset of germination, and K is the rate of increase (Roche *et al.* 1997).

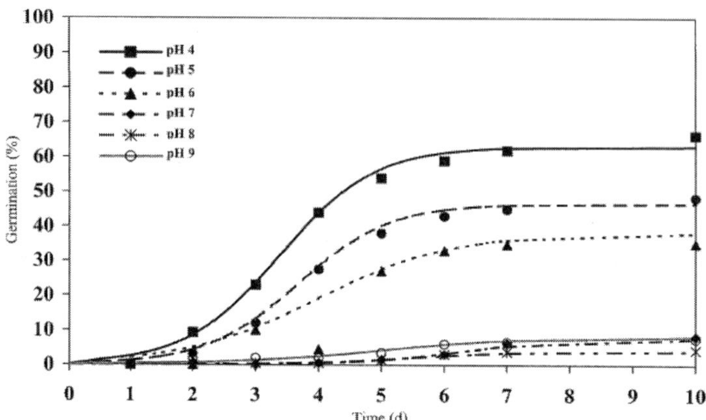

Fig. 6. Influence of solution pH on crowfootgrass germination averaged across temperature regimes, modeled using the equation $y = M (1 + \exp[-K(t - L)])^{-1}$ with estimated parameters (and standard errors) as follows:

pH	M	K	L	R^2
4	63.1 (2.7)	1.34 (1.29)	3.4 (0.2)	0.99
5	46.7 (2.0)	1.37 (0.44)	3.7 (0.2)	0.99
6	38.2 (2.2)	0.95 (0.17)	4.0 (0.4)	0.99
7	8.6 (0.3)	1.21 (0.21)	6.3 (0.2)	0.99
8	4.1 (0.2)	1.2 (0.23)	5.5 (0.3)	0.99
9	8.3 (0.5)	0.88 (0.21)	5.0 (0.4)	0.99

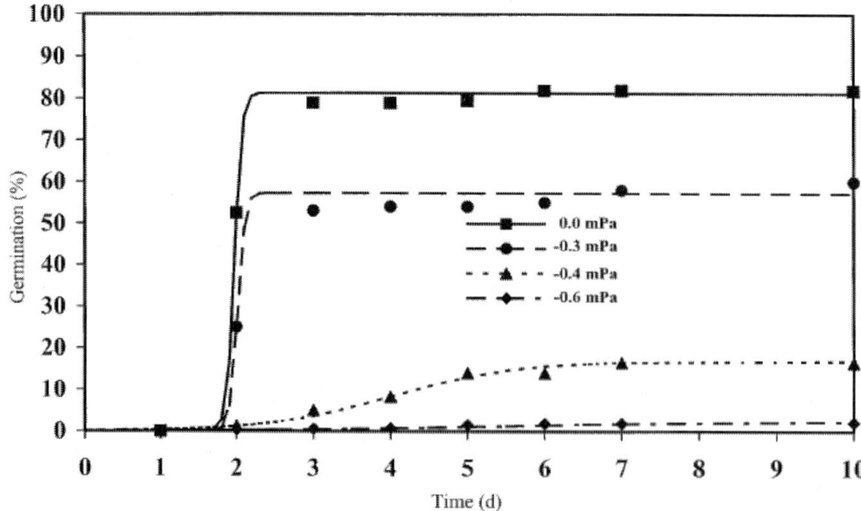

Fig. 7. Influence of water stress on crowfootgrass germination averaged across temperature regimes, modeled using the equation $y = M (1 + \exp[- K(t - L)])^{-1}$ with estimated parameters (and standard errors) as follows:

mPa	M	K	L	R^2
0.0	81.3 (2.3)	19.8 (4.8)	2.0 (0.2)	0.75
– 0.3	57.3 (3.3)	19.6 (0.3)	2.0 (0.3)	0.83
– 0.4	16.9 (8.4)	1.3 (1.2)	4.1 (0.3)	0.69
– 0.6	2.5 (0.5)	0.67 (0.6)	5.5 (1.5)	0.89

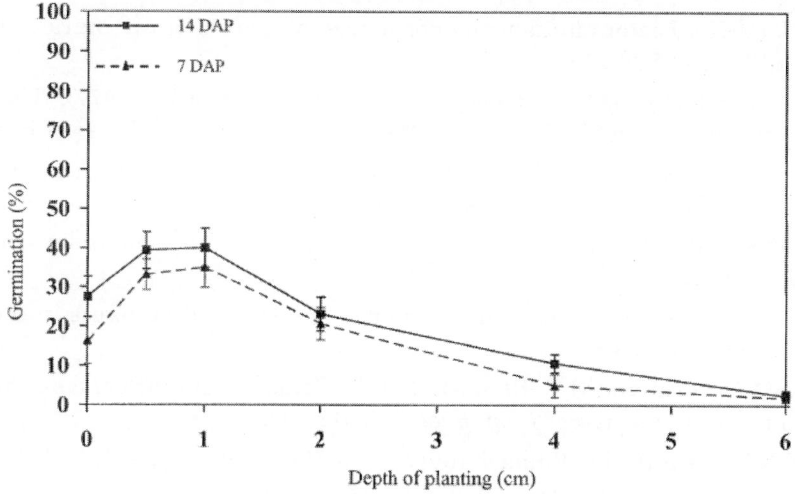

Fig. 8. Cumulative emergence of crowfootgrass seed buried 0, 0.5, 1, 2, 4, or 6 cm 7 and 14 d after planting. Vertical bars represent standard errors of the mean.

Decreasing germination of crowfootgrass seed with increasing burial depth in populations from Nigeria and India, and the preponderance of germination occurred at burial depths of 1.5 cm or less was observed (Kumar *et al.*,1971; Sharma and Chivinge, 1982).

REFERENCES

Adu, A.A., A.R. Yeo and O.T. Okusanya. (1994). The response to salinity of a population of *Dactyloctenium aegyptium* from a saline habitat in southern Nigeria. *J. Trop. Ecol.*, 10: 219-228.

Asif Saeed, M. and Waheed Sabir, A. (1993). Dormancy behaviour of *Dactyloctenium aegyptium* and effect of temperature, pH and pre–soaking on seed germination. *Pakistan J.Agric. Res.* 14 (4).

Bridges, D.C., C.K. Kvien, J.E. Hook and C.R. Stark, Jr. (1994a). An analysis of the use and benefits of pesticides in U.S.-Grown peanut: I Southeastern Production Region. Tifton, GA: National Environmentally Sound Production Agriculture Laboratory. 42 p.

Bridges, D.C., C.K. Kvien, J.E. Hook and C.R. Stark, Jr. (1994b). An analysis of the use and benefits of pesticides in U.S.-Grown peanut: III Virginia-Carolina Production Region. Tifton, GA: National Environmentally Sound Production Agriculture Laboratory. 39 p.

Burke, I.C., W.E. Thomas and F.S. Jane (2003). Influence of environmental factors on after-ripened crowfootgrass (*Dactyloctenium aegyptium*) seed germination. *Weed Sci.*, 51: 342–347.

Ferrell, J.A., Murphy, T.R. & Webster, T.M. (2006). Using pre-emergence herbicides to improve establishment of centipede grass (*Eremochloa ophiuroides*) from seed. *Weed Technology* 20(3): 682-687.

Gardiner, C.J., J.G. McIvor and A. Jansen. (1993). Survival of seeds of tropical grassland species subjected to bovine digestion. *Journal of Applied Ecology,* 30(1): 75-85.

Gupta, K.C. (1973). Factors influencing dormancy in seeds of crowfootgrass. *Biochem. Physil. Pflanz.* BPP., 164:582-587.

Holm, L.G., Plucknett, D.L., Pancho, J.V. & Herberger, J.P. (1977). The World's Worst Weeds: distribution and biology. East-West Centre/University Press of Hawaii.

Holm, J.R., J.V. Pancho, J.V. Herberger and D.L. Plucknett (1979). A geographical Atlas of World weeds. John Wiley and Sons: New York. p. 119.

Hoveland, C.S. and G.A. Buchanan. (1973). Weed seed germination under simulated drought. *Weed Sci.,* 21:322-324.

Jenik and K.A. Longman. (1965). The dormancy and survival of plants in the humid tropics. *J. W. Afr. Sci. Ass.,* 10:64.

Kumar, A., M.C. Joshi and B.V. Ramesh. (1971). Some factors influencing the germination of seeds in two desert grasses. *Trop. Ecol.*, 12:202-208.

Murthy, K.N.K., Murali, K., Ramachandra, C. & Rajashekarappa, K.S. (2007). Effect of pre-emergent herbicides on weed control, growth and yield of transplanted onion (*Allium cepa* L.). *Research on Crops* 8(1): 204-208.

Nishimoto, R. K. and L. B. McCarty. (1997). Fluctuating temperature and light influence seed germination of goosegrass (*Eleusine indica*). *Weed Sci.* 45:426–429.

Okusanya, O.T. and A.A. Sonaike. (1991). Germination behaviour of *Dactyloctenium aegyptium* from two localities in Nigeria. *Physiol. Plantarum.*, 81:489-494.

Peters, J., Ed. (2000). Association of Official Seed Analysts Tetrazoleum Testing Handbook Contribution #29, 1st Revision. Lincoln, NB: Assoc. Off. Seed Anal. pp. 19–21.

Popay, A.I. (1974). Investigations into the behaviour of the seeds of some tropical weeds. 1. Laboratory germination tests. *East African Agricultural and Forestry Journal,* 40: 31- 43.

Qadir, S.A. and H.N. Shahzadi. (1969). Seed germination of common cultivated trees, shrubs and some wild grass. *Pak. J. Forestry,* 19:195-220.

Roche, C.T., D.C. Thill and B. Shafii. (1997). Estimation of base and optimum temperatures for seed germination in common crupina (*Crupina vulgaris*). *Weed Sci.,* 45:539–533.

Sajjad Hussain, Sadar Uddin Siddiqui, Shahida Khalid, Atif Jamal, Abdul Qayyum and Zahoor Ahmad. (2007). Allelopathic potential of senna (*Cassia angustifolia* vahl.) on germination and seedling characters of some major cereal crops and their associated grassy weeds. *Pak. J. Bot.*, 39(4): 1145-1153.

Sharma, B.M. and A.O. Chivinge. (1982). Contributions to the ecology of *Dacytloctenium aegyptium* (L.) P. Beauv. *J. Range Man.*, 35: 326-331.

Simpson, G.M. (1990). Seed dormancy in grasses. Press Syndicate of the University of Cambridge: New York. p. 35.

Taylorson, R.B. and M.M. Brown. (1977). Accelerated after-ripening for overcoming seed dormancy in grass weeds. *Weed Sci.,* 25: 473-476.

Watson, L. and M. J. Dallwitz. (1992). The Grass Genera of the World. Wallingford, U.K.: CAB International. pp. 276–277.

Wiggins, I.L. and D.M. Porter. (1971). Flora of the Galapagos Islands, Stanford University Press, Noticias de Galapagos, vol. 44 1986 Stanford, Calif. 998pp.

THORN APPLE *(Datura* spp.*)*

Common name : Thorn apple

Botanical name : Datura spp.

Tamil name : Oomathai

Hindi name : Dhatura

Family : Solanaceae

Origin and distribution : It is native to Mexico and has been introduced from there throughout the tropics and subtropics. It is probably grown throughout tropical Africa as an ornamental. It is documented as a naturalized weed in many African countries, but is probably seriously under-reported.

Botany

Habitat : It is locally abundant in waste places, roadsides, brushwood, open grassland, teak forest and on river banks, from sea-level up to 1000 m altitude. It tolerates various soil types, but prefers clayey or loamy soils. Often with a double and purplish corolla, is widely cultivated in gardens.

Habit : Annual shrubby herb (Fig. 1).

Stem : Stem short-hairy to glabrous.

Leaf : Leaves large, entire, sinuate or deeply toothed, lanceolate, unequal, up to about 10 cm long or more.

Floral biology

Flower : Flowers pedicillate, solitary, terminal, white in color; calyx: sepals 5, gamosepalous, persistent forming a cup shaped structure at the base of mature fruit; corolla: petals 5, gamopetalous, infundibuliform, twisted, tips of the teeth acuminate; androecium: stamens 5, epipetalous; gynoecium: bicarpellary, syncarpous; ovary superior, obliquely placed, each loculous contains numerous ovules; style long; stigma capitate (Khan and Hussain, 1960) (Fig. 2).

Fig. 1. Plant

Fig. 2. Inflorescence

Flowering and pollination : Flower from June to July. Pollination by hawk moths and to be largely self fertile.

Fruit : Fruit thorny capsule, dehisces septifragally to form four valves to liberate numerous seeds (Wallis, 1960) (Fig. 3).

Fig. 3. Fruit

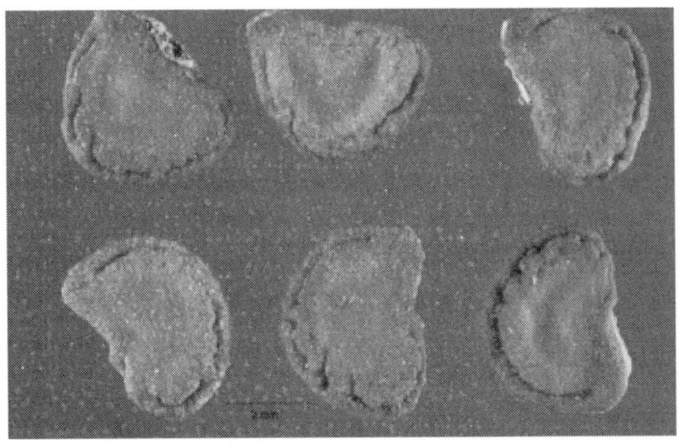

Fig. 4. Seed

Seed biology

Seed : Seeds flattened, reniform, black or dull dark brown and marked by shallow reticulate depressions (Fig. 4).

Propagation : By seeds

Seed dispersal : It splits open when ripe, dispersing the seeds. Another means of dispersal is by the fruit spines getting caught in the fur of animals, who then carry the fruit far from the mother plant.

Seed dormancy and germination : According to Dorado *et al.* (2009), datura seeds posses some type of embryo dormancy. In *Datura stramonium*, the dormancy is due to hard seed coat (Sarfaraz Khan Marwat *et al.*, 2005).

Seed storage : Long-term storage under IPGRI preferred conditions. It can lie dormant underground for years and germinate when the soil is disturbed.

Effect on crop : The plant is now considered an invasive species in several locations. For example, because of the similarity of its life cycle to that of cotton, it is a pest in cotton fields. It is also a potential seed contaminant.

Control : Digging and pulling out plants before seed is set. Contact herbicides such as diquat, fatty acids pelargonic acid or glufosinate ammonium should easily control this weed. Systemic herbicides such as glyphosate (Round Up or Tumble weed) will also be effective and are the material of choice where mature plants are to be eliminated.

Medicinal uses : Relieving asthma, cough, tuberculosis and bronchitis by smoking the dried leaves, roots or flowers as a cigarette or in a pipe. These 'asthma cigarettes' have been shown to be very effective in some cases.

In Africa, other traditional uses of the leaves include the treatment of haemorrhoids, boils, sores, skin diseases, rheumatism, headache, toothache, cholera, parasites such as ringworm and guinea worm, and as an abortifacient and anaesthetic.

In Ethiopia, the seed oil is used for massaging painful body parts. It is also widely used as a pesticide. Seeds mixed with sorghum flour are used as poison bait for rats. Plant extracts have been used successfully against caterpillar pests in forestry, cotton pathogens and domestic vermin such as fleas and jigger, and to reduce insect damage in stored wheat and maize.

Past work done on Seed Science and Technology

1. Seed dormancy and germination

In *Datura ferox*, the LFR of seed germination involves both an increase in the expansive force of the embryo and a decrease in the mechanical resistance of the micropylar portion of the endosperm (Sanchez *et al.,* 1986; de Miguel and Sanchez, 1992). Endosperm softening involves degradation of the main cell-wall component (1β, 1-4 mannan) by endo- β -mannanase and β–mannosidase (Sanchez *et al.*, 1990). The mobilization of reserve proteins and extensive ultra-structural modifications in other parts of the cells accompany the changes in the walls. This pre-radicle protrusion syndrome is almost completely confined to the micropylar portion of the endosperm (Sanchez *et al.*, 1990; Mella *et al.*, 1995; Sanchez and de Miguel, 1997).

Presence of dormancy in *Datura* spp is because of: (1) a significant effect with fluridone + GA (in seeds collected in 2005, germination reached 41% after this treatment, compared with 10.5% in the control) was observed; (2) the cumulative germination did not reach 100% (the maximum germination was 73% in naturally chilled seeds collected in 2005) at any treatment; and (3) the low percentage of germination in the seeds collected in 2006, even after pricking where a maximum percentage germination of 14% was observed. These results are suggesting that, under field conditions, seeds of *Datura* spp are released from dormancy in two steps, the first one during the spring of the first year and the second step during the subsequent winter. The high sensitivity to light perceived by phytochromes has long been known to be involved in dormancy mechanisms of *Datura* spp and a minimum period of burial would be needed to reduce or eliminate the light requirement and induce seed germination (Jose Dorado *et al.*, 2009) (Table 1).

Table 1. Cumulative germination data (%) observed at 20°C in seeds collected in 2004, 2005, and 2006 in *Datura* spp.

	***Datura* spp**				
	2004	**2004 NC[a]**	**2005**	**2005 NC**	**2006**
Control	2±1.7*	5±2.6	11±3.6	6±6.0	0±0.0
Fluridone + GA	7±4.2	13±6.1	41±14.8	N.O	4±3.0
Pricking	30±7.1	53±5.9	68±4.7	73±10.3	14±11.0

[a]Abbreviations: NC, natural chilling of seeds during 2 months; fluridone + GA, mixture of 42 mM of 95% fluridone and 100 mM gibberellin; pricking, seeds were pricked manually using a scalpel; N.O., germination was not observed because of the contamination of the dishes.
* Standard deviation on the basis of four replications.

Sarfaraz Khan Marwat *et al.* (2005) indicated that in untreated seeds of *Datura stramonium* L., maximum germination was observed at 30°C, suddenly decreased at 32°C and absolutely negative results were obtained at 25°C and 35°C (Table 2). With the removal of testa, greatest germination was observed at 30°C after 10 days (Table 3). It is concluded that optimum temperature for the germination of *Datura stramo-nium* L. seeds is 30°C. With the removal of testa the germination increased. So, it is concluded that hard testa of *Datura stramonium* L. seed slows down germination.

Table 2. Results showing the percentage of germination of untreated seeds of *Datura stramonium* L. at different temperatures in the laboratory.

No. of Days	**Temperature**			
	25°C	**30 °C**	**32°C**	**35 ° C**
2	-	-	-	-
4	-	28.33	8.33	-
6	-	36.66	10.00	-
8	-	36.66	11.66	-
10	-	36.66	18.33	3.33

Table 3. Results showing the percentage of germination of mechanically scraped seeds of *Datura stramonium* L at various temperatures in laboratory.

No of Days	Temperature			
	25°C	30°C	32°C	35°C
2	5.00	11.66	3.33	13.33
4	13.33	18.33	8.33	16.66
6	16.66	30.00	18.33	16.66
8	18.33	35.00	21.66	21.66
10	18.33	40.00	33.33	23.33

Arben Mehmeti *et al.* (2011) reported that germination percentage reached at the end of the six-week period of non-stratified seeds was slightly different between the populations, with highest rates for populations originating from the western part of Kosovo (around 90%), and lowest for population from the eastern part (around 85%). Germination percentage of non stratified populations were much higher than those with stratification over all tested temperature (Fig. 5).

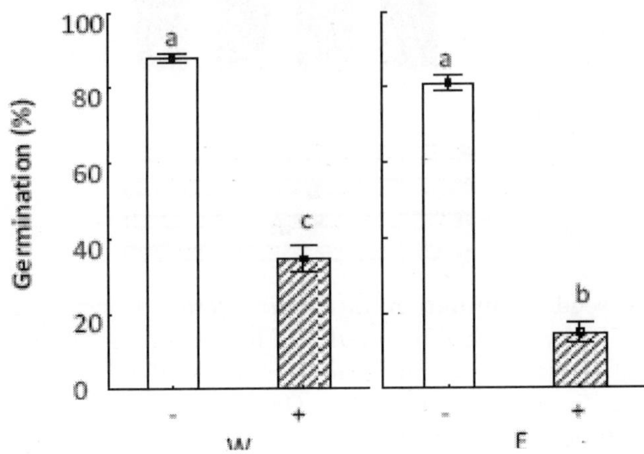

Fig. 5. Percentage seed germination mean over all tested temperatures of populations of *D. stramonium* at the end of a six week period in climate chamber experiment. Seeds (□: non-stratified + cold-stratified) originated from two sub – regions (W = western part, E= eastern part) of Kosovo. Data are means ± SE for five replicates (50 seeds per replicate). Means sharing the same letter are not significantly different (P> 0.05; Tukey's HSD).

2. Effect of light on germination

In the seeds of several species, germination is promoted by a LFR and inhibited by several hours of exposure to continuous FR given even after a pulse of FR is no longer able to cancel the effect of R (Negbi and Koller, 1964; Hartmann, 1966;

Bewley and Black, 1982). This effect of continuous FR is known as a high-irradiance response (HIR) and is most efficient between 710-720nm (Frankland and Taylarson, 1983). Interestingly, during de-etiolation, VLFR, LFR and HIR inhibits germination (Casal *et al.,* 1998) and this failure has generally not been tested for seed germination.

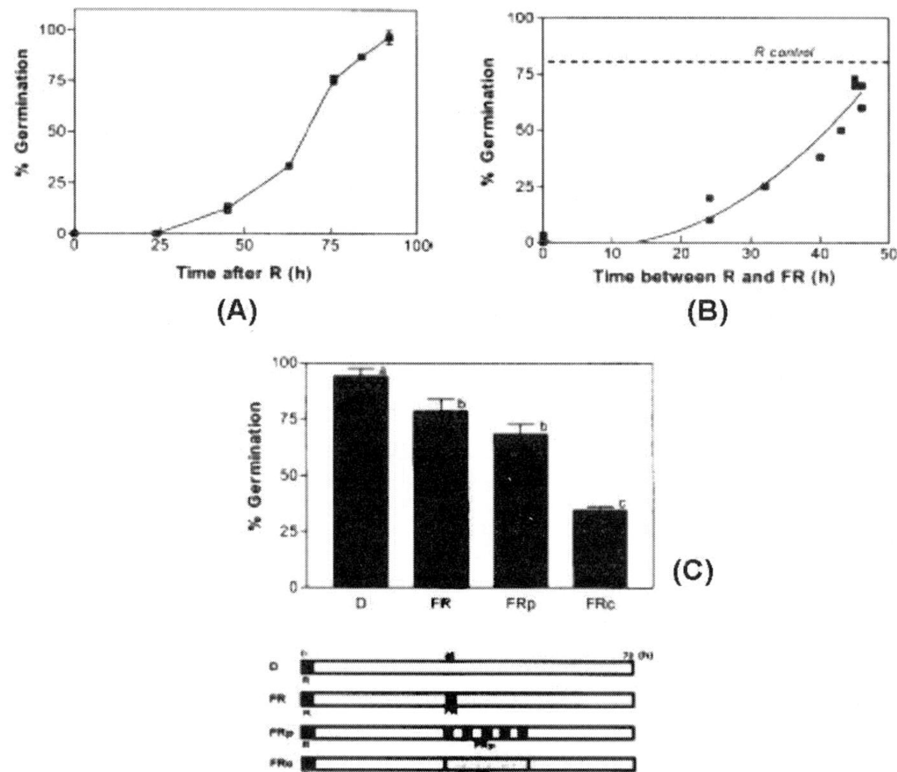

Fig. 6. The HIR of seed germination in *Datura ferox*. Germination of decoated seeds incubated at 20-30°C is promoted by a pulse of R (A). A pulse of FR cancels the effect of R when given immediately after R, but gradually loses effectiveness when a dark period is interposed between R and FR pulses (B). Germination of decoated *Datura ferox* seeds given a R pulse and incubated for 45h in darkness (20-30°C) is inhibited by 6h of FR (FRc) (15 mol m^{-2} s^{-1}), but not by hourly FR pulses (FRp) (3 min h^{-1}, 300 μmol m^{-2} s^{-1}) or a single FR pulse followed by 5.5 h of darkness (FR) (30 min h^{-1}, 15 μmol m^{-2} s^{-1}). Some seeds remained as controls not exposed to any FR (D). All the seeds were further incubated in darkness to complete 72 h after the initial R pulse (25°C). Data are means and SE of 4-5 independent experiments. Different letters indicate significant differences, $P < 0.05$.

Sanchez *et al.* (1986) studied the influence of phytochrome on endosperm softening and cellulase activity on light-stimulated *Datura ferox* seeds. Endosperm softening preceded the earliest signs of radicle protrusion, and there was good correlation between the % of seeds with soft endosperm at 48 h after red light (R) and germination at 96 h after R. Cellulase activity was stimulated by R and the increase in activity preceded, by more than 24h, radicle protrusion and endosperm softening. The effect of R was reversed by far red (FR), but by delaying the irradiation with FR until cellulase activity had increased significantly, it was observed that removing Pfr did more than just stop any further increase, the level of cellulase activity decreased in about 24 h close to the dark controls. Cellulase activity was decreased by a FR irradiation even when more than 60% germination had escaped from reversion. These results indicate that phytochrome influence on cellulase is not an indirect consequence of the stimulus of germination and that the continuous presence of Pfr is required for the cellulase activity to remain high.

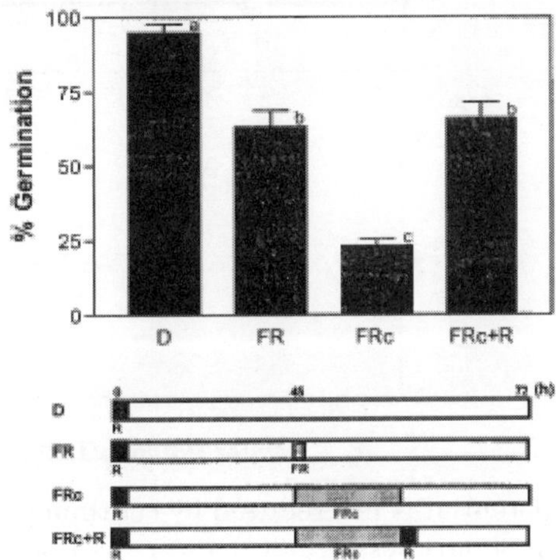

Fig. 7. The inhibition of germination imposed by continuous FR is cancelled by a subsequent R pulse. Decoated seeds were given a R pulse followed by 45 h of darkness (20-30°C) and then exposed to a single pulse of FR (FR), 6h of FR (FRc), 6h of FR followed by a R pulse (FRc + R) or no FR (D). All the seeds were further incubated in darkness (25°C) to complete 72 h after the initial R pulse. Data are means and SE of 3-5 independent experiments. Different letters indicate significant differences, $P<0.05$.

Formation of Pfr sets in motion several processes that end in seed germination. In seeds with coat-imposed dormancy, some of these processes takes place in the embryo, while others are located in the micropylar portion of the endosperm. The time required for Pfr to complete its intervention is different for each one. In *Datura ferox* the escape time for promotion of embryo growth potential is approximately 24h (de Miguel and Sanchez, 1992), whereas for endosperm softening and germination it is longer than 45 h (de Miguel and Sanchez, 1992, Fig. 6B). Even after the apparent completion of Pfr action, prolonged FR (6h) can inhibit germination (Fig. 6C). This effect of FR, which goes beyond converting existing Pfr into Pr, is known as HIR. It is shown that in *Datura ferox,* continuous FR is effective as late as 45h after an inductive R pulse. At this time, extensive physiological and structural changes have already taken place. The activity of mannan hydrolases has been high for more than 15h, the mobilization of the protein bodies and the formation of vacuoles in the endosperm cap are well underway and endosperm softening is clearly perceptible (Mella *et al.*, 1995; Sanchex and de Miguel, 1997). Despite these changes, 6h of continuous FR blocked germination and reimposed the need for a high level of Pfr to restart the processes (Fig. 7). A likely target for this last-minute action was embryo growth potential. However, continuous FR did not affect embryo growth potential (Fig. 8).

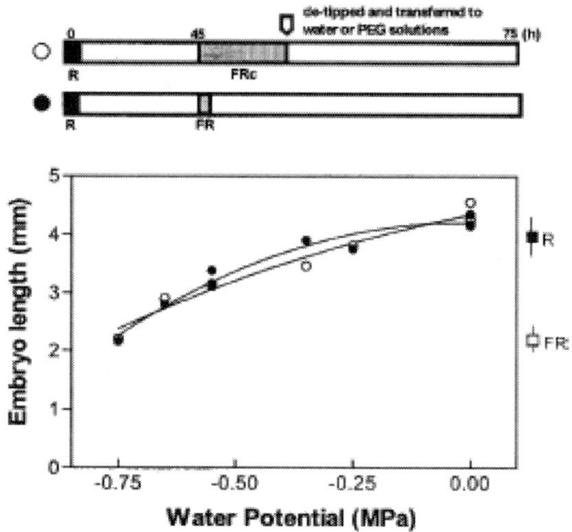

Fig. 8. Embryo growth potential is not reduced by continuous FR. Seeds pretreated with a R pulse followed by 45 h of darkness (20-30°C) were exposed to 30 min FR followed by 5.5 h of darkness(•) or 6 h FR (O). At this point, all the seeds were de-tipped under green safelight and transferred to polyethylene glycol solutions of the indicated water potential. The length of the embryos was measured after 24h of further incubation in darkness (25° C). Each datum point is the average of 10 embryos. The length of the embryo in seeds given only a R (■) or FR (□) pulse followed by 51h of darkness is indicated.

Sanchez and de Miguel (1997) shown that endosperm weakening, a necessary step in the light induction of germination in *Datura ferox* seeds, is related to the hydrolysis of cell wall mannose polysaccharides. It is demonstrated that the activities of two enzymes involved in mannan degradation (ß-mannanase and ß-mannosidase) are strongly promoted in the micropylar endosperm, after Pfr formation but long before radicle protrusion. An increase in enzyme activities occurred in dissected seed parts at similar rates to those in whole seeds, but only when at least a portion of the embryonic axis was present during incubation. De-embryonation of seed parts prevented the Pfr-induced increase in hydrolase activities. Exogenous GA_3 promoted mannan-degrading enzyme activities both in the micropylar region of far red-treated whole seeds and those of isolated and deembryonated endosperm sections. The response to Pfr and GA_3 was far more pronounced in the micropylar portion than in the rest of the endosperm. Pfr promotion of hydrolase activities and germination was blocked by paclobutrazol, an inhibitor of gibberellin synthesis. A close correlation was found between ß-mannanase activity measured 45 h after Pfr formation and germination scored 24 h later. These results support the hypothesis that in D. ferox, endosperm softening is promoted by Pfr via the production in the embryonic axis of a factor, probably a gibberellin, which moves to the endosperm where it stimulates cell wall-mannan hydrolytic enzyme activities.

Botto *et al.* (1998) studied the dormancy pattern of buried seeds of *Datura ferox* by testing germination responses to red light (R), far-red light (FR) or darkness after exhumation at regular time intervals. The seeds were deeply dormant at the time of burial in winter. After three months (spring), exhumed seeds began to respond to an R pulse. Summer temperatures did not induce secondary dormancy. During winter of the second year, seed germination after an R pulse increased further; however, germination after an FR pulse or in darkness also increased and reached values close to R. Changes in seed germination after an R or an FR pulse and in darkness were also studied, combining different depths of burial (0.5, 5 and 10 cm), presence of vegetation (with or without vegetation canopies), and date of exhumation (4 or 7 months after burial). The seeds placed at 0.5 cm depth did not germinate in any light treatment. A very high sensitivity to light, indicated by the promotion of germination by an FR pulse compared to darkness, was exhibited only in seeds buried at 5 cm in a soil without vegetation. A typical R/FR reversible response was observed in seeds exhumed at 5 and 10 cm in almost all situations of burial; except in a soil without vegetation, at 10 cm depth, in late summer when even dark controls showed high germination. It is concluded that buried seeds of D. ferox are released from dormancy in two steps, the first step during the spring of the first year and the second step during the subsequent winter. This process is not obviously controlled by summer temperatures. The acquisition of a very high sensitivity to light depends on vegetation cover and depth of burial, and thus it is predicted to be affected by agricultural practices.

de Miguel *et al.* (2000) investigated whether the red light pulse and prolonged far-red light act on the same or different processes during germination of *Datura ferox* seeds. Forty-five hours after the inductive red light pulse, germination could not be blocked by one pulse or six hourly pulses of far-red light, but was significantly reduced by 6 h of continuous far-red light. The pulse of red light increased embryo growth potential and the activities of β–mannanase and β–mannosidase extracted from the micropylar region of the endosperm. Continuous far-red light had no effect on embryo growth potential or β–mannosidase activity, but severely reduced the activity of β–mannanase. The effect of far-red light had the features of a high-irradiance response of phytochrome. Both germination and β–mannanase activity were restored by a pulse of red light given after the end of the continuous far-red treatment. It is concluded that the low-fluence response and high-irradiance response modes of phytochrome have antagonistic effects on seed germination and that the control of β–mannanase activity is one process where this antagonism is established.

Rodolfo A. Sánchez *et al.* (2002) observed that phytochrome-induced softening of the micropylar endosperm of Datura ferox seeds is followed by radicle protrusion and preceded by a significant increase in the extractable activity of β-mannanase and β-mannosidase (6 and 3 times over the far-red light controls, respectively) and a decrease in the mannan content of the micropylar endosperm cell walls. This relationship between phytochrome-induced germination, endosperm softening and mannan degradation was studied in Datura ferox seeds subjected to a range of water potentials during the presence of P_{fr}. Low water potential inhibited the decrease in mannose-rich cell-wall insoluble polysaccharides, the increase in activity of β-mannosidase and endosperm softening. A good correlation was found between activity of β-mannosidase measured at 45 h after exposure to a red light pulse and germination counted 24 h later when, with different external water potentials, germination varied between 0 and 90%. In contrast, no inhibitory effect of low water potential on *in vitro* β-mannanase activity was detected. It is suggested that the in vivo action of β-mannanase might be limited by accumulation of manno-oligosaccharides caused by inhibition of β-mannosidase. These data are consistent with the concept of a link between mannan mobilization and endosperm softening. Part of the effect of low water potential on P_{fr}-induced germination may be mediated by inhibition of β-mannosidase activity, with consequences for mannan hydrolysis and endosperm softening.

3. Effect of Germination

Safaraz Khan Marwat *et al.* (2005) found maximum germination percentage in leaf mold soil, in both pot and field experiments. This is due to its properties which provide most favorable environment for germination (Table 4).

Table 4. The percentage of germination of *Datura stramonium* L/ seeds in the pot and field experiment in different soil conditions

Type of soil	Germination (%) at pot	Germination (%) at field
Clayey soil	40	60
Loamy soil	42	40
Sandy soil	56	64
Control soil	16	48
Leaf Mold (Soil)	60	72

4. Seed burial

Benvenuti and Macchia (1995) carried out trials in order to investigate the effect of hypoxia on seed germination of *Datura stramonium* L. in petri dishes and when buried at various depths in soil. Hypoxia was found to cause a decrease in germination capacity and germination rate. This inhibition was partially alleviated by daily exchange of hypoxic gas surrounding the seeds during incubation. Similarly, seed scarification allowed maintenance of a higher germination capacity under conditions of low oxygen availability, showing that the seed coat was only partially gas permeable. Oxygen deficiency led to a decrease in respiratory capacity. However, this was probably compensated for by induction of fermentation metabolism. The possibility of removing the final products of fermentation exerts a decisive influence on seed germination, especially in an environment such as soil, where their diffusion into the surrounding environment is restricted. Thus daily nitrogen flushing partially eliminated this inhibition, even under conditions of low external oxygen availability. It was therefore, postulated that the main depth–derived inhibition was not caused directly by oxygen deficiency but by the increasing difficulty in eliminating toxic fermentation products, which was found to be proportional to the degree of hypoxia. Finally, incubation for several days under completely anaerobic conditions induced secondary dormancy. This was probably due to an ecological adaptation mechanism that prevents germination under conditions that are unfavourable for survival.

Benvenuti (2003) performed laboratory trials to test germination and emergence characteristics of jimsonweed (*Datura stramonium* L.) seeds buried in 10 different soil types (with or without the control of soil external gas environment) with pronounced sandy or clay texture. Germination inhibition due to burial depth was found to be directly proportional to clay content and inversely proportional to sand content. Measurement of soil air permeability showed a close relation between gas exchange potential and depth-mediated germination inhibition. Comparative analysis of the germination response in nonsoil and soil hypoxia suggested that inhibition is caused not so much by hypoxia *per se* as by the presence of fermenting metabolites that could not easily be eliminated due to decreased respiratory activity. *In situ* inspection of buried seeds also revealed that the increased time required for emergence in day soils is primarily due to increased mean germination time rather than greater difficulty in seedling penetration upwards through the soil before emergence. Partial removal of germination inhibition of buried seeds was facilitated by elevated air oxygen availability, but only with sandy soils, showing that inhibition is closely linked to soil ability to induce gas exchange with external air. At excessive burial depth (12 cm), seeds exhibited induction of secondary dormancy independent of soil texture. In conclusion, these experiments demonstrated that soil physical properties have a strong effect on buried-seed ecology and consequently on seedbank dynamics in the agroecosystem.

REFERENCES

Arben Mehmetia, Adem Demaja and Rainer Waldhard. (2011). Germination traits of *Datura stramonium* L. in Kosovo. 41st Annual Meeting of the Ecological Society of Germany, Austria and Switzerland, Oldenburg, Germany, 5-9 Septenber.

Benvenuti, S. and M. Macchia. (1995). Effect of hypoxia on buried weed seed germination. *Weed Res.,* 35(5): 343-351.

Benvenuti, S. (2003). Soil texture involvement in germination and emergence of buried weed seeds. *Agronomy J.,* 95(1):191-198.

Bewley, J. D. and M. Black. (1982). The physiology and biochemistry of seeds in relation to germination. New York. Springer Veriag, 1982.

Botto, J. F, R. A. Sanchez and J. J. Casal. (1995). Role of phytochrome B in the induction of seed germination by light in Arabidopsis thaliana. *J. of Plant Physiology*, 146:307-312.

Casal, J. J., R. A. Sanchez and J. F. Botto. (1998). Modes of action of phytochromes. *J. Exp.t Bot.*, 49: 127-138.

de Miguel L., M.J. Burgin, J.J. Casal and R.A. Sanchez. (2000). Antagonistic action of low-fluence and high-irradiance modes of response of phytochrome on germination and ß-mannanase activity in *Datura ferox* seeds. *Journal of Experimental Botany*, Vol. 51, No. 347, pp.1127-1133.

de Miguel, L and R. A. Sanchez. (1992). Phytochrome-induced germination, endosperm softening and embryo growth potential in *Datura ferox* seeds: sensitivity to low water potential and time to escape to FR reversal. *J. of Experimental Botany*, 43: 969-974.

Frankland, B. and R. B. Taylarson. (1983). Light control of seed germination. **In:** Shropshire W, Mohr H, ed. *Encyclopedia of plant physiology*. New York: Springer Veriag, 428-456.

Halmer, P. and J. D. Bewley. (1979). Mannanase production by the lettuce endosperm. *Planta*, 144: 333-340.

Hartmann, K. M. (1966). A general hypothesis to interpret 'high energy phenomena' of photo-morphogenesis on the basis of phytochrome. *Photochemistry and Photobiology*, 5: 349-366.

Jose Dorado, Cesar Fernandez-Quintanilla and Andrea C. Grundy. (2009). Germination patterns in naturally chilled and nonchilled seeds of Fierce Thornapple (*Datura ferox*) and Velvetleaf (*Abutilon theophrasti*). *Weed Science* 57:155–162.

Khan, A. H and S. M. Hussain. (1960). *Datura stramonium* L. and its cultivation in Abbot-abad, *Pak. J. Sc.,* 12(3): 137- 141; Republished in 1962, in *Pak. J.* For. P.F.I. Peshawar; 12: 90-92.

Mella, R. A, S. Maldonado and R. A. Sanchez. (1995). Phytochrome-induced structural changes and protein degradation prior to radicle protrusion in Datura ferox seeds. *Canandian J. of Bot.,* 73: 1371-1378.

Negbi, M. and D. Koller. (1964). Dual action of white light in the photocontrol of germination of *Oryzopsis miliacea. Plant Physiology*, 39: 247-253.

Rodolfo A. Sánchez, Lucila de Miguel, Carlos Lima and Rosa M. de Lederkremer. (2007). Effect of low water potential on phytochrome-induced germination, endosperm softening and cell-wall mannan degradation in Datura ferox seeds. Seed Science Research, 12(3):pp 155-163.

Sanchez, R. A. and L. de Miguel. (1997). Phytochrome promotion of mannan-degrading enzyme activities in the micropylar endosperm of *Datura ferox* seeds requires the presence of the embryo and gibberellins synthesis. *Seed Sci. Res.*, 7: 27-33.

Sanchez, R. A., L. de Miguel and O. Mercuri. (1986). Phytochrome control of cellulose activity in *Datura ferox* L. seeds and its relationship with germination. *J. Expt. Bot.*, 37; 1574-1580.

Sanchez, R. A., L. Sunell, J. Labavitch and B. A. Bonner. (1990). Changes in endosperm cell walls of two *Datura* species before radicle protrusion. *Plant Physiology*, 93, 89-97.

Sarfaraz Khan Marwat, Fazal ur Rehman and Saifullah khan. (2005). Germination of seeds of *Datura stramonium*. under different conditions (temperature & soil). *Gomal University Journal of Research,* 21: 45-49.

Wallis, T. E. (1960). A Text Book of Pharmacognosy. 4th ed. J and A. Churchil Ltd. 104 Gloucester place, W.I.London; pp. 220, 221.

SIRUPULLATI (*Desmodium triflorum*)

Common name : Three-flowered beggarweed.

Botanical name : Desmodium triflorum

Tamil name : Sirupullati

Hindi name : Kudaliya, motha

Family : Fabaceae

Origin and distribution : Old World tropics and subtropics (Wagner *et al.*, 1999). Pantropical, and northward into the southern USA. In much of the Pacific the species was presumably introduced and has become naturalized eastward to the Societies and Hawai'i (Smith, 1985).

Botany

Habitat : In Hawai'i, naturalized and common in dry, disturbed sites near the coast and on dry slopes and lava flows, also in lawns, pastures, waste places and along roadsides, 2-440 m (Wagner *et al.,* 1999). In Fiji, abundantly naturalized at elevations from near sea level to 600 m in lawns, waste places, and villages, on grassy hillsides, and along roadsides and forest tracks (Smith, 1985).

Habit : Terrestrial, annual, prostrate herb, up to 50 cm long, rooting at nodes (Fig. 1).

Stem : Stems rounded, solid, glabrous. Stipules present, triangular, glabrous or hairy.

Leaf : Leaves compound, trifoliolate, alternate spiral, stalked, obovate, less than 2cm long wide, glabrous on both sides, margin entire, apex emarginated, base rounded, pinnately veined.

Floral biology

Flower : Flowers bisexual, grouped together in an axillary, few-flowered, stalked head, petals 5, red, blue or purple (Fig. 2).

Fig. 1. Plant

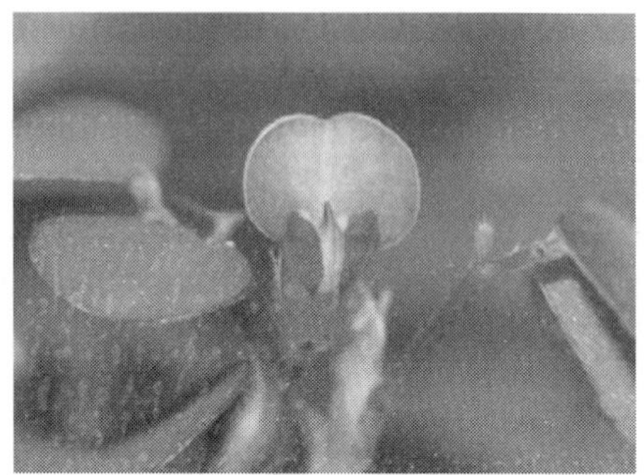

Fig. 2. Inflorescence

Fruit : Fruit an articulated pod, 17 mm long, about 2.3 mm broad, 3-7-jointed, indehiscent (Fig. 3).

Fig. 3. Fruit

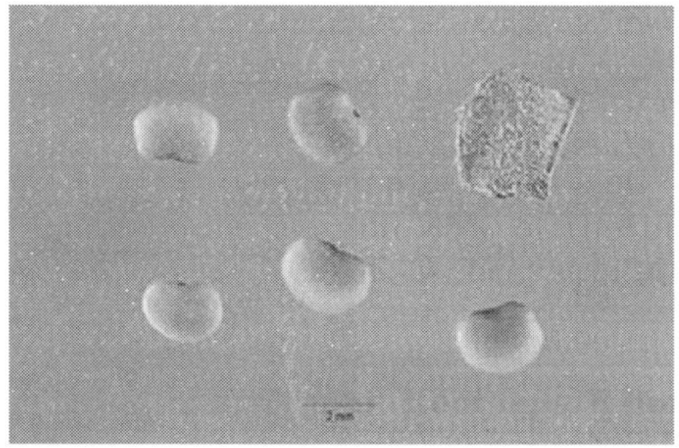

Fig. 4. Seed

Seed biology

Seed : Seeds ovoid to rounded in outline, reniform, smooth surface, olive, brown, or black (Fig. 4)

Propagation : By seed

Seed dispersal : Seed is dispersed by grazing animals

Seed dormancy and germination : Desmodium trifolium posseses the hard seed coat dormancy. Guo Qiang *et al.* (2010) collected the seeds at different maturing stages and treated with four chemical or physical methods. The combination of method and timing for optimal germination were 0.5% NaOH for 10 min. (58% seeds germinated), 0.5% H_2SO_4 for 3 min. or 5% H_2SO_4 for 1 min. (48%) and physical scarification for 5 min. (58%), all of which were higher than the control (38%). Germination rate in full ripe stage (55%) was higher than that in cribbing maturing (47%) and over mature stage (42%).

Seed storage : Jha and Jha (2006) expressed that freshly harvested seeds failed to germinate even after acid scarification and incubation under different light qualities, whereas nine-month old seeds exhibited germination of 23% without acid scarification or nutrient treatment. The legumes contained yellowish brown (Yb) and brick-red (Br) seeds in their pods in the ratio of 4:1. The Yb seeds had hard seed covering very slowly permeable to water. Storage of seeds under room conditions in sealed polythene pouches up to three years profoundly inhibited germination in Br seeds, but had little effect on Yb seeds. The nine-month-old Br seeds had maximum germination (38%) when submerged in water for 24 h before sowing, whereas germination in Yb seeds of the same age was 77%. A germination of 84% was observed when seven-eight-month old seeds were scarified with sulphuric acid for 10 min and incubated under red light for a week.

Effect on crop : The replacement of a mat forming grass Chrysopogon acicultatus in one of the Lever Brothers estates in the Solomon Islands, by the legume *Desmodium triflorum* had a marked effect on copra yield (Carrad, 1977). In 1929 the grass had spread to most parts of the Pepesala estate. Cultivation started to replace it with *Desmodium triflorum* and by 1934 the grass was practically replaced. The yield increase was certainly significant, three factors may have been responsible: nitrogen-fixation by the legume, a less competitive species or the cultivation effect resulting in increased aeration and nutrient release, or a combination of all three (Carrad, 1977).

Control measure : Cultural control: Hand weeding.

Chemical control: Pre-emergence prometryne alone or with simazine, alachlor plus 2,4-D, diuron or fluometuron plus surfactant. Foliar treatment with 2,4-D or 2,4,5-T should be effective .

Medicinal uses : The roots are reputedly carminative, tonic and diuretic and used in bilious complaints. The leaves are galactagogue and ground with cow's milk, they are given daily in the morning. They are also given to children to treat diarrhoea due to indigestion and also in convulsions. The main actions include anti-spasmodic, sympathomimetic, central nervous system stimulation, curare-mimetic activity and diuretic.

REFERENCES

Carrad, B. (1977). Cattle and coconuts: a study of copra estates in the Solomon Islands. Draft report for South Pacific Commission, ANU, Canberra, Australia.

Guo-Qiang, H., M. A. Zong-ren, L. Fang-chao, W. Jian-yi. (2010). Effect of four different treatments on hard seed and germination of legume species of *Desmodium triflorum. Pratacultural Science,* 17.

Jha, S. N. and P. K. Jha. (2006). Germination responses of some forage species of eastern terai, Nepal. *Ecoprint (An international Journal of Ecology),* 13: 49-60.

Smith, A. C. (1985). Flora Vitiensis nova: a new flora of Fiji. National Tropical Botanical Garden, Lawai, Kauai, Hawaii. Volume 3. 758 pp.

Wagner, Warren L. Herbst and S. H. Derral R.Sohmer. (1999). Manual of the flowering plants of Hawaii. Revised edition. Bernice P. Bishop Museum special publication. University of Hawai'i Press/Bishop Museum Press, Honolulu. 1919 pp.

FALSE AMARANTH (*Digera muricata* L.)

Common name : False Amaranth

Botanical name : Digera muricata (L.) Mart. (Syn. *D. arvensis* Forsk.).

Tamil name : Thoyya keerai

Family : Amaranthaceae

Origin and distribution : It is widespread in eastern tropical Africa (from Sudan and Ethiopia south to Tanzania), Madagascar and tropical and subtropical Asia (from Yemen to Afghanistan, Pakistan, India, Malaysia and Indonesia).

Botany

Habitat : Common weed of drylands, roadsides and waste places.

Habit : Erect or prostrate, annual herb becoming perennial with wide spreading branches, prostrate below (Fig. 1).

Stem : Slender, glabrous.

Leaf : Leaves variable, ovate elliptic or rounded, acute, base rounded or cuneate.

Fig. 1. Plant

Fig. 2. Flower

Floral biology

Flower : Inflorescence a long-pedunculate (up to 14 cm long), axillary, spike-like bracteate raceme up to 30 cm long, each bract subtending a subsessile partial inflorescence with a central fertile flower and 2 sterile lateral flowers. Flowers pink or greenish, in axillary peduncled spikes; bracteate, bracts subulate, persistent; central flowers of the spike fertile, lateral ones sterile; perianth calycine, membranous, 4-5 lobed, lobes oblong, erect, the two outer larger, 5-9 nerved; stamens 5, free; ovary superior, 1-celled, 1-ovuled, compressed, truncate, style short, filiform, stigma recurved (Fig. 2).

Fruit : Fruit a subglobose, hard, indehiscent nutlet. 2mm in diameter, ridged, enclosed by the persistent perianth and falling together with the sterile flowers and bracteoles (Fig. 3).

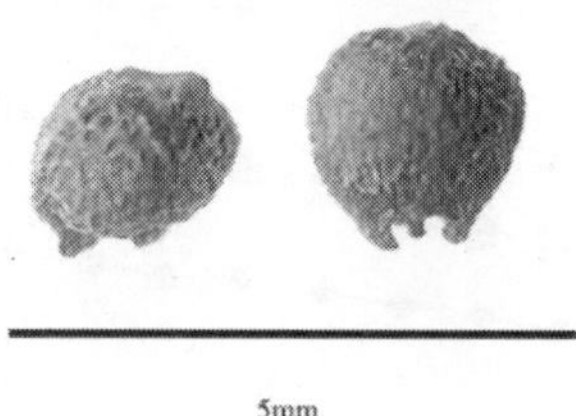

Fig. 3. Fruit

Seed biology

Seed : The seed mass of *Digera muricata* varied from 0.0038 to 0.0132 g (range = 0.0094g). The mean seed weight of population was 0.0095 (Aziz and Shaukat, 2010).

Flowering : Flowering: August-September.

Effect on crop : Digera arvensis is host weed for the caterpillars in the tobacco and cruciferous crops. Important weed in cotton, sugarcane.

Control : In the case of cotton, CMU (3-(p- chlorophenyl)-1,1-dimethyl urea) pre-emergence @ 0.5-1.0 kg ha-1 2-3 weeks after transplanting.

Uses : Leaves and young shoots of *Digera muricata* are locally used as a vegetable, e.g. in Africa (Ethiopia, Kenya) and in India. *Digera muricata* is considered a famine food. The whole plant is also commonly grazed as a forage, particularly by sheep and goats. In Senegal, *Digera muricata* is used internally against digestive system disorders and in India seeds and flowers are used to treat urinary disorders.

Past work done on Seed Science and Tecnology

According to Seemi Aziz and Syed Shahid Shaukat (2010), seed size had considerable effect on final percentage germination (Fig. 4) and large-sized category gave significantly greater germination percentage than those placed in medium-size class followed by small-size seeds (Fig. 4). Root: shoot ratio was significantly higher ($p<0.01$) for seedlings from large-sized seeds followed by medium and small seeds. Higher amounts of dry weights (g) of both root and shoot were observed in the plants emerged from large-sized seeds in all three species. Mortality seems to be higher in plants developed from small-sized seeds followed by medium and large seeds. The highest amount of reproductive output in large-seeded individuals followed by medium and small seeded individuals was observed (Table. 1).

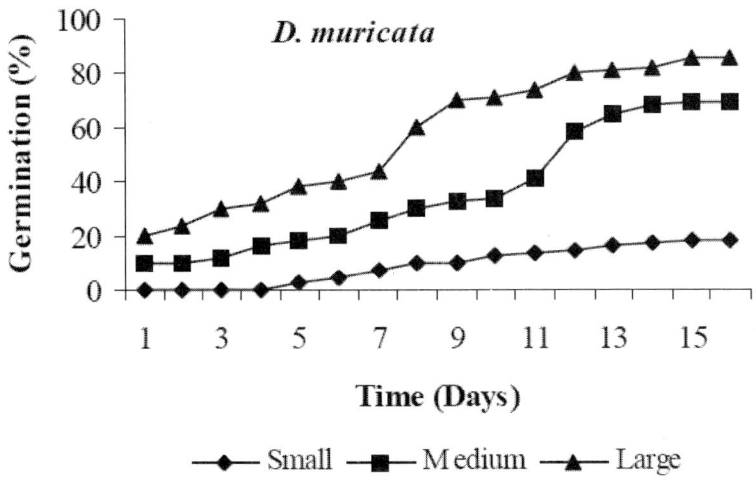

Fig. 4. Cumulative germination percentages over time of *Digera muricata* seeds belonging to three size categories.

Table. 1 The effect of seed size on the reproductive structures of *Digera muricata*. (mean ± std. error)

Species	Seed size	No. of fruits/plant	No. of seeds/plant
D. muricata	Large	20.0 ± 0.50***	20.5 ± 0.50***
	Medium	16.0 ± 0.63**	16.0 ± 0.63**
	Small	8.0 ± 0.53 n.s.	8.0 ± 0.53*

Note: Level of significance; * = p<0.01, ** = p<0.01, *** = p<0.001; n.s. = Non-significant

Muhammad Uzair *et al.* (2009) observed that the highest number of seedlings (40±11) was recorded in 20kg ash treatment with *D. muricata,* while lowest number of seedlings appeared in 20kg sludge treatment. Higher amount of sludge reduced the seed germination of this species while higher amount of ash promoted the germination and this effect was significantly different (Table 2). Though maximum height was recorded in control, while maximum fresh weight and dry weight were recorded in 20kg sludge however, no significant differences were recorded between the treatments.

Table 2. Effect of industrial sludge and ash on the mean number of seedling, height, fresh weight and dry weight of *Digera muricata* (L.) Mart.

Treatments	Number of seedlings	Height (cm)	Fresh weight (g)	Dry weight (g)
Control	ab	a	a	a
	22 ± 3	28.36 ± 1.84	641.71 ±_184.09	116.37 ± 30.66
10 kg sludge	ab	a	a	a
	26 ± 1	26.67 ± 0.73	1217.96 ± 591.45	179.73 ± 68.46
20 kg sludge	ab	a	a	a
	8 ± 4	21.16 ± 10.60	641.71 ± 184.09	255 ± 156.60
10 kg ash	b	a	a	A
	19 ± 5	24.97 ± 1.12	870.69 ± 192.00	175.37 ± 101.59
20 kg ash	ab	a	a	a
	40 ± 11	27.94 ± 1.46	703.92 ± 96.48	129.64 ± 18.22
LSD	18.95	15.68	2145.25	319.01

REFERENCES

Seemi Aziz and Syed Shahid Shaukat. (2010). Effect of seed mass variations on the germination and survival of three desert annuals. *Pak. J. Bot.*, 42(4): 2813-2825, 2010.

Muhammad Uzair, Moinuddin Ahmed and Kanwal Nazim. (2009). Effect of industrial waste on seed bank and growth of wild plants in dhabeji area, Karachi, Pakistan. *Pak. J. Bot.*, 41(4): 1659-1665.

LARGE CRABGRASS (*Digitaria sanguinalis*)

Common name : Large crabgrass

Botanical name : Digitaria sanguinalis (L.) Scop.

Tamil name : Arisipul

Family : Poaceae

Origin and distribution : It is native to the Iberian Peninsula, France, the Apennine Peninsula, the Balkan Peninsula, Eastern Europe, Hungary, Turkey, the Caucasus, Central Asia, Myanmar, China, Korea, Japan, Africa, North and South America, Australia and New Zealand and is naturalized in North and Central Europe.

Botany

Habitat : Disturbed areas of prairies, weedy meadows, edges of degraded wetlands, areas along roads and railroads, lawns and gardens, vacant lands, fields, grassy paths, and miscellaneous waste areas.

Habit : Annual grass, usually red- to violet-splotched(Fig. 1).

Stem : Initially prostrate, then curvily ascending, sparsely haired at the nodes, growing upto 50cm high.

Leaf : Leaf blades 5-15cm long and 3-12mm wide. Leaf blade green to purple, both sides with silky, shiny hairs; often reddish with central strip and pale at the margin. Ligule membranous, white, 1-2mm long, truncate. Auricles absent. Sheaths green to reddish violet, with long blister-like hairs, especially at the sheath base.

Floral biology

Inflorescence : Panicle with 4-10 finger-like spike-like racemes, each 2-16cm long, not all originating from a single point but with one or more 1-2cm below the others (Fig. 2).

Fig. 1. Plant

Fig. 2. Inflorscence

Flower : Spikelets elliptic, plano-convex, about 3mm long, paired, on short, unequal pedicels; each spikelet has a single fertile floret, lower glume minute, up to 1mm long, upper glume half to two thirds the length of the spikelet, hairy. Lemmas as long as the spikelet, the lower green, hairy and minutely rough on the nerves towards the tip; upper lemma glabrous, smooth, pale green or light brown (Fig. 3).

Fruit : Caryopsis enclosed in 2 sterile flowering-glumes, 2-4 mm long, greenish-gray, halved-fusiform.

Seed biology

Seed : Shiny, yellowish-brown, 2-3mm long (Fig. 4).

Fig. 3. Flower

Fig. 4. Seed

Flowering, pollination and fruit set : Crabgrass flowers in mid to late summer and the seeds are set soon after in a cluster of spikes that resemble fingers.

Propagation : By seeds

Seed dispersal : It is dispersed by human activity.

Life cycle : Annual grass. Though it is annual it exhibits perennial growth. A single plant is capable of producing 150 to 700 tillers and 150,000 seeds. Seeds germinate best from early spring to late summer. Crabgrass seeds are dormant for a short period of time after they shed from plants. Crabgrass continues to grow until midsummer when days become shorter, and vegetative growth slows as plants enter their reproductive stage. Plants that emerge early in the season and have a long period of vegetative growth are much larger and more competitive than plants that germinate late in the season.

Seed dormancy and germination : Dormant for several weeks, even under ideal conditions for germination (Toole and Toole, 1941 *in* Gardner, 1996). Typical dormancy lasts 2.5-6 months in *Digitaria sanguinalis* seeds (Laudien and Koch, 1972 *in* Gardner, 1996). Freshly harvested seed took a minimum of 196 days to germinate, versus 7-14 days for stored 1 year-old seed (Toole and Toole, 1941 *in* Gardner, 1996). Toole and Toole (1941) conducted preliminary tests for *Digitaria sanguinalis* and found that when placed under favourable germination conditions, after initial dormancy of fresh seed diminished and germination started, it progressed slowly over several months, yet steadily, until nearly all of the viable seed (95-98%) had germinated. When alternating temperatures were provided, approximately the same percentage germinated, yet the pattern of germination was altered. Prechilling the seed at 37°F (3°C) shortened the time before germination initiated along with increasing the rate of germination. As the seed ages, temperature requirements and germinations percentages change for *Digitaria sanguinalis* seeds (Gardner 1996). Less than 50% of the seed produced from the following year germinates the following spring, but will germinate in subsequent seasons and years (Kowalsick, 2001). According to Steinmaus *et al.* (2000), the germination rate increased linearly with temperature between 15°C and 30°C for *D. sanguinalis* seeds. King and Oliver (1994) increased the germination percentage of large crabgrass [*Digitaria sanguinalis* (L.) Scop.] seed by elevating the soil temperature from 15 to 25°C. Dormancy was broken when *Digitaria sanguinalis* seeds were kept 16 months in dry storage, or 2 months in moist sand (Gardner, 1996).

Seed storage : Viability is only few years. *Digitaria sanguinalis* seeds can remain viable in soil for at least 3 years (Elmore and Cudney, 1998)

Effect on crop : It is an aggressive weed in some subtropical crops, mainly sugarcane, corn and soya. In European corn fields it is less competitive than other warm season grasses; it mainly escapes control by late emergence in corn. It is used as a pasture grass in some areas of North America. Because of its long germination period from May to September (germination temperature minimum 20°C) (68°F) large crabgrass is a troublesome weed in corn. Tillering starts after emergence of the fourth leaf. Tillers and leaves increase under suitable conditions such as drought and low temperatures.

Control : Establishing a dense and healthy stand of turfgrass is the best way to control crabgrass and other annual weeds, including grasses and broad-leaf weeds. The proper mowing height and frequency, fertilization and irrigation are part of the weed control program and should be practiced throughout the growing season. The best time for preemergence (Prodiamine at 0.8 kg ha^{-1}, oryzalin at 2.2 kg ha^{-1}, dithiopyr at 0.3 kg ha^{-1}, pendimethalin at 1.7 kg ha^{-1}, oxadiazon at 3.4 kg ha^{-1} (Jack Johnson, 1997)) application of herbicides is late April or early May. Postemergence (Sethoxydim at 0.11 kg/h and Dithiopyr at 0.28 kg/ha (Jack Johnson, 1997)) herbicides can be used when crabgrass is in the 2- to 5-leaf stage. Repeated applications may be required depending upon the treatments.

Uses : The seeds, most notably those of fonio, can be toasted and ground into a flour, which can be used to make porridge or fermented to make beer. Foniohas been widely used as a staple crop in parts of Africa. It also has decent nutrient qualities as a forage for cattle.

Past work done on Seed Science and Technology

Rakesh *et al.* (2000) stated that exposure of weed seeds to 4 or 12°C for 7 d, followed by transfer to greenhouse (32/25°C) reduced subsequent germination percentage of hemp sesbania and johnsongrass, relative to that of control seeds maintained in the greenhouse (Table 1). However, germination percentage of barnyardgrass was not reduced. This may indicate that cold treatment of hemp sesbania and johnsongrass seed induced them into secondary dormancy.

Table 1. Effect of a 7-d temperature on subsequent weed seed germination within 3 weeks after transfer to a greenhouse (32°C day/25°C night).

Temperature regime	Species		
	Hemp sesbania	Barnyardgrass	Johnsongrass
		Emergence (%)	
4° C →32/25° C	76	92	64
12° C →32/25° C	70	90	58
32/25° C (Control)	94	96	80
LSD ($P \leq 0.05$)	8	NS	14

[NS] Nonsignificant

Heating also caused the seeds on the soil surface to germinate, but further growth was retarded. When the heating pads were removed, all seedlings appeared to be etiolated, because of the presence of the opaque heating pad and the insulating material.

After an additional 96 h of cold treatment, all seedlings exposed to –12°C were killed (Table 2). Although the seedlings exposed to 4°C were still alive, growth was minimal. Thus 4°C was too low for rapid growth, but not lethal. These results were more distinct after the seedlings were transferred to the greenhouse, where seedlings placed in the cold room at –12°C failed to grow, while those kept at 4°C developed some green pigmentation after 2–3 d.

Table 2. Seedlings emerged under cold room conditions after concomitant heat treatment (32°C for 96h) or in a greenhouse, of a soil seeded at 2.5 cm depth and seedlings killed after subsequent chilling for 7 d.

Temperature regime	Species		
	Hemp sesbania	Barnyardgrass	Johnsongrass
		Emergence (%)	
4° C →32/25° C	56	68	44
12° C →32/25° C	44	78	36
32/25° C (Control)	78	94	80
LSD ($P < 0.05$)	16	16	16
		Mortality (%)[y]	
4° C→ 32° C →° C	21	18	28
-12° C→ 32° C→ -12° C	100	100	100

32/25° C (Control)	---	---	---
LSD (P ≤ 0.05)	10	18	10

[x] Cold room temperature (where seeded soils were chilled), concomitant heat treatment temperatures and reversal to cold room temperatures.
Control represents ambient greenhouse temperature (32°C day/25°C night).

[y] Percentage of emerged seedlings that died after a second cold treatment.
Control treatments were not subjected to cold treatment.

Exposure to 4°C resulted in 21% control of hemp sesbania, 18% control of barnyardgrass, and 28% control of johnsongrass from a depth of 2.5 cm. Greenhouse conditions provided adequate sunlight for the etiolated seedlings to develop pigmentation. Death of some seedlings from this treatment was perhaps due to their poor adaptability to cold stress

Some seeds of all three species subjected to –12 °C were viable after the two successive cold treatments (Table 3). About 20%, 30%, and 5% of hemp sesbania, barnyardgrass, and johnsongrass seeds, respectively, failed to germinate following artificial heating. Thus, either the duration of artificial heating was inadequate or physiological or environmental thresholds necessary for seed germination were not attained. Total germination percentages for trays kept in the greenhouse throughout the study were 82%, 102%, and 88%, for hemp sesbania, barnyardgrass, and johnsongrass, respectively (Table 3).

Table 3. Effects of exposing an artificial seed bank to greenhouse temperatures (32/25°C) after subjecting the seed bank to three temperature regimes, on residual germination and total germination.

Temperature treatment	Species					
	Hemp sesbania	**Barnyard grass**	**Johnson grass**	**Hemp sesbania**	**Barnyard grass**	**Johnson grass**
	Residual germination (%)			Total germination and control (%)[a]		
4°C	12	8	8	68 (32)	76 (37)	52 (54)
–12°C	18	32	4	62 (78)	110 (68)	40 (95)
32/25°C (Control)	4	8	8	82	102	88
LSD (P≤0.05)	8	10	6	–	–	–

*Total percentage of germination was calculated as [(seedlings surviving after second cold treatment + residual germination)/Total number of seedlings germinated in untreated soils] *100% Percentage of control given in parenthesis were calculated as, {1-[(Seedlings surviving after second cold treatment + Residual germination)/Total number of seedlings germinated in control soils]}*100%

Taylorson and Hendricks (1979) reported that seed dormancy of *D. sanguinalis* was strikingly overcome by ethanol (Table 4).

Table 4. Effect of ethanol on percent germination of several species of annual grasses[a]

Species	Germination (%)			
	Control (H2O)		Ethanol (50 μl)	
	– R	+ R	– R	+ R
Panicum capillare	3	58	84	84
Digitaria sanguinalis	4	6	23	76
Digitaria ischaemum	0	2	20	49
Echinochloa crus-galli	0	0	17	72
Eragrostis ciliamensis[a]	0	1	0	6
Setaria faberi[a]	16	15	45	34

a. Maximum effective amount was of ethanol 10 μl which is approximately equivalent to an 0.1 M solution

King and Oliver (1994) conducted experiments to evaluate the influence of temperature and water potential on water uptake, germination, and emergence of large crabgrass in order to predict emergence in the field. Water uptake of seed soaked in polyethylene glycol solutions of 0 to -1400 kPa underwent an initial imbibition phase followed by a lag phase and subsequent increase in water content when radicles emerged from the seed. Maximum germination at 15°C was 12% at 0 kPa and 60% at 25°C at 0 to -200 kPa osmotic potential. In the growth chamber, large crabgrass emergence from soil began 2 to 3 d after planting at 30 or 35°C and within 9 to 10 d at 15°C. Maximum emergence of 77% occurred at 25°C and at a soil water potential of -30 kPa. Emergence percentage decreased as water potential decreased or as temperature increased or decreased. A logistic equation described emergence of large crabgrass at each combination of temperature and soil water potential at which emergence occurred, and a predictive model was developed and validated by field data. In the field, there was little or no emergence at soil temperatures below 15°C or water potentials below -50 to -60 kPa. The model predicted the time of onset of large crabgrass emergence and the time to reach maximum emergence to within 2 to 4 d of that recorded in field experiments. The model also predicted the correct number of flushes of emergence occurring in the field in three of four experiments.

REFERENCES

Elmore, C.L. and D.W. Cudney. (1998). Bermudagrass. The University of California, Statewide Integrated Pest Management Project, UC Pest Management Guidelines.

Gardner, D. (1996). Crabgrass (*Digitaria* spp.) taxa. Outline for: Agronomy 517; Weed Biology and Ecology, Spring Semester, 1996.

Jack Johnson, B. (1997). Preemergence and postemergence herbicides for large crabgrass (*Digitaria sanguinalis*) control in centipedegrass (*Eremochloa ophiuroides*). *Weed Technology, Vol.,* 11: 144-148.

King, C.A. and L.R. Oliver. (1994). A model for predicting large crabgrass (*Digitaria sanguinalis*) emergence as influenced by temperature and water potential. *Weed Sci.,* 42: 561–567.

Kowalsick, T. (2001). Crabgrass control. Cornell University, Cornell Cooperative Extension, Suffolk, New York.

Laudien, H. and W. Koch. (1972). Some comments on the ecology and distribution of *Echinochloa crus-galli* (L.), *Digitaria sanguinalis* (Scop.), and *Setaria* spp. in the Federal Republic of Germany. Notiziaria sulle Malattie delle Piante, 86: 149-162.

Rakesh S. Chandran and Megh Singh. (2000). Depletion of an artificial weed seed bank during the dormant period via heating and subsequent chilling of soil. *Hortscience,* 35(6):1092–1094.

Steinmaus, S.J., S.P. Timonthy and S.H. Jodie. (2000). Estimation of base temperature for nine weed species. *J. Exp. Bot.,* 51: 275-286.

Taylorson, R.B. and S.B. Hendricks. (1979). Overcoming dormancy in seeds with ethanol and other anesthetics. *Planta,* 145: 507-510.

Toole, E.H. and V.K. Toole. (1941). Progress of germination of seed of *Digitaria* as influenced by germination temperature and other factors. *Journal of Agricultural Research,* 63: 65-90.

PROSTRATE GOMPHRENA (*Gomphrena decumbens*)

Common name : Prostrate Gomphrena, Prostrate Globe Amaranth, Coastal Globe Amaranth

Botanical name : Gomphrena celosioides C. Martius. Syn. *G. decumbens* auct. non Jacq.

Tamil name : Neervadamalli

Family : Amaranthaceae (Amaranth family)

Origin and distribution : Native to America, but naturalized in India and become a widely distributed weed in the tropics and subtropics

Botany

Habitat : Common in dry, wet, cultivated and garden land, pastures and roadsides.

Habit : Prostrate, trailing, much branched annual herb; Leaves and stem covered with long hairs (Fig. 1).

Fig. 1. Plant

Stem : Stem is somewhat round and densely clothed with ad pressed or spreading, long white hairs in the young parts. Nodes are much swollen and sometimes densely covered with the hairs.

Leaf : Leaves opposite, spathulate with base tapering and clasping the stem partly.

Floral biology

Flower : Inflorescences are white flower-heads tinged with pink or red, spherical to short-cylindric, 1-1.3 cm across. Bracts have toothed crests. Flowers white, in dense spikes, each subtended by a bract and a pair of bracteoles, membr-anous; perianth 5-lobed, lobes lanceolate, densely woolly; stamens 5, filaments connate below into a long tube cleft at the top; ovary subglobose, 1-celled, superior; style short, stigmatic surface yellow (Fig. 2)

Fig. 2. Flower

Flowering and pollination : Flowering year-round. Pollination by bees as pollinators (Bush, 1995).

Seed biology

Fruit and seed : Utricle enclosed by hard perianth; seeds compressed, ovoid, 1.5 mm. in diameter, brown, faintly reticulate, shining (Fig. 3).

Fig. 3. Seeds

Propagation : By seed

Seed dispersal : The seed is spread readily by its many sharp burrs which attach to animals, motor vehicle tyres etc.

Seed dormancy and germination : Dehusking (removal of seed covering structures) promoted germination. Seed coat dormancy is clearly present in *Gomphrena celosioides* as the removal of the hull resulted in almost complete germination (Wambugu and Nyamongo, 2010).

Seed storage : Longevity is for 1 – 2 years

Effect on crop : Allelopathic substances are most soluble in methanol, yielding the most inhibitory effect on soybean growth. The extracts of either stems and roots of weeds caused a significantly decreased germination and seedling growth of soybean (Wanphen Srijankul, 1996).

Control measure : Apply 2,4-D ester 800 g/l in young seedlings only.

Uses : Antimalarial and is substitute for quinine. It is also used for land conservation. Good for yarning and laxative.

Past work done in Seed Science and Technology

Seed dormancy and germination

Wambugu and Nyamongo (2010) reported that the effect of temperature on seed germination of species of Amaranthaceae was found to be significant with most species performing better at alternating temperatures of 35°C/20°C as compared with constant ones (Fig. 4).

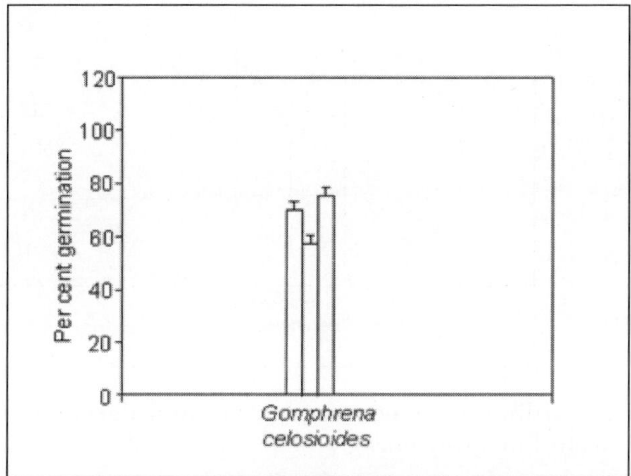

Fig. 4. Average percent germination (± SE) for seeds of Gomphrena subjected to three temperature regimes: 20°C (black bars); 25°C (grey bars); and 35°C/20°C (white bars).

Dehusking (removal of seed covering structures) affected seed germination highly significantly, promoting germination in *Gomphrena celosioides* and *Aerva lanata*. Seed coat dormancy was clearly present in *Gomphrena celosioides* as the removal of the hull resulted in almost complete germination (Fig. 5).

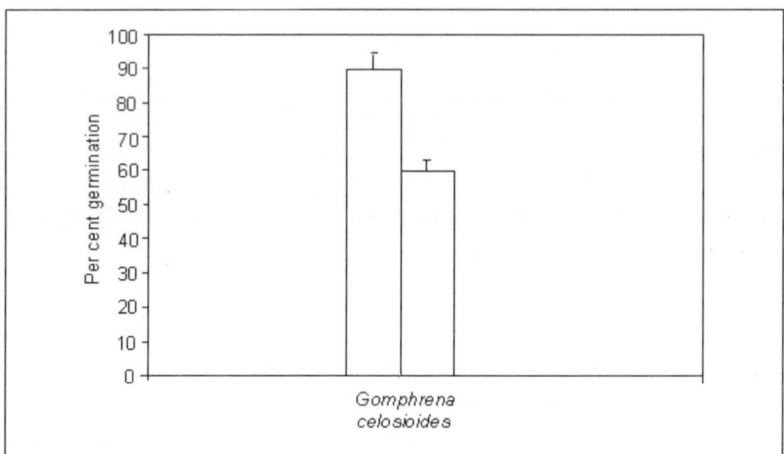

Fig. 5. Effect of seed covering structure on seed germination and dormancy of Gomphrena of *Amaranthaceae*: dehulled (back bars) and non-dehulled (grey bars).

Application of chilling or KNO_3 treatment as dormancy release mechanisms, while causing a positive effect in *G. celosioides*, reduced germination in others (Fig. 6 and 7).

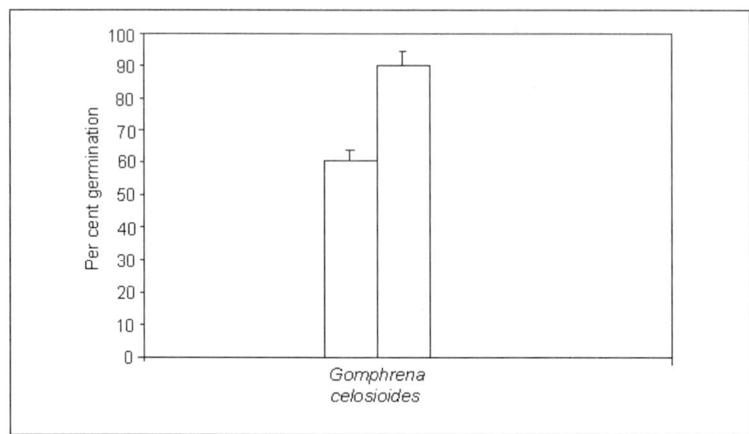

Fig. 6. Effect of moist chilling on seed germination and dormancy of Gomphrena chilling (black bars) and no chilling (grey bars).

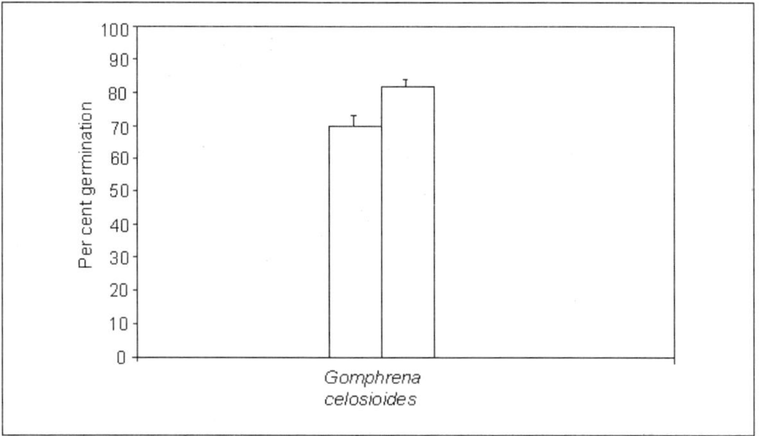

Fig. 7. Effect of potassium on seed germination and dormancy of Gomphrena: No KNO_3 (black bars) and KNO_3 applied (grey bars).

REFERENCES

Bush, M. B. (1995). Neotropical plant reproductive strategies and fossil pollen representation. *The American Naturalist.* 145(4): 594-609.

Wambugu, P. W. and D. O. Nyamongo. (2010). Seed dormancy and germination testing protocol for various economically useful plant species in the amaranthaceae family. *J. of New Seeds*, 11(4): 412-421.

Wanphen Srijankul. (1996). Allelopathic effects of some weeds on growth and yield of soybean (*Glycine max* (L.) Merrill.) cv. SJ.4. Thesis. Kasetsart Univ., Bangkok. p. 67-79 .

BROOMRAPE (*Orbanche cernua* Loef L.)

Common name : Broomrape

Botanical name : *Orbanche cernua* Loefl.

Tamil name : Pokayilai-kalan.

Family : Orobanchaceae

Origin and distribution : The Mediterranean region is considered to be one of the centres of origin of Orobanche species. The species are distributed Worldwide from temperate climates to the semi-arid tropics

Botany

Habitat : Broomrapes (*Orobanche* spp.) are obligate, chlorophyll-lacking holo root parasites that parasitize many dicotyledonous species (Parker and Riches 1993; Foy *et al.,* 1989). Broomrape plant lives directly on their hosts by attaching strong haustoria to their roots, penetrating the tissues, and absorbing the food gathered by the host plants for their own development (Manschadi *et al.,* 2001; Al-Khateeb *et al.,* 2005).

Habit : An erect, parasitic annual herb, growing upto 6-12 inches (Fig. 1).

Stem : Stem fleshy, unbranched, pale brown with scales at the base.

Leaf : A few acute scale leaves.

Floral biology

Flower : Flowers in spikes or racemes each with a scale-like bract, calyx 4-fid or bipartite, corolla 2 lipped tube curved dialated above, white, lobes of corolla pale blue or violet; stamens 4, didynamous; ovary 2-3 carpels, 1 celled (Fig. 2). The flowers are hermaphrodite (have both male and female organs).

Fig. 1. Plant

Fig. 2. Flower

Flowering : These plants generally flower from late winter to late spring

Fruit : Fruit a 2-valved capsule with many ovoid seeds (Fig. 3).

Seed biology

Seed : Seeds ovoid in shape (Fig. 4 A,B,C) and very tiny (200 x 250 μm).

Propagation : By seeds

Seed dispersal : Dispersed through sunflower seeds and by wind. It is a potential seed contaminant.

Life cycle : Perennial. The life cycle of *O. cernua* is completed in about 3 months after planting. In the presence of tobacco plants, orabanche seeds germinate in the second week after planting and infect the roots in the third week. Orabanche shoot emerges from the sixth week onwards. Flowering is completed by the seventh week. Seed formation and capsule drying are completed in the eleventh and twelth week, respectively. Seeds upon germination, the small broomrape seed develop a tube-like radicle that attaches to the host root surface. After attachment to the host root, the radicle develops a haustorium, which penetrates the root and forms connections to the vascular system of the host plant (Parker and Riches, 1993; Hassan, 2000; Kubo *et al.*, 2009). The portion of the parasite remains outside of the root tissue then develops into a tubercle that initiates crown roots and a floral meristem that produces a floral spike (Foy *et al.*, 1989; Parker and Riches, 1993). The small broomrape flower stalks emergences from the tubercle about 4 to 5 months after initial parasitic attachment to red clover (Lins *et al.,* 2005).

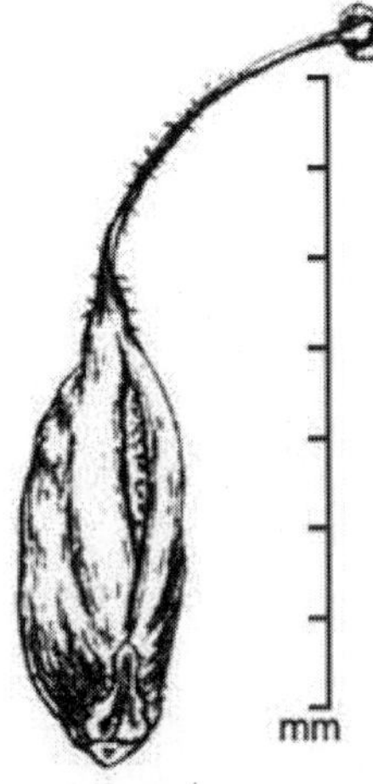

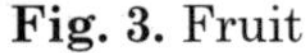
Fig. 3. Fruit

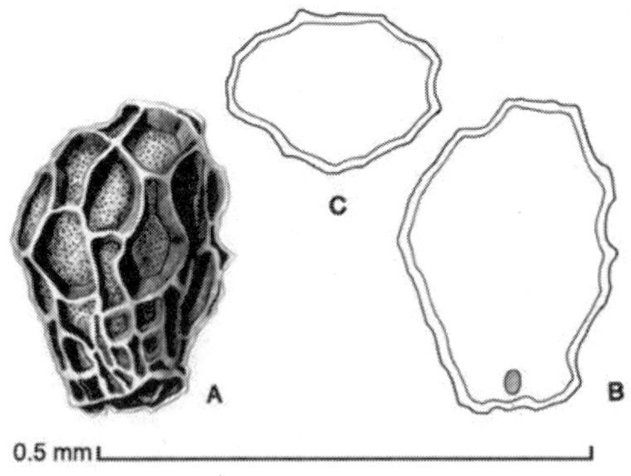

A, Seed; B, longitudinal section of seed showing embryo; C, transection of seed

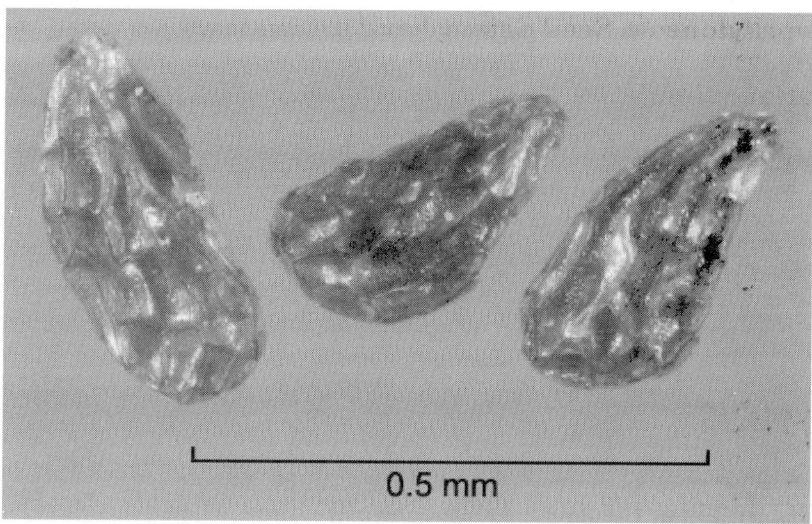

Fig. 4. Seed

Seed dormancy and germination : Orabanche produces a vast number of seeds (Saghir *et al.,* 1973a) and are capable of staying dormant in the soil for many years waiting for the germination stimulant being exuded from the plant root (Saghir *et al.,* 1973b). *Orobanche* is dormant and require a period of after-ripening or so called post-harvest ripening period (Riches, 1989). Seed germination occurs when ripened seeds are preconditioned by exposure to warm moist conditions for several days followed by exogenous chemical signals produced by host roots and some non-hosts (germination stimulant) (Worsham, 1987).

Seed storage and burial : Most of the seeds in the soil will not be reached by the stimulant, but will remain viable for upto 15 years, forming a seed reservoir for the next cropping seasons (Maass, 2001).

Effect on crop : The damage caused by the parasites *Orobanche* on field and vegetable crops is significant in the Near East, South and East Europe and in various republics of the former Soviet Union. It causes yield losses ranging from 5-100% (Linke *et al.,* 1989). At an infestation of 7.3 shoots per shoots per host plant, growth inhibition and yield loss in tobacco was 50%. *Orabanche* represents an increasing threat to several vegetable and fruit crops (Al-Khazraji *et al.,* 1987) in a wide range of cultivated lands, leading to yield losses of up to 100% in the host plant in several places worldwide (Sauerborn, 1991).

Control : Prevention of formation of the inoculum potential form the basis of *Orobanche* control. Regular weekly hand pulling of tender *Orobanche* shoots before they set seed, has reduced the original stand by 85% after 2 years and by 96% after 4 years. Soil application of ethylene dibromide (EDB) @ 2-5ml m^{-1} in heavy clay or light medium soil significantly reduced *Orobanche* infestation.

Uses : Root is cooked or roasted then, eaten.

Previous work done on Seed Science and Technology

1. Germination stimulants

The parasite produces thousands of viable seeds that can remain dormant in the soil for several years and seed dormancy is only broken by the presence of host plant root in the vicinity of the seed.

In nature, germination of *Orobanche* seed takes place only when the seed comes in contact or close proximity to the root of a host plant (Sauerborn, 1991; Kreutz, 1995).

The seeds need to be exposed to a moist environment (called preconditioning) for several days at a suitable temperature (optimum 15 to 20°C) before seeds respond to germination stimulants (Kebreab and Murdoch, 1999; Morozov *et al.*, 2000). Bar Nun and Mayer (1993), reported the synthesis of certain new proteins during preconditioning and subsequent germination of *Orobanche* seeds.

Small broomrape seeds require the presence of germination stimulant, a chemical signal for germination (Foy *et al.,* 1989). Germination stimulants include alectrol and orobanchol, which are analogues of strigolactones and have been isolated from the root exudates of the host red clover (Yokota *et al.,* 1998).

According to Dhanapal and Struik, (1996) germination phase of *Orobanche cernua* LoefL. is a critical period in the life cycle of this parasitic weed. By stimulating the germination through chemicals in the absence of hosts or through natural stimulants by exposing seeds to trap crops, the seed bank can be reduced. Better results were obtained when germination was tested in the presence of seedlings of host plants under glasshouse conditions. GR 24 at 0.1 and 1.0 mg kg^{-1} was found most effective in stimulating the germination of the seeds followed by gibberellic acid at 10 and 20 mg kg^{-1}. Without addition of chemicals, trap crops strongly increased germination. Especially greengram (*Vigna radiata*) and sunhemp (*Crotalaria juncea*) were effective.

Previous *in vitro* studies (Barghouthi *et al.*, 2000), showed that some of the tested bacteria (*Enterobacter* sp. QUBC20 and *Serratia marcescens* QUBC6) may enhance *in vitro* germination of *Orobanche* seeds in the presence of the growth regulator GR24. GR24 is a synthetic strigolactone analogue (Mangnus *et al.,* 1992). Strigolactones and flavonoids are secondary plant exudates that trigger secondary signals in parasitic plants such as *Striga* and *Orobanche* (Steinkellner *et al.*, 2007). Strigolactones play a role in regulating above-ground plant architecture and in communication with other underground organisms (Umehara *et al.,* 2008; McSteen, 2009). They act as secondary signals which rhizotropically direct the parasitic germ tube to finding host root (Sauerborn *et al.,* 2007). Other bacterial isolates (*Stenotrophomonas maltophilia* QUBC61, and unidentified isolates QUBC67, and QUBC68) showed no significant inhibition of radical elongation. Zermane *et al.,* (2007) have shown that *Pseudomonas fluorescens* Bf7-9 can suppress the pre-emergence of *O. foetida* Poir. and *O. crenata* when tested in pot experiments.

Serghini *et al.* (2001) supported the hypothesis of a host defense mechanism capable of preventing germination or development of the germinated parasitic germination or development of the parasitic seed. Thus, in the presence of isolated sunflower roots, some germinated *Orobanche cernua* seeds stopped growing, turned brown and died, the percentage of seeds turning brown being higher in the presence of isolated roots from the resistant sunflower variety (Fig. 5B). Isolated sunflower roots from both the resistant and the susceptible variety stimulated *Orobanche cernua* seed germination (Fig. 5A), but not that of *Orobanche ramose.* This effect may be accounted for by host excretion of toxic compounds.

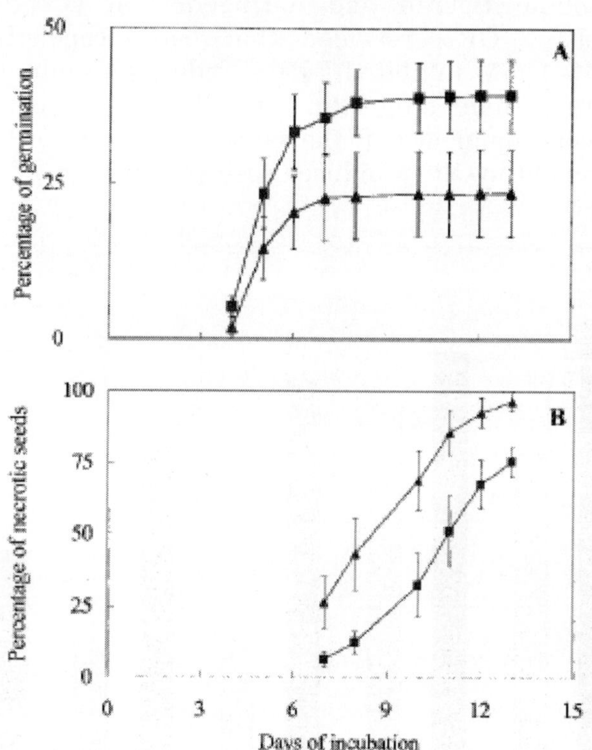

Fig. 5. Characterization of *O. ceruna* seed germination on agar in the presence of resistant cortes (▲) or susceptible Agrosur (■) sunflower roots. (A) Germinated seeds of *O.ceruna* population J6 Arburg; values expressed as a percentage of the total seeds. (B) Germinated seeds with browning symptoms, values expressed as percentage of the germinated seeds. Data are means of five independent replicates ± SE.

The absence of tubercle formation on the roots of the resistant variety, but not in the presence of the susceptible one, was observed using the tray attachment bioassay. In the resistant sunflower variety, browning of the attached *Orabanche* radical and of the sunflower root around the attachment point was clearly visible. This has been associated with the induced synthesis of phytoalexins or other toxic compounds whose accumulation may block radical penetration through the host root, preventing connection to the host vascular tissue and causing parasite death. The data presented here clearly indicated that, in sunflower, the induced synthesis, accumulation and excretion of toxic compounds seems to be an important defensive mechanism for preventing parasitic plant infection. The data obtained here suggest that the sunflower coumarins scopoletin, its glucosyl derivative, scopolin and ayapin may also be part of the sunflower defence strategy against parasitic *Orabanche,* by acting as allelochemicals (preventing *O. cernua* seed germination or killing germinated seeds) or phytoalexins (preventing root penetration and connection to the vascular system). The evaluation on the effect of scolpoletin, ayapin, scopolin, chlorogenic acid, isoliquiritigenin and naringenin on *Orobanche cernua* seed germination induced by GR_{24} revealed that both scopoletin and ayapin at concentrations of 0.01–1 mM inhibited GR_{24} – induced germination. Ayapin was a more potent inhibitor than scopoletin at all the concentrations tested. A combination of the two coumarines, at the same final concentration, led to a higher germination inhibition of the GR $_{24}$- induced seed germination (Fig. 6).

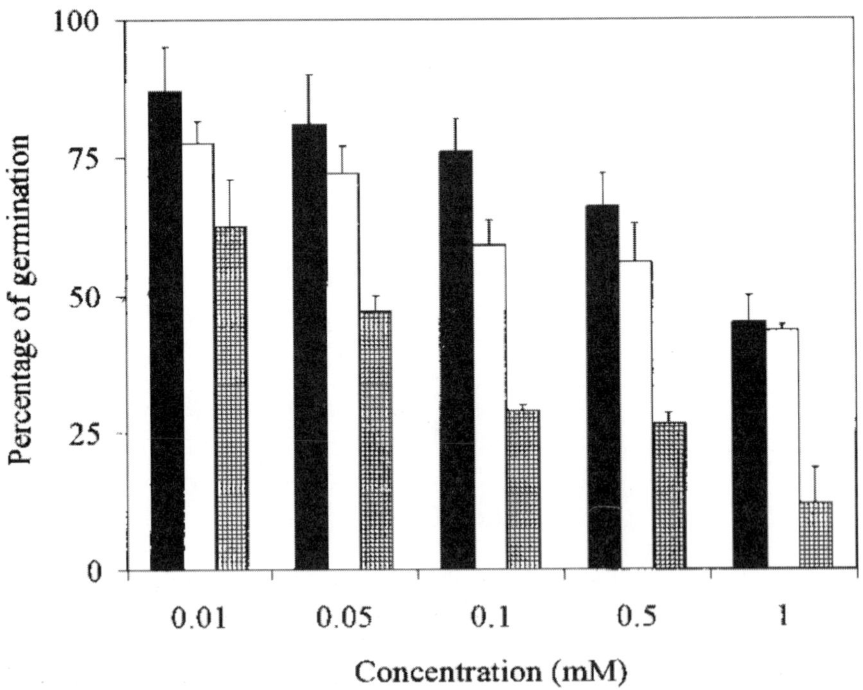

Fig. 6. Inhibition of the GR_{24} – induced *O. cernua* seed germination by scopoletin and ayapin. After conditioning, seeds were incubated for 4 d in darkness at 25°C in the presence of GR_{24} alone or with scopoletin (■), ayapin (□) or scopoletin+ ayapin (▩) at different concentrations. Data correspond to the J6 Arburg population and are expressed as a percentage of germination of the total seeds. Germination in MES buffer without GR_{24} was lower than 5%, and with GR_{24} reached 93 %. Values are means of four independent replicates ± SE.

Hassan *et al.* (2010) exposed seeds of *Orobanche minor* to 50, 75, 100 and 150 mM NaCl solutions during or after conditioning period (for 7 days) and induced to germinate by a synthetic germination stimulant GR24. Seed germination was decreased significantly with the increase in salt solution concentration during conditioning. Significant reduction in *O. minor* seed germination was observed at highest NaCl level (150 mM). It reduced germination by 92% as compared to the corresponding control (Table 1). Germination response of *O. minor* seeds conditioned in water and then treated with mixture of GR24 and NaCl 1:1 (v/v) displayed that germination of *O. minor* seeds was inhibited with increasing NaCl concentrations. Seeds treated with 150 mM, exhibited reduced germination by 77% as compared with control (Fig. 7). Of all NaCl concentration tested, 100 mM displayed the most inhibitory effect and significantly affect germination and radicle elongation of parasitic weed.

Table 1. Effect of NaCl on *O. minor* seeds germination in response to GR24 (during conditioning)

Germination						
	NaCl (mM)					
GR 24 (μM)	Water	50	75	100	150	Average
3.4 μM	84.6	48.8	44.2	25.4	24.0	45.30
0.34 μM	62.9	31.8	26.7	30.0	18.0	33.86
0.034 μM	58.9	17.4	16.4	18.2	5.2	23.23
Average	68.79	32.66	29.13	24.52	18.73	

LSD interaction = 6.516, LSD GR24 = 13.033, LSD NaCl = 9.216

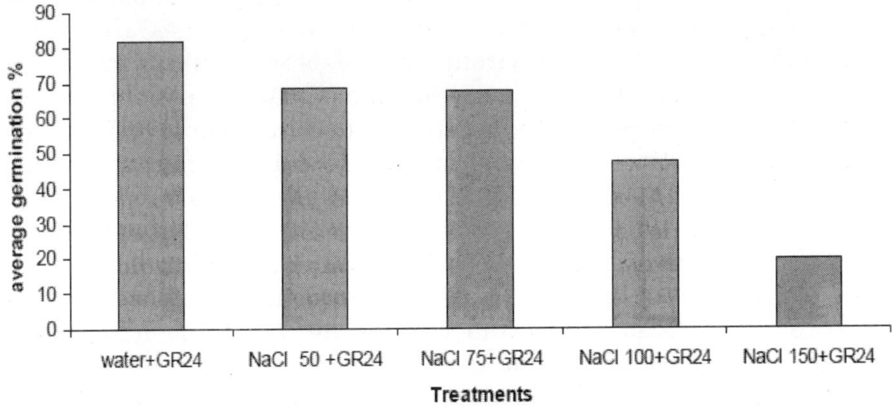

Fig. 7. Effect of different level of NaCl mixed with GR_{24} on *O. minor* germination (after conditioning). Vertical bar indicates LCD

2. Salt salinity and germination

Maps of the general distribution and existence of *Orobanche* spp. around the World indicate the absence of these parasitic plants in regions characterized by saline soil, as confirmed by Abu-Irmaileh (1998) who observed the absence of these parasites in the high salt soils of the region south to the Dead Sea in Jordan. In case of green vascular plants, salt stress is probably more critical during their seed germination (Al Karaki, 2000), through induced plasmolysis and/or permeation of toxic salt ions into their embryos (Tobe *et al.,* 1999). The effect of salinity on the seeds of *Orobanche* spp. during their preconditioning period and later on after being exposed to the germination stimulant exuded by their host and some non-host root systems remains not clearly understood. In Jordan, Abu Irmaileh (1998) observed that no *Orobanche* infections were found on tomato in the area around the Dead Sea and the southern Jordan valley where soil salinity is reported to reach 16.4 ds/m.

Al-Khateeb *et al.* (2005) stated that there was a significant reduction in seed germination along with the increasing salt concentration (0, 25, 50, 75, and 100 mm) resulting in 35.2, 32.5, 23.6, 14.3 and 9.2 germination, respectively. Similar results of 29.4, 21.3, 20.5 and 17.4% germination were also obtained when *O. cernua* seeds exposed to 0.0, 1.0, 1.25 and 1.5 Molar (M) levels of NaCl for 9 hours respectively. With no salt preconditioned seeds showed heavier proteins profile bands at the range of 6.5-14.2 Kilo Dalton (KDa) than dry seeds. Seeds from the 0.75 mM NaCl treatment showed similar profile to the water preconditioned ones, plus an extra band at the range of 29-36 KDa. However, the protein profile of 1.0 and 1.5 M NaCl treated seeds showed weaker bands with the absence of the 29-36 KDa band.

3. Biological control

Plant resistance to pafactorialrasitic plants, both non-host and host-specific, seems to be a complex multifactorial process. This interplay is dependent on the host (species, varieties and populations), the parasite (species and races or population), and biotic and abiotic environmental factors (Cubero *et al*., 1994). Taking into account the complex life cycle of the parasite, resistance may be proposed. A potential strategy might be directed at interrupting specific developmental stages such as germination, haustorial induction, attachment, penetration and connection to the vascular system, development, emergence, and flowering (Jorrin *et al*., 1999). The earlier stages may be the most efficient targets for resistance, and low production of germination stimulants as a resistance or avoidance mechanism is well documented (Butler, 1995). In addition to germination, alternative mechanisms operate in varieties resistant to *Striga* and *Orabanche*. It has been reported that once parasite seeds have germinated and become attached to the host root, either root tissue penetration and connection to the vascular tissue or tubercle development are prevented in resistant hosts (Lane *et al*., 1997; Goldwasser *et al*., 1997). Other typical plant defence response responses against pathogenic microorganisms (Jackson and Taylor, 1996) are also induced in response to parasitic plant infection.

These include host cell browning around the penetration site (Dorr *et al.*, 1997; Lane *et al.*, 1997; Goldwasser *et al.*, 1997), phytoalexin accumulation (Wegmann *et al.*, 1991; Lane *et al.*, 1997; Westwood *et al.*, 1998), lignifications and cell wall phenolic deposition (Antonova, 1994; Ish-Shalom Gordon *et al.*, 1994; Mayer *et al.*, 1997; Goldwasser *et al.*, 1999), and pathogenesis – related protein induction (Joel and Portnoy, 1998). These inducible mechanisms, as for phytopathogens (Jackson and Taylor, 1996), may result in thew restriction of infection structures to small areas immediately surrounding the penetration site or prevention of their connection to the vascular system.

Table 3. Pathogenicity of isolated fungi on seeds of *Orabanche ramose*

Isolated fungi	**Disease ratings of fungal infection on *O. ramosa***	
	Seed germination (%)	**Loss in seed texture[a]**
Alternaria alternata	0	100
Fusarium sp.	0	100
Dendrophoma sp.	0	100
Rhizoctonia sp.	0	100
Cheatomium sp.	25	50
Pullularia sp.	78	-
Aspergillus sp.	78	-
Penicillium sp.	78	-
Control	78	0

[a] Loss in integrity of mounted seed (0=no loss to 100=complete loss) when pressed on microscopic slide.

According to Hameed *et al.* (2001), species of the genus *Orobanche* are parasitic flowering plants, holoparasites, which cling to the roots of green plants. Their tiny seeds (200 × 250 µm) germinate in response to chemical stimuli produced by host and some non-host plants. Successful contact with their host leads to development of haustoria for obtaining water and food. The shoots above the ground expose flowers and disseminate seeds. The samples of *Orobanche ramosa, O. crenata, O. cernua,* and *O. egyptiaca* were collected from different localities in Jordan showed one of the following disease symptoms: dry rot at the base of the stem; general deterioration and expanded lesion from base upward; soft tissue maceration of stem; and black rot of flower parts with incomplete maturation of the ovary and seeds. Isolation from diseased stems and seeds indicated that several fungi were isolated, mainly, *Fusarium* spp., *Alternaria alternata, Rhizoctonia* sp., *Dendrophora* sp., *Chetomium* sp., and an ascomycetus fungus with a perithecium. Pathogenicity tests showed that *Fusarium* spp. and *Alternaria alternata* attacked healthy living tissue of *Orobanche* spikes. These fungi caused lesions of black soft rot and complete deterioration within 5-7 days (100%). They also attacked *Orobanche* seeds, arresting their germination and causing maceration of non-germinated and germinated seeds after 5-7 days of incubation (Table 3). Table 4 shows that the crude culture filtrate may contain some inhibitory factor(s) against the Orobanche seed germination. Germination was reduced by about 80% compared to the control. Meanwhile, *Dendrophora* sp. and *Chetomium* sp. caused limited lesion at first, but were able to colonize the tissue as it ages and senesces. This study showed the presence of a potential endogenous pathogenic fungi in Jordan, which can be investigated as a biological control for *Orobanche*.

The insect *Phytomyza orobanchia* Kalt. is responsible for reducing seed production in *O. crenata* Forsk. to 50% (Amesellem *et al.,* 2001).

Sameer Barghouthi and Mazen Salman (2010) indicated that *Pseudomonas aeruginosa* QUBC1, *P. fluorescens* QUBC3, *Bacillus atrophaeus* QUBC16, and *B. subtilis* QUBC18 significantly inhibited radical elongation (P50.01) of both *O. aegyptiaca* and *O. cernua* relative to control radicles, whereas *Microbacterium hydrocarbonoxydans* QUBC11 and *Ochrobactrum anthropi* QUBC13 showed less inhibitory effects. Other bacterial isolates had no inhibitory effects (Table 5).

Table 4. Effects of crude culture filtrates on *Orobanche* seed germination[a]

Isolated fungi	Germination of *orabanche* seeds (%) on seed lawn, poured technique	
	O.ramosa	
Fusarium *sp.*	36	3.7
Alternaria alternata	56	7.7

[a]Applied, poured technique on seed lawn of *Orobanche* on water agar.

Table 5. Inhibition of radical lengths of *O. cernua* or *O. aegyptiaca* seeds treated with bacteria relative to seed length and to control radicals.

Bacterium		Radical length to relative to seed		Radical length, % of control (% inhibition)	
	CFU x 10^8 ± SD[1] mL^{-1}	*O. aegyptiaca*	*O. cernua*	*O. aegyptiaca*	*O. cernua*
Control (no bacteria)	0.00	3.07 ± 1.45[a]	2.23 ± 0.82[a]	100 (0.00)	100 (0.00)
P. fluorescens QUBC3	11.3 ± 0.6	1.07 ± 0.60[d]	1.36 ± 0.59[cd]	34.8 (65.2)	61.0 (39.0)
O. anthropi QUBC13	6.7 ± 0.4	1.78 ± 0.84[c]	1.88 ± 0.66[ab]	58.0 (42.0)	84.3 (15.7)
M. hydrocarbonoxydans QUBC11	4.7 ± 1.0	2.36 ± 0.62[b]	1.49 ± 0.50[bc]	76.9 (22.1)	66.8 (33.3)
S. epidermidis QUBC21	2.1 ± 0.4	1.14 ± 0.44[d]	1.23 ± 0.32[cd]	37.1 (62.9)	55.2 (44.8)

P. aeruginosa QUBC1	1.7 ± 0.5	0.64 ± 0.61^{ef}	1.01 ± 0.66^{cd}	20.1 (79.9)	45.3 (54.7)
B. atrophaeus QUBC16	0.75 ± 1.2	0.40 ± 0.39^{f}	0.72 ± 0.48^{d}	13.1 (86.9)	32.3 (67.7)
B. subtilis QUBC18	0.17 ± 0.04	0.93 ± 0.84^{de}	1.15 ± 0.78^{cd}	30.3 (69.7)	51.6 (48.4)

[1]Highest (top) to lowest (bottom) bacterial CFU mL^{-1} obtained at $A_{600nm} = 0.22\ mL^{-1}$

The establishment of the inhibitory effect of the most promising isolates, *B. atrophaeus* QUBC16 and *P. aeruginosa* QUBC1 on radical elongation of both Orobanche spp. is a step towards utilizing such bacteria as biocontrol agents against *O. aegyptiaca*, *O. cernua*, and potentially other *Orobanche* species.

REFERENCES

Abu Irmaileh B. E. (1998). Effect of salinity on *Orobanche* germination and establishment. *In:* current problem of *Orobanche researches,* proceeding of the fourth international workshop on *Orobanche*, ed. by K. Wegmann, Musselman, L. and D. Joel. Bolgaria.

Al Karaki, G. (2000). Germination of tomato cultivars as influenced by salinity. *Crop Res.*, 19: 225-229.

Al-Khateeb, W. M., K. M. Hameed and R. A. Shibli. (2005). Effect of salinity on *Orobanche cernua* seed germination. *Plant Pathol. J.*, 21:391-394.

Al-Khazraji, T. O., K. M. Hameed and A. R. Saghir. (1987). Effects of inoculum density of *Orobanche* seeds in soil on tomato and tobacco. *Zanco,* 5:73-84.

Amsellem Z., Kleifeld Y., Kerenyi Z., Hornok L., Goldwasser Y., Gressel J. (2001). Isolation, identification and activity of mycoherbicidal pathogen from juvenile broomrape plants. Biological Control 21(3), 274- 284.

Antonova, T. S. (1994). Biochemical aspects of the development of new virulent forms in the Moldavian population (race C) of *Orobanche cumana* Wallr. against the background of resistant SJ. Biology and management of *Orobanche. Proceedings of the 3rd Interna-tional Workshop on Orobanche and related Striga research.* Amsterdam: Royak Tropical Institute, 290-292.

Bar Nun, N. and A. Mayer. (1993). Preconditioning and germination of *Orobanche* seeds: respiration and protein synthesis. *Phytochemistry,* 34: 39-45.

Barghouthi, S., F. Faqieh and A. Badrieh. (2000). Effect of bacteria on broomrape seed germination. Third International Weed Science Congress, IWSC, 6-11 June 2000, Foz Do Iguassu, Brazil. Corvallis, OR: International Weed Science Society, Oregon State University, p. 183.

Bulter, L. G. (1995). Chemical communication between the parasitic weed *Striga* and its crop host a new dimension in allelochemistry. In: Inderijit, Dakshini KMM, Einhellig FA, (eds.) Allelopathy: organisms, processes and applications. ACS Symposium series 582: 158-168.

Cubero, J. I. (1994). Breeding work in Spain for *Orobanche* resistance in faba bean and sunflower. *In* Pieterse, A.H, Verkleij, J.A.C. & Borg, ter Borg, S.J. eds. *Proc. 3rd Int. Workshop on Orobanche and related Striga Research.* Royal Tropical Institute, Amsterdam, The Netherlands, 465-473.

Dhanapal, G. N. and P. C. Struik. (1996). Broomrape (*Orobanche cernua*) control before attachment to host through chemically or biologically manipulating seed germination. *Netherlands Journal of Agricultural Science,* 44: 279-291.

Dorr, I., A. Staack and R. Kollmann. (1994). Resistance of *Helianthus annus* to *Orobanche*-histological and cytological studies. In: Pieterse AH, Verkleij JAC, ter Borg SJ, eds. Biology and management of *Orobanche*. Proceedings of the 3rd International Workshop on *Orobanche* and related Striga Research. Amsterdam: Royal Tropical Institute, 276-289.

Foy, C. L., R. Jain and R. Jacobsohn. (1989). Recent approaches for chemical control of broomrape (*Orobanche* spp.). *Rev. Weed Sci.*, 4:123–152.

Goldwasser, Y., Y. Kleifeld, D. Plakhine and B. Rubin. (1997). Variation in vetch (*Vicia* spp.) response to *Orobanche aegyptiaca. Weed Science*, 45: 756-762.

Hameed, K. M., I. M. Saadoun and Zaineb Al-Shyab. (2001). Potential biological control of *Orobanche* by fungi isolated from diseased specimens in Jordan. *Plant Pathol. J.* 17(4): 189-195.

Hassan, E. (2000). Efficacy of broomrape *(Orobnche ramosa L.)* control methods on faba bean *(Vicia faba L)* growth and development. M.Sc. Thesis, Jordan University of Science and Technology, Irbid, Jordan.

Hassan, M.M., Y. sugmuto, A.G.E. Babiker, Y. Yamauchi, M.G.E. Osman, S. O.Yagoub. (2010). Effect of NaCl on Orobanche spp and Striga hermonthica seeds germination during and after conditioning. *Bioscience Res.,* 7(1), 26-31.

Iah-Shalom-Gordon, N. Y. Cohen and R. Jacobson. (1994). Ultra-structural differences in roots of sunflower.

Jackson, A.O. and C.B. Taylor. (1996). Plant-microbe interactions: life and death at the interface. *The Plant Cell*, 8: 1651-1668.

Joel, D.M. and V.H. Portnoy (1998). The angiosperous root parasite. *Orobanche* L. (Orobanchaceae) induces expression of a pathogenesis-related (Pr) gene in susceptible tobacco roots. *Annals of Botany* 81: 779-781.

Jorrin, J., K. Serghini, A. Perez de Luque, F.A. Macias, J.C. Garcia Galindo, L. Garcia-Torres and M. Castejon–Munoz. (1999). Plant resistance to parasitic angiosperms: a biochemical point of view. In: Wegmann K, Musselman L. Joel D, (eds.) Current problems of *Orobanche* research, proceedings of Fourth International Workshop on Orobanche. Albena: Institute for wheat and sunflower 'Dobroudja', 43-49.

Kebreab, E. and A.J. Murdoch. (1999). A quantitative model for loss of primary dormancy and induction of secondary dormancy in imbibed seeds of *Orobanche spp. Exp. Bot.,* 50: 211-219.

Kreutz, C.A.J. (1995). Orobanche. The European broomrape species: 1. Central and Northern, Europe. Natuurhistorisch Genootschaap in Limburg, Maastricht, The Netherlands.

Kubo, M., H. Ueda, P.Y. Park, M. Kawaguchi and Y. Sugimoto. (2009). Reactions of *Lotus japonicus* ecotypes and mutants to root parasitic plants. *Journal of Plant Physiology,* 166(4): 353-362.

Lane, J.A., D.V. Child, G.C. Reiss, V. Entcheva and J.A. Beiley. (1997). Crop resistance to parasitic plants. In: Crute IR, Holub EB, Burdon JJ, (eds.) The gene-for-gene relationship in plant-parasite interactions. London: CAB International, 81-97.

Linke, K.H., J. Sauerborn, and M. C. Saxena. (1989). *Orobanche* field guide. University of Hohenheim, Germany/ICARDA, Aleppo, Syria (also available in Arabic).

Lins, R.D., J. BC. Colquhoun, M. Cole and C. A. Mallory-Smith. (2005). Post-emergence small broomrape (*Orobanche minor*) control in red clover. *Weed Technol.*, 19: 411–415.

Maass, E. (2001). Spontaneous germination in *Striga*. *In* Fer, A. Thalouarn, P., Joel, D.M., Musselman, L.J., Parker, C. & Verkleij, J.A.C., eds. *Proc. of the 7th Int. Parasitic Weed Symposium*. Nantes, France. pp. 129.

Mangnus, E.M., F.J. Dommerholt, R.L.P. de Jong and B. Zwanenburg. (1992). Improved synthesis of strigol analogue GR24 and evaluation of the biological activity of its diasteroisomers, *Journal of Agricultural and Food Chemistry*, 40: 1230-1235.

Manschadi, A.M., J. Sauerborn and H. Stutzel. (2001). Quantitative aspects of *Orobanche crenata* infestation in faba beans as affected by a biotic factors and parasite soil seed bank. *Weed Res.,* 41:311-324.

Mayer, M.J., J. Steel, D.V. Child, J.A. Hargreaves and J.A. Bailey. (1997). Early stages of infection of maize (*Zea mays*) and *Pennisetum setosum* roots by the parasitic plant *Striga hermonthica. European Journal of Plant Pathology*, 103: 815-827.

McSteen, P. (2009). Hormonal regulation of branching in grasses. *Plant Physiology*, 149:46-55.

Morozov, I.V., C.L. Foy and J.H. Westwood. (2000). Small broomrape (*Orobanche minor*) and aegyptian broomrape (*Orobanche aegyptiaca*) parasitization of red clover (*Trifolium pratense*). *Weed Tech.,* 14: 312-320.

Parker, C. and C.R. Riches. (1993). Parasitic weeds of the world. *Biology and control*, Wallingford: CAB International.

Riches, C.R. (1989). The biology and control of *Alectra vogelii* Benth. (Scrophulariaceae) *in Botswana*. University of Reading, UK. (Ph.D. thesis).

Saghir, A.R., C.L. Foy, K.M. Hameed, C.R. Drake and S.A. Tolin. (1973a). Herbicide effects on parasitism of tomato by Hemp Broomrape. *Weed Sci.,* 21: 253-257.

Saghir, A.R., C.L. Foy, K.M. Hameed, C.R. Drake and S.A. Tolin. (1973b). Studies on the biology and control of *Orobanche ramose* L. *Proc. Eur. Weed Res. Coun. Symp. Parasitic Weeds*, Malta. 197: 106-116.

Sameer Barghouthi and Mazen Salman. (2010). Bacterial inhibition or *Orobanche aegyptiaca* and *Orobanche cernua* radical elongation. *Biocontrol Science & Technology,* 20(4):423-435.

Sauerborn, J. (1991). Parasitic flowering plants, ecology and management, Germany: Verlag Josef Margraf Scientific Books.

Sauerborn, J., D.M. Stovera and J. Hershenhornb. (2007). The role of biological control in managing parasitic weeds, *Crop Protection*, 26: 246-254.

Serghini K., A. Pérez-De-Luque, M. Castejòn-Munoz, L. Garcìa-Torres and J. V. Jorrìn. (2001). Sunflower (*Helianthus annuus* L.) response to broomrape (*Orobanche cernua* Loefl.) parasitism: induced synthesis and excretion of 7-hydroxylated simple coumarins. *J. Exp. Bot.,* 52: 2227-2234.

Steinkellner, S., V. Lendzemo, I. Langer, P. Schweiger, T. Khaosaad, J.P. Toussaint and H. Vierheilig. (2007). Flavonoids and strigolactones in root exudates as signals in symbiotic and pathogenic plant – fungus interactions, *Molecules,* 12, 1290-1306.

Tobe, K., L. Zhang and K. Omasa. (1999). Effect of NaCl on seed germination of five nonhalophytic species from a Chinese desert environment. *Seed Sci.& Technol.,* 27: 851-863.

Umehara, M., A. Hanada, S. Yoshida, K. Akiyama, T. Arite, N. Takeda-Kamiya, H. Magome, Y. Kamiya, K. Shirasu, K. Yoneyama, J. Kyozuka and S. Yamaguchi. (2008). Inhibition of shoot branching by new terpenoid plant hormones. *Nature*, 455 (7210): 195-200.

Wegmann, K., E. von Elert, H. J. Harloff and M. Stadler. (1991). Tolerance and resistance to Orobanche. In: Wegmann K, Musselam L, (eds.) Progress in Orabanche research. Proceedings of the International Workshop on *Orobanche* Research, Tubingen: Eberhard-Karls-University, 318-321.

Westwood, J. H., X. Yu, C. L. Foy and C. L. Cramer. (1998). Expression of a dense-related 3-hydroxy-3-methylglutaryl CoA reductase gene in response to parasitization by *Orobanche* spp. *Molecular Plant-Microbe Internations*, 11: 530-536.

Worsham, A. D. (1987). Germination of witchweed seeds. In Musselman, L.J., (ed.) *Parasitic weeds in agriculture. Vol.* 1. *Striga.* CRC Press, Boca Raton, Florida, USA, 45-61.

Yokota, T.H., K. Sakai, K. Okuno, K. Yoneyama and Y. Takeuchi. (1998). Alectrol and oroban-chol, germination stimulants for *Orobanche minor* from its host red clover: *Phytochem.*, 49: 1967–1973.

Zermane, N., T. Souissi, J. Kroschel and R. Sikora. (2007). Biocontrol of broomrape (*Orobanche crenata* Forsk. and *Orobanche foetida* Poir.) by *Pseudomonas fluorescens* isolate Bf7-9 from the faba bean rhizosphere. *Biocontrol Science* and *Technology*, 17: 483-497.

INDIAN SORREL (*Oxalis corniculata* L.)

Common name : Indian Sorrel

Botanical name : Oxalis corniculata L.

Tamil name : Paliakiri

Hindi name : Amrul

Family : Oxalidaceae (Wood Sorrel Family)

Origin and distribution : This species is cosmopolitan in its distribution and its place of origin is unknown. *Oxalis corniculata* L., creeping or yellow woodsorrel (Oxalidaceae), is a spreading herbaceous species with cosmopolitan distribution throughout temperature and tropical zones (Eiten, 1963; Holm *et al.,* 1977). It is predominantly weedy in habit, occurring in atleast 17 crops in 44 countries worldwide (Holm *et al.,* 1977).

Botany

Habitat : Common weed occurring in moist situations and garden land. In the northern hemisphere, *O. corniculata* is confined to gardens, lawns and greenhouses (nurseries), where it is serious pest (Eiten, 1963; Lourteig, 1979).

Habit : A creeping herbaceous perennial weed with long-stalked trifoliate leaf (Fig. 1).

Stem : Stem is creeping.

Leaf : Leaves radicular cauline, usually digitate and leaflets obcordate.

Floral biology

Flower : Yellow, umbellate on slender long peduncles; sepals 5, fused, imbricate; petals 5, free; stamens 10 in two whorls of 5 each; ovary superior, pentacarpellary, syncarpous; ovules many attached to axile placenta (Fig. 2).

Fig. 1. Plant

Fig. 2. Flower

Flowering, pollination and fruit set : Flowering period is March to October.

This species is predominantly inbreeding, but has retained the capacity to outcross; this presumably derived breeding system contributes to success as an aggressive and colonizing weed in disturbed habitats (Ornduff, 1972; Lovett Doust *et al.*, 1981).

Fruit : Fruit is a loculicidal capsule (Fig. 3).

Seed biology

Seed : Seeds are oval, flat, rough, reddish brown sometimes with gray spots and about 1/25 inch long (Fig. 4).

Propagation : Through seed. A major factor contributing to weediness in *O. corniculata* is prolific reproduction. This stoloniferous herb reproduces vegetatively by rooting at nodes and also by seeds which are dispersed forcefully from capsules away from the parent plant (Robertson, 1975; Holm *et al.*, 1977).

Seed dispersal : Noteworthy for its explosive discharge of the contained. There are about 10 to 50 seeds per pod, with a potential for more than 5,000 seeds per plant. When seedpods mature, they rupture, and seeds are forcefully expelled, landing up to 10 feet from the plant. Because seeds are rough, they can stick to machinery, plastic pots, irrigation tubing and clothing.

Fig. 3. Fruit

Fig. 4. Seed

Seed dormancy and germination : Seeds require light for germination. Optimum seed germination occurs between 60° and 80°F, although it can occur at lower temperatures. Seeds can germinate any time of year in California, but most plant establishment takes place in fall. It isn't known how long seeds remain viable in the soil. Moist, hot conditions inhibit seed germination; for example, 4 hours of moist heat at 97°F decreases germination by 96%, while 8 hours stops it altogether.

Seed storage : Holt (1987) exposed that when seeds were stored dry in laboratory conditions, three of four seed lots examined retained over 80% germination capacity until 8 months; 50% capacity remained after 15 months.

Effect on crop : Seed reproduction and dispersal are the primary means of invasion and establishment in greenhouses, where *O. corniculata* is a persistent weed of container-grown horticultural plants.

Control measure : First, consider whether control can be achieved using non-chemical means such as digging out or suppressing with mulch. Contact weedkillers or glyphosate-based products, which have low persistence in the soil, may suffice.

Medicinal uses : It is anthelmintic, anti-inflammatory, astringent, depurative, diuretic, emmenagogue, febrifuge, lithontripic, stomachic and styptic. It is used in the treatment of influenza, fever, urinary tract infections, enteritis, diarrhoea, traumatic injuries, sprains and poisonous snake bites. An infusion can be used as a wash to rid children of hookworms. The plant is a good source of vitamin C and is used as an antiscorbutic in the treatment of scurvy. The leaves are used as an antidote to poisoning by the seeds of *Datura* spp., arsenic and mercury. Yellow, orange and red to brown dyes are obtained from the flowers. The boiled whole plant yields a yellow dye. An infusion of leaves is used to remove opacities of the cornea and is dropped into the eyes for itching lids. A decoction of leaves is used as a gargle.

Past work done in Seed Science and Technology

Germination requirement

Holt (1987) conducted a study to investigate the physiological responses of greenhouse-produced *Oxalis corniculata* seeds to light, temperature, moist heat treatment, aging and season of production. Fresh seeds exhibited over 90% germination and required low levels of light (5 μmol m^{-1} s^{-1}, 400-700 nm) to germinate. Seeds germinated over a broad, yet seasonally-dependent range of incubation temperatures. Seeds produced in winter had the narrowest temperature range of germination (15 to 25°C) and the lowest germination per cent (44% at 2 week) at optimum temperature (17°C); seeds produced in summer had the widest temperature range of germination (10 to 30°C) and the highest germination percent (93% at 2 week) at optimum temperature (17°C). Incubation at non-optimum temperatures between 5 and 40°C suppressed or slowed the rate of germination until seeds were placed at optimum temperature, where full germination subsequently occurred. Moist heat treatment at temperatures over 40°C resulted in varying degrees of inhibition of subsequent germination. These results indicate that the seasonal temperature and daylength effects on maternal plants in the green-house environment are major determinants of seed germination characteristics of *O. corniculata.*

REFERENCES

Eiten, G. (1963). Taxonomy and regional *variation* of Oxalis section Corniculatae. 1. Introduction, keys and synopsis of the species. *Am. Midl. Nat.*, 69: 257 -309.

Holm, L.G., J.V. Plucknett and J.P. Herberger. (1977). The world's worst weeds-distribution and biology. The Univ. Press, Hawaii, Honolulu.

Holt, J.S. (1987). Factors affecting germination in greenhouse-produced seeds of *Oxalis corniculata,* a perennial weed. *Amer. J. Bot.*, 74(3): 429-436.

Lourteig, A. (1979). Oxalidaceae extra-austroamericanae.ll. Oxalis L. section Corniculatae DC. *Phytologia,* 42: 57-198.

Lovett Doust, L., J. Lovett Doust and P.B. Cavers. (1981). Fertility relationships in closely related taxa of Oxalis, section Corniculatae. *Can. J. Bot.,* 59: 2603-2609.

Ornduff, R. (1972). The breakdown of trimorphic incompatibillity in Oxalis section Corniculatae. *Evolution*, 26: 52-65.

Robertson, K.R. (1975). The Oxalidaceae in the southeastern United States. *J. Arnold Arbor.* Harv. Univ. 56 223-239.

INDIAN PURSLANE (*Portulaca oleracea* L.)

Common name : Indian Purslane

Botanical name : Portulaca oleracea L.

Tamil name : Karie keerai, pasarai or paruppu keerai.

Hindi name : Lunia

Family : Portulacaceae

Origin and distribution : Native to Eurasia. Tropical and subtropical in distribution. Mexico and Australia are considered to be centres of its diversity. The cultivated forms probably originated in the old World.

Botany

Habitat : Orchards, vineyards, crop fields, landscaped areas, gardens, roadsides and other disturbed sites.

Habit : Fleshy annual herbs, usually succulent (Fig. 1).

Fig. 1. Plant

Fig. 2. Flower

Stem : The main stem is short and erect bearing a number of lateral branches which lie prostrate on the ground; they grow to a length of 4-20 inches and are green or reddish in colour.

Leaf : The leaves are alternate at the base, sub-opposite or opposite at the ends of branches and crowded round the flowers. A well developed leaf is 1 inch by ½ inch; tip is rounded; the base is tapering and the margins are red or green. The lower surface of the leaf is pale and is covered with glistening dots. The leaf-stalk is short.

Floral biology

Flower : Flowers are yellow without stalks, from 2-6 in number and are collected together in clusters at the ends of branches. Sepals are 2 and fleshy; they are united at the base and are free above. Petals are 5 and are yellow in colour. Stamens vary in number from 8-20 and surround the ovary. The ovary is 1-celled, many ovuled and is half inferior; stylar branches are from 3-6 (Fig. 2).

Flowering, pollination and fruit set : Flowering in June - November. Seedlings can mature and produce seeds in about 5-8 weeks. Capsules mature about 1-3 weeks after flowering.The primary sources of pollination of purslane are bees, insects and wind (Hanson, 2004). In the Philippines up to 10000 and in Norih America up to 243 000 seeds are produced per plant (Waterhouse, 1994).

Fruit : Fruit is dry, dehiscing transversely by the upper part enclosed by the sepals, coming off as a lid leaving the lower part in the plant (Fig. 3).

Fig. 3. Fruit

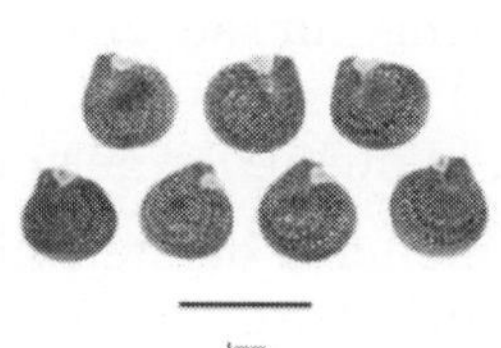

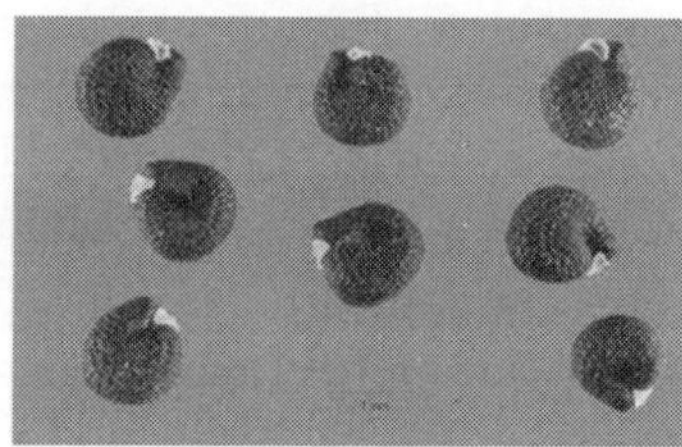

Fig. 4. Seed

Seed biology

Seed : Seeds are minute (0.5 mm dameter), blackish brown in colour and are covered with concentric lines of tubercles (Fig. 4).

Propagation : Reproduces by seed and sometimes by stem fragments.

Root fragment regeneration—The fleshy stems of common purslane can remain moist and viable for several days after cultivation and hoeing and reroot forming new plants when the gardens or fields are irrigated (Cudney and Elmore, 1999). Having this capability, allows the plant to persist even with the use of mechanical cultivation.

Seed dispersal : Most seeds fall near the parent plant.

Seed dispersal by humans: Humans are a large source of seed dispersal among the purslane. Human cultural practices often times break the seeds off of the plant without any awareness (Kiyoko and Cavers, 1980).

Seed dispersal by animals: Animals are another large source of dispersion of purslane seeds (Mitich, 1997). Animals, such as birds, will eat the seeds and deposit them in other areas where the new seed would be allowed to germinate and grow.

Seed dispersal by water: Precipitation and the use of irrigation also allows the seeds to be broken off of the plant and washed into other arable areas (Mitich,

1997). This method of dispersal is dependent upon weather patterns and cultural watering practices of humans.

Seed storage and burial : Each plant has potential to produce thousands of seeds and it is said that their germination power can last seven years. Buried seeds survive upto 40 years (Cudney and Elmore, 1999). Ozer (1972) found that the seeds of *P. oleracea* buried in the soil can remain viable for 30 years or more, while Egley and Chandler (1983) reported that the some weeds buried in soil can still be viable after 5.5 years, but the viable seed ratio was less than 1%.

Seed dormancy and germination : Some seeds survive ingestion by birds. Germination occurs erratically spring through summer, especially as day time/night time temperatures approach 35°/20°C and after soil disturbance. Newly matured seeds require light to germinate, but 1 –year old seeds can germinate under dark conditions. Fresh purslane seed does not exhibit dormancy and has the ability to germinate immediately, allowing for several generations per year (Kiyoko and Cavers, 1980). Certain cultural practices disturb the seeds that are left in the soil and bring them to the surface, allowing for germination (Cudney and Elmore, 1999).

Effect on crop : It was ranked 9th of the world's worst weeds, being recorded in 45 crops in 81 countries (Holm *et al.*, 1977). It caused up to 80% yield loss in sugarbeet in additive field competition experiments.

Control : The purslane sawfly, *Schizocerella pilicorni* and a leafminer weevil, *Hypurus bertrandiperris* are two accidentally introduced biological control agents that have become widespread in California. Hand weeding, mulching, spraying of glyphosate-bades herbicide over the purslane to kill it.

Uses : In Greek popular medicine, purslane is used as a remedy for constipation and inflammation of the urinary system. In the Dominican Republic, all parts of *Portulaca oleracea* are used in treatments for internal parasites. In North India it is known to act as a liver tonic and is used in diseases of the liver. In traditional Chinese medicine, it is used to treat infections or bleeding of the genito-urinary tract as well as dysentery. The fresh herb may also be applied topically to relieve sores and insect or snake bites on the skin.

Past work done on Seed Science and Technology

1. Allelopathic effects

Alireza Dadkhah and Ali Mohammad Asaadi (2010) reported that germination was progressively inhibited with increasing concentration of foliar dry powder aqueous extract of *Eucalyptus camaldulensis*. Germination percentage of *Portulaca* was not affected by different concentration of foliar aqueous extract of *Eucalyptus camaldulensis* (Fig. 5). This indicated that *Eucalyptus* allelochemicals were not inhibitor on *Portulaca* germination. Although, seed germination and germination rate of *Portulaca* were not affected by aqueous extract concentrations, growth of young seedlings of *Portulaca* was severely reduced especially at the higher aqueous extract concentration. Root length of *Portulaca* decreased by 61.9 and 83.3% at low (10 g l^{-1}) and high (20 g l^{-1}) concentrations of aqueous extract of Eucalyptus, respectively (Fig. 6). Shoot length of *Portulaca* also decreased by 32 and 68% at low (10 g l^{-1}) and high (20 g l^{-1}) concentrations of aqueous extract of Eucalyptus compared to control (Fig. 7).

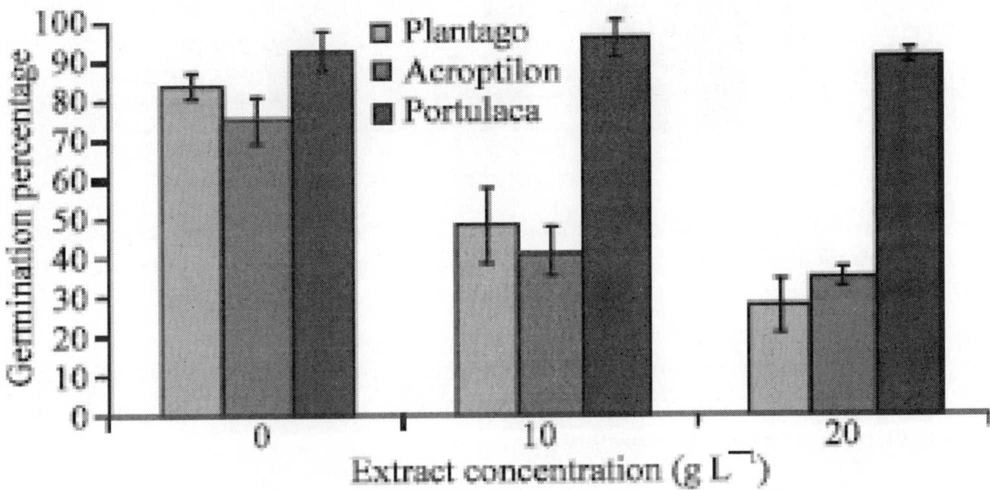

Fig. 5. Germination percentage of studied weed seeds in the presence of various concentrations of aqueous extracts of *Eucalytus camaldulensis*. Each reported value represents the mean ± S.D of 6 replications.

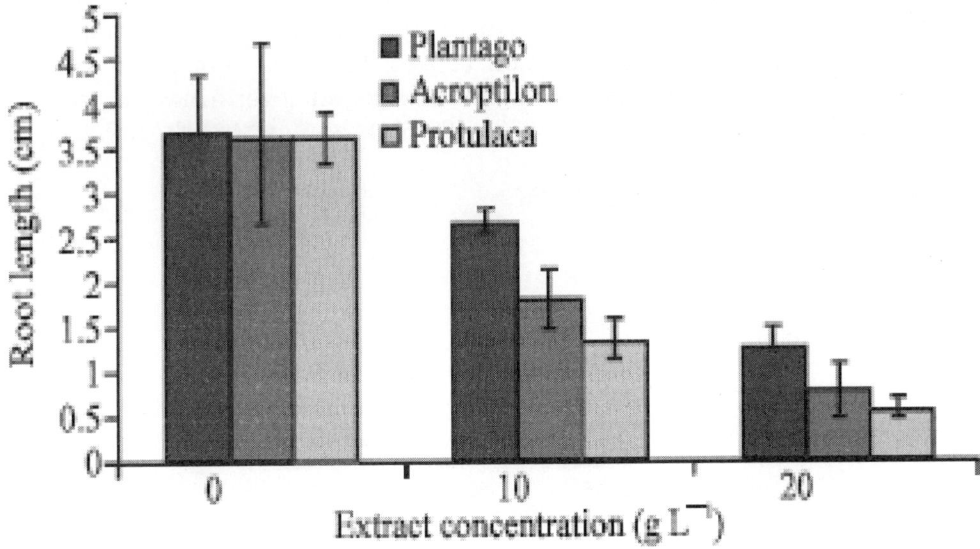

Fig. 6. Root length of studied weed species in the presence of various concentrations of aqueous extracts of *Eucalyptus camaldulensis*. Each reported value represents the means ± S.D of 6 replications.

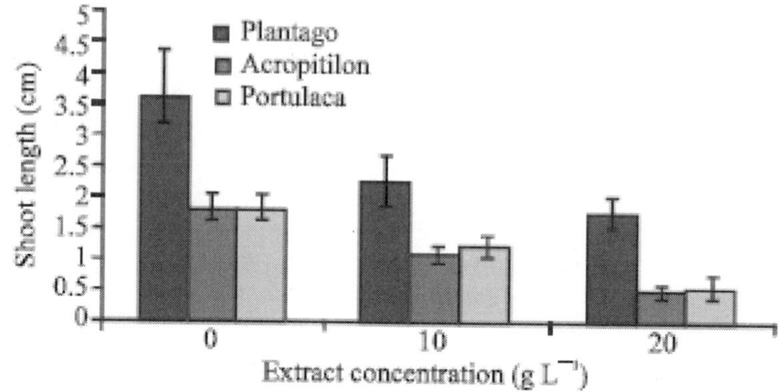

Fig. 7. Shoot length of studied weed species in the presence of various concentrations of aqueous extracts of *Eucalyptus camaldulensis*. Each reported value represents the means ± S.D of 6 replications

The inhibitory effect of *Eucalyptus* on seed germination and seedling of weeds may be related to the presence of allelochemicals including phenolic contents and volatile compounds in its foliage. Furthermore, the toxicity might be due to synergistic effect rather than single one.

2. Germination biology

Germination response to temperature and light

Chauhan and Johnson (2008) stated that germination of freshly harvested seed of *P. oleracea* was influenced by temperature and light, and its interaction ($P < 0.05$) of these (Fig. 8). In the dark, germination (1.1 to 2.2%) was not influenced by the temperatures tested. However, in the light/dark regime, germination was significantly lower at 25/15°C (58.9%) and 35/25°C (60.6%) than at 30/20°C (69.4%).

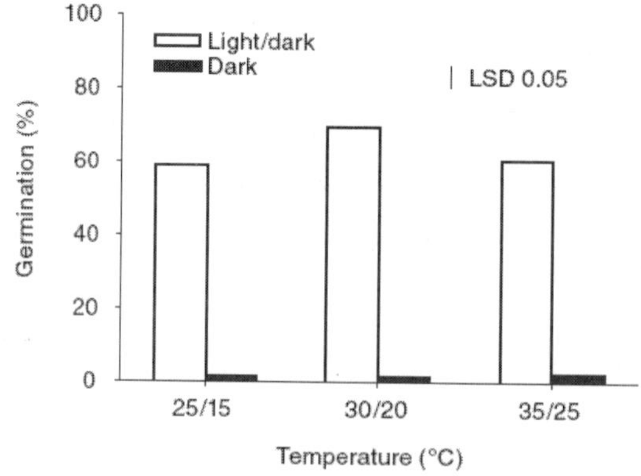

Fig. 8. Effect of temperature (25/15, 30/20, and 35/25°C alternating day/night temperatures) and light (light/dark and dark) on germination of freshly harvested seed of *Portulaca oleracea*.

Germination decreased linearly with decreasing osmotic potential ($G = 66.7 + 88.9x$, $r^2 = 0.98$) (Fig. 9). Germination decreased from 71% to 10% as osmotic potential decreased from 0 to −0.6 MPa; germination was completely inhibited at osmotic potential of −0.8 MPa.

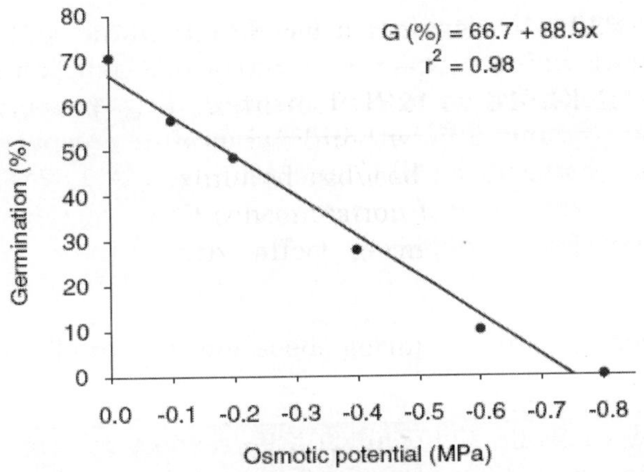

Fig. 9. Effect of osmotic potential (MPa) on germination of *Portulaca oleracea* seed incubated at 30/20°C in a 12 h photoperiod for 14 days. Line represents a linear model fitted to the data.

The data on seedling emergence of *P. oleracea* were fitted to an exponential decay curve (E (%) = 63.2 x e−4.8x, $r^2 = 0.99$). Seedling emergence decreased sharply with increased seed burial depth (Fig. 10). Seedling emergence was greatest (63%) for seed present on the soil surface, but decreased progressively as depth increased. No seedlings emerged from seed buried at a depth of 2 cm.

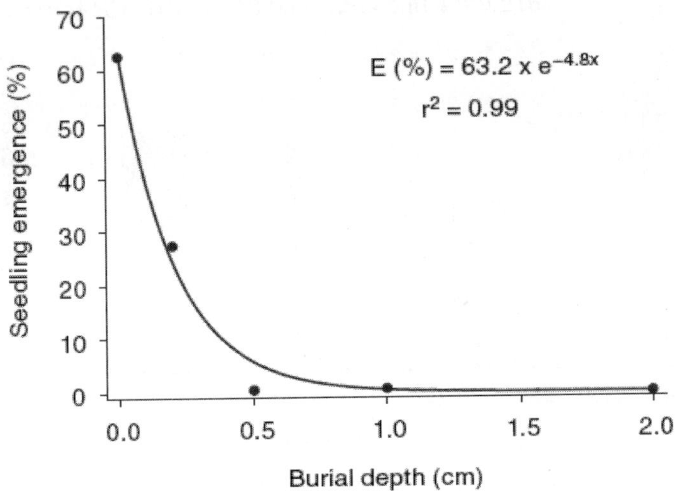

Fig. 10. Effect of seed burial depth on seedling emergence of *Portulaca oleracea.* Line represents an exponential decay model fitted to the data.

3. Germination response to seed burial and solarisation

El-Keblawy and Al-Ansari have found that seed dormancy and germination requirements of the cosmopolitan *Portulaga oleracea* weed are differed among Canadian and UAE populations.

Elmore indicated that *P.oleracea* in California could be controlled with high temperature for a short duration or lower temperature for a longer duration. Upto 8 h did not reduce the germination of that moist *P.oleracea* seeds at 45°C, but four hours reduced it by 43 and 90% at 50°C and 60°C, respectively. In the UAE however, despite that the surface soil temperature reached over 60°C for more than four hours every day, seeds of *P.oleracea* were able to germinate after 26 days of solarisation (El-Keblawy *et al.*, 2004).

Ali El-Keblawy and Al Hammadi (2007) indicated that final germination for non-solarized (control) seeds of UAE, Egyptian and Canadian were 78.0±2.6, 90.1±4.1 and 86.0±3.8%, respectively. Soil solarization for 45 days had resulted in the deterioration of the germination of these seeds to 10.7, 4.1 and 2.0% respectively, at 2.5 cm and to 24.0, 23.6 and 20.2%, respectively at 15 cm depth. This means that the response of the seeds of different source differ greatly according to the depth (Table.1).

Table 1. Effects of depths of seed burial, solarization duration and seeds sources on the final germination percentage (means ± SE) for stored seeds of *Portulaca oleracea*

Depth (cm)	Duration (days)	Seed source			
		UAE	Egypt	Canada	Overall
2.5	15	22.00 ± 2.60	25.60 ± 4.10	10.20 ± 1.06	19.30 ± 2.60
	30	11.80 ± 3.70	7.30 ± 2.15	6.00 ± 1.10	8.40 ± 2.30
	45	10.70 ± 1.30	4.10 ± 1.64	2.00 ± 2.00	5.60 ± 1.60
	Overall	14.80 ± 2.50	12.30 ± 2.60	6.10 ± 1.40	11.10 ± 2.20
7.5	15	40.20 ± 3.70	28.00 ± 3.60	34.10 ± 4.80	34.10 ± 4.00
	30	25.36 ± 2.43	20.46 ± 2.96	26.59 ± 9.40	24.14 ± 5.34
	45	24.40 ± 3.60	4.20 ± 1.60	9.00 ± 1.00	12.60 ± 2.00
	Overall	30.00 ± 3.30	17.60 ± 2.80	23.20 ± 5.10	23.60 ± 3.80
15	15	50.20- ± 5.10	46.90 ± 3.20	45.60 ± 5.40	47.60 ± 4.60
	30	41.20 ± 3.70	39.20 ± 2.50	28.00 ± 4.80	39.50 ± 3.20
	45	24.00 ± 3.70	23.60 ± 4.20	20.20 ± 3.60	22.60 ± 3.80
	Overall	38.50 ± 3.70	36.60 ± 3.30	34.60 ± 4.80	36.50 ± 3.90

Overall depath	15	37.50 ± 3.70	33.50 ± 3.70	29.90 ± 3.70	33.60 ± 3.70
	30	26.10 ± 3.70	22.30 ± 2.50	23.50 ± 5.10	24.00 ± 3.50
	45	19.70 ± 3.70	10.60 ± 2.50	10.40 + 2.20	13.50 ± 2.50
	Overall	27.80 ± 3.70	22.10 ± 2.90	24.60 ± 3.70	24.80 ± 3.20

4. Effect of temperature

Freshly harvested seed of *Portulaca oleracea* (purslane) did not germinate in darkness between 10 to 40°C; however, seed stored for one year had temperature dependant dark germination (Singh, 1973).

REFERENCES

Ali El- Keblawy and Fatima Al Hammadi. (2007). Seed source, depth of burial and solarization duration affect the efficiency of soil solarization in control of *Portulaca oleracea* weed. *Global journal of environmental Research,* 1(2): 86-91.

Alireza Dadkhah and Ali Mohammad Asaadi. (2010). Allelopathic effects of *Eucalyptus camaldulensis* on seed germination and growth seedlings of *Acroptilon repens, Plantago lanceolata* and *Portulaca oleracea. Research journal of Biological sciences,* 5 (6): 430-434.

Bhagirath S. Chauhan and David E. Johnson. (2008). Germination biology of *Portulaca oleracea* L. Sixteenth Australian Weeds Conference. Crop and Environmental Sciences Division, International Rice Research Institute, Los Baños, Philippines.

Cudney, D. and C. Elmore. (1999). University of California. Statewide Integration Pest Management Program.

Egley, G. H. and J. M. Chandler. (1983). Longevity of weed seeds after 5.5 years in the Stoneville 50-year buried seed study. *Weed Sci.,* 31: 264-270.

El-Keblawy, A., H. Dhaheri, S. S. Al Mazrooei, N. A. Ghaili, E. M. Al Hedhani, A. M. R. Al Salmi, R. R. M. Ali and H. S. Al Ameri. (2004). Using soil solarization in weed control under the environment of the UAE. Proceeding of 5th Annual Conference for Research Funded by UAE University, Al-Ain, UAE.

Hanson, G. (2004). University of Idaho. Common Purslane. *Portulaca oleracea.*

Holm, L.G., D. L. Plucknett, J. V. Pancho and J. P. Her- berger. (1977). The world's worst weeds: distribution and biology. (University of Hawaii Press, Honolulu, Hawaii USA).

Kiyoko, M. and P B. Cavers. (1980). The biology of Canadian weeds. 40. *Portulaca oleracea* L. *Can. J. Plant Sci.,* 60:953-963.

Mitich, L. W. (1997). Common purslane *(Portulaca oleracea). Weed* Tech., 11(2):394-397.

Ozer, Z. (1972). Yabanci ot tohumlarinin yasam muddetleri. Ataturk Universitesi Ziraat Fakultesi Dergisi, 3: 233-239.

Singh, K. P. (1973). Effect of temperature and light on seed germination of two ecotypes of *Portulaca oleracea* L. *New Physiologist* 72 : 289-295.

Waterhouse, D. F. (1994). Biological control of weeds: Southeast Asian prospects. ACIAR Monograph No. 26, 302pp.

RHYNCHOSIA (*Rhynchosia minima* (L.) DC)

Common name : Rhynchosia, Burn mouth vine

Botanical name : Rhynchosia minima (L.) DC

Tamil name : Chitt-avarai

Hindi name : Kulata, kulthi

Family : Leguminosae - Papilionoideae (Faboideae)

Origin and distribution : It is an almost cosmopolitan plant on the heavier textured soils of the tropics and subtropics. It is widespread in Sudan and East, Syrian Arab Republic to India (on black cotton soils in Mysore State), Nepal, southern United States (South Carolina to Florida and Texas) to Mexico, Panama, Venezuela, Argentina, Australia.

Botany

Habitat : A weed adjacent to cultivated areas.

Habit : Twining annual herb; branches are many, thin and wiry growing to a length of 3-5 feet (Fig. 1).

Fig. 1. Plant

Fig. 2. Flower

Stem : The stem is rather thick, striate, four-sided, hollow and densely clothed with short hooked hairs.

Leaf : Leaves trifoliate, leaflets subcoriaceous or membranous, obtuse, conspicuously gland – dotted beneath, as broad as long, deltoid in the lower half; stipules minute, lanceolate, caducous.

Floral biology

Flower : Flowers yellow, short peduncled, 6-12 flowers, elongated axillary racemes, bracts minute, linear; calyx pubescent, teeth exceeding, the tube acuminate, lowest tooth setaceous; corolla papilonaceous, twice the calyx; stamens diadelphous; ovary superior, 1-celled, 2-ovuled, style incurved, stigma capitates (Fig. 2).

Flowering and pollination : Flowering: August-January. Cross-pollination. Bees are pollinators.

Fruit : Pod long, glabrescent, turgid, slightly recurved, 2- seeded (Fig. 3).

Fig. 3. Fruit

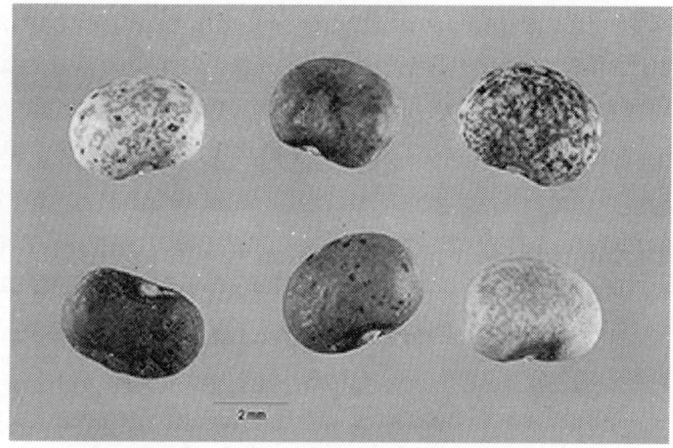

Fig. 4. Seed

Seed biology

Seed : Seeds one coloured, black or brown, 3 mm long with short hilus (Fig. 4).

Propagation : By seed.

Dormancy and germination mechanism : Constant high temperature pre-treatment given to the seeds of *Rhynchosia minima,* beside breaking hard seedcoat dormancy, tends to stimulate germination as well as seedling growth (Mullick and Chatterji, 1970).

Seed storage and burial : Seeds having 8% germination following 18 years open storage at room temperature (Ewart, 1908).

Effect on crop : Host for yellow mosaic virus.

Control : Glyphosate is mixed with fluroxypyr to control

Uses : Manure, fodder, diarrhoea, dysentery; genital stimulants/depressants; laxatives, vermifuges miscellaneously poisonous or repellent, anus, haemorrhoids.

Past work done on Seed Science and Technology

1. Seed dormancy and germination

Rangaswamy and Lakshmi Nandakumar (1985) reported that, under laboratory conditions, *Rhyncosia minima* seeds germinated only if the hilum was pricked and that, in seeds sown on moistioned filter paper, a prominent brownish-black halo developed around the pricked area, presumably from leaching of phenolics. Krishnamurthy *et al.* (1975) identified galic acid, protocatechuic acid, prodelphinidin, and hydroquinone diacetate in *R. minima* seeds.

Rangaswamy and Lakshmi Nandakumar (1985) identified that the seed coat of *Rhynchosia minima* is composed of (1) a surface deposit of waxy material interfused with a lipoidal substance, β-sitosterol; (2) a subjacent 3-μm adcrustation of hemicellulose-cellulose complex, which is shown to be a separate entity; and (3) a layer of palisade cells in which the secondary walls are impregnated with arabinan and the lumen contains tannin and phenolic compounds. These features constitute a gradative barrier to water uptake by seed. Seeds failed to germinate unless the impermeable seed coat surface or hilum was damaged.

According to Mullick and Chatterji (1970), exposing the seeds of *Rhynchosia minima* to constant high temperature treatment, beside breaking hard seedcoat dormancy, tends to stimulate germination as well as seedling growth. Favourable results have been obtained by exposing to high temperature at 60°C to 90°C for 12 to 24 hours.

Shahid Shaukat and Nazia Burhan (2000) expressed that seeds have water impermeable seed coats and thus they have physical dormancy. Physical scarification with sand paper was more effective in breaking seed dormancy than chemical scarification with hydrochloric acid (Fig. 5).

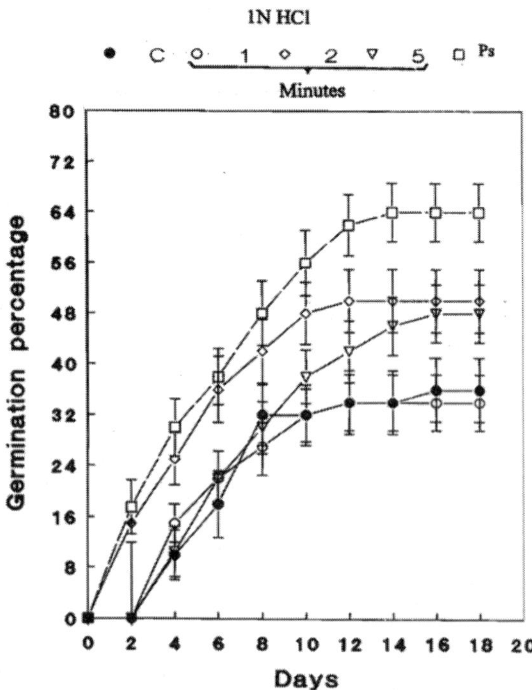

Fig. 5. Effect of physical and chemical scarification on germination of *Rhyncosia minima*. C. control; 1, 2, 5 min. in HCl treatment: Ps. Physical scarification

Dry heat treatment (50°C and 70°C) markedly increased the germination percentage (Fig. 6). Under field conditions dry heat during summer apparently breaks seed dormancy thereby allowing the seeds to germinate in considerable numbers after rain showers.

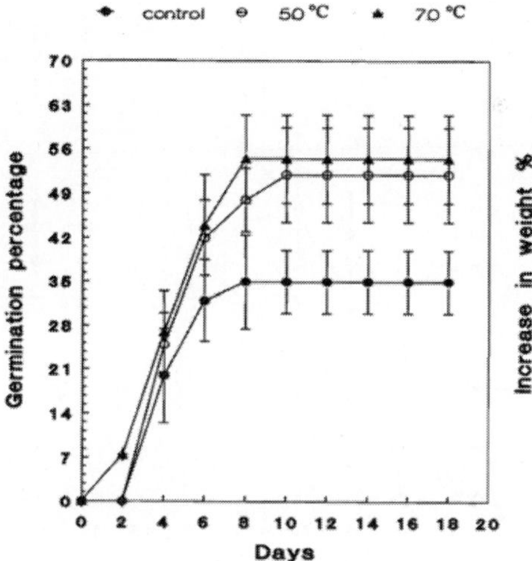

Fig. 6. Effect of dry heat treatment on germination of *Rhyncosia minima* at 50 and 70° C

Greater germination percentage was observed in light compared to dark (Fig. 7). Final germination percentage varied with temperature (25°-40°) with the optimum at 35° (Fig. 8). Significant inhibition of germination occurred at 6, 7, and 8 bar NaCl salinity (Fig. 9). While in polyethylene glycol (PEG 4000) germination was reduced at 7 and 8 bar (Fig. 10). At iso-osmotic concentrations, PEG 4000 caused lesser inhibitory effect compared to NaCl salinity.

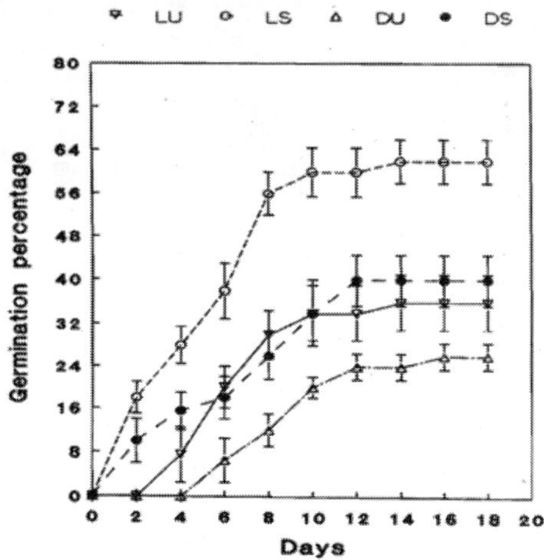

Fig. 7. Effect of light and dark on germination of *Rhyncosia minima*. LU, unscarified seeds in light; LS, scarified seeds in light; DU, unscarified seeds in dark; DS, scarified seeds in dark.

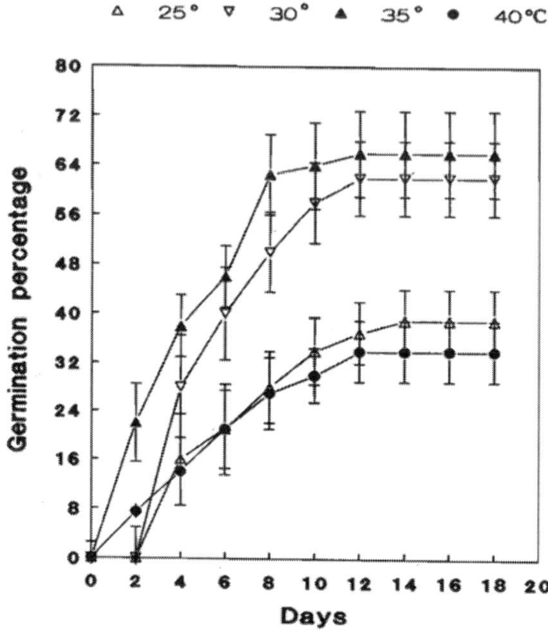

Fig. 8. Effect of temperature regimes on germination of *Rhyncosia minima*

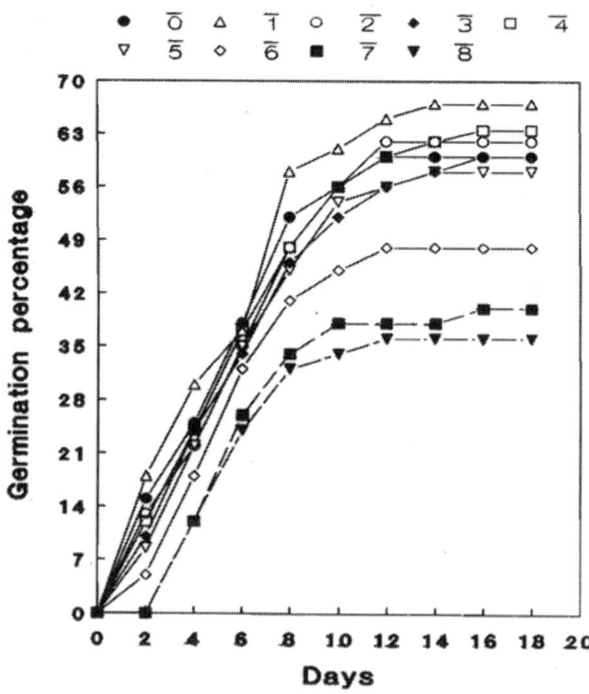

Fig. 9. Effect of salinity levels on germination of *Rhyncosia minima*

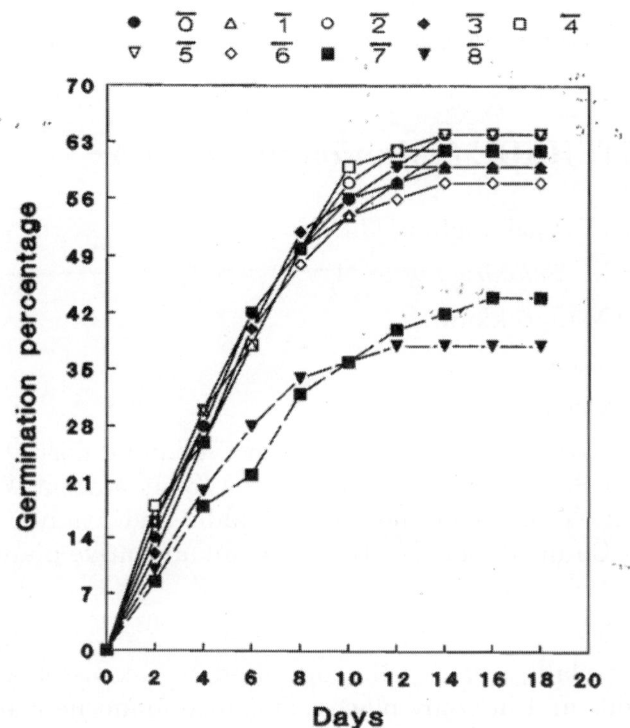

Fig. 10. Effect of moisture stress (created by PEG-4000) on germination of *Rhyncosia minima*

Albert Madueno-Molina *et al.* (2006) found that the use of concentrated sulfuric acid for 30 minutes and the heat treatment at 70°C for 5 hours promoted the highest germination percentage.

REFERENCES

Alberto Madueño-Molina, Diego García-Paredes, Jesús Martínez-Hernández, Candelario Rubio-Torres, Ana Navarrete-Valencia and y José Bojórquez-Serrano. (2006). Seed germination of wild Bean, *Rhynchosia minima* (L.) DC., under different seed treatments. *Bioagro Vol.*18 (2).

Ewart, A.J. (1908). On the longevity of seeds. *Proceedings of the Royal Society Victoria*, 21: 1-210.

Krishnamurthy, H.G, L. Krishnaswami and N.S. Rangaswamy (1975). Hydroquinone diacetate from Rhynchosia minima. *Phytochemistry.* 14: 2518-2519

Mullick, P. and U.N. Chatterji (1970). Germination regulating mechanism in arid-zone plants: high temperature stimulus and germination of Rhynchosia seeds. *Plant Science,* 76(2): 39:46.

Rangaswamy, N. S. and Lakshmi Nandakumar (1985). Correlative studies on seed coat structure, chemical composition and impermeability in the legume *Rhynchosia minima.* *Bot. Gaz.,* 146(4): 501-509.

Shahid Shaukat, S. and Nazia Burhan (2000). Fecundity, seed characteristics and factors regulating germination of *Rhynchosia minima* (L.) D.C. *Pak. J. Bot.,* 32(1): 211-226.

BLACK NIGHTSHADE (*Solanum nigrum* L.)

Common name : Black nightshade

Botanical name : Solanum nigrum L.

Tamil name : Manatakkali

Hindi name : Mokoi

Family : Solanaceae

Origin and distribution : Native of Europe, Asia and northern Africa. European part of the Former Soviet Union, the Caucasus, Central Asia, Western Siberia, the Far East. Distributed also in Scandinavia, Middle and Atlantic Europe, Mediterranean, Iran, India, China, Japan, North Africa; an adventive plant in North America.

Botany

Habitat : This plant prefers rich, cultivated, well-moistened soils, but can even grow on saline soils and in stony places. It is also found near roads, houses, waste dumps and bush thickets along rivers.

Habit : An erect, much branched, unarmed, annual herb (Fig. 1).

Stem : Branchlets zig-zag.

Leaf : Leaves thin, glabrous ovate – lanceolate, sinuate toothed or entire.

Floral biology

Flower : Regular (actinomorphic). Corolla white, fused, wheel-shaped, 5-lobed, 6–14 mm (2.2–5.8 in.) wide. Calyx fused, campanulate, deeply 5-lobed. Stamens 5, anthers in a conical group. Gynoecium composed of 2 fused carpels. Inflorescence a lax, usually 3–8-flowered cyme (Fig. 2).

Fruit : Fruit is globose berry 0.6 cm diameter, globular, black but sometimes brownish-green or red or yellow (Fig. 3).

Seed biology

Seed : Seeds discoid minutely pitted (Fig. 4).

Life cycle : Maximum productivity is up to 280,000 seeds. May be biennial particularly in milder conditions, where it may live from 2-5 years.

Flowering, pollination and fruit set : This plant flowers in June-September, bears fruit in July-November. Flowering occurs 5-9 weeks after germination and continues until the plant dies. Early germinating plants take several weeks more to flower compared to late germinating plants.

Fig. 1. Plant

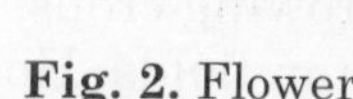

Fig. 2. Flower

Fig. 3. Fruit

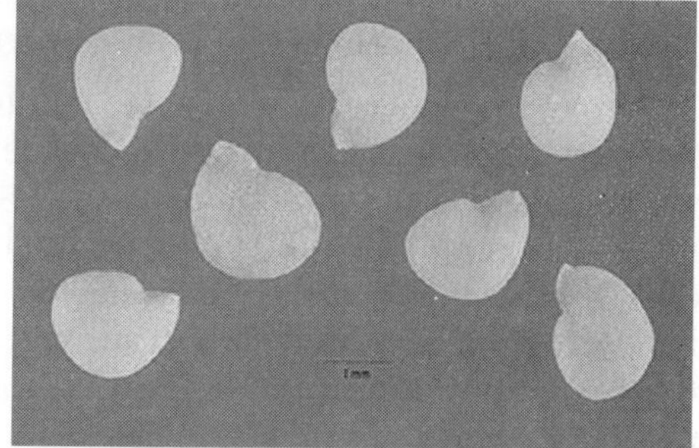

Fig. 4. Seed

Seed dispersal : The berries fall from the plants when mature: occasionally remain on the plant. Often the seeds germinate *in situ* when the prevailing environment conditions are optimal.

The ripe berries are extremely succulent, often brightly colored and/or aromatic, and are attractive to a variety of birds and animals. They are therefore, effectively dispersed by birds all over the world, as well as by various animals such as rodents, lizards, rabbits, cattle and man. There is evidence that the seeds pass through the digestive tracts of both birds and animals undamaged (Burgert *et al.*, 1973; Weller and Phipps, 1978/79). The seeds can also be dispersed by water (Burgert *et al.*, 1973).

Propagation : Propagation is mainly by seed; seedling development is epigeal. Shoot cuttings may also be used as propagules though plants propagated in this way branch, spread and yield less than those propagated by seed (Mwafusi, 1992).

Plantlets have also been regenerated from *S. nigrum* mesophyll chloroplasts (Wang and Xia, 1983).

Seed dormancy and germination : The mechanism of seed dormancy is associated with impermeable seed coat (Suthar *et al.,* 2009). Minimum temperature for seed germination is 10-12°C, maximum is 34-36°C and optimum is 24-26°C. Seeds germinate from depths of 0.5-1 cm. Alternating temperatures are required to break dormancy and lower light intensities or darkness are generally more favourable for germination. There is vigorous recruitment from soil stored seed in burnt and unburnt areas after rain. Levels of primary seed dormancy among populations, enabling it to maximise its chance of survival by regulating germination timing to favourable conditions. It is therefore, best to time weed control measures to larger seedling flushes.

Seed storage : Seedbank persistence, 5+ years.

Effect on crop : S. nigrum is a segetal weed of mainly tilled crops; it is rarely found in close-growing crops, such as grain and fodder crops. This plant is especially abundant in cotton fields. Host to nematodes, fungi and viruses. Produces alkaloids, glycosides and accumulates nitrates. This weed can be a host plant for *Synchitrium endobioticum*, the pathogen of potato wart.

Control : Control measures include thorough main and pre-sowing soil treatments, harrowing of tilled crops, eradication of *S. nigrum* before flowering, mowing of this weed in uncultivated places. Prevent seed set for several years. Hand weeds small infestations. In bushland situations, manually remove plants before flowering. For large infestations, starane 1 l ha^{-1} (20 ml/10 l) applied when actively growing in summer, will provide reasonably selective control. 2, 4-D amine (500 g l^{-1}) 1 l ha^{-1} or 20 ml/10 l can also be used for control of young plants in early summer.

Uses : Black nightshade is used for skin diseases, rheumatism, and gout. Juice of the herb is given in chronic enlargement of the liver. It can cure ear and eye diseases. It is sometimes prescribed to remove the effect of old age.

In India, black nightshade berries are grown and eaten by local people. The dish is more common in Northern Tamil Nadu, Southern Andhra and Southern Karnataka.

In Ethiopia, ripe fruits of black nightshade are eaten by children during normal times, while all the affected people eat fruits of the plant during famines. The leaves are also collected by women and cooked in salt water. They are then consumed like any other vegetable.

It is known that farmers in the Konso Special Woreda use it as source of food, until their maize crops are ready for harvesting, since the plant matures before maize. The Welayta people of the adjacent Semien Omo Zone do not remove this weed appearing in their gardens. Instead, they cook and eat the leaves of the plant.

Past work done on Seed Science and Technology

1. Seed dormancy and germination

Suthar *et al.* (2009) observed 72% germination was when the seeds were scarified with needle-pin while with sand paper, 78% seed germination was observed followed by 33% germination with mortar pestle scarification (Table 1). It was observed that sand paper scarification method was the best for breaking dormancy. The macroscleried layer of mature seed coat is mainly responsible for imparting impermeability to the seeds.

Table 1. Effects of mechanical scarification on seed germination of *Solanum nigrum*

Method		Germination (%)
Unscarified seeds		52 ± 1.5
Sacrification with	Needle-pin	72 ± 1.3
	Sand paper	78 ± 0.9
	Mortar-pestle	33 ± 1.0

Scarification with diluted sulphuric acid gave 48% of germination at 1.0 min. and then it decreased with increasing in time. Similarly, germination was the highest, 38% at 30 sec., with concentrated sulphuric acid and decreased later on (Table 2). It was observed that chemical scarification with both dilute and concentr-ated sulphuric acid showed adverse effect as compared to that of control (unscarified seeds), which could be due to adverse effect caused by the acid to the embryo.

Table 2. Effect of chemical scarification with dilute and conc. H_2SO_4 on seed germination of *Solanum nigrum*

	Method	Germination (%)
Unscarified seeds		53 ± 1.8
Scarified with	0.15 min	23 ± 1.2
a. Dilute H_2SO_4	0.30 min	34 ± 1.8
	1.0 min	48 ± 2.8
	2.0 min	32 ± 1.6
b. Conc. H_2SO_4	0.15 min	34 ± 1.2
	0.30 min	38 ± 0.8
	1.0 min	20 ± 2.6
	2.0 min	10 ± 2.6

Low imbibition was observed in unscarified seeds as compared to sand paper scarified seeds at all pH treatments. But, in both seeds, there was no significant variation in imbibition percentage in acidic pH as well as alkaline pH range. The rate of germination of scarified and unscarified seeds also showed variation. The maximum percentage germination occurred in petriplate irrigated with only distilled water (neutral pH). The percentage rate germination was low at acidic as well as alkaline pH in both the sets of scarified and unscarified seeds (Table 3). It was observed neutral pH plays an important role in increasing germination of *S. nigrum*.

Table 3. Effect of pH on imbibition and seed germination of *Solanum nigrum*

		pH values							
Parameter	**Seeds**	**4.1**	**5.1**	**6.1**	**6.5**	**D/W**	**8.1**	**9.2**	**10.3**
% of imbibition (6 hrs)	Scarified	8.39	8.31	8.59	8.43	8.51	8.13	8.21	8.17
	Unscarified	6.83	6.31	6.91	6.83	6.91	6.78	6.78	6.61
% of germination	Scarified	71 ± 1.2	73 ± 0.8	75 ± 1.4	78 ± 1.2	82 ± 1.6	76 ± 3.2	73 ± 0.6	63 ± 1.2
	Unscarified	31 ± 0.6	38 ± 0.8	40 ± 1.2	38 ± 1.2	48 ± 1.6	36 ± 0.8	28 ± 1.8	22 ± 2.0

Both scarified and unscarified seeds showed an increase in rate of imbibition with increase in time interval. After soaking, the scarified seeds for 5.5 h, the maximum imbibition was observed and it remained constant afterwards. Similarly, the rate of imbibition of unscarified seeds was low and maximum imbibition was observed at 6.0 h and then remained constant (Table 4).

Table 4. Rate of imbibition on scarified and unscarified seeds of *S. nigrum*

	Percentage imbibition in gm	
Time (hrs)	**Unscarified seeds**	**Scarified seeds**
0.5	5.31	5.81
1.0	5.38	5.93
1.5	5.48	6.46
2.0	5.50	6.49
2.5	5.53	6.54
3.0	5.58	6.58
3.5	5.60	6.63
4.0	5.81	6.71
4.5	5.93	7.13
5.0	6.31	7.41
5.5	6.68	7.42
6.0	6.73	7.42
6.5	6.74	7.42
7.0	6.74	7.42

The maximum percentage of seed germination was observed in all the three sets in sand paper scarified seeds as compared to unscarified seeds. In scarified seeds, the maximum percentage of seed germination was observed in the seeds, sown one cm deep in the soil followed by seeds, sown superficially and 2.0 cm deep in the soil. Similar trends were observed in case of unscarified seeds with low values. Thus, a suitable depth of sowing of the seeds is 1.0 cm below the soil surface (Table 5).

Table 5. Effect of sowing depth on seed germination of *Solanum nigrum*

	% Germination	
Sowing depth	**Scarified**	**Unscarified**
Superficial	68 ± 7	38 ± 3
1 cm	78 ± 8	52 ± 4
2 cm	54 ± 4	44 ± 2

Unscarified seeds showed lesser percentage of germination as compared to sand paper scarified seeds at all temperature treatments. The maximum percentage of germination was observed at 27°C followed by 32°C in scarified seeds in which the lowest percentage seed germination was observed at low temperature (3% after 5 days of incubation at 5°C) as well as at high temperature (2% after 5 days of incubation at 60°C). Variation was observed in the germination rate of scarified and unscarified seeds and the maximum percentage germination was noted on 4th day. The rate of percentage germination was low for first two days and on last day (Table 6). Therefore, the temperature (27°C) increased the germination of *S. nigrum* scarified seeds.

Table 6. Effect of temperature on seed germination of *Solanum nigrum*

		% Germination / day				
Temp (C)	**Seeds**	**1**	**2**	**3**	**4**	**5**
5	Scarified	-	-	7 ± 0.8	9 ± 0.8	3 ± 0.3
	Unscarified	-	-	1 ± 0.9	3 ± 0.2	2 ± 0.2
10	Scarified	-	-	8 ± 0.6	7 ± 0.2	4 ± 0.4
	Unscarified	-	-	2 ± 0.4	2 ± 0.6	1 ± 0.6
27	Scarified	-	4 ± 0.8	28 ± 0.6	38 ± 1.3	14 ± 1.8
	Unscarified	-	1 ± 0.3	8 ± 1.2	6 ± 0.3	4 ± 0.8
32	Scarified	-	-	14 ± 1.2	22 ± 0.8	18 ± 0.2
	Unscarified	-	-	8 ± 0.3	4 ± 0.3	3 ± 0.4
40	Scarified	-	-	6 ± 0.6	18 ± 0.7	3 ± 0.3
	Unscarified	-	-	5 ± 0.2	3 ± 0.3	1 ± 0.7

50	Scarified	-	-	8 ± 0.2	3 ± 0.4	7 ± 0.3
	Unscarified	-	-	3 ± 0.1	4 ± 0.8	2± 0.2
60	Scarified	-	-	3 ± 0.2	8 ± 0.7	2 ± 0.2
	Unscarified	-	-	1 ± 0.3	3 ± 0.3	2 ± 0.4

2. Seed storage and burial

Ozer (1972) reported that seeds of *Solanum nigrum* L. buried in the soil can remain viable for 30 years or more. Ilhan Uremis and Nezihi Uygur (2005) recorded that seeds recovered 7 years after burial and seed longevity was determined using by germination tests and TTC test. Burial depth, 15 and 30 cm had no significant effect on seed viability. Seed longevity was 5.38% for *S.nigrum*.

Suthar *et al.* (2009) observed the highest percentage of seed germination in the month of June followed by July, August, October and December in case of sand paper scarified seeds and June in case of unscarified seeds, while there was not much variation in other months (Table 7). The scarified seeds showed higher germination as compared to that of unscarified seeds at every month.

Table 7.

	Germination (%)	
Months	**Scarified seeds**	**Unscarified seeds**
Oct.	68 ± 1.2	43 ± 0.8
Nov.	63 ± 1.3	42 ± 0.3
Dec.	68 ± 1.2	46 ± 0.4
Jan.	62 ± 0.8	43 ± 0.8
Feb.	58 ± 0.7	38 ± 0.7
Mar.	53 ± 0.3	37 ± 1.3
Apr.	68 ± 0.8	43 ± 0.8
May	63 ± 1.1	46 ± 0.9
June	73 ± 0.8	54 ± 1.5
July	71 ± 1.2	52 ± 1.1
Aug.	69 ± 0.3	50 ± 0.7
Sept.	49 ± 0.2	38 ± 0.9

REFERENCES

Burgert, K.L., O.C. Burnside and C.R. Fenster. (1973). Black nightshade leaves its mark. Nebr. Univ. Coll. Agric. Q. Serving Farm, 20: 8-10.

Ilhan Uremis and F. Nezihi Uygur. (2005). Seed viability of some weed species after 7 years of burial in the Cukurova region of Turkey. *Asian Journal of Plant Sciences,* 4: 1-5.

Mwafusi, C.N. (1992). Effects of propagation method and deflowering on vegetative growth, leaf yield; phenolic and glycoalkaloid contents of three black nightshade selections used as vegetables in Kenya. MSc Thesis, University of Nairobi.

Ozer, Z. (1972). Yabanci ot tohumlarinin yasam muddetleri. Ataturk Universitesi Ziraat Fakultesi Dergisi, 3: 233-239.

Suthar, A.C., V.R. Naik and R.M. Mulani. (2009). Seed and seed germination in *Solanum nigrum* Linn. *American-Eurasian J. Agric. & Environ. Sci.,* 5 (2): 179-183.

Wang G. and Z. Xia. (1983). Regeneration of plantlets from *Solanum nigrum* L. mesophyll protoplasts. *Acta Botanica Sinica,* 25(2):111-114.

Weller, R.F. and Phipps, R.H. (1978/1979). A review of black nightshade (*Solanum nigrum* L.) *Prot. Ecol.,* 1: 121-139.

JOHNSON GRASS (*Sorghum halepense* (L.) Pers.)

Common name : Johnson grass

Botanical name : Sorghum halepense (L.) Pers.

Tamil name : Solam

Hindi name : Baru, chinna, bajara, jangli-jowar

Family : Poaceae

Origin and distribution : Believed to be of Mediterranean origin (Meredith, 1955), but introduced very early to India (Bor, 1960); now widespread through the subtropics. Called Johnson grass after Colonel Johnson who first grew it in Alabama, United States.

Botany

Habitat : A weed of arable land, along irrigated canals, roadsides and waste places

Habit : Annual grass (Fig. 1).

Stem : Culms erect, or geniculately ascending, or decumbent; 50-300 cm long. Culm-nodes pubescent.

Leaf : The 2 ft (0.6 m) long, lanceolate leaves are arranged alternately along a stout, hairless, somewhat upward branching stem and have distinct, white midribs.

Floral biology

Flower : Inflorescence: Inflorescence is a panicle with branches tipped by a raceme.

Fig. 1. Plant

Flower: Spikelets heteromorphic. Sessile spikelet dorsally compressed, 4-5.5 mm long, 1.5-2.3 mm wide, shiny, straw-yellow to dark reddish brown, of one fertile floret and one basal sterile lemma. Spikelet callus pubescent; pedicels with silky hairs. Lower glume coriaceous, length of spikelet, its margins enclosing upper glume, 2-keeled near apex, apex dentate, 3-fid; upper glume coriaceous. Sterile lemma hyaline; fertile lemma hyaline, 2-4 mm long; palea present or absent or minute. Pedicellate spikelet(s) dorsally compressed, glumes chartaceous, 4-6.5mm long, containing empty or staminate lemmas (Fig. 2).

Fig. 2. Inflorescence; Flower

Flowering, pollination and fruit set : It is primarily self-pollinated. Some pollination is effected by wind, especially when plants are ≤ 425 feet (130 m) apart. Depending on the climate, flowering begins roughly two months after growth commences and continues throughout the growing season. Most of the year's rhizome growth takes place after flower production (Warwick and Black, 1983). Hundreds of seeds are produced on each panicle throughout the summer flowering period (Monaghan, 1979; Warwick and Black, 1983).

Seed biology

Seed : Caryopsis reddish brown, 2.5-3 mm long, dorsally compressed (Fig. 3).

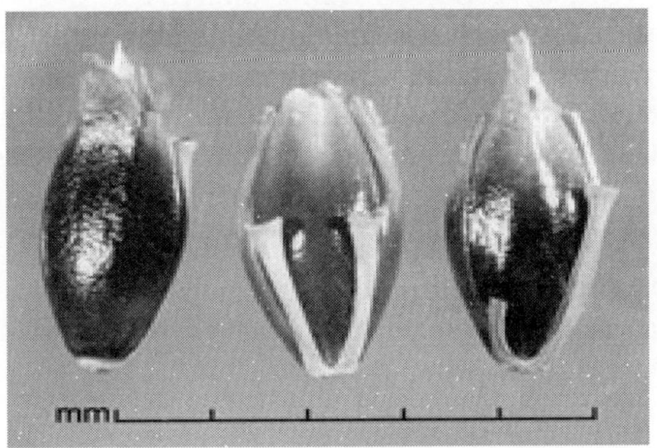

Fig. 3. Seed

Propagation : Reproducing from rhizomes and seeds.

Seed dispersal : Holm *et al.* (1991) noted that crops seeds are the primary means of *Sorghum halepense* spread. *Sorghum halepense* seeds are dispersed by infructescence shatter dropping seed near the parent plant or by water, wind, ingestion by birds and cattle, catching onto animal fur, contamination in grain and hay and contamination of farm equipment then shared by farmers (California Department of Food and Agriculture, Encycloweedia, 2002; Holm *et al.*, 1977 *in* Warwick and Black, 1983; Holm *et al.*, 1991; McWhorter, 1971b *in* Warwick and Black 1983; Youngclaus 1972 *in* Warwick and Black, 1983) *Sorghum halepense* can also spread by means of its extensive rhizomes. If these fragments are moved by vehicles or machinery, humans or animals, or water, further spread of *Sorghum halepense* can occur. The use of *Sorghum halepense* as a forage encourages further spread of the species (Chambers and Hawkins, 2002).

Seed dormancy and germination : Sorghum halepense seed is dormant when dispersed, dormancy differing with ecotype (Bennett, 1973 *in* Warwick and Black, 1983; Taylorson and McWhorter, 1969 *in* Warwick and Black, 1983, Monaghan, 1979 *in* Warwick and Black, 1983). Less than 10% germination was observed at 68°F (20°C), indicating most of the seeds of 44 ecotypes studied were highly dormant (Taylorson and McWhorter, 1969 *in* Holm *et al.*, 1991, Taylorson and McWhorter, 1969 *in* Warwick and Black, 1983).

Dormancy is seemingly mechanical; the seed coat contains tannins causing a reduced permeability to water (Bennett, 1973 *in* Warwick and Black, 1983, Taylorson and McWhorter, 1969 in Warwick and Black, 1983). Most *Sorghum halepense* seeds do not germinate the year they are produced, although they germinate readily the subsequent year (Holm *et al.*, 1977 *in* Newman, 1993). The required afterripening period lasts an average of 4-5 months (California Department of Food and Agriculture, *EncycloWeedia*, 2002) which mostly overcomes dormancy (Anderson, 1968 *in* Warwick and Black, 1983).

After-ripened seeds with glumes still attached were observed to be associated with residual dormancy (Anderson, 1968 *in* Warwick and Black, 1983; Harrington, 1923 *in* Warwick and Black, 1983). Harrington, 1916, 1917 *in* Holm *et al.,* 1991) showed that removal of the glume could give 95% germination during some experiments with dormant *Sorghum halepense* seeds. About 20-40% of *Sorghum halepense* seed is enclosed by its glumes, thus having this imposed dormancy. As previously stated, in California *Sorghum halepense* seed germination can occur from early spring through early fall, with optimum germination temperature approximately 75°F (24°C). Temperatures below 57°F (14°C) will typically inhibit germination.

Depending on *Sorghum halepense* biotype, germination rates have been increased after chilling followed by short periods of high temperatures, when fluctuating temperatures occur, with exposure to light, and with nitrate levels increased. Taylorson and McWhorter (1969 *in* Holm *et al.*, 1991) found that pretreatment of seeds by keeping them at 50°F (10°C) for 2 weeks, then followed by temperatures of 104°F (40°C) for 2 hours broke dormancy.

Seed storage and burial: The spikelet bracts (glumes) tightly surround and enclose *Sorghum halepense* seeds, protecting them from decomposition in the soil for several years. When the glumes are removed, freshly harvested seed can germinate immediately. 60-75% *Sorghum halepense* seed buried in soil during trials in Mississippi remained viable after 2.5 years (Egley and Chandler, 1978 *in* Warwick and Black 1983).

In California, viability after 5 years was 50%, although one study showed a diminished viability of 2% after 6 years in soil (Looker, 1981 *in* Warwick and Black, 1983). After passing through the digestive tracts of animals, and after 7 years in dry storage, *Sorghum halepense* seed has reportedly stayed viable (Holm *et al.,* 1977 *in* Warwick and Black, 1983). Some estimation suggests that some of the seed remains viable for up to 15 years.

Effect on crop : Johnsongrass is very invasive and can reduce corn and soybean yields over 30% and 40% respectively, even with normal control efforts. In addition to *Sorghum halepense'* s competitive ability is the production of allelopathic substances. Friedman and Horowitz (1970) showed a retardation effect in barley, mustard and wheat as well as in other Johnson grass plants by both an exudate from Johnson grass and decaying plant matter. An allelopathic compound from Johnson grass, dhurrin, inhibited seed germination and seedling development in several plant species (Warwick and Black, 1983).

Johnson grass is a desirable forage crop and erosion stabilizer.

Control : Torching and burning, mowing and grazing, tilling and ploughing and herbicide application can help to control this weed. Repeated tilling and ploughing has resulted in the control of Johnson grass in some locations (Warwick and Black, 1983). A single application of the herbicide glyphosate results in an 85% reduction in Johnson grass

Uses : An excellent grass for hay and pasture in the southern United States from Alabama to Texas. Palatable and nutritious when fed early, comparable to timothy as hay, and when properly fertilized and cut in the boot stage, ranks next to alfalfa and ahead of oat and soybean hays for feeding dairy cows. The seeds are eaten in times of scarcity. Reported to be cyanogenetic, demulcent, depurative, diuretic, poison and tonic. It is a folk remedy for blood and urinary disorders (Duke and Wain, 1981).

Past work done on Seed Science and Technology:

1. Seed dormancy and germination

Taylorson (1975) observed that a 10°C dark prechilling of Johnsongrass [*Sorghum halepense* (L.) Pers.] seeds, when terminated by a 2-hr, 40°C temperature shift, potentiates about 40% germination at 20°C in darkness (Table. 1). Irradiation of the seeds before, during and at the end of prechilling with far red light reduces the subsequent germination, although red irradiation after the far red can overcome some of the inhibition. However, either brief red or far red irradiation given immediately after the temperature shift inhibits subsequent germination by one-third to onehalf. The results suggest that the far red-absorbing form of phytochrome is a factor in the prechill-induced dark germination and that phytochrome participates in the inhibition of germination by irradiations immediately after the temperature shift.

Table 1. Effect of R and FR irradiation after a prechill and temperature shift on germination of Johnsongrass seeds

Light treatment		Irradiance (μ 20 cm^{-2})	Germination (%)
Quality	Duration		
	0 (Control)		33 ± 3
R	1 sec	23	31 ± 3
R	4 sec	23	28 ± 2
R	16 sec	23	28 ± 1
R	1 min	23	25 ± 2
R	4 min	23	21 ± 2
R	16 min	23	23 ± 2
R	5 min	192	21 ± 1
FR	5 min	377	21 ± 1
R + FR	5 + 5 min		24 ± 6

Seeds were prechilled in darkness at 10°C for 14 days, given a 2-hr, 40°C temperature shift, irradiated as indicated and returned to 20°C dark for 5 days to germinate.

Huang and Hsiao (2006) observed that seeds of Johnsongrass [*Sorghum halepense* (L.) Pers.] germinated to higher percentages (20–30% higher) when incubated at 28 and 35°C than at 10 or 22°C. After-ripening was accelerated by dry storage of these seeds at 50°C. Seeds pre-chilled at 6°C for 2–4 weeks followed by incubation at 28°C germinated 40–60%. Light effects on germination were related to incubation temperatures; inhibitory at 22°C; no response at 28°C; and stimulatory at 35°C. Effects of gibberellin A_3 (GA_3) also varied depending on incubation temperature, sodium hypochlorite (NaOCl) immersion and light conditions. Immersion of dry seeds in either 700 mM NaOCl, 900 mM H_2O_2 or concentrated H_2SO_4 before incubation in water was effective in breaking dormancy. This result suggests the modes of action of H_2SO_4 in the termination of dormancy may be similar to those of NaOCl and H_2SO_4.

2. Effect of water stress

Water stress during seed development has been found to reduce dormancy of *Sorghum halepense* seed through modifications in glumes properties which resulted in an enhancement of their permeability to oxygen diffusion (Benech-Arnold *et al.*, 1992).

Singh and Singh (2009a) reported that decreasing water potential from -0.1 to -0.2, -0.4 and -0.8 MPa decreased the germination of weed species over control. Germination of johnson grass was significantly reduced at -0.2 MPa and only 23% seed germinated at -0.4 MPa (Table 2). Water stress of -0.1 and -0.2 MPa delayed the germination of johnson grass by one and two weeks, respectively.

Table 2. Effect of osmotic potential and duration on per cent germination of johnson grass seed*

Duration (DAT)	Control	-0.1 MPa	-0.2 MPa	-0.4 MPa	-0.8 MPa	-1.2 MPa
7	45	29	16	4	0	0
14	56	50	31	8	0	0
21	66	66	31	15	0	0

* Means followed by same letter within species do not significantly differ under Student-Newman-Kelus (P=0.5). DAT = Days after transplanting.

Singh and Singh (2009b) reported that increasing the temperature from 15 to 20 and 30°C increased germination of weed species. Germination was reduced drastically at -0.1 and -0.5 osmotic potential compared to control treatments. There was no germination of weed species at -1.0 Mpa osmotic potential. Germination of johnsongrass was negligible below 30°C, but increased water stress decreased germination greatly (Fig. 4). An osmotic stress of -0.1 decreased the germination of johnsongrass by 49% over no water stress at 30°C and no germination occurred with increasing water stress.

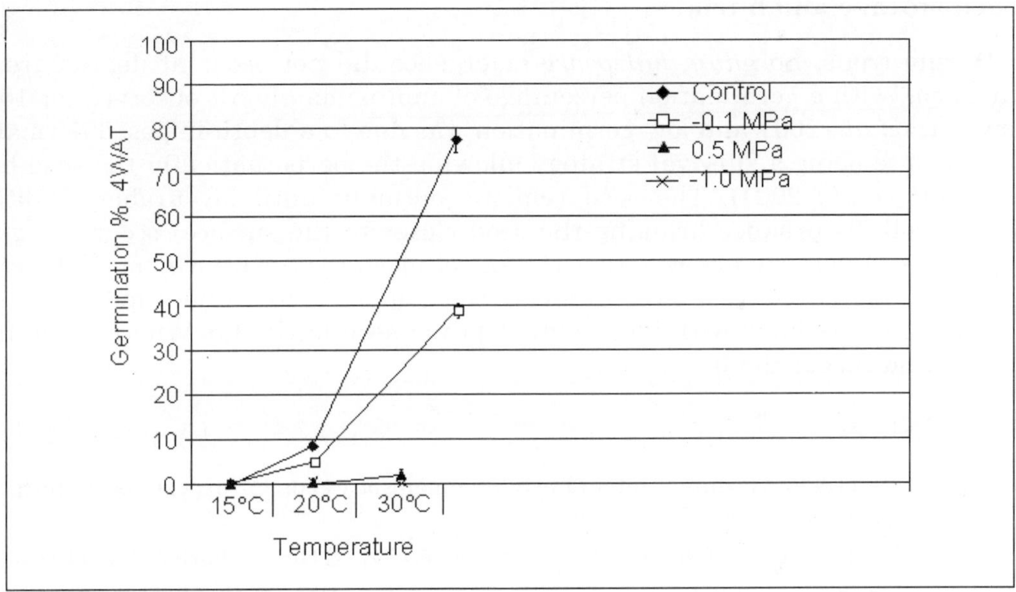

Fig. 4. Effect of temperature and water potential on germination of johnsongrass. Bars indicate standard error of means (SEm).

3. Effect of Temperature

Seed germination rate of johnson grass has been found to increase with temperature upto 36°C and then declined at 40°C (Holshouser *et al.*, 1996). Germination of Johnson grass was 20-30% higher when incubated at 28 and 35°C than at 10 or 22°C (Huang and Hsiao, 1987). Incubating seeds at 35°C had stimulatory effect on germination of johnson grass, whereas it was inhibitory when incubated at 22°C and no effect at 28°C. Highest germination of johnson grass at 35°C and no germination at 15°C suggest that seeds were in dormant stage at lower temperature regimes. Strong correlation of soil temperature and germination of johnson grass have been reported earlier (Benech-Arnold *et al.*, 1988; 1990).

Singh and Singh (2009b) studied the effect of temperature and found that Johnson grass had no germination at 15°C temperature, but higher temperature was less inhibitory. Johnson grass recorded 61% germination at 45°C compared to only 8% germination at 20°C. Increase in temperature from 20°C increased the germination of johnson grass and highest germination was recorded at 35 followed by 40°C.

4. Effect of pH

Singh and Singh (2009b) reported that johnson grass and strangler vine had no germination at pH 3, but similar germination was observed at pH 5, 7, 9 and 11.

5. Seed storage and burial

During trials, *Sorghum halepense* emergence did not occur at depths greater than 10 cm, with a germination percentage of approximately 5% observed for 10 cm (Benvenuti *et al.,* 2001 and low germination was due to a depth-imposed dormancy. This response being a survival strategy allowing the perpetuation of the seed bank (Benvenuti *et al.*, 2001). The seed remains dormant until favourable conditions (possibly soil disturbance bringing the seed closer to the surface) occur. *Sorghum halepense* seedlings can emerge from 15 cm, yet most emerge from 7 cm (Holm *et al.*, 1991). The emergence was lower in a clay soil compared to a fine, sandy loam soil, and emergence declined with increasing depth of seed burial (McWhorter, 1971a in Warwick and Black, 1983).

REFERENCES

Anderson, R. N. (1968). Germination and establishment of weeds for experimental purposes. W.F. Humphrey Press Inc., New York. 236.

Benech-Arnold, R. L., C. M. Ghersa, R. A. Sanchez and A. E.Garcia-Fernandez. (1988). The role of fluctuating temperatures in the germination and establishment of *Sorghum halepense*. Regulation of germination under leaf canopies. *Funct. Ecol.* **2** : 311-318.

Benech-Arnold, R. L., C. M. Ghersa, R. A. Sanchez and P. Insausti. (1990). Temperature effects on dormancy release and germination rate in *Sorghum halepense* (L.) Pers. Seeds: a quantitative analysis. *Weed Res.* 30 : 81-89.

Bennett, H. W. (1973). Johnsongrass, carpetgrass and other grasses for the humid south. *In*: Heath, M.E., D.S. Metcalfe, and R.F. Barnes (eds.). Forages. Iowa State University Press, Ames, Iowa, 286-293.

Benvenuti, S., M. Macchia and S. Miele. (2001). Quantitative analysis of emergence of seedlings from buried weed seeds with increasing soil depth. *Weed Science,* 49(4): 528-535.

Bor, N. L. (1960). The grasses of Burma, Ceylon, India and Pakistan, London, Pergamon press.

California Department of Food and Agriculture, EncycloWeedia. (2002). JohnsongFrass (Sorghum halepense (L.) Pers.), Shattercane (Sorghum bicolor (L.) Moench). Healy, E.A., S. Enloe, J.M. DiTomaso, B. Roberson, N. Dechoretz, S. Schoenig, P. Akers, L. Butler, and J. Garvin (eds.). Non-Cropland Weed group, UC Extension Service, Weed Science Program, Department of Vegetable Crops, The University of California. Davis, CA. 95616. website: http://pi.cdfa.ca.gov/weedinfo/sorghum2.html.

Chambers, N. and T. O. Hawkins. (2002). Invasive plants of the Sonoran Desert, a field guide. Sonoran Institute, Environmental Education Exchange, National Fish and Wildlife Foundation, with funding from many other organizations. Tucson, Arizona. 120.

Duke, J. A. and K. K. Wain. (1981). Medicinal plants of the world. Computer index with more than 85,000 entries. 3 vols.

Egley, G. H. and J. M. Chandler. (1978). Germination and viability of weed seeds after 2.5 years in a 50-year buried seed study. *Weed Science,* 26: 230-239.

Friedman, T. and M. Horowitz. (1970). Phytotoxicity of subterranean residues of three perennial weeds. *Weed Research,* 10: 382-385.

Harrington, G. (1916). Germination and viability of johnsongrass seed. *Proceedings of the Association of Official Seed Analysts of North America,* 9: 24-28.

Harrington, G. (1917). Further studies on germination of johnsongrass seeds. *Proceedings of the Association of Official Seed Analysts of North America,* 10: 71-76.

Harrington, G.T. and W. Crocker. (1923). Structure, physical characteristics and compositon of the pericarp and integument of Johnsongrass seeds in relaton ot its physiology. J. Agric. Res. 23:193-222.

Holm, L. G., D. L. Plucknett, J. V. Pancho and J. P. Herberger. (1977). The world's worst weeds. University Press of Hawaii, Honolulu, Hawaii, 609.

Holm, L. G., D. L. Plunknett, J. V. Pancho and J. P. Herberger. (1991). The world's worst weeds. Distribution and biology. Krieger Publishing Company, Malabar, Florida. 609.

Holshouser, D. L., J. M. Chandler and H. I. Wu. (1996). Temperature dependent model of non-dormant seed germination and rhizome bud break of johnson grass (*Sorghum halepense*). *Weed Sci.* 44 : 257-265.

Huang, W. Z. and A. I. Hsiao. (1987). Factors affecting seed dormancy and germination of Johnsongrass, *Sorghum halepense* (L.) Pers. *Weed Res.* 27: 1-12.

Looker, D. (1981). Johnsongrass has an Achilles heel. *New Farm* 3: 40-47.

McWhorter, C. (1971a). Carbohydrate metabolism of johnson grass as influenced by seasonal growth and herbicide treatments. *Weeds,* 9: 563-568.

McWhorter, C. G. (1971b). Growth and development of johnsongrass ecotypes. *Weed Science,* 19: 141-147.

Meredith, D. (1955). The grasses and pastures of South Africa, Johannesburg, Central News Agency Ltd.

Monaghan, N. (1979). The biology of johnson grass (*Sorghum halepense*). *Weed Research,* 19: 261-267.

Newman, D. (1993). The nature conservancy element stewardship abstract for *Sorghum halepense*. The Nature Conservancy. 1815 North Lynn St., Arlington, Virginia. Website: http://tncweeds.ucdavis.edu/esadocs/documnts/sorghal.html

Singh S. and M. Singh. (2009a). Effect of temperature and water potential on germination of twelve weed species. *Indian J. Weed Sci.,* 41 (3 & 4): 135-145.

Singh S. and M. Singh. (2009b). Effect of temperature, light and pH on germination of twelve weed species. *Indian J. Weed Sci.,* 41 (3 & 4): 113-126.

Taylorson, R. B. (1975). Inhibition of prechill-induced dark germination in *Sorghum halepense* (L.) Pers. seeds by phytochrome transformations. *Plant Physiol.,* 55: 1093-1097.

Taylorson, R. B. and C. G. McWhorter. (1969). Seed dormancy and germination in ecotypes of johnsongrass. *Weed Science* 17: 359-361.

Warwick, S. and L. Black. (1983). The biology of Canadian weeds - *Sorghum halepense*. *Canadian Journal of Plant Science,* 63: 997-1014.

Youngclaus, W. A. (1972). Johnsongrass is everybody's business. *California Weed Conference Proceedings,* 24: 8-11.

CORN SPURRY (*Spergula arvensis*)

Common name : Corn spurry

Botanical name : Spergula arvensis

Tamil name : Thadikeerai

Family : Caryophyllaceae

Hindi name : Jangli dhania, bandhania, mun-muna, pittpapra

Origin and distribution : Native to Europe. Distributed in almost all of western Europe, northern Africa, northern America, China, India, Japan, northern and central regions of the European part of the former USSR, the Caucasus, western and eastern Siberia, the Far East.

Botany

Habitat : Waste ground, cultivated sites, disturbed sites and railroads.

Habit : Forms irregular clumps. Can grow in a mound up to 12 inches high and 24 inches wide. Mostly seen growing between, around, and through other weeds (Fig. 1).

Stem : Stem - single or much-branched, 10 - 50 cm tall, bright green, finely hairy and sometimes slightly sticky.

Leaf : Leaves are 2-5 cm long, very narrow, flat or slightly rounded in cross-section. The leaves of young plants are in a rosette-like cluster resembling grass. On older plants, leaves are in whorls of 6 to 30 or more at each node.

Floral biology

Flower : Flowers are perfect, have 5 white petals, 10 yellow stamens and are about 1 cm in diameter. There are 5 sepals behind the petals, which are slightly offset (Fig. 2).

Flowering and pollination : Flowering occurs during September and May. Plants hermaphrodite or andromonoecious or dioecious or gynodioecious. Pollination entomophilous. Spurry is predominantly a self-pollinating plant, though cross pollination occurs regularly. The flowers remain open only for a short period, and when the weather is bad, they remain closed altogether. When the weather is favourable, the flowers open at 10-11 a.m. and close at 3-4 p.m. At the bases of the anthers, nectar is excreted by the flower, which causes a spurry crop in flower to extend a typical sickly smell. The anthers are curved backwards to such an extend that the insects, in their attempt to reach the nectar, touch the anthers with one side of their body and the stigma with the other side, so that cross pollination can occur.

Fruit : Boll is wide ovate, opens with 5 folds or does not open (Fig. 3).

Seed biology

Seed : Seeds per plant are from 1000-10,000. Seeds are small, 1-1.8 mm in diameter, black, with narrow brownish margin (Fig. 4).

Propagation : By seed.

Fig. 1. Plant

Fig. 2. Flower

Fig. 3. Fruit

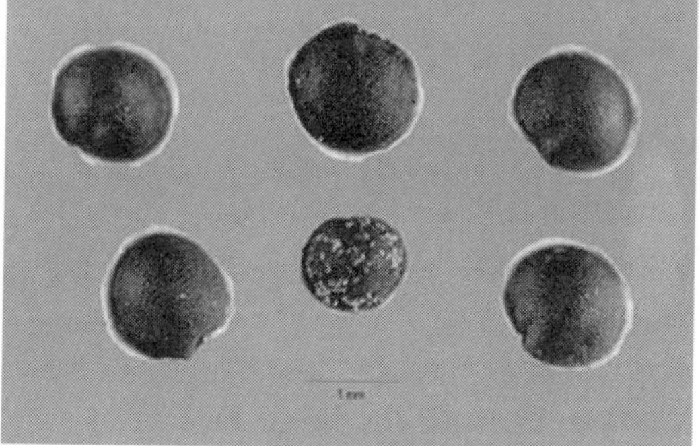

Fig. 4. Seed

Seed dispersal : Spreading by seed. Corn spurry seed has been a contaminant of grass, clover and cereal seeds. Short distance seed dispersal is possible in mud on the tyres of farm vehicles. The seeds have been found in cattle, horse, deer and pig droppings. Seedlings have been raised from the excreta of various birds.

Life cycle : Seeds are not thought to last for very long in the soil. Germination is mainly in the spring, but some also occurs in the autumn. In California, where it was introduced in the mid-nineteenth century (Robbins, 1940), the species grows as a winter annual, germinating in fall and flowering in winter and spring. In Europe and in most other parts of the world, *Spergula arvensis* grows as a summer annual, producing seeds in summer and early fall (Birch, 1957; New, 1958, 1961; Holm *et al.*, 1977). Plants can develop ripe seeds within 8 to 10 weeks of emerging and 2 generations may occur in a year.

Seed dormancy and germination : Seed collected 5, 10 and 15 days after flowering gave 4, 8 and 12% germination respectively. Dormancy is broken by raising temperatures in spring and does not depend on exposure to cold winter temperatures. Light, nitrate and desiccation following imbibition, all stimulate seed germination. The dormancy could also be broken by eithera warm or cold stratification (Karssen *et al.,* 1988). Some seed germinates in the autumn in the wild while some germinates in the spring.

Seed storage and burial : Seeds can remain viable in soil for at least 5 years. Seed numbers decline faster in cultivated soil. Seed recovered from excavations and dated at 30 years old is reported to have germinated. In dry storage, seed longevity is over 15 years.

Effect on crop : Severe growth inhibition observed when some vegetable crops were infested with corn spurry (*Spergula arvensis* L.). Interference by a natural population of the weed reduced the shoot weights of English pea (*Pisum sativum* L.) and collard (*Brassica oleracea* L.) by 93% and 72%, respectively (Harrison *et al.*, 1997).

Control : Inter-row cultivation, harrowing before crop shoots emerge can control this weed. Control is by thorough liming or chalking. Small seedlings are susceptible to flame weeding. Apply Preside (30g ha^{-1}) before the 4th leaf stage. Few insects and pathogens are known to attack corn spurrey, but exposure to an arbuscular-mycorrhizal fungal inoculum has been shown to reduce the biomass of the weed.

Uses : Seed can be dried and ground into a meal, then used with flour for making bread. The seed is rich in oil. A famine food, it is only used when all else fails. The seed contains saponins so some caution is advised. The plant has been used as a diuretic.

Past work done on Seed Science and Technology

1. Seed dormancy and germination

Spergula exhibits a genetically – based seed coat polymorphism in which two primary seed types, papillate (P) and nonpapillate (NP) are produced (New, 1958). Individual plants bear seeds of only one type; therefore, populations may be composed of one or more morphs. In British populations, population morph ratios form a stable clinal pattern from NE to SW Britain, in which NP plants predominate in northern populations and P plants in southern populations (New,

1958). New suggested this clinal pattern was associated with differences in germination response to temperature; she found that NP seeds germinated faster and to a greater degree at 13°C than P seeds, whereas at 21°C, P seeds germinated more rapidly, with a similar final percent germination (New, 1958). Nonpapillate plants also showed a reduction in the proportion of fertile capsules produced per plant at 21°C (New, 1958).

Jones and Hall (1979) studied the effect of ethylene on quantitative and qualitative aspects of respiration in *Spergula arvensis* L. seeds. Oxygen consumption and CO_2 evolution from seeds increased 60 h after addition of ethylene to the dry seed; these increases were considered to be a function of the number of seeds that were germination, rather than a direct effect of ethylene. The respiratory quotient (RQ) on day 1 (0.5) was lower than on subsequent days, indicating an absence of anaerobic processes in the early stages of germination and the use of a different respiratory substrate to that oxidized on days 2–4. Measurements of C_4/C_1ratios showed that ethylene did not break dormancy by initiating a switch from the Embden-Meyerhof-Parnas Pathway (EMPP) to the pentose phosphate pathway (PPP). On the contrary, some evidence was obtained for a switch from the PPP to the EMPP after 3–6 h in pre-imbibed and non pre-imbibed seeds treated with ethylene.

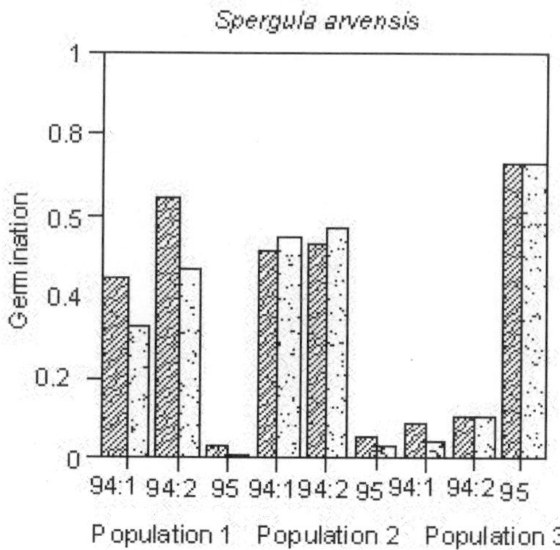

Fig. 5. Germination (proportion) of seeds collected at the same site in 1994 (94:1 and 94:2) and in 1995. Recently harvested (94:1 and 95) and one-year-old dry-stored (94:2) seeds were tested for germination in light (stippled columns) and darkness (striped columns).

According to Jones and Hall (1979), ethylene and light synergise in promoting seed germination in *Spergula arvensis* L. Increasing carbon dioxide concentration above atmospheric levels did not affect germination, but removal of carbon dioxide resulted in an inhibition of germination in the light in non-preimbided seeds. This inhibition was more pronounced in the absence than in the presence of ethylene, suggesting that ethylene can partially overcome the carbon dioxide requirement of the seed. With preimbibed seeds, removal of carbon dioxide had no significant effect on germination, indicating that the phase of sensitivity to carbon dioxide occurred during the first 12 h of inhibition. The ethylene-sensitive phase of seed germination was found to occur during the first 24 h of imbibition, but subsequent to the carbon dioxide-sensitive phase.

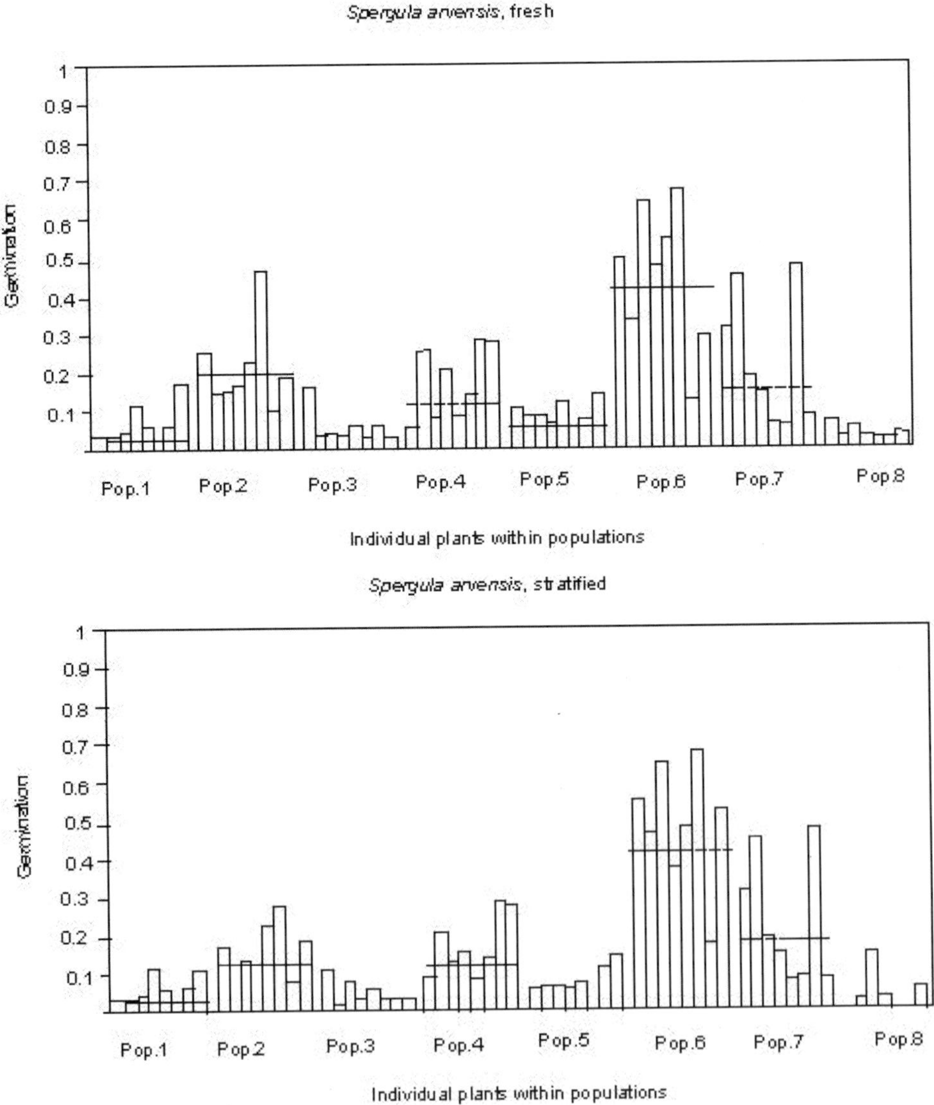

Fig. 6. Germination (proportion) of *Spergula arvensis*seeds from eight individual plants from each of eight populations. Dotted lines indicate mean germination within populations.

Wagner (1988) stated that *Spergula arvensis* L. exhibits a generically based seed coat polymorphism in which two primary seed types, papillate (P) and nonpapillate (NP) are produced. This study examined germination response of the two morphs and seedling emergence in pacific coast populations. Germination behaviour was complex and variable among populations and between seed morphs. Seeds from some populations exhibited greater temperature sensitivity and/or differentiation between morphs. Freshly produced seeds from grassland and ruderal populations were strongly dormant at the time of dispersal, in contrast to seeds from cultivated or summer – wet populations. In San Francisco Bay area populations, germination and seedling emergence was largely simultaneous, following the first substantial fall rains.

Andersson and Milberg (1998) reported that seeds from the two years (1994 and 1995) differed in dormancy level in *Spergula arvensis*. Germination percentage differed significantly among populations. Furthermore, dormancy level in seeds from different mother plants also varied (Fig. 5, 6). Variations at the three levels tested (year, population and mother plant) indicate that these species have a random pattern of variation in dormancy level. It is concluded that variation in seed dormancy among mother plants, populations and years must be taken into account when testing the germination characteristics of a species and also when attempting to model weed seed bank dynamics.

Milberg *et al.* (2000), compared the germination in light and darkness after cold stratification of seeds of 54 species known or suspected to accumulate persistent seed banks. Germination became less dependent on light with increasing seed mass. This pattern was clear in a direct correlation of individual species data as well as when considering phylogenetically independent contrasts. The latter analysis suggests that light response and seed mass coevolved.

Stephan Goudey *et al.* (2006) reported that depending on the applied concentration, nitrate can either stimulate or inhibit germination of dormant seeds of *Sinapis arvensis* L. (wild mustard). Seed NO^{-}_{3} levels that corresponded to the maximal germination frequencies recorded, ranged from 0.3 to 4.4-nmol seed^{-1} for applied NO^{-}_{3} concentrations between 2.5 and 20 mol m^{-3}. Germination was significantly lower in seeds containing more than 5 nmol NO^{-}_{3}. Although the presence of NO^{-}_{3} within the seed was required to promote germination, seed NO^{-}_{3} levels were 5 to 15 times less than levels calculated from the volume of solution taken up by the seeds. Seed NO^{-}_{3} levels also responded in a passive fashion to the external NO_3^- concentration when the seeds were incubated on filter paper or in soil. In a representative soil containing 26 mg NO^{-}_{3}-N per kg dry weight and 8 to 16% water by weight, the amount of NO^{-}_{3} taken up by the seeds reached inhibitory levels. The results suggest that levels of inorganic nitrogen (NO^{-}_{3}, NH^{+}_{4}) in managed soils may play an important role in regulating germination of dormant *S. arvensis* seeds.

REFERENCES

Andersson, L. and P. Milberg. (1998). Variation in seed dormancy among mother plants, populations and years of seed collection. *Seed* Science Research 8, 29-38.

Birch, W.R. (1957). The seeding and germination of some Kenya weeds. *J. Ecol.* 45, 85-91.

Harrison, H.F., J. K. Peterson and M. Snook. (1997). Studies on the Allelopathic Effect of Corn Spurry (*Spergula arvensis* L.) on Cole Crops and English Pea. *HortScience*. Vol. 32(3), 430.

Holm, L., D. Plucknett, J. Pancho and J. Herberger. (1977). The world's worst weeds: distribution and ecology. University Press of Hawaii, Honolulu. 609 pp.

Jones, J.F. and M.A. Hall. (1979). Studies on the requirement for carbon dioxide and ethylene for germination of *Spergula arvensis* seeds. *Plant Science Letters*, 16(1): 87-93.

Karssen, C. M., M.P.M. Derkx and B.J. Post. (1988). Study of seasonal variation in dormancy of *Spergula arvensis* L. seeds in a condensed annual temperature cycle. *Weed Research,* 28: 449-457.

Milberg, P., L. Andersson and K. Thompson. (2000). Large seeded species are less dependent on light for germination than small – seeded ones. *Seed Science Research*. 10: 99-104.

New, J.K. (1958). A population study of *Spergula arvensis*. Part 1. Two clines and their significance. *Annals of Botany*, 22: 457-477.

New, J.K. (1961). Biological fluora of the British Isles. No.76. *Spergula arvensis*. *Journal of Ecology*, 49: 205- 215.

Robbins, W.W. (1940). Alien plants growing without cultivation in California. Bull. 637. Calif. Agr. Exp. Sta., Berkeley, Calif.

Stephen Goudey, J. Saini and M.S. Spencer. (2006). Role of nitrate in regulating germination of *Sinapis arvensis* L. (wild mustard). *Plant cell and environment* 11(1): 9-12.

Wagner, L.K. (1988). Germination and seedling emergence in *Spergula-arvensis*. *American Journal of Botany* 75, 465-475.

ASIATIC WITCHWEED (*Striga lutea*)

Common name : Asiatic witchweed

Botanical name : Striga asiatica (L.) Kuntze Syn. *S. lutea* Lour.

Tamil name : Pallipoondu, kollaippalli, chirakachitam, chirakacitappuntu

Family : Scrophulariaceae

Origin and distribution : Africa and Asia (Werth *et al.*, 1984)

Botany

Habitat : A common obligate root parasitic flowering weed on roots of sorghum and sugarcane. This weed is known for its abundant seed production and the seeds remain viable for a long period. Once a field is infested with this weed, it would be very difficult to control.

Habit : An erect scabrous, hirsute, branching semi parasitic herb occurring on roots of graminaceous hosts (Fig. 1).

Fig. 1. Plant

Stem : Stems are erect, rarely branched.

Leaf : Leaves alternate, lower opposite, linear, entire up to 3.7 cm long, narrow and rough

Floral biology

Flower : Flowers in slender spikes with long bracts and minute bracteoles; calyx tubular 5 – lobed tube with 5-15 ribs; corolla tubular, slender, the limb spreading, upper lip 2-fid, lower 3 –lobed; stamens 4, didynamous; ovary superior, bicarpellary, syncarpous with many ovules (Fig. 2).

Flowering, pollination and fruit set : Flowering occurs in September-October. Each plant can produce 500,000 seeds. Flowers develop about 3 weeks after emergence. Viable seed is produced within 2 weeks of flowering

Fruit : Fruit is a capsule (Fig. 3).

Fig. 2. Flower

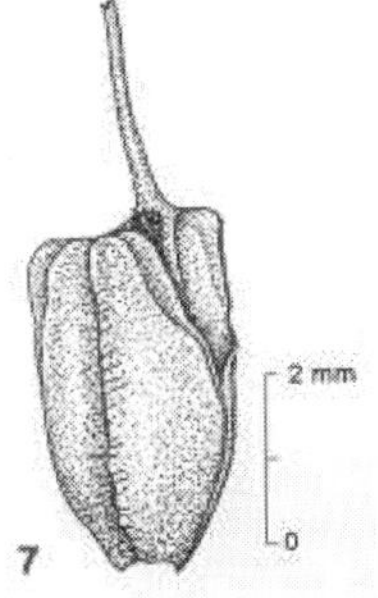

Fig. 3. Fruit

Seed biology

Seed : Seeds are brown, oval, dust-like, 0.2 mm long. Seed surfaces striated, over-layed with a reticulate pattern (Fig. 4). It is the smallest of any one seed, measuring only 0.20 – 0.50 mm long.

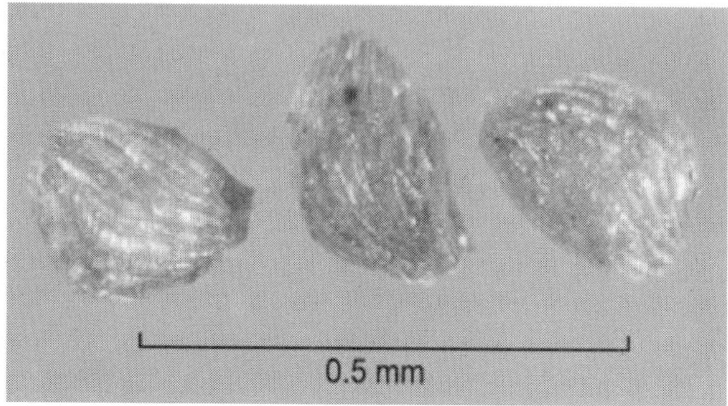

Fig. 4. Seed

Propagation : By seeds.

Seed dispersal : Witch weed seeds spread easily by wind, water and soil *via* animal vectors. The chief means of dispersal, however, is through human interaction, by means of machinery, tools, and clothing. In Africa, seeds are frequently dispersed with planting material of which they are contaminants (Berner *et al.,* 1994).

Life cycle : After germination, produces 2-3 long radical form haustoria on nearly host roots and continue their growth. They suck water, nutrient and food material from them. Each *Striga* produces a million seeds after 30 days of its emergence. It grows under soil for 20 days and then it emerges out as green plant. At this stage, it is a partial parasite. A bell-like swell forms where the parasitic roots attach to the roots of the host. The pathogen colonizes underground where it may spend the next 4–7 weeks before emergence, when it rapidly flowers and produces seeds. A minimum of 60 days is required from seed germination to seed production (http://www.cdfa.ca.gov).

Seed dormancy and germination : Seeds germinate in the presence of host root exudate and develop haustoria which penetrate host root cells. Host root exudate contain strigolactones, signaling molecules that promote *Striga* seed germ-ination. Seeds require an after ripening period of 6 weeks under warm conditions to 40 weeks under freezing conditions. Dormant seeds survive freezing [–15°C (5° F)] for at least 49 days and can remain viable under field conditions for up to 14 years or more.

Germination is complex and requires about a 1-3 week conditioning period at a suitable temperature regime [20-40°C (68-104° F), optimum 35°C (95° F)] under moist conditions, followed by a chemical signal from a nearby root of a host plant. Proximity of host root to seed must be within a few millimetres. Under these conditions, seeds germinate within 24 hours. After 3 weeks of conditioning without a chemical signal, germination ability of seeds decrease, and some seed may pass into a secondary dormancy. Light exposure or wet soils inhibit germination. Irregular or light rainfall appears to promote seed germination and plant vigour.

Seed storage and burial : The seeds remain viable in the soil for over 10 years.

Effect on crop : Parasitic plant that attacks agricultural crops, including corn, sorghum, sugarcane and rice (Werth *et al.,* 1984). It is capable of significantly reducing yields, in some cases wiping out the entire crop. Host plant symptoms such as stunting, wilting, and chlorosis are similar to those seen from severe drought damage, nutrient deficiency and vascular disease.

Control : Trap cropping is a conventional practice of controlling *Striga.* Trap crop induce *Striga* seeds to germinate, but *Striga* fail to make haustoria on them and perished. Cotton, sunflower and soybean are effective trap crops for striga.

Dicamba kills *Striga* effectively than 2,4-D. Fenac (1-3 kg/ha) as pre-planting application control the *Striga* in maize and sugarcane. Trifluralin and nitraline are used to control *Striga* weed in cotton and soybean (http://www.agriinfo.in/). Successful control depends on eliminating the seed reserves of *Striga* spp. in soil and preventing parasitism. A proven method of eliminating these seed reserves is soil injection of ethylene gas. This method was used successfully in the United States to control *Striga asiatica* (Berner *et al.* (1999).

Past work done on Seed Science and Technology

1. Suicidal germination (Germination induction)

To germinate, seeds of *Striga* and other parasitic species require after-ripening pretreatment (conditioning) in a moist warm environment for 2–14 days and subsequent exposure to an exogenous stimulant. The natural germination stimulants so far reported are thought to be sesquiterpene lactones and are collectivelyna med strigolactones (Butler, 1995).

Induction of germination in the absence of or away from a host root, termed suicidal germination, had been considered to have great potential for *Striga* control (Worsham, 1987).

Seeds of *Striga* has special germination requirements. To germinate, the seed should be kept in a warm moist environment for several days prior to its exposure to exogenous germination stimulants. A group of sesquiterpene lactones, collectively called strigolactones (Fig. 5) are potent germination stimulants for both *Striga* and *Orabanche* (Butler, 1995; Joel *et al.,* 1995). Strigol, the first strigolactone, was isolated from the root exudates of a false host, cotton (*Gossypium hirsutumb* L.) (Cook *et al.,* 1966; Cook *et al.,* 1972). Later, Siame *et al.*(1993) identified strigol in the root exudates of the genuine hosts of *Striga*, sorghum (*Sorghum bicolour* (L.) Moench), maize (*Zea mays* L.), and proso millet (*Pennisetum glaucumb* (L.) R. BR.) (Siame *et al.,* 1993). Yasuda *et al.* (2003) have recently isolated strigol from an aseptic root culture of *Menispermum dauricum* DC., a Chinese medicinal, non-host plant. Strigol is therefore, produced by both host and non-host plants. In contrast to strigol, all of the other natural strigolactones characterised to date have only been isolated from host plants: sorgolactone from sorghum (Hauck *et al.*, 1992), alectrol from cowpea (Yokota *et al.,* 1998) and orobanchol from red clover (Muller *et al.*, 1992). It is likely that these strigolactones would also be produced by non nost plants.

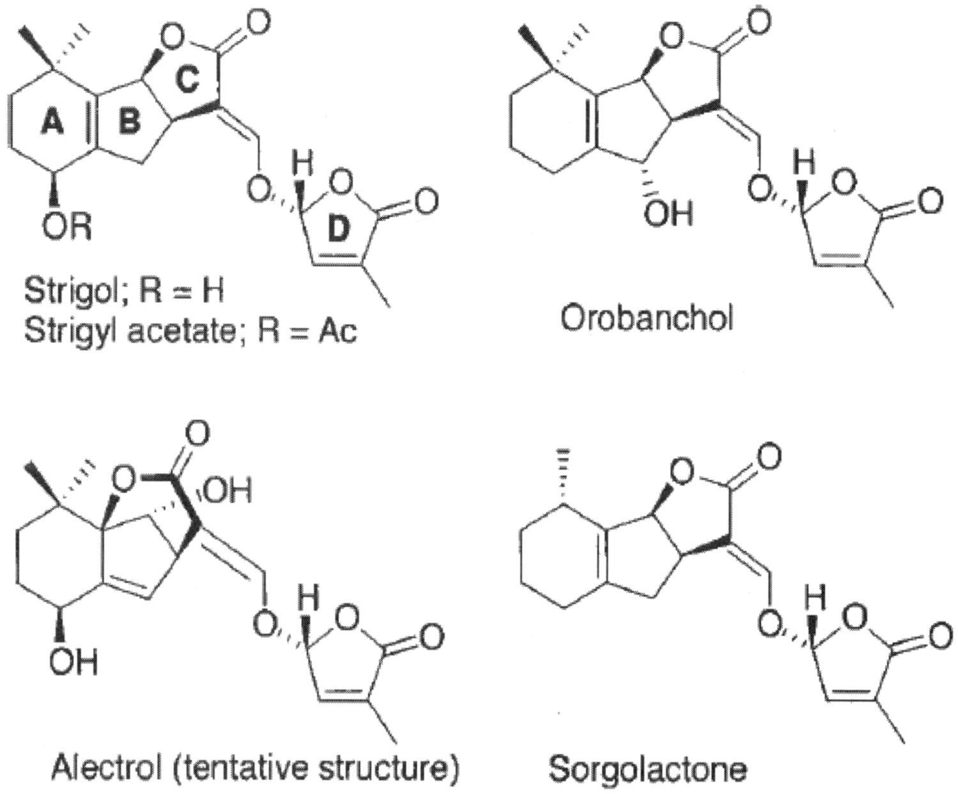

Strigol; R = H
Strigyl acetate; R = Ac

Orobanchol

Alectrol (tentative structure)

Sorgolactone

Fig. 5. Naturally occurring germination stimulants for weedy root parasites.

Brown and Edwards (1945) incubated *Striga lutea* seeds at 22°C for twenty-one days on moist filter paper, transferring them to another culture vessel provided with filter paper moistened with a solution of thiourea and allylthiourea, incubating for a further period of some days at 22°C and finally exposing the culture to 34°C for 24–48 hours. Thus the experimental process embodied three separate phases, each necessitated by a particular condition; the first by the fact that the parasite seed reacts vigorously to the host stimulant only after exposure to moisture for some days; the second by the effect of immediate exposure to the higher temperature with the substitute stimulant, which is that of inhibiting germination; and the third by the fact that neither with the normal nor with the substitute stimulant does extensive germination occur at the lower temperature, however prolonged is the period of incubation.

Egly (1971) reported that scarification of the seed envelope promoted germination in the absence of the stimulant. With adequately (2-week) pretreated seed, a cut or puncture through the aleurone at the radicular end induced typical germination in which the radicle elongated, but did not penetrate the intact envelope over the radicle. Incubation in oxygen did not promote germination regardless of the site of seed scarification. Ten per cent carbon dioxide reduced the germination of punctured seed. Seed germinated equally well when scarified over the radicle and incubated in air, nitrogen, darkness, or light. Brief treatments with sulphuric acid also induced germination. The softening effect of stimulatory acid treatments upon the aleurone was first evident 6h later and only occurred at the radicular end of the seed. Indol-3yl-acetic acid inhibited and ethylene stimulated germination of scarified and non-sacrified seed. Gibberellic acid(GA_3) had no apparent effect upon nonscarified seed but promoted germination of scarified seed.

Inadequately (1-day) or excessively (12- to 16-week) pretreated seed germinated poorly or not at all when treated with the germination stimulant, ethylene, GA_3, or sulphuric acid. Some seed germinated slowly when scarified over the radicle, but the germination rates and totals were less than those of scarified seed pretreated for 2 weeks. Additions of stimulant, ethylene or GA_3, acelerated the germination rates of scarified but inadequately pretreated seed. Of the treatments tested, only GA_3, increased the slow germination of the excessively pretreated scarified seed.

Results indicated that the aleurone restrained radicle elongation. Scarification over the radicle removed the restraint and permitted radicle emergence. However, the ability of the radicle to elongate, as influenced by time of seed pretreatment and exogenous stimulant, GA_3 or ethylene, determined whether or not the scarified seed germinated.

Patterson *et al.* (1982) stated that ethylene-stimulated germination of witch-weed seed first occurred after 13, 10, 6, 3, and 2 days of conditioning in moist sand at day/night temperatures of 20/14, 23/17, 26/20, 29/23, and 32/26°C, respectively. Maximum germination percentages in these regimes were 0.5, 3, 20, 24, and 37%, respectively. No germination occurred at 17/11°C. Witchweed seed survived in sand frozen for 7 weeks at -7 or -15°C and subsequently germinated in response to ethylene or in the presence of corn (*Zea mays* L., 'DeKalb B73 × Mo.17H') roots. The parasites emerged from the soil and flowered when maintained at 29/23°C after the termination of the freezing treatments. In other experiments, witchweed parasitized corn and/or sorghum [*Sorghum bicolor* (L.) Moench 'DeKalb E-59^{+}'] root systems in a sandy loam under 26/17, 26/20, 29/20, 32/23, and 32/26°C day/night regimes. Witchweed emerged from the soil with 26/20, 29/20, 32/23 and 32/26°C and flowered with 26/20, 32/23, and 32/26°C day/night regimes. Underground development and subsequent emergence of the parasites were substantially reduced with day/night temperatures below 29/20°C. Winter soil temperatures and growing season soil and air temperatures are unlikely to limit the spread of witchweed into important corn- and sorghum-producing areas of the United States.

Seed germination requires an exogenous stimulant that initiates ethylene production within the *Striga* seed (Babiker *et al.*, 1993; Logan and Stewart, 1991) or directly provides ethylene (Eplee, 1975). Without adequate moisture and an external germination stimulant the seeds remain dormant. Once germinated, the *Striga* seedling must attach to a host root within 3–5 days or the seedling dies (Worsham, 1987). Stimulating germination of *Striga* spp. seeds in the absence of a host plant results in the death of the parasite seedlings and reduction of seed reserves in the soil, thus preventing parasitism on subsequent crops.

Several methods of stimulating germination of *Striga* spp. seeds in the absence of host plants have been used or suggested for control of the parasites in agricultural systems. The most effective of these has been soil injection of ethylene gas in an eradication program for *S. asiatica* from two southeastern states in the United States (Eplee, 1975; Sand and Manley, 1990). However, ethylene gas is pressurized, flammable, and requires specialized storage and application equipment for soil injection.

A suggested alternative has been to introduce synthetic analogues of strigol—"strigol" is the trivial name applied to the germination stimulant isolated from cotton (*Gossypium hirsutum* L.) and identified by Cook *et al.* (1966, 1972)—into *Striga*-infested soil to stimulate the parasite's seed germination (Johnson *et al.*, 1976). However, the instability of these analogues in soil precludes their usefulness (Babiker *et al.*, 1987). Strigol analogues are also expensive, and a process to inexpensively produce large quantities of soil-stable analogues has not been developed.

A practical alternative for *Striga* spp. control for African farmers is rotation of nonhost cultivars, selected for effectiveness in producing germination stimulants, with host crops (Doggett, 1984; Berner *et al.*, 1996a). However, this method may take several seasons of cultivation to a selected nonhost cultivar to eliminate parasite seed reserves from farm fields (Eplee, 1975; Berner *et al.*, 1996b).

The pathovar glycinea of the bacterium species *Pseudomonas syringae* has been found to produce ethylene both *in planta* and *in vitro* (Sato *et al.*, 1987; Weingart and Volksch, 1997). This bacterium is widely distributed geographically and causes halo blight on soybeans. However, the genes for ethylene production are carried on plasmids within the bacterium (Nagahama *et al.*, 1994). If levels of ethylene production by strains of this pathovar are sufficient to stimulate seed germination of *Striga* spp., then strains of *P. syringae* pv. glycinea, or other bacteria capable of producing sufficient levels of ethylene, i.e., non-pathogenic strains carrying the plasmid(s) with the ethylene production gene, might be useable as soil amendments to eliminate *Striga* spp. seed reserves in soil and prevent parasitism.

Igbinnosa and Okonkwo (1992) tested seeds of witchweed seeds with 18 growth regulators and chemicals and revealed that the cytokinins [zeatin, kinetin and 6-benzylaminopurine (BAP)], ethylene and sodium hypochlorite stimulated germination of cowpea witchweed seeds at various concentrations. Witchweed seeds preconditioned in different concentrations of sodium hypochlorite for 3 wk had germination rates of 25 to 89% after incubation in distilled water, 10^{-5} M solution of kinetin, 50 mg l^{-1} of 2-chloroethylphosphonic acid, and exudates from cowpea roots. Witchweed seeds preconditioned in distilled water and subsequently incubated in distilled water did not germinate.

Babiker *et al.* (1993) reported that 800–1000 nl ethylene/li^{-1} headspace gas was necessary to achieve greater than 65% germination of *S. asiatica* seeds.

According to Berner *et al.* (1999), bacterium *Pseudomonas syringae* pathovar glycinea synthesizes relatively large amounts of ethylene. Efficacy of stimulating germination of seeds of *Striga* spp. was compared among bacteria, volatile compounds, root exudates and synthetic stimulants. Seeds of three *Striga* spp. were tested over a 10-month period. No seed germination was ever observed with sterile water. When compared across *Striga* spp. the bacterial strains were consistently better stimulators of germination of seeds of the parasites than ethylene gas or root pieces of a *Vigna unguiculata* cultivar known to stimulate germination of parasite seeds. The strains were as effective in germinating *S. aspera* and *S. gesnerioides* seeds as a synthetic germination stimulant (Table 1). This results showed that ethylene-producing bacteria are highly effective in promoting seed germination in *Striga* spp. suggest that these bacteria may provide a practical means of biological control of *Striga* spp. in Africa and other locations.

Table. 1 Summary of germination of *Striga* spp. seeds stimulated by strains of *Pseudomonas syringae* pv. *glycinea* (Psg), ethylene gas, a synthetic germination stimulant (GR24), and roots of a cowpea (*Vigna unguiculata*) cultivar

Striga species	**(% seed germination)**						
	Psg strains			**Ethylene (2.5 mg)**	**GR24 (10mg/liter)**	**Cowpea roots (1 g)**	**Mean**
	16/83	**19/84**	**8/83**				
S. aspera	10.3	10.0	10.1	7.1	12.8	8.1	9.7
S. gesnerioides	14.2	14.5	13.1	14.1	13.4	13.8	13.8
S. hermonthica	14.6	14.8	13.6	2.0	44.1	7.2	16.0
Mean	13.1	13.1	12.3	7.7	23.4	9.7	13.2

Least squares means of actual percentage of germination that was adjusted, through analysis of covariance, using viability of seeds of the three species in the three experiments as a covariate in the analysis.

Sugimoto *et al.* (2001) revealed that root culture of *Menispermum dauricum* DC. (Menispermaceae), a broad-leaved herbaceous plant (Sugimoto, 1999), was most potent in inducing germination of conditioned *Striga hermonthica* seeds.

Table. 2. Percent *Striga* seed germination as induced by *C. argentia* selected trap crops

Crop/plant	% germination of *Striga* seeds			
	Distance 1 (3 mm)	**Distance 2 (9 mm)**	**Distance 3 (15 nm)**	**Mean**
C. aregentia	1.48	2.58	5.67	3.24
Cotton	14.48	11.4	4.6	10.16
Cowpea (Ebealat)	0.35	0.85	0.93	0.71
Seredo (improved sorghum)	5.33	6.78	5.65	5.92
Inoke (susceptible sorghum) (positive control)	15.25	9.33	7.55	10.71
Distilled water (negative control)	0.0	0.0	0.0	0.0
Mean	6.15	5.16	4.07	
s.e.d[a]	1.5			
s.e.d[b]	1.7			
C.V. (%)	60			

[a] Differences between treatments/crops

[b] Differences between distances

Sugimoto (2002) found that root culture filtrate of *Stephania cepharantha* (Menispermaceae), obtained from established cultures (Sugimoto, 2002), was also active in inducing germination of conditioned *Striga hermonthica* seeds.

According to Olupot *et al.* (2003), *C. argentia* could induce suicidal germination in *Striga* seed by as much as 68% compared to cotton which was taken to be the standard. The crops differed significantly in their abilities to stimulate *Striga* seed germination. The susceptible sorghum variety stimulated *Striga* germination much more than all other crops and was closely followed by cotton though the difference between the two was not significant. In this study, it has been demonstrated that *C. argentia* can also induce *Striga* seeds to germinate. It is suggested that the differences in percent germination of *Striga* seeds induced by different crops is due to different amounts of germination stimulant produced (Table. 2).

Yasuda *et al.* (2003) isolated and identified the major germination stimulant, (+)-strigol, from *M. dauricum* root culture filtrate (Fig. 6). They isolated 140 mg of (+)-strigol from 6 l of *M. dauricum* root culture filtrate, on the basis of peak area monitored in the preparative HPLC. Assuming that the activity found in the culture filtrate is attributable only to (+)-strigol, its concentration is roughly 1 mg l^{-1}. This indicates that the recovery rate of the stimulant in the present experiments was as low as 2%. Preliminary experiments showed that conditioned *Striga hermonthica* seeds germinated and formed haustoria in proximity to *M. dauricum* roots.

Sato *et al.* (2005) demonstrated that cotton roots exude strigol and strigyl acetate, but no other known strigolactones such as orobanchol and acectrol. The production of strigol was detected even in the root exudates collected during the first 24h of incubation and reached a maximum 5-7 days later. The average exudation of strigol and strigyl acetate during the incubation period was *ca*, 15 and 2 pg/plant/day, respectively, indicating that strigol mainly contributed to germination stimulation by the cotton root exudate. Strigyl acetate was more active toward *Striga* germination and, it is likely that germination stimulation by the cotton root exudate was attributable to strigol. The biological importance of strigyl acetate is not yet clear, although strigyl acetate might be a storage form or biosynthetic precursor of strigol. Cotton is a false host of *Striga* and is used as a trap crop which induces the germination of *Striga,* but is not paralized, and consequently results in the suicidal germination of *Striga* seeds. Host plants of *Striga* such as maize, pearl millet and sorghum have been reported to produce strigol as one of the germination stimulants.

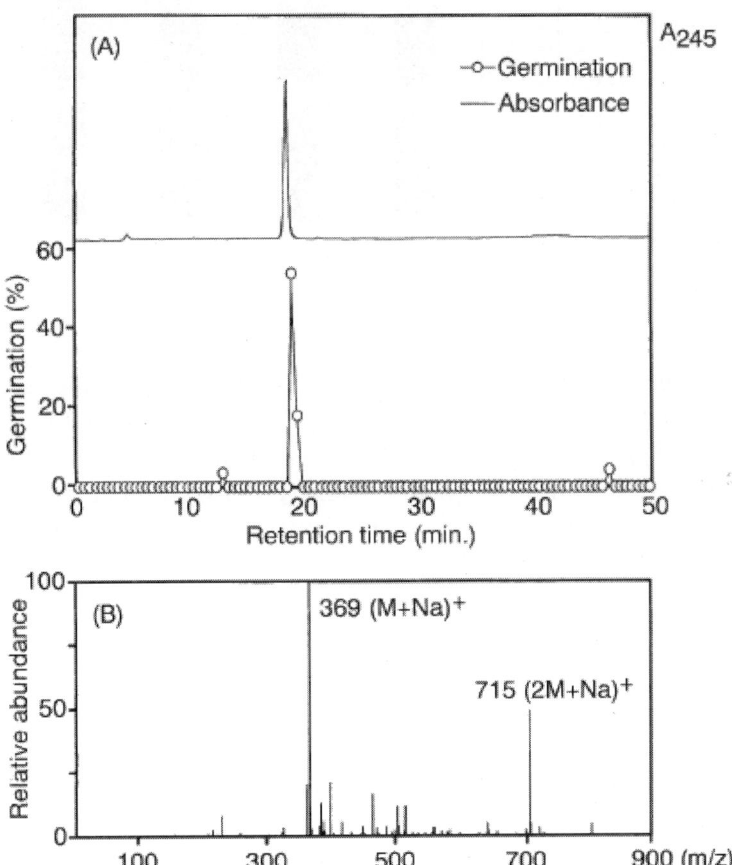

Fig. 6. LC-ESI-MS analysis of germination stimulant isolated from *M. dauricum* root culture filtrate. Elution profile by analytical HPLC (A) and mass spectrum (B) stimulant.

Sato *et al.* (2005), as shown in Fig. 7, strigyl acetate was 10-1000 times less active than strigol toward *O. minor* seed germination. Since strigol exhibited distinct germination stimulation at a concentration as low as 10^{-11} M, the estimated concentration of strigol (but not of strigyl acetate) was high enough to stimulate seed germination of *O. minor*. Although strigol and strigyl acetate may decompose more rapidly in the soil than in the glass wool used in this study, it is likely that these strigolactones induced the seed germination of root parasitic weeds when the seeds were in the vicinity of the root system in the soil.

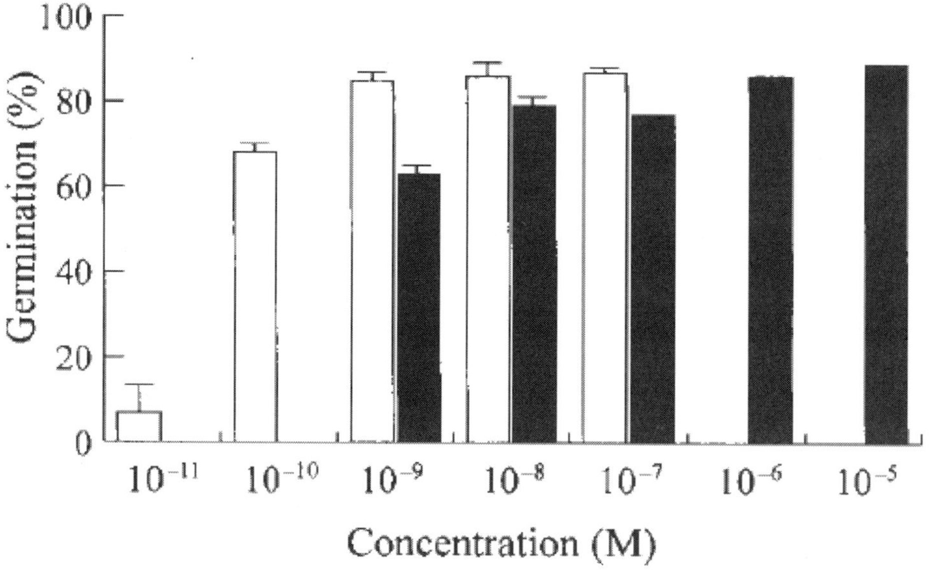

Fig. 7. Stimulatory activity of strigol and strigyl acetate toward *O. minor* seed germination. □, strigol; ■, strigyl acetate.

Hassan *et al.* (2010) reported that *Striga* seed, previously conditioned in presence of water induced the highest germination (21%) in response to root extracts of sorghum (Table 3). All concentrations of NaCl decreased *S. hermonthica* seed germination in response to sorghum stimulant in comparison with the corresponding aqueous control. However, the highest concentrations of NaCl (100 mM) reduced germination significantly. It reduced germination by 52% as compared to their control (Table 3).

Table 3. Effect of NaCl of *S. hermonthica* seeds germination

Treatment	Germination (%) NaCl (mM)				
	Water	25	50	100	Means
Sorghum root macerate	21	12.8	13.7	10.6	14.5

SE ± 3.1

2. Seed storage and burial

Berner *et al.* (1999) reported that in a single growing season each *Striga* plant can produce 500,000 seeds (Saunders, 1933), which may remain viable for 14 years in soil (Bebawi *et al.*, 1984). Viability of the seeds of *S. aspera* and *S. hermonthica* used decreased over the course of the study (Table 4). The germination decreased from 53.6 (8 months) to 34.0 per cent (18 months) in the case of *S. aspera*, while in *S. hermonthica* decreased from 72.3 (6 months) to 59.2 per cent (16 months). In the case of *S. gesnerioides*, the germination decreased to 0.7 (37 months) from 36.0 per cent (5 months).

Table 4. Viability of seeds from collections of *Striga aspera, Striga gesnerioides*, and *Striga hermonthica*[a]

Striga species	Seed age (months)	Mean viability (%)[b]	Standard error
S. aspera	8	53.6	8.04
	14	40.9	7.16
	18	34.0	6.8
S. gesnerioides	31	12.3	2.46
	37	0.7	0.42
	5	36.0	5.40
S. hermonthica	6	72.3	6.07
	12	63.7	5.96
	16	59.2	4.98

[a] Viability determined by tetrazolium chloride staining.

[b] Means based on four replications, each conducted on different days.

3. Effect on crop

Striga species in Uganda, severely affects sorghum yield, the third most important cereal crop in the country. Yield loses at individual farms are estimated at 60–100% (Anon., 1997).

Yield losses on cereals attributable to infection by these parasites may reach 100% (Lagoke *et al.,* 1991) while losses in cowpea reach 83% (Aggarwal and Ouedraogo, 1989). It is estimated that, on average, a 40% loss in crop production in Africa is due to *Striga* spp., and this accounts for an annual loss in agricultural revenue of $7 billion (Robson and Broad, 1989).

Olupot *et al.* (2003) reported that inter-planting *C. argentia* into sorghum reduced *Striga* emergence by an average of 55% in a season (Table 5) and increased the yield of a susceptible sorghum variety in the field by 35% compared to the sole sorghum treatment (Table 6). Results of the screen house experiment showed that inter-planting *C. argentia* into sorghum at a ratio of 2:1 suppressed *Striga* best by as much as 48% and resulted in the highest yield increase (100%) compared to the sole sorghum treatment.

Table 5. The effect of inter-planting *Celosia argentia* and cowpea into sorghum on *Striga* emergence

Treatment	*Striga* emergence[a] (plants/m^2)			
	1999a	1999b	2000a	Mean
Cowpea/sorghum	8.6[b]	4.4[b]	33.0[b]	15.3
C. argentia/sorghum	13.5[b]	7.1[b]	34.4[b]	18.3
Sorghum (sole)	24.6	21.1	76.7	40.8
C.V. (%)	26.3	12	6.1	

[a] Analysis based on transformed data. Means presented as original data.

[b] Significantly low, L.S.D. (5%).

Table 6. The effect of inter-planting *Celosia argentia* and cowpea into sorghum on sorghum yield

Treatment	Grain yield (kg/ha)			
	1999a	1999b	2000a	Mean
Cowpea/sorghum	413	163	134	236.7
C. argentia/sorghum	697	186	168	350.3
Sorghum (sole)	674	153	93	306.7
s.e.d.	102.5	46.2	30.1	
C.V. (%)	31.6	82.3	48.4	

REFERENCES

Aggarwal, V.D. and J.T. Ouedraogo. (1989). Estimation of cowpea yield loss from *Striga infestation. Trop. Agric.*, 66:91-92.

Anonymous. (1997). *Striga* survey in Eastern Uganda. Serere Agricultural and Animal production Research Institute/National Agricultural Research Organisation.

Babiker, A.G.T., A.M. Hamdoun, A. Rudwan, M. Mansi and H.H. Faki. (1987). Influence of soil moisture on activity and persistence of the strigol analogue GR24. *Weed Res.,* 27: 173–178.

Babiker, A.G.T., G. Ejeta, L.G. Bulter and W.R. Woodson. (1993). Ethylene biosynthesis and strigol-induced germination of *Striga asiatica. Physiol. Plant,* 88: 359-365.

Berner, D.K., K.F. Cardwell, B.O. Faturoti, F.O. Ikie and O.A. Williams. (1994). Relative roles of wind, crop seeds and cattle in the dispersal of *Striga* species. *Plant Disease,* 78:402-406.

Bebawi, F.E, R.E. Eplee, C.E. Harris and R.S. Norris. (1984). Longevity of witchweed (*Striga asiatica*) seed. *Weed Sci.*, 32: 494-497.

Berner, D.K., M.O. Alabi, U. Di-Umba and F.O. Ikie. (1996a). Proposed integrated control program for *Striga hermonthica* in Africa. *In* "Advances in Parasitic Plant Research" (M. T. Moreno, J. I. Cubero, D. Berner, D. Joel, L. J. Musselman, and C. Parker, (Eds.), pp. 817–825. Junta de Andalucia, Direccion General de Investigacion Agraria, Cordoba.

Berner, D.K., R.J. Carsky, K.E. Dashiell, J.G. Kling and V.M. Manyong. (1996b). A land management based approach to integrated *Striga hermonthica* control in sub-Saharan Africa. *Outlook Agric.,* 25: 157–164.

Berner, D.K., N.W. Schaada and B. Volksch. (1999). Use of ethylene-producing bacteria for stimulation of *Striga* spp. seed germination, *Biol. Control,* 15: 274-282.

Brown, R. and M. Edwards. (1945). Effects of thiourea and allylthiourea on the germination of the seed of *Striga lutea. International weekly Journal of Science,* 155: 455-456.

Butler L.G. (1995). Chemical communication between the parasitic weed *Striga* and its crop host. A new dimension in allelochemistry. In: Inderjit K.M., Dakshini M., Enhelling F.A., (eds). Allelopathy, Organisms, Processes and Applications. Washington, DC: American Chemical Society. pp. 158–166.

Cook, C.E., L.P. Whichard, B. Turner, M.E. Wall and G.H. Egley. (1966). Germination of witchweed (*Striga lutea* Lour.): isolation and properties of a potent stimulant. *Science,* 154: 1189-1190.

Cook, C.E., L.P. Whichard, M.E. Wall, G.H. Egley, P. Coggon, P.A. Luhan and A.T. McPhail. (1972). Germination stimulants. II. The structure of strigol- a potent seed germination stimulant for witchweed (*Striga lutea* Lour.). *J. Am. Chem. Soc.,* 94: 6198-6199.

Doggett, H. (1984). *Striga:* Its biology and control—an overview. *In* "*Striga* Biology and Control" (E. S. Ayensu, H. Doggett, R. D. Keynes, J. Marton-Lefevre, L. J. Musselman, C. Parker, and A. Pickering, (Eds.), pp. 27–36. Int. Counc. Sci. Union, Paris, France and the Int. Dev. Res. Ctr., Ottawa.

Egley, G.H. (1971). Influence of the seed envelope and growth regulators upon seed dormancy in witchweed (*Striga lutea* Lour). *Annals of Bot.* 36: 755-770.

Eplee, R.E. (1975). Ethylene: A witchweed seed germination stimulant. *Weed Sci.*, 23: 433-436.

Hassan, M.M., Y. Sugmuto, A.G.E. Babiker, Y. Yamauchi, M.G.E. Osman and S.O. Yagoub. (2010). Effect of NaCl on *Orobanche spp* and *Striga hermonthica* seeds germination during and after conditioning. *Bioscience Research,* 7(1): 26-31.

Hauck, C., S. Muller and H. Schildknecht. (1992). A germination stimulant for parasitic flowering plants for *Sorghum bicolor*, a genuine host plant. *J. Plant Physiol.,* 139: 474-478.

Igbinnosa, I. and S.N.C. Okonkwo. (1992). Stimulation of germination of seeds of cowpea withchweed (*Striga gesnerioides*) by sodium hypochlorite and some growth regulators. *Weed Science*, 40: 25-28.

Joel, D.M., J.C. Steffens and D.E. Matthews. (1995). Germination of weedy root parasites. In "Seed Development and Germination", (eds.) J. Kigel and G. Galili, Marcel Dekker, Inc., New York, 567-597.

Johnson, A.W., G. Roseberry and C. Parker. (1976). A novel approach to *Striga* and *Orobanche* control using synthetic germination stimulants. *Weed Res.,* 16: 223–227.

Lagoke, S.T.O., V. Parkinson and R.M. Agunbiade. (1991). Parasitic weeds and control methods in Africa. In " Combating Striga in Africa" S.K. Kim, (Ed.), pp. 3-14. Proc. Int. Workshop organized by IITA, ICRISAT and IDRC, 22-24 August 1988 IITA, Ibadan.

Logan, D.C. and G.R. Steward. (1991). Role of ethylene in the germination of the hemiparasite *Striga hermonthica. Plant Physiol*., 97: 1435-1438.

Muller, S.C. Hauck and H. Schildknecht (1992). Germination stimulants produced by *Vigna unguiculata* Walp cv Sanders Upright. *J. Plant Growth Regul*., 11: 474-478.

Nagahama, K., K. Yoshino, M. Matsuoka, M. Sato, S. Tanase, T. Ogawa and H. Fukuda. (1994). Ethylene production by strains of the plant-pathogenic bacterium *Pseudomonas syringae* depends upon the presence of indigenous plasmids carrying homologous genes for the ethylene-forming enzyme. *Microbiology,* 140: 2309 –2313.

Olupot, J.R., D.S.O. Osiru, J. Oryokot and B. Gebrekidan. (2003). The effectiveness of *Celosia argentia* (*Striga "chaser"*) to control Striga on sorghum in Uganda. *Crop Protection*, 22: 463–468.

Patterson, D.T., R.L. Musser, E.P. Flint and R.E. Eplee. (1982). Temperature responses and potential for spread of witchweed (*Striga lutea*) in the United States. *Weed Science*. 30: 87-93.

Robson, T.O. and H.R. Broad. (1989). "*Striga* – improved management in Africa" FAO Plant Production and Protecton Paper 96, Rome.

Sand, P.F. and J.D. Manley. (1990). The witchweed eradication program survey, regulatory and control. *In* "Witchweed Research and Control in the United States" P. F. Sand, R. E. Eplee, and R. G. Westbrooks, (Eds.). *Weed Sci. Soc. Am*., Champaign, 141–150.

Sato, M., K. Urushizaki, K. Nishiyama, F. Sakai and Y. Ota. (1987). Efficient production of ethylene by *Pseudomonas syringae* pv. *glycinea* which causes halo blight in soybeans. *Agric. Biol. Chem.,* 51: 1177–1178.

Sato, D., A.A. Awad, Y. Takeuchi and K. Yoneyama. (2005). Confirmation and quantification of strigolactones, germination stimulants for root parasitic plants *Striga* and *Orobanche,* produced by cotton. *Biosci. Biotecchnol. Biochem*., 69(1): 98-102.

Saunders, A.R. (1933). "Students in Phanerogamic Parasitism, with particular reference to *Striga lutea* Lour." South Africa Dept. of Agric. Sci. Bull. 128, The Government Printer, Pretoria.

Siame, B.P., Y. Weerasuriya, K. Wood, G. Ejeta and L.G. Bulter. (1993). Isolation of strigol, a germination stimulant for *Striga asiatica*, from host plants. *J. Agric. Food. Chem*., 41: 1486-1491.

Sugimoto, Y. (1999). Menispermum species (moonseed vines): *In vitro* culture, and the production of dauricine and other secondary metabolites. In: Bajaj, Y.P.S. (Ed.), *Biotechnology in Agriculture and Forestry,* 43: 290–309.

Sugimoto, Y., H.A. A. Babiker, T. Saisho, T. Furumoto, S. Inanaga and M. Kato. (2001). Chlorinated alkaloids in *Menispermum dauricum* DC. root culture. *J. Org. Chem*., 66: 3299–3302.

Sugimoto, Y. (2002). *In vitro* culture and the production of secondary metabolites in Stephania. In: Nagata, T. and Ebizuka, Y. (Eds.), *Biotechnologyin Agriculture and Forestry, Vol.,* 51: 281–305.

Weingart, H. and B. Volksch. (1997). Ethylene production by *Pseudomonas syringae* pathovars in vitro and in planta. *Appl. Environ. Microbiol.,* 63: 156–161.

Werth, C.R., J.L. Riopel and N.W. Gillespie. (1984). Genetic uniformity in an introduced population of witchweed (*Striga asiatica*) in the United States. *Weed Sci.,* 32: 645-648.

Worsham, A.D. (1987). Germination of witchweed seeds. *In* Musselman, L. J. (ed.) *Parasitic Weeds in Agriculture. Vol. 1. Striga.* CRC Press, Inc., Boca Raton, Florida, U.S.A., 45-62.

Yasuda, N.Y. Sugimoto, M. Kato, S. Inanaga and K. Yoneyama. (2003). (+)- Strigol, a witchweed seed germination stimulant, from *Menispermum dauricum* root culter. *Phytochemistry,* 62: 1115-1119.

Yokota, T., H. Sakai, K. Okuno, K. Yoneyama and Y. Takeuchi. (1998). Alectrol and orobanchol, germination stimulants for *Orobanche minor,* from its host red clover. *Phytochemistry,* 49: 1967-1973.

BLACK PIG WEED (*Trianthema portulacastrum*)

Common name : Black pig weed, desert horse-purslane, giant pig weed, horse purslane

Botanical name : Trianthema portulacastrum L

Tamil name : Saranai

Hindi name : Santhi, patharchatta

Family : Aizoaceae

Origin and distribution : Found throughout the tropical and subtropical countries.

Botany

Habitat : A common weed in dry land and garden lands, gregarious causing considerable hinderance to normal cultivation.

Habit : Prostrate, profusely branching, glabrous, annual herb (Fig. 1).

Stem : Stem is round, striate, glabrous below and hairy above.

Leaf : Leaves oppositve, petioled, petiole sheathing at the base, obovate, in unequal pairs.

Floral biology

Flower : Flowers solitary, sessile, sheathed by the base of the petiole; calyx 5 – lobed, lobes obtuse, green above and pink inside; cuspidate. Petals 0; stamens 10-15, inserted near the top of the calyx tube; ovary superior, 1-2 celled, free, style 1, subulaten (Fig. 2).

Fruit : Capsule scarious below, top mitriform, coriaceous, 3-5 seeds with concentric mucriculate lines (Fig. 3).

Seed biology

Seed : Kidney-shaped, in spiral ended by a beak, 2 mm in diameter. The tegument is tuberculate, reddish and brown to black (Fig. 4).

Propagation : By seeds and cuttings.

Seed dispersal : Seed dispersal is achieved by several methods: one seed is dispersed in the detached cap of the capsule, which can float and the other seeds are either dispersed individually from the capsule or remain on the annual, parent plant where they will germinate and establish new plants.

Life cycle : The production of flowers and seeds of *Trianthema portulacastrum* starts 20–30 days after germination of the seeds.

Fig. 1. Plant

Fig. 2. Flower

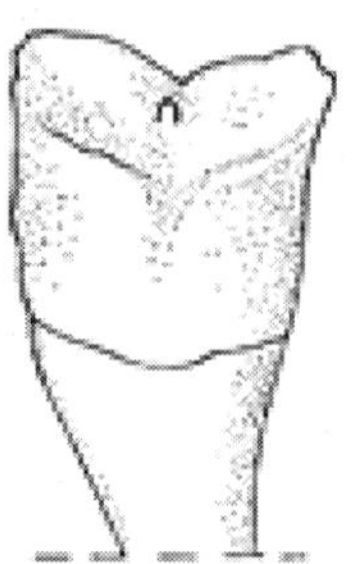

Fig. 3. Fruit

Fig. 4. Seed

Seed dormancy and germination: Seeds do not exhibit dormancy and fresh seeds can germinate when incubated at 35°C to the extent of 54 to 56% (Balyan and Bhan, 1986). It was found that eighth day was the optimum period for testing seed germination of *Trianthema portulacastrum* (Umarani and Selvaraj Jacqueline, 1994).

Seeds germinated from 20 to 45°C. Maximum germination of 90% was observed at 35°C. Germination was highest between the 4th and 8th days of incubation. Maximum seedling emergence resulted when seeds were planted at 1 cm and gradually declined with the increased seeding depth. Seeds lying on the surface gave very poor seedling emergence (Balyan and Bhan, 1986b).

Seed storage : Germination increases during 7 and 8 months air-dry storage at room temperature (Balyan and Bhan, 1986b). The seeds are orthodox in nature.

Seeds stored in soil gave significantly higher germination than there shared in the laboratory.

Effect on crop : It is found almost throughout the India as a weed in cultivated and wastelands. And it is considered as a number one problematic terrestrial weed by virtue of its infestation in various agricultural and vegetable crops especially during the rainy seasons. It could reduce crop yields by 32%, losses may be even more, if not properly managed.

Control : Cultural control: It may be difficult to control by tillage.

Biological control: Gibbago trianthemae, the causal agent of leaf spot on *T. portulacastrum*, is isolated from diseased plants collected. Plants sprayed with conidia are killed within 9 days. In host-range studies, the fungus is pathogenic only to *T. portulacastrum*. This fungus may be a useful agent for the biological control of this species.

Chemical control: Propanil @ 2 kg ha-1 applied two weeks after rice emergence or application of 2, 4-D @ 500 g/ha or almix @ 4g/ha.

Uses : The plant, dried out and reduced to powder, is reportedly used against throat troubles and anti-fungal agent. A decoction of the roots is reputedly used as an emmenagogue, and in larger doses it is abortive. Recent studies have investigated potential hepatoprotective properties. The leaves are diuretic and applied in the treatment of oedema, jaundice, strangury and dropsy. A decoction of the herb is used as a vermifuge and is useful in rheumatism; it is considered an antidote to alcoholic poisoning. The fleshy nature of leaves makes them suitable for use as a wound-dressing.

Past work done on Seed Science and Technology

1. Seed production

A mature horse purslane plant can produce as many as 3330 fruits per plant and 6-10 seeds/fruit (Prakongvongs, 1999).

Balyan and Bhan (1986 a) reported that the first three seeding dates resulted in significantly higher seedling emergence than the August 30 date. Leaf number, leaf area, dry matter, and number of branches per plant at the peak growth stages of this weed were significantly higher for the July 20 sowing than for the other sowing dates, followed closely by the June 30 sowing. Similarly, number of nodes per plant, seeds per plant, and seed weight per plant were also significantly higher for the July 20 sowing, followed closely by the June 30 sowing. Phenological behavior of the plants indicate that: (a) maximum seedling emergence takes place during June and July, (b) rapid and vigorous growth occur during July and August (rainy season) when conditions for growth are optimum, and c) production of flowers and seeds starts 20 to 30 days after sowing.

2. Allelopathic effects/ Germination inhibition

Hussain *et al.* (1987) reported that *Trianthema portulacastrum* contains caffeic, chlorogenic, p-hydroxybenzoic, p_ coumaric and ferulic acids which may inhibit or promote seed germination of other plants.

Randhawa *et al.* (1999) stated that germination of *Trianthema portulacastrum* was slightly suppressed while root and shoot length was promoted by maize extract. On the contrary, germination, root and shoot length of maize were suppressed by *Trianthema* extract (Table 1 & 2).

Table 1. Effect of different concentrations of maize water extract on root and shoot length (cm) of *Trianthema portulacastrum*

Treatment	Root length (cm)	Shoot length (cm)
C_0 (Control)	2.21	1.34
C_1 (25%MWE)	2.15	1.39
C_2 (50%MWE)	2.44	2.10
C_3 (75%MWE)	3.08	2.25
C_4 (100 %MWE)	3.38	2.89

Table 2. Effect of different concentrations of *Trianthema* water extract on root and shoot length (cm) of maize

Treatment	Root length (cm)	Shoot length (cm)
C_0 (Control)	3.82	2.99
C_1 (25%MWE)	3.64	2.87
C_2 (50%MWE)	3.43	2.48
C_3 (75%MWE)	3.12	2.04
C_4 (100 %MWE)	2.26	1.82

Concentrated sorghum water extract (100%) reduced the germination of *Trianthema portulacastrum* by 15 to 20% (Randhawa *et al.*, 2002). Cheema *et al.* (2002) reported that sorghum water extracts reduced weed growth by 28%, but this inhibition level still not effective for profitable crop production. For better and eco-friendly control of horse purslane, the effectiveness of sorghum water extracts needs to be improved. Duke *et al.* (2000) stated that mixture of two or more allelopathic aqueous extracts caused more inhibition of weeds due to synergistic mode of action. Likewise in another field study Cheema *et al.* (2003) found that application of sorghum + eucalyptus + sunflower water extracts combination furnished >70% weed reduction than sorghum water extract alone in wheat, probably due to the presence of and synergistic effects of different allelochemicals in these species.

Randhawa *et al.* (2002) revealed that sorghum water extract at higher concentration (100%) reduced the germination of *Trianthema portulacastrum* by 15 to 20%. Root and Shoot length of *Trianthema portulacastrum* was also significantly suppressed at higher concentrations (i.e. 75 and 100%) of sorghum water extract. While lower concentration (25%) of the extract promoted shoot length of *Trianthema portulacastrum* (Table 3).

Table 3. Allelopathic effect of sorghum water extract on the germination and seedling growth of *Trianthema portulacastrum*

Parameters	SWE (0%)	SWE (25%)	SWE (50%)	SWE (75%)	SWE (100%)
Germination percentage	75 ab	80a	75ab	70bc	60c
Root length (cm)	3.01 ab	3.15a	2.70b	2.40c	2.31c
Shoot length (cm)	2.24 b	2.82a	2.20b	2.11b	2.02bc

SWE = Sorghum water extract

Mahmood *et al.* (2010) reported that in the laboratory bioassay, sorghum+ sunflower water extract combination at higher concentration (100%) completely inhibited germination of horse purslane. In pot experiment, the foliar application of sorghum + sunflower water extract combination at higher concentration (100%) greatly suppressed growth traits (shoot and root length) of horse purslane seedling and this combination reduced shoot dry weight by 66% over control. Inhibitory effect was proportional to the concentrations of the extracts and higher concentration had the stronger inhibitory effect (Table 4).

Muhammad Saeed *et al.* (2011) revealed that *Helianthus annuus* water extract reduced germination (10.31%), shoot length (0.25 cm) and root length (0.28 cm) of horse purslane as compared to control which showed germination (86.72%), shoot length (1.81 cm) and root length (2.43 cm). The degree of percent germination inhibition of horse purslane was in the order of *Helianthus annuus*> *Parthenium hysterophorus*> *Datura alba*> *Sorghum bicolor*> *Eucalyptus camadulensis*> *Trianthema portulacastrum*> Control.

Table 4. Allelopathic effect of different plant water extracts on germination, shoot and root length of horse purslane

Treatments	Germination (%)	Shoot length (cm)	Root length (cm)
Plant water extracts			
W_0 = Control (Distilled water)	92.08 a †	1.70 a	2.52 a
W_1 = Sorghum water extract	35.00 b	0.56 b	0.88 b
W_2 = Sorghum + mulberry (WEs)	20.83 d	0.30 e	0.50 e
W_3 = Sorghum + brassica (WEs)	21.67 d	0.44 c	0.66 d
W_4 = Sorghum + sunflower (WEs)	13.33 e	0.24 f	0.37 f
W_5 = Sorghum + maize (WEs)	24.17 cd	0.37 d	0.64 d
W_6 = Sorghum + rice (WEs)	27.92	0.45 c	0.70 c

† = means not sharing a letter in common differ significantly at p 0.05; WEs = Water extracts

Table 5. Effect of different aqueous extracts on horse purslane.

Treatments	Germination (%)	Shoot length (cm)	Root length (cm)
Control	86.72 a	1.81	2.43 a
Trianthema portulacastrum L.	31.95 b	0.63 b	0.79 b
Parthenium hysterophorus L.	17.82 d	0.38 d	0.41 e
Datura alba L.	18.67 d	0.39 d	0.55 d
Helianthus annuus L.	10.31 e	0.25 e	0.28 f
Sorghum bicolor L.	21.15 d	0.49 c	0.57 cd
Eucalyptus camadulensis L.	25.00 c	0.50 c	0.61 c
$LSD_{0.05}$	3.47	0.04	0.05

Means not sharing a letter in common are differ significantly at $P_{0.05}$

2. Seed dormancy and germination

According to Lee *et al.* (2011), germination response of both populations (IRRI: IR population; university of the Philippines; UP population) was greater at 30/20°C and 35/25°C day/night temperatures than they were at 25/15°C alternating day/night temperatures. Germination of both populations was greater in the light/dark regime than in darkness. In dark, depending on the temperature, seed germination of the UP population ranged from 37 to 62%, whereas seed germination of the IR population was <20%. Exposure to 5 min. at 117 and 119°C for the IR and UP populations, respectively, reduced germination to 50% of maximum germination. Osmotic potential of −0.26 MPa inhibited germination to 50% of the maximum for the UP population, whereas the corresponding value for the IR population was −0.37 MPa. Seeds placed on or near the soil surface had maximum emergence, and emergence declined with increase in seed burial depth. Seedling emergence of the UP and IR populations was 74% and 13%, respectively, for seeds placed on the soil surface. For both populations, no seedlings emerged from a soil burial depth of 6 cm or more. Germination and emergence responses to light and seed burial depth differed between the two populations of horse purslane. Residues on the soil surface of upto 6 Mg ha^{-1} did not influence seedling emergence of either populations. Knowledge gained in this study could contribute to developing components of integrated weed management strategies for horse purslane.

REFERENCES

Balyan, R.S. and V.M. Bhan. (1986a). Emergence, growth and reproduction of horse purslane (*Trianthema protulacastrum*) as influenced by environmental conditions. *Weed sci.,* 34: 516-519.

Balyan, R.S. and V.M. Bhan. (1986b). Germination of horse Purslane (*Trianthema portulacastrum*) in relation to temperature, storage conditions and seeding depths. *Weed Sci.,* 34: 513-515.

Cheema, Z.A., A. Khaliq and M. Mubeen. (2003). Response of wheat and winter weeds to foliar application of different plant water extracts. *Pakistan J. Weed Res.*, 9: 89–97.

Cheema, Z.A., M. Asim and A. Khaliq. (2002). Sorghum allelopathy for weed control in cotton (*Gossypium arboreum* L.). *Int. J. Agric. Biol.*, 2: 37–41.

Duke, S.O., F.E. Dayan, I.G. Romagni and M.A. Rimando. (2000). Natural products as a source of herbicides: Current status and future trends. *Weed Res.*, 40: 99–111.

Hussain, F., Yasmin and F. Rehman. (1987). *Trianthema portulacastrum* Linn. Exhibits allelopathy: Advances in weed Sci. Proc. PakIndo Weed Control Workshop, pp. 325-334, Islamabad.

Lee, J, Bhagirath S. Chauhan and David E. Johnson. (2011). Germination of fresh horse purslane (*Trianthema portulacastrum*) seeds in response to different environ-mental factors. *Weed Science* 59 (4): 495-499.

Mahmood, A., Z.A. Cheema, A. Khaliq and A.U. Hassan. (2010). Evaluating the potential of allelopathic plant water extracts in suppressing horse purslane growth. *Int. J. Agric. Biol.*, 12: 581–585.

Muhammad Saeed, Muhammad Ashfaq and Bakhtiar Gul. (2011). Effect of different allelochemicals on germination and growth of horse purslane. *Pak. J. Bot.*, 43(4): 2113-2114.

Prakongvongs. (1999). Germination and growth of horse purslane (*Trianthema portulacastrum* L.). Proc. the 17th Asian Pacific Weed Sci. Soc. Conf., November, 22-27. Bangkok, Thailand.

Randhawa, M.A., Z.A. Cheema and Muhammad Anjum Ali. (2002). Allelopathic effect of sorghum water extract on the germination and seedling growth of *Trianthema portulacastrum. Int. J. Agri. Biol.,* 4 (3): 383-384.

Umarani, R. and A. Selvaraj Jacqueline. (1994). Studies on dormancy and germination of *Trianthema portulacastrum. Indian Journal of Weed Science*, 26 (3 & 4): 82- 88.

PART - III
SEED BIOLOGY OF WETLAND WEEDS

R. Revathi[1] and V. Anbukkarasi[2]

1. PG scholar, Department of Seed Science and Technology,
2. Research Associate, Horticulture College and Research Institute,
Tamil Nadu Agricultural University, Coimbatore – 641 003,
E-mail: revsjagu@gmail.com/ anbuhort@gmail.com

BLISTERING AMMANIA (*Ammania baccifera* L.)

Common name: : Neermel neruppu

Botanical name : *Ammania baccifera* L.

Tamil name : Cirukalluruvi, Kalluppaikkati, Malaimurikam, Neermel Neruppu, Pakuputtipati, Tipputu

Hindi name : Dadamari, dadmari

Family : Lythraceae

Origin and distribution : Sub tropical Asia to tropical Australia, tropical Africa.

Botany

Habitat : A very variable species of rice fields and other wet places up to 1200 m.

Habit : Herb (Fig. 1).

Stem : The stems are somewhat 4-angled with numerous, ascending branches.

Leaf : Leaves are oblong, oblanceolate or narrowly elliptic, about 3.5 cm long, those on the branches very numerous, small, 1 to 1.5 cm long, with narrowed base and pointed or somewhat rounded tip.

Floral biology

Flower : Flowers are small, about 1.2 mm long, greenish or purplish, borne in dense axillary clusters (Fig. 2).

Flowering and fruiting : Flowering in May to October, fruiting in July to November.

Fruit : Capsules are nearly spherical, depressed, about 1.2 mm in diameter, purple, and irregularly circumsciss above the middle (Fig. 3).

Seed biology

Seed : Seeds are black in colour (Fig. 4).

Propagation : Through seed.

Agricultural importance : May become very harmful in rice fields, especially when fields run dry due to lack of water.

Uses : Plant used in Ayurvedha and siddha. The leaves or the ashes of the plant, mixed with oil, are applied to cure herpetic eruptions. The fresh, bruised leaves have been used in skin diseases as a rubefacient and as an external remedy for ringworm and parasitic skin affection.

Fig. 1. Plant

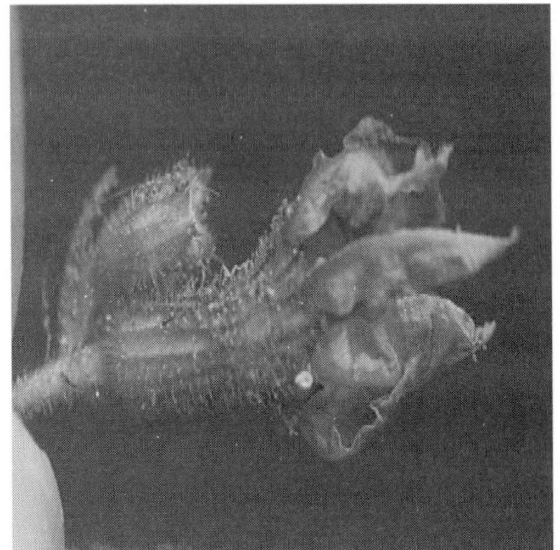

Fig. 2. Flower

Fig. 3. Fruit

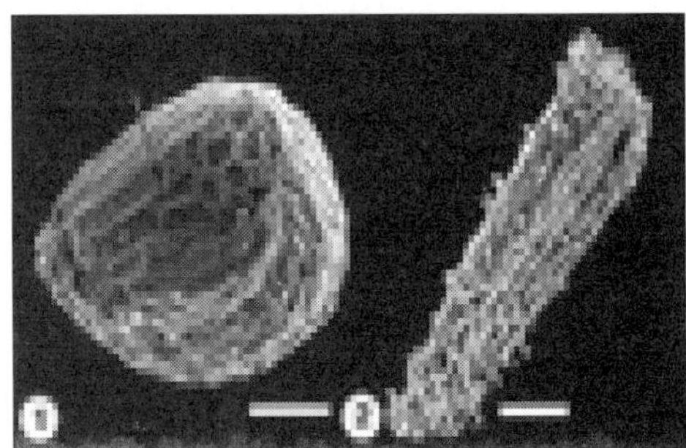

Fig. 4. Seed

Control : 2, 4-D and other growth-regulating herbicides like oxadiazon, a mixture of propanil with oryzalin, nitralin or butachlor are used.

Soil seed bank : Crop rotation significantly affects the density and species composition of weed seedbank. By rotation of wheat with maize and soybean for 2 years, seedbank density decreased 27.16% and 44.44% respectively compared with rice - wheat.

LONG LEAVES BARLERIA (*Asteracantha longifolia*)

Common name : : Kulikhara, kokilaksah, long Leaves barleria

Botanical name : Asteracantha longifolia

Tamil name : Nirumuli

Hindi name : Talimkhana

Family : Acanthaceae

Origin and distribution : This plant is widely distributed throughout tropical and sub-tropical regions of India and other parts of the world including Phillippines, Srilanka, Burma, Malaya and Nepal.

Botany

Habitat : Generally the plant is found on wet places.

Habit : This is an annual herb or under shrub, it grows to a height of about 60 cm (Fig. 1).

Stem : The plant has a number of fasciculated, usually unbranched sub glandular stems, each 60 to 120 cm tall. The stems have nodules hispid with long hairs.

Leaf : Leaves are sessile, oblong- lanceolate or linear lanceolate, spines yellowish brown and 2 to 3 cm (sometimes more) long. These are found in whorls of 6 at each node. The two outer leaves of the whorl remain much larger than the four inner ones. Each leaf is greenish brown in colour. These are acute, entire and hairy.

Floral biology

Flower : Flowers are yellowish brown to blue in colour. These usually occur in apparent whorls of eight in four pairs at each node. Bracts occur on nodes. These are usually 2.5 cm long with long and white hairs. Corolla is about 3 cm long, widely two lipped, tube about 1.6 cm long, abruptly swollen at top. Stamens are four, didynamous with the second pair larger. Calyxes are four- partite with upper sepal 1.6 to 2 cm long, broader than the upper three. Calyxes are linear, lanceolate, and coarsely hairy on the back and with hyline ciliated margin (Fig. 2).

Fruit : Fruits are two celled, linear, and oblong, compressed about 8cm long, and pointed 4 to 8 seeded capsules. These have single layer of epidermis covered with striated cuticle, followed by 5 to 10 layered, thick walled, oval to hexagonal, lignified sclerenchymatous cells (Fig. 3).

Flowering and fruiting : Submerged plants do not flower, but flowering and fruiting take place when plants emerge out of water. It appears that different temperatures and light conditions might be playing some role in the ecophysiology of this plant species.

Seed biology

Seed : Seeds like the eye of a cuckoo. Seeds are ovate, flat or compressed, 0.2 to 0.25 cm long and 0.1 to 0.1 to 0.15 cm wide, hairy, appearing smooth when soaked in water immediately get coated with mucilage, light brown, taste slightly bitter and odour not distinct (The Ayurvedic Pharmacoepia of India, 1999) (Fig. 4).

Fig. 1. Plant

Fig. 2. Flower

Fig. 3. Fruit

Fig. 4. Seed

Propagation : Through seeds.

Life cycle : Asteracantha longifolia grows well and completes its life cycle mostly when water recedes and then, it takes terrestrial mode of life.

Seed dispersal : Dispersed by water and wind.

Seed dormancy and germination : The seeds of *Asteracantha longifolia* prefer germinating in light. Germination was also favoured in blue and red lights, whereas total darkness delayed this process. The optimum temperature for germination of seeds was 29°C in continuous white light. The seeds did germinate in total darkness as well, but the percentage of germination remained poor and with high temperatures beyond 30°C, the pace of germination became slow. Higher temperatures suppressed the seedling growth both in continuous white light and total darkness. Blue and red lights promoted hypocotyls growth, whereas radicle was inhibited (Sen and Chawan, 1970).

Medicinal uses : In Ayurveda, the leaves, seeds and the roots are used for treatment of rheumatism, urinary tract infection, inflammation, jaundice, hepatic obstruction, pain, etc. Powdered seeds taken with milk is said to increase libido.

Control : Physical : Hand weeding is difficult due to the presence of thorns and a woody root.

Chemical: Dicamba, glyphosate, picloram and triclopyr are very effective (Parsons and Cuthbertson, 1992).

REFERENCES

Parsons, W.T. and E.G. Cuthbertson (1992). Noxious weeds of Australia, Inkata Press, Melbourne and Sydney.

Sen, D.N. and D.D. Chawan (1970). Role of light and temperature in relation to seed germination and seedling growth of *Asteracantha longifolia* Nees. *Osterr. Bot. Z.* 118(3): 226-232.

The Ayurvedic Pharmacoepia of India (1999). Government of India, Ministry of Health and Family Welfare, Department of Indian System of Medicine & Homeopathy, New Delhi.

INDIAN PENNY WORT (*Centella asiatica*)

Common name : : Indian penny wort

Botanical name : Centella asiatica

Tamil name : Vallarai, Yoshanavalli, Chandaki, Pindeeri

Hindi name : Brahmi

Family : Apiaceae or Umbelliferae

Origin and distribution : It is a native of Madagascar. It is found in swampy areas of India, commonly found as a weed in crop fields and other waste places throughout India up to an altitude of 600 m, in tropical and subtropical regions. The plant is reported to occur also at higher altitudes of 1550m in Sikkim and 1200 m in Mount Abu. It grows widely in regions of east India, China, Japan and Australia, Sri Lanka and South Africa.

Botany

Habitat : It is commonly found in rice fields along the bunds of irrigation channels.

Habit : Perennial herb (Fig. 1).

Root : The plant consist of vertical root stalk.

Stem : Stem is prostrate slender, creeping with long stolens and nearly glabrous or hairy on young parts.

Leaf : Leaves are cordate or hastate or orbicular or reniform or sub entire or palmately lobed consisting of long petiole and small stipules. Crowded leaves can be seen at nodes, consisting of very long stalks and sheathing leaf bases. Leaf blades are dentate, crenate with thick radiate veins and dark green in colour. Leaves are glabrous on both surfaces.

Floral biology

Flower : Flowers are small, sessile and dark pink in colour. They arise as simple umbels of 3 – 6 flowers at the ends of slender peduncles arising from the axils of leaves and much shorter than petioles supported below by an involucre of 2-boat shaped membranous persistent bracts (Fig. 2).

Fig. 1. Plant

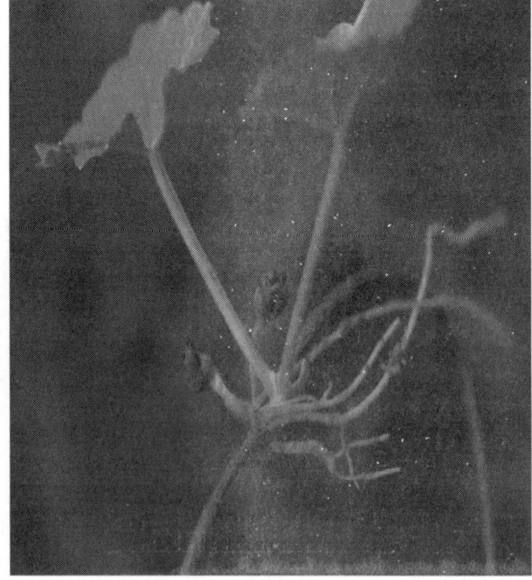

Fig. 2. Flower

Flowering and fruiting : Flowering in June to October; fruiting in July to November.

Fruit : Clusted at joints. Carpels oblong, sub–cylindric curved and less in length much laterally compressed, readily separating into 2 indehiscent halves (mericarps) united by a very narrow plane of junction. Vittae are seen in the furrows of each mericarp. The fruits when bruised have an aromatic odour, pungent, nauseous and bitter taste (Fig. 3).

Seed biology

Seed : Solitary in each mericarp, pendulous embryo, laterally compressed (Fig. 4). *Centella asiatica* had 15 seeds per ramet which are brownish in colour and slightly kidney-shaped. The average length and breadth of seeds were 2.84 and 1.81mm, respectively. The 100 seeds was 0.13g (Devkota and Jha, 2010).

Seed dispersal : Through water.

Fig. 3. Fruit

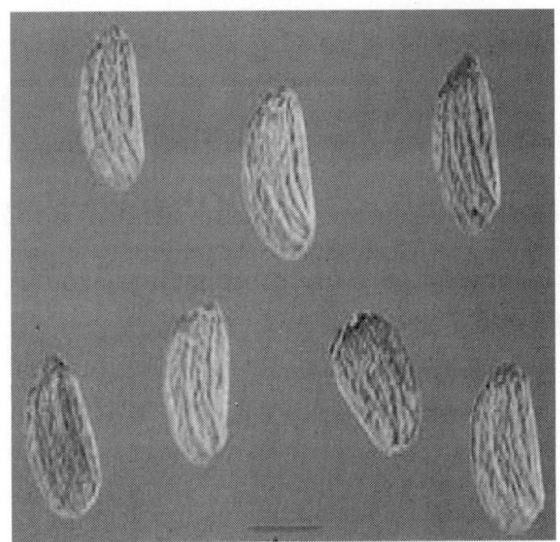

Fig. 4. Seed

Propagation : Centella asiatica reproduces both vegetatively and sexually (Brock *et al.*, 2003). Severe competition exists between ramets and seedlings. Seedling mortality is high in natural settings. Ramets play an important role in population maintenance in nature (Singh *et al.*, 1999).

Seed dormancy and germination : Freshly harvested seeds of *C. asiatica* did not germinate even after gibberellic acid (GA_3) treatment and exposure to different treatments with light qualities, while two-three months old seeds exhibited germination (82%) without pre-treatment at warm environment (25 -30°C).

Seed storage : TTC test indicated 82% viability in freshly harvested seeds of *C. asiatica*. Seeds viability deteriorated as duration of storage increased and they became non-viable after storage for thirty months.

Seeds offer genetic variation and ease of portability and storage as compared to vegetative propagules apart of resilience to draught (Brock *et al.*, 2003).

Medicinal uses : Gotu kola has been used externally for burns, psoriasis, prevention of scar formation following surgery, recovery from an episiotomy following vaginal delivery of a newborn and treatment of external fistulas.

Extract of *Centella* exerted anti-inflammatory and antifilarial effects and is used as memory enhancer and wound healer (Shukla *et al.*, 1999). Clinical trials have also shown that it can help those with chronic venous insufficiency (Brinkaus *et al.*, 2000).

Control : Mechanical control: Refrain from watering the areas where Centella has grown for up to two weeks. Dig up and remove the plants at the roots by hand or if there is a large area, removal can be done by machine.

Chemical control: MSMA @ 0.4 kg ha^{-1} applied 25-30 d after emergence provides effective control.

Past work done in Seed Science and Technology

According to Devkota and Jha (2010), freshly harvested seeds of *C. asiatica* did not germinate even after GA_3 treatment and incubation under different light qualities, whereas two-three months old seeds exhibited germination (82%) without pre-treatment at warm environment (25-30°C). Germination of seeds pretreated with 60°C water were significantly lower than germination of seeds pre-treated with 10% HNO_3 and seeds treated with different concentration of GA_3. Among the different treatments of GA_3, soaking seeds in 10 ppm resulted in highest germination and was followed by 20 ppm (Fig. 5).

The germination was slightly higher in red light and inhibited in blue and far red lights. The seeds were defective on germination in dark conditions (Fig. 6).

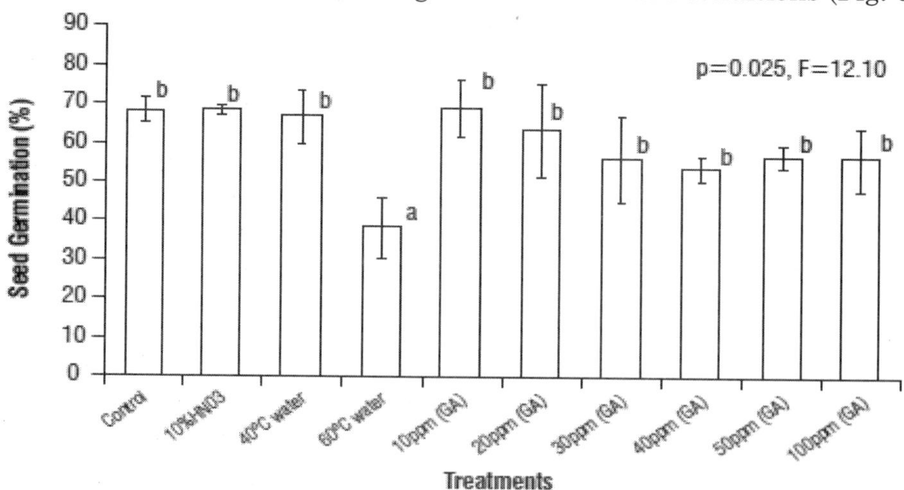

Fig. 5. Effects of different pre-treatments on seed germination of *C. asiatica* after six weeks (bars marked with the same letter are not significantly different from the control, p>0.05).

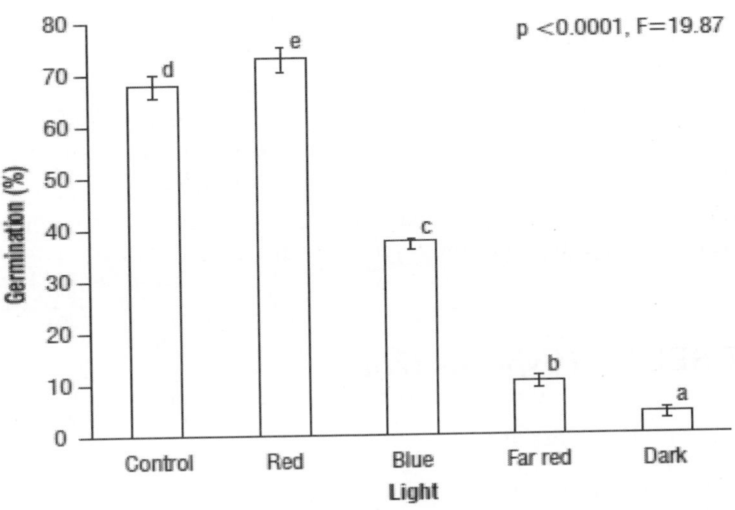

Fig. 6. Final germination percentages of *Centella asiatica* seeds in different coloured light (observation at sixth week).

At salinities up to 3500 ppm, NaCl germination rate ranged 70-73%. The germination was inhibited by 57 and 89% at 4500ppm and 5500ppm NaCl, respectively and completely abolished at 6500 ppm salinity (Fig. 7).

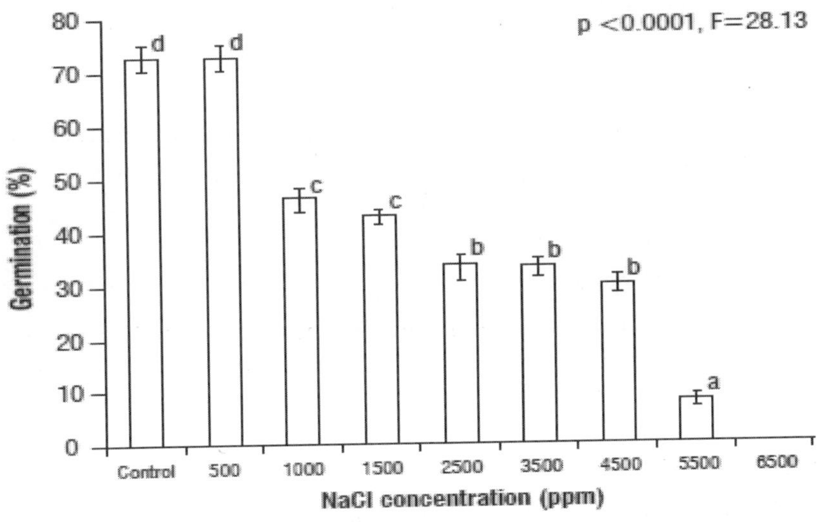

Fig. 7. Germination of seeds of *Centella asiatica* in different concentration of NaCl (bars marked with the same letter are not significantly different at p>0.05).

REFERENCES

Devkota, A. and P.K. Jha (2010). Seed germination responses of the medicinal herb *Centella asiatica*. *Brazilian Journal of Plant Physiology*, 22(1):143-150.

Brock, M.A., D.L. Nielsen, R.J. Shiel, J.D. Green and J.D. Langley (2003). Drought and aquatic community resilience: the role of eggs and seeds in sediments of temporary wetlands. *Freshwater Biol.*, 8: 1207-1218.

Singh Pratibha, U.P. Singh and J.S. Singh (1999). The effect of leaf extracts of *Centella asiatica* and *Andrographis paniculata* on spore germination of some fungi. *Journal of Plant Protection in the Tropics,* 12 (2): 106-112. ISSN 0127-6883.

Shukla, A., A.M. Rasik and B.N. Dhavan (1999). Asiaticoside induced elevation of antioxidant levels in healing wounds. *Phytotherapic Res.,* 13:50-54.

RICE FLAT SEDGE (*Cyperus iria*)

Common name : Rice flat sedge

Botanical name : Cyperus iria

Synonym : Cyperus panicoides Lam., *Cyperus resinosus* Hochst., *Cyperus santonici* Rottb.

Tamil name : Oosikorai

Family : Cyperaceae

Origin and distribution : Tropical and temperate Asia, Australia, tropical east Africa naturalized in (sub-) tropical America

Botany

Habitat : C. iria is a principal weed in paddy fields throughout the world and is also common in upland fields in many countries (It is reported in 17 crops and in 22 countries).

Habit : Annual sedge (Fig. 1).

Roots : A fibrous root system.

Leaves : Dark green, shiny, linear in outline but tapering to a point, and three-ranked. Individual leaves may be from 3 to 8 mm wide and have rough margins toward the leaf tips. A membranous sheath occurs at the leaf bases.

Floral biology

Flowers : Terminal seed heads occur at the ends of the triangular stems. Seed heads may reach 8 inches in length, are open, and are composed of several dense spikes. Individual spikes are, in turn, made up of many goldfish-brown spikelet's that are approximately 5 to 13 mm long and 1 to 2 mm wide. Seed heads also have 3 to 7 unequal bracts surrounding the inflorescence, the shortest one being much longer than the overall inflorescence (Fig. 2).

Flowering and fruiting : Throughout the year.

Seed biology

Fruit : A three-angled brown achene (Fig. 3).

Fig. 1. Plant

Fig. 2. Flower

Fig. 3. Fruit

Propagation : By seeds. It is a prolific seed producer and spreads quickly.

Seed dormancy and germination : Seed is dormant, but can germinate about 75 days after shedding. Germination of *C. iria* was stimulated by light and warm fluctuating temperatures. Germination of *C. iria* was influenced to a greater degree by increasing salt and water stress. Seeds sown on the soil surface gave the greatest percentage of seedling emergence, and no seedlings emerged from seeds buried in soil at depths of ≥1 cm. Flooding, although not continuous or deep, had a suppressive effect on emergence and growth of *C. iria* (Chauhan and Johnson, 2009).

Germination of *C. iria* was stimulated at temperature between 20 and 36°C. The germination was inhibited with osmotic potentials increasing from -0.4 to 1.0 MPa.

Seed longevity and burial : Bhatt and Singh (2007) studied the soil seed bank dynamics of weed flora in upland and low land paddy cultivation areas. They found that the mean total number of individuals of weeds that germinated in soil samples of upland site was 517 compared to 503 at lowland site. The highest number of weeds (186) was of the family Cyperaceae (37% of the species). The proportion of this family was 35.6% at the upland site.

Agricultural importance : Concentrated leachates from the roots, shoots and inflorescences of *C. iria* adversely affect germination, root and shoot growth of rice seedlings.

Chemical control : C. iria is controlled by butachlor and is moderately susceptible to oxadiazon @ 0.75-1.0 kg ha^{-1} applied after harrowing and sowing of rice. It is also controlled by early post-emergence commercial mixtures of butachlor and propanil, or 2, 4-D @ 0.5 to 0.8 kg ha^{-1}, MCPA @ 0.4 kg ha^{-1}, propanil @ 2 kg ha^{-1} applied 20-30 days after-emergence or almix @ 4 g ha^{-1}.

Past work done on Seed Science and Technology

Jagat Devi Ranjit *et al.* (2007) showed that the weed seed accumulation of *C. iria* was highest in the 5-10cm soil depth (Table 1)

Table 1. Response of sedge weed seed bank density to soil depth at before wheat planting (BWP), after wheat harvest (AWH) and after rice harvest (ARH).

Treatments	BWP	AWH	ARH
Soil depth			
0-5 cm	173	104	1028
5-10 cm	255	221	87
10-15 cm	103	113	85
15-20 cm	82	73	55

Effect of water stress: There is large variation in the germination of weed species under different moisture regimes. Germination of *Cyperus iria* was inhibited by 50% at 0.5 MPa, whereas it required only 0.1 MPa water potential for similar reduction in *C. difformis* (Chauhan and Johnson, 2010).

REFERENCES

Bhatt, M.D. and S.P. Singh (2007). Soil seed bank dynamics of weed flora in upland and lowland paddy cultivation areas of far western Nepal. *Scientific World*, 5(5):54-59.

Chauhan, B.S. and D.E. Johnson (2009). Ecological studies on *Cyperus difformis, Cyperus iria* and *Fimbristylis miliacea*: three troublesome annual sedge weeds of rice. *Ann. Appl. Biol.*, 155: 103-112.

Jagat Devi Ranjit, Rungsit Suwanketnikom, Sombat Chinawong, Sutevee Suprakarn, Isara Sooksathan and Sunanta Juntakool (2007). Weed seed bank response to soil depth, tillage and weed management in the mid hill ecology, *Kasetsart J. (Nat. Sci.),* 41: 17-33.

Chauhan, B.S. and D.E. Johnson (2010). The role of seed ecology in improving weed management strategies in the tropics. *Adv. Agron.*, 105: 222-262.

UMBRELLA SEDGE (*Cyperus difformis*)

Common name: Small-floweed, Umbrella sedge

Botanical name : Cyperus difformis

Tamil name : Vattakorai

Hindi name : Dila, motha

Family : Cyperaceae

Origin and distribution : Introduced from Central America. Distributed to Europe across to Japan, South Africa, Indian continent, south east Asia and Malesia to the Pacific.

Botany

Habitat : Usually a weed of wet places. The plant normally grows in flooded or in very moist soils. It is primarily a weed of paddy or flooded rice. It is frequently found in small pools, along rivers, canals, and streams, in open wet places, and in grassy swamps. It grows best in rich, fertile soils, but can grow in poorer sandy or clay soils of unused lands or in fallow rice fields.

Habit : Annual sedge, submersed or more commonly emergent (Fig. 1).

Roots : Numerous, fibrous, reddish.

Stems : Rather weak or tender, triangular, slightly winged, smooth, 1 to 3mm thick.

Leaf : Linear, few, weak, rather abruptly acuminate, smooth or slightly rough at the top, usually shorter than the plant is high (often two-thirds of the height), 2 to 5 mm wide, sometimes reduced to sheaths.

Floral biology

Flower : Flowers are greenish brown to purplish and cluster densely to form round, stalked or stalkless flower heads, each having two to three modified leaves below it (Fig. 2).

Fig. 1. Plant

Fig. 2. Flower

Flowering : Flowering takes place from May through November.

Fruit : The fruit is egg shaped, pale greenish brown, glossy, and very tiny less than 1/25 of an inch (1 mm) (Fig. 3).

Seed biology

Seed : Thousands of small seeds are produced (Fig. 4).

Fig. 3. Fruit

Fig. 4. Seed

Propagation : By seeds. Having heavy seed production and massive seedling densities, the plant quickly covers the ground and becomes dominant.

Life cycle : In tropical areas, *C. difformis* apparently can flower and produce seeds all year long, if the soil moisture is sufficient. One plant could complete the vegetative cycle in one month and produce a new generation of weeds during the second month, producing 50,000 seeds with about 60% germination.

Through high seed production and short life cycle, this weed can spread rapidly to become a dominant weed in a rice field.

Seed dormancy and germination : C. difformis seeds do not have dormancy (Kim *et al.*, 1987). Germination of *C. difformis* was influenced to a greater degree by increasing salt and water stress. In all three species, seeds sown on the soil surface gave the greatest percentage of seedling emergence, and no seedlings emerged from seeds buried in soil at depths of ≥1 cm. Intermittent flooding to shallow depths, however, was less effective in controlling *C. difformis*; deep flooding was needed to suppress growth of *C. difformis* seedlings. When the flooding was delayed to 21 days after sowing, there was little growth reduction in all three species (Chauhan and Johnson, 2009). According to Mortimer *et al.* (2005), small seeded *C. difformis* exhibited physiological polymorphism in germination response to flooding.

Agricultural importance : C. difformis is a widespread and serious weed of paddy rice. This weed appears to be more of a competitor for water and nutrients than for light.

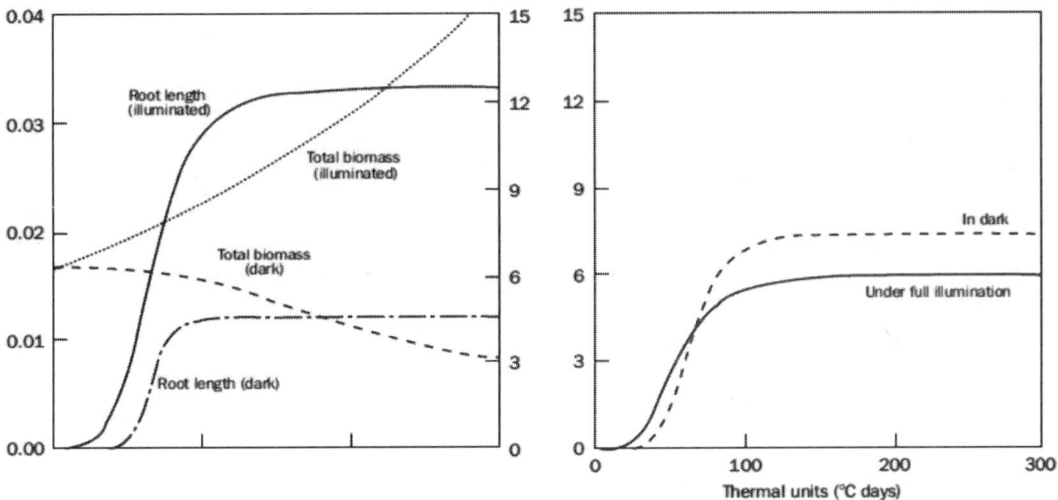

Fig. 5. Seedling growth responses in *Cyperus difformis* in response to flooding. Germinated seeds were placed in water columns containing nutrients to a depth of 50 mm and either illuminated (light/dark) on a 12-h cycle or maintained in total darkness at a temperature of 30/20°C day/night. PAR = 450 μE m^{-2} s^{-1}. Repeated destructive harvests were taken over the time course, enabling fit of the functions shown.

Past work done on Seed Science and Technology

Mortimer *et al.* (2005) illustrated dramatic differences in resource mobilization and partitioning in *cyperus difformis* (Fig. 5). In *C. difformis*, net assimilation of biomass occurred below a 45-mm water column in full illumination, but biomass was not partitioned into photosynthetic tissue that emerged above the water surface, by 300 thermal units (TU). Root length (a partial surrogate of root biomass) extension was noticeable immediately after germination under both light regimes, suggesting that carbohydrate assimilate was partitioned to roots in the first instance. Maximum shoot height was achieved in the dark regime over 300 TU. Emergence of leaf tissue from illuminated water columns was not achieved until 700 TU but plants returned to lighted conditions from after 300 TU in darkness did not grow. With respect to response to flooding regimes, with *C. difformis* maintaining a longer aquatic stage before emergence from the water column because of underwater photosynthesis.

Jagat Devi Ranjit *et al.* (2007) showed that the weed seed accumulation of *C. difformis* was highest in the 5-10cm soil depth (Table 1).

Table 1. Response of sedge weed seed bank density to soil depth at before wheat planting (BWP), after wheat harvest (AWH) and after rice harvest (ARH).

Treatments	BWP	AWH	ARH
Soil depth			
0-5 cm	1220	861	1534
5-10 cm	1014	884	621
10-15 cm	758	541	528
15-20 cm	643	459	268

REFERENCES

Chauhan, B.S. and D.E. Johnson (2009). Ecological studies on *Cyperus difformis, Cyperus iria* and *Fimbristylis miliacea*: three troublesome annual sedge weeds of rice. *Annals of Applied Biology*, 155 (1): 103–112.

Jagat Devi Ranjit, Rungsit Suwanketnikom, Sombat Chinawong, Sutevee Suprakarn, Isara Sooksathan and Sunanta Juntakool (2007). Weed seed bank response to soil depth, tillage and weed management in the mid hill ecology. *Kasetsart J. (Nat. Sci.)* 41: 17 - 33.

Kim, Soon Chul and Keith Moody (1987). Germination of two rice cultivars and several weed species. *Korean J. Weed Sci.*, 9(2): 116-122.

Mortimer, A. M., O. Namuco and D.E. Johnson (2005). Seedling recruitment in direct-seeded rice: weed biology and water management. Paper presented at the World Rice Research Conference, 4-7 Nov 2004, Tokyo and Tsukubu, Japan.

BARNYARD GRASS (*Echinochloa crus-galli*)

Common name : Barnyard grass, dutch kut grass, cockspur grass.

Botanical name : *Echinochloa crus-galli* (L.) P. Beauv.

Tamil name : Kudirai vali, chettha karai

Hindi name : Kayada, sawank

Family : Poaceae.

Origin and distribution : Native from tropical Asia, it has spread and now occurs widely in tropical and subtropical regions and extending from lat 50° N to 40° S. It is the most widespread weed of rice throughout South and Southeast Asia.

Botany

Habitat: It occurs in low and medium altitudes, favoring open sunny places and wet soils, and can continue to grow when partially submerged. It is a common weed in swamps and aquatic places. It also grows well in drier soils, but is shorter and has fewer tillers, panicles and seeds. According to Barret and Seaman (1980), Barret (1982) and Barret and Wilson (1983), Crus-galli persist as a weed of shallow-water areas and paddy banks.

Habit : Terrestrial, tuffed, erect herb (Fig. 1).

Roots : A fibrous root system.

Stems : Usually erect, thick, without hairs (glabrous), often branched at the lower nodes, and may be tinted red to maroon at the base.

Fig. 1. Plant

Fig. 2. Flower

Leaves : Rolled in the shoot, smooth and without ligules or auricles. Leaves range from 4 to 20 inches in length and may be 5-30 mm wide. Leaves have a distinct white midvein that becomes keeled toward the basal potions of the leaf. A few short hairs may occur at the leaf bases.

Floral biology

Infloersence : Terminal panicle ranging from 4 to 16 inches in length. Panicles may be green to purple in colour and are comprised of individual spikelet that may develop a 2 to 10 mm long terminal (Fig. 2).

Flowers : The flowers are sessile and bisexual (Fig. 3).

Fig. 3. Flower, spikelet, opened spikelet showing (left to right) second glume, fertile floret, awned sterile floret and first glume.

Flowering and pollination : Echinochloa crus-galli plants are self-compatible and highly autogamous. In common with many grasses, cleistogamous flowers are produced late in the season when spikelets do not emerge from the surrounding leaf sheaths. Anthesis of individual spikelets occurs within a single day, but the flowering period of inflorescences can last for several days with spikelets at the top of the inflorescence entering anthesis before those at the base. The mating system of *E. crus-galli* involves a high degree of self-fertilization with occasional outcrossing mediated by wind. Predominant selfing results in a high degree of homozygosity within populations and relatively low levels of heterozygosity at polymorphic loci

Seed biology

Seed : Seed is caryopsis and average seed weight is 1.73 mg. The seed is light in weight when compared to other species (Fig. 4).

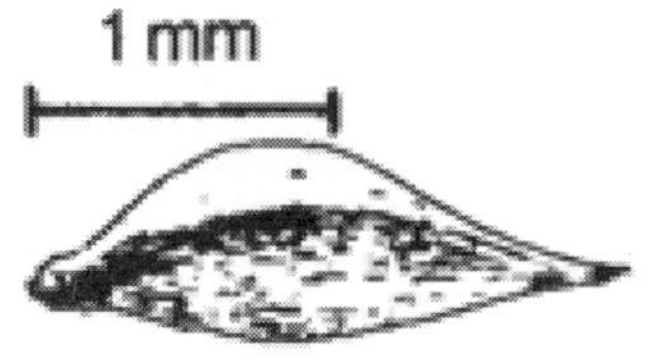

Fig. 4. Seed

Propagation : By seeds.

Life cycle: E. crusgalli takes 42-64 days to complete its life cycle. The seeds are often mixed with rice seed. The seed production can vary from 2000 to 40000 per plant (Holm *et al.*, 1977)

Seed dispersal : Approximately 50 volume of seeds remained as float after 4-5 d in water (Barrett and Wilson, 1983) indicating that dispersal of seeds by water is likely, particularly in flooded habitats. Viable *E. crus-galli* seed has also been reported in irrigation water (Wilson, 1979). In addition granivorous birds, especially blackbirds (*Agelaius* spp.) and waders are important dispersal agents, particularly in rice field habitats (Smith and Shaw, 1966; Barrett and Seaman, 1980). Probably the most important contemporar), agent of long distance dispersal is man, since barnyard grass seed is a common contaminant of many seed crops and weed seeds are inadvertently introduced to fields during the sowing of the crop.

Seed dormancy and germination : Freshly collected seeds of *E. crus-galli* exhibit innate dormancy or enforced dormancy, the duration of which varies considerably (Rahn *et al.*, 1968; Barrett and Wilson, 1983). The dormancy is due to the pericarp and epidermis (Arai and Miyahara, 1963). Rahn *et al.* (1968) demonstrated that fresh seeds had only 0.3-l.4% germination. Storage of seeds for 4 and 8 months increased germination to 19 and 44%, respectively. Of a sample of *E. crus-galli* seeds planted at various depths at the beginning of November, only 4.7% germinated over the first 2 months, whereas 25.4% germinated during the sixth and seventh months after planting (Horng and Leu, 1978). The most effective methods of breaking dormancy were removal of outer seed covering or pericarp, freezing and thawing for 4 d, exposing moist seeds to 120°F (49°C) for 5 h, soaking them in acetone for 20 min, in concentrated H_2SO_4 for 8 min, or in water for 4 d (Rahn *et al.*, 1968). Use of 0.2% KNO_3 as a moistening agent gave slightly higher germination (Rahn *et al.*, 1968). The optimum chilling temperature for breaking dormancy was l-10°C for *E. crus-galli* var. *praticola* (Watanabe, 1978). Miyahara (1975) concluded that primary dormancy was broken by low winter temperatures, alternating temperatures in spring and spring flooding in paddy fields.

Dormant *E. crus-galli* seeds can be stimulated to germinate by contact with solutions of n-propanol (Taylorson, 1988).

Caryopsis of barnyardgrass contained in a single spikelet were dormant at maturity, but dormancy was overcome when the spikelets were dry after ripened at 23°C or were treated with high temperatures, acid scarification, or punctured with a scalpel (Sung *et al.*, 1987).

Scarification of seeds in gut passage of waterfowl accelerated the germination (De Vlaming and Proctor, 1968; Figuerola *et al.*, 2005), which uplifts the mechanical dormancy, making the seed coat permeable to water (Santamaria *et al.*, 2002).

The seeds were wrapped in nylon fabric, buried 20 cm under the grass sward and exhumed at monthly intervals. The germination of seeds under light conditions at a constant temperature of 25°C fluctuated between 0% and 96%. Percentage of germination increased faster in secondary than primary dormant seeds at both constant 25°C and 5°C (Honek *et al.*, 1999).

Control measures : Biological control: *E. crusgalli* is attacked by a number of (mostly polyphagous) insects, fungi and viruses. The North America weevil *Listronotus humilis* might be of interest as biological control. Its larvae eat the growing points and young tissues of *E. crusgalli* seedlings.

Chemical control: In India for rice crop, pre-emergence application of butachlor @1.5 kg a.i/ha, anilophos @400 g/ha, pretilachlor @1.0 kg/ha, pendimethalin @1.5 kg/ha.

Uses: A warm-season bunchgrass used as cattle fodder and sometimes cultivated for this purpose. Also suited for ensilage, but not for hay. Usually fed green. Grass also used for reclamation of saline and alkaline areas, especially in Egypt. Grain of some varieties used as food in times of scarcity and sometimes used for adulterating fennel.

Reported to be preventative and tonic, barnyard grass is a folk remedy for treating carbuncles, hemorrhages, sores, spleen trouble, cancer and wounds. The shoots and/or the roots are applied as aseptic to wounds. The plant is a tonic, acting on the spleen.

Past work done on Seed Science and Technology

Seed dormancy and germination

Effect of temperature

Takabayashi and Nakayama (1981) found that germination of *E. crus-galli* var. *praticola* seeds buried 5 cm deep in sand was high in February and May, but low in summer, a trend they attributed to secondary dormancy. Arai and Miyahara (1963) showed that storage of dormant seeds in submerged soil at 5°C overcome dormancy. Storage of seeds in nitrogen accelerated release of seeds from dormancy much faster at 30°C than at 5°C. Barrett and Wilson (1983) conducted germination tests on 9 and 18 month-old seeds of 18 populations of *E. crus-galli* var. *crus-galli* and I1 populations of *E. crus-galli* var. *oryzicola.* They found that the decay of dormancy in var. *crus-galli* was less rapid than in var. *oryzicola* following dry storage and burial in soil (Table 1).

The germination of *E. crus-galli* seeds was not reduced by exposure to 60 and 80°C temperatures for 1h but exposure to 100, 120 and 130°C for the same length of time significantly lowered percent germination (Maun, 1977). The rate of germination was reduced only at 120°C and 130°C.

The minimum, optimum, and maximum temperatures for germination of *E. crusgalli* were determined to be 13°C, 20-30°C and 40°C, respectively (Brod 1968; Roche and Muzik, 1964).

Table 1. Percentage germination following 9 and 18 months of dry storage of seeds of *Echinocola crus-galli* vars. *crus-galli*

		9 months	18 months
		Average germination (%)	Average germination (%)
Taxon	***n* populations**	**± SD**	**± SD**
Crus-galli	18	25.7±24.9	96.7±2.7
Oryzicola	11	92.9±7.3	94.7±2.8

Shifts to a higher temperature, often for only 2 h in darkness, can break dormancy of seeds of several species, including *Nigella, Rumex*, and *Echinochloa crus-galli* (Isikawa, 1957; Taylorson and Dinola, 1989; Taylorson and Hendricks, 1972; Vicente *et al*., 1968), and buds of *Iris* spp (Halevy *et al*., 1964). In most cases, however, high temperature and light are required for breaking dormancy (Toole *et al*., 1955, 1957).

Effect of pH, moisture and salt stresses

The seeds of *E. crus-galli* germinate over a wide range of temperatures, from 13 to 40°C (Roche and Muzik, 1964; Rahn *et al*., 1968). The optimum temperature for germination as reported by Watanabe (1981) was 5°C alternating with 30°C. The rate of germination of barnyard grass seed is markedly affected by temperature. Vengris *et al.* (1966a) showed that seeds germinated in only 5 d under high temperature conditions in mid-summer as compared to 11-12 d in early spring or fall when temperatures were relatively low. When temperature was not limiting (held at 20°C) the germination of *E. crus-galli* var. *praticola* seeds (wintered in soil) was entirely dependent on exposure to sunlight or to fluorescent light (Watanabe, 1977). Under ideal germination conditions in a glasshouse, seeds of var. *oryzicola* from each of four locations in California germinated significantly faster than those of var. *crus-galli* (Barrett and Wilson 1983).

Aslam *et al.* (1987) reported that seeds from an *E. crusgalli* population from Pakistan did not germinate at salinity of 1.5% NaCl, indicating that the Ohio seed source from a saline habitat was more salt-tolerant at the germination stage than the seeds from Pakistan collected in a nonsaline location.

According to Rahman and Irwin (1990), seeds of *Echinochloa crusgalli* from both scarification and stratification + light treatments responded similarly to salinity treatments (Table 2). Highest germination percentages were found in the distilled water controls in both the stratification + light and scarification treatments. Germination decreased in the 1.5% NaCl concentration by 33.5% and 30.0% from controls for each treatment, respectively. No seeds germinated in the 2.0% NaCl treatment. Percent seed germination in both treatments decreased significantly ($P < 0.05$) with increasing salt concentration from 0% to 1.5%. Stratified seeds incubated in darkness did not germinate in any of the treatments (Table 2).

Table 2. The effect of salinity on germination percentages (means SE) of *Echinochloa crus-galli* seeds after scarification and stratification + light temperatures. (Each Value ± 1 SE).

	Scarification	Stratification + light	Stratification –light
0.0% NaCl	85.0 ± 0.91	87.0 ± 1.47	0
0.5% NaCl	72.5 ± 1.04	74.5 ± 1.85	0
1.0% NaCl	57.5 ± 1.04	55.0 ± 0.82	0
1.5% NaCl	55.0 ± 0.91	53.5 ± 0.65	0
2.0% NaCl	0	0	0

Germination of *Echinochloa crus-galli* was significantly reduced at -0.5 MPa (Boyd and Van Acker, 2004).

Effect of light

Seeds in which dormancy had been broken (1 yr old) germinated best with continued exposure to light (60%) as compared with continuous dark (6%) (Rahn *et al.*, 1968).

Effect of maturity

More mature (dark grey or brownish and shiny) seeds produced significantly higher germination than immature (light grey) seeds (Rahn *et al.*, 1968); however, some seeds harvested at milk or dough stage may be viable.

Combined effects

Dong-Sheng Tang *et al.* (2010) reported that in *E. crus-galli* var. *crusgalli* was higher than *A. retroflexus* and *D. adscendens*. Combined application of red light and N-compounds significantly promoted seed germinations. In this weed, the best concentrations for N-compounds were 50 mM (KNO_3), 5 mM (NH_4Cl), 50 mM (NH_4NO_3, NH_4NO_3) and 25 mM ($NaNO_2$). Similar to *A. retroflexus*, NH_4NO_3 and $NaNO_2$ provided maximum germination percentages (Fig. 4). Nitrite induced seed germination of 64.67%, which was lower than nitrate (76.67%) (Fig. 5).

Summary

Exposure to approximately 440 W of 660.5nm light for 5 min (Watanabe, 1977); chemical treatment with urea and potassium nitrate (Moursi *et al.*, 1979a); treatment with GA after storage of seeds at low temperature (Rizk *et al.*, 1978); physical scarification and H_2SO_4 or NaOH treatment (Yamasue *et al.*, 1977); sudden changes in temperature (Averkin, 1978a); exposure to freezing temperatures for 6 or 10 week (Moursi *et al.*, 1979b); and ultrafreezing in liquid nitrogen to - 196°C for 5 min (Jordan *et al.*, 1983). Conversely, percent germination was found to decrease when seeds were submerged under 20 cm of water (Panggabean, 1976), exposed to freezing temperatures for 20 week (Moursi *et al.*, 1979b) or when the density of the soil in which they were sown reached 1.3 g cm^{-3} (Averkin, 1978b).

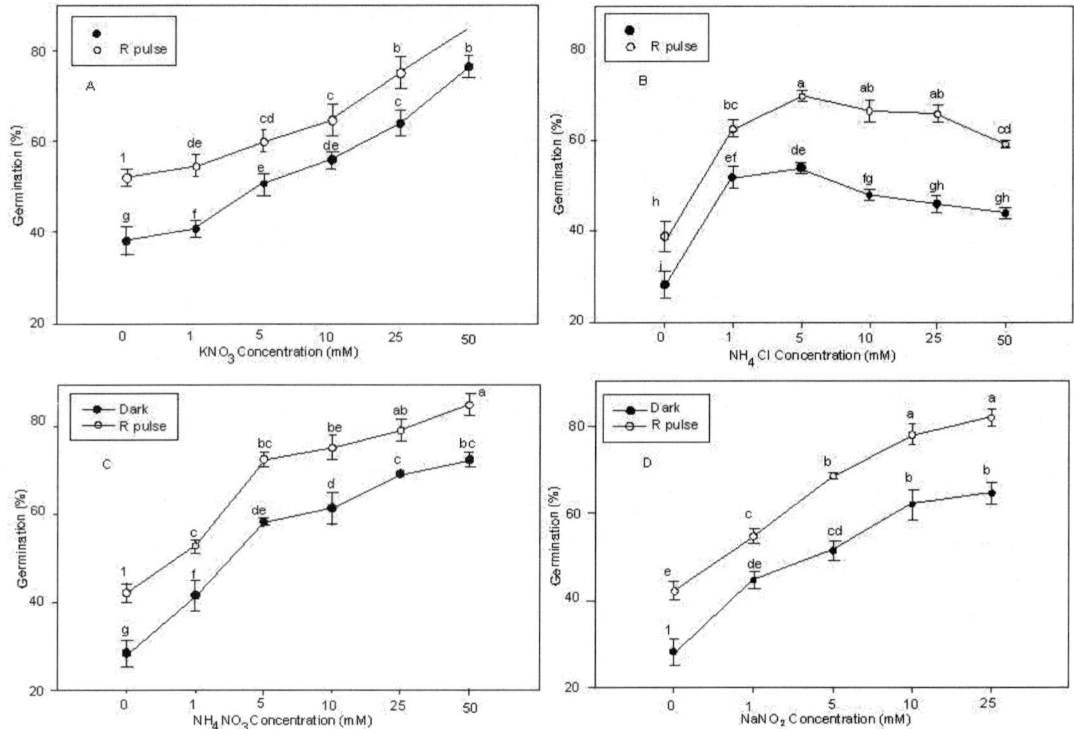

Fig. 5. Germination (%) of *Echinochloa crus-galli* seeds in response to N-compounds and red light. D represents dark condition, while R pulse indicates red light irradiation for 10 min.

Seed longevity and burial

Rahn *et al.* (1968) found no evidence of deterioration in viability of *E. crus-galli* seeds after storage for 3 yr in a glass jar in lab (dry) or loam soil at two depths (8 and 25 cm). Similarly, Barrett (1983) observed no reduction in viability of 6- to 8-yr-old seed of *E. crus-galli* var. *crus-galli* stored dry at room temperature. No reduction in germination of seeds buried in submerged soil for 30 months was observed by Roche and Muzik (1964); however, seeds buried at 10- and 20-cm depths under normal soil conditions for the same period lost considerable viability. Dawson and Bruns (1975) buried seeds of *E. crusgalli* at 2.5, l0 and 20 cm depths in irrigated and nonirrigated sandy loam soil. They showed that seeds exhumed from l0 to 20cm depths had highest germination in the second year after burial. The seeds buried for 13 yr had 3% viabllity but those buried for l5 yr were nonviable. Seeds buried at 20 cm remained viable for a longer period than at l0 cm probably because of greater induced dormancy (Roche and Muzik, 1964). Germination was more rapid in sandy loam rather than loamy soil and soil compacted by tamping and surface watering in a greenhouse produced higher germination (Rahn *et al.* 1968).

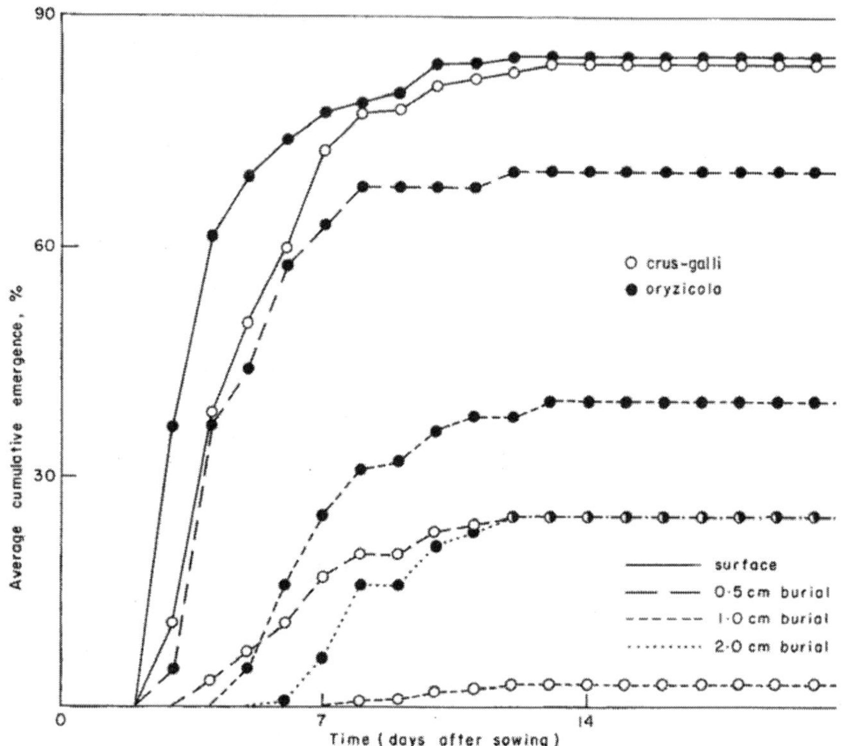

Fig. 6. Seedling establishment of *Echinochloa crus-galli* and *oryzicola* from different burial depths (surface, 0.5, 1.0, 2.0 cm) in saturated soil under glasshouse conditions.

Dawson and Bruns (1962) buried seeds at depths ranging from 0 to 15 cm. The seeds germinated equally well at all depths but the maximum number of seedlings emerged from 1 to 2 cm depths. The maximum depth from which some seedling emergence occurred was 10 cm. The emergence occurred first from shallower depths. The surface-lying seeds germinated, but produced few seedlings probably because of a lack of moisture. In saturated soils, however, seed burial caused a reduction in the rate of emergence even at 0.5 to 2 cm depths (Barrett and Wilson 1983). Seedling emergence was significantly greater at all burial depths (0, 0.5, 1.0, 2.0 cm) in var. *oryzicola* as compared to var. *crus-galli*. No seedling emergence in var. *crus-galli* occurred from 2 cm depth (Fig. 6). In *crus-galli*, 13% of the seeds flooded at 26°C produced seedlings which emerged from water after 4 weeks. At 18°C, emergence was delayed until after 5 weeks, when approximately twice the number of seedlings emerged that at 26°C. A single *crus-galli* seedling emerged at 10°C after 6 weeks of immersion (Fig. 7). Seedling emergence of var. *oryzicola* was also significantly greater than var. *crus-galli* from soil flooded to 9 and 18 cm water depth (Barrett, 1983).

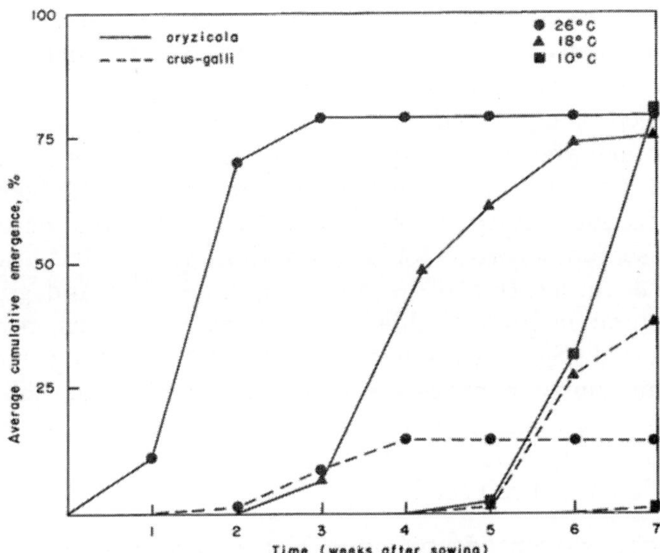

Fig. 7. Seedling emergence of *Echinochloa crus-galli* vars. *crus-galli* & *oryzicola* from soil flooded to a depth of 8 cm at three temperature regimes (10°, 18°, 26°C) in growth chambers.

Wevers and Staas-Ebregt (1980) reported that emergence was positively correlated with soil water content up to the field capacity but negatively correlated with sowing depth. They associate emergence with the energy reserves of the seed in relation to the product of mechanical soil resistance and sowing depth.

Only a small fraction of the seeds (variable between local populations) are capable of germination at dispersal or following a short period of storage (Honek and Martinkova, 1996; Martinkova and Honek, 1997). Dormancy in the rest of the seeds may be removed by prolonged exposure of dry seed at elevated temperature (after ripening) or exposure of moist seed at low temperature (stratification). In fields of the temperate zone where dispersed seed remains on the soil surface or is incorporated into the soil by tillage, dormancy is mostly terminated by stratification. Although dormancy release of some buried seeds may be completed by the autumn, in most seeds, dormancy termination takes place during winter and early spring (Martinkova and Honek, 2000). Under sparsely sown crops, the seedlings then emerge in late April to early May. The non-germinated seeds persist in the soil seed bank, often for many years (Burnside *et al.*, 1996), and may germinate in response to complex stimuli (Yoshioka *et al.,* 1998). Germination occurs at temperatures between 13 and 40°C (Brod, 1968; Maun and Barrett, 1986; Rahn *et al.*, 1968; Roche and Muzik, 1964; Torma and Hodi, 2002).

The 1-yr-old seeds have just started afterripening or lost dormancy. In the 8-yr-old seeds, afterripening increased germination percentage for ca. 2.5 yr (Martinkova and Honek, 2004) and buried seeds each year underwent dormancy/nondormancy cycles with maxima in April to June and minima in September to November (Honek *et al.*, 1999; Martinkova and Honek, 2000).

Martinkova *et al.* (2006) investigated the germination in 1-yr and 8-yr-old seed lots of barnyardgrass stored dry at 25°C (after ripening) and buried in the field (stratification). The after ripened seed lost dormancy within 2 yr after dispersal, whereas buried seed passed through annual dormancy/nondormancy cycles. Germination percentage and rate in 8-yr-old after ripened and each of the stratified seed lots varied with temperature. Optimum temperatures for germination were between 27 and 31°C, and the range of adequate temperatures increased with seed age. A common base temperature for germination was 11.7°C. In 1-yr-old after ripened material, the proportion of germinating seed (5%) and germination rate were similar at all temperatures (Table 3). Barnyardgrass thus, revealed a plastic strategy of germination. Stratification during the first winter removed dormancy and allowed germination at a narrow range of temperatures. This constraint on optimum germination temperature decreased with increased seed age. Seedling emergence, thus, may vary according to whether seed population originates from the previous year or from old soil seed banks.

Table 3. Germination percentage (mean ± SE) at five constant temperatures in seeds of different age and storage conditions

Temperature	Seed			
	1-yr old		8-yr old	
	Afterripened	Buried	Afterripened	Buried
C				
17	2.6 ± 0.8	0.4 ± 0.3	35.2 ± 4.8	5.2 ± 2.8
20	2.9 ± 0.7	9.0 ± 1.7	49.4 ± 4.9	36.4 ± 3.1
25	1.0 ± 0.4	22.0 ± 2.8	68.2 ± 4.4	64.6 ± 1.8
30	2.0 ± 0.6	56.6 ± 4.5	65.6 ± 3.3	60.2 ± 2.3
35	3.4 ± 0.6	32.8 ± 9.2	42.4 ± 2.5	58.4 ± 3.0
Mean	2.4 ± 0.4	24.2 ± 9.8	52.2 ± 6.4	45.0 ± 11.1

REFERENCES

Arai, M and M. Miyahara (1963). Physiological and ecological studies on barnyardgrass (*Echinochloa crus-galli* Beauv. var. *oryzicola* Ohwi). VI. On the elongation of plumule through soils after germination. *Proc. Crop Sci. Soc. Jpn.* 31:367-370.

Aslam, Z., M. Salim, R.H. Quresh and G.R. Sandhu (1987). Salt tolerance of *Echinochloa crus-galli. Biol. Plant.* 29: 66-69

Averkin, G.V. (1987a). The seed germination of some weed species as affected by the temperature in the Novosibirsk region. *Byull. Vses. Naucho-Issled. Inst. Rastenievod.* 78: 67-68.

Averkin, G.V. (1987b). The germination of weed seeds in different layers of the ploughed horizon in relation to soil density. *Byull. Vses. Naucho-Issled. Inst. Rastenievod.* 78: 73-76.

Barrett, S.C.H. (1982). Genetic variation in weeds. In: Biological control of weeds with plant pathogens. R. Charudattan & H. Walker (Eds). John Wiley and Sons Inc., New York, pp.73-98.

Barrett, S.C.H. (1983). Crop mimicry in weeds. *Econ. Bot.*, 37: 255-282.

Barrett, S.C.H. and B.F. Wilson (1983). Colonizing ability in the *Echinochloa crus-galli* complex (barnyard grass). II. Seed biology. *Can. J. Bot.*, 61: 556-562.

Barrett, S.C.H. and D. Seaman (1980). The weed flora of Californian rice fields. *Aquat. Bot.*, 9: 351-376.

Boyd, N.S. and R.C. Van Acker (2004). Seed germination of common weed species as affected by oxygen concentration, light and osmotic potential. *Weed Sci.*, 52: 589-596.

Brod, G. (1968). Studies on the biology and ecology of barnyard grass, *Echinochloa crus-galli. Weed Res.*, 8: 115-127.

Burnside, O. C., R. G. Wilson, S. Weisberg and K. G. Hubbard (1996). Seed longevity of 41 weed species buried 17 years in eastern and western Nebraska. *Weed Sci.* 44:74–86.

Dawson, J.H. and V.F. Bruns (1962). Emergence of barnyardgrass, green foxtail and yellow foxtail seedlings from various soil depths. *Weeds*, 10: 136-139.

Dawson, J.H. and V.F. Bruns (1975). Longevity of barnyardgrass, green foxtail and yellow foxtail seeds in soil. *Weeds Sci.*, 23: 437-440.

De Vlaming, V. and V.W. Proctor (1968). Dispersal of aquatic organisms: viability of seeds recovered from the droppings of captive killdeer and mallard ducks. *Am. J. Bot.,* 55, 20–26.

Dong-Sheng Tang, Muhammad Hamayun, Abdul Latif K, Zabta Khan Shinwari, Yoon-Ha Kim, Sang-Mo Kang, Joon-Hee Lee, Chae-In Na, Yasmi Nawaz, Kee-Kyung Kang and In-Jung Lee (2010). Germination of some important weeds influenced by red light and nitrogenous compounds. *Pak. J. Bot.*, 42(6): 3739-3745.

Figuerola, J., L. Santamaria, A.J. Green, I. Luque, R. Alvarez and I. Charalambidou (2005). Endozoochorous dispersal of aquatic plants: does seed gut passage affect plant performance. *Am. J. Bot.* 92, 696–699.

Halevy, A.H., J. Shoub and D. Rakati-Aaylon (1964). The effects of storage temperature and growing conditions on intermediate size bulbs of wedgewood iris. *Israel J. Agr. Res.* 14:11-17.

Holm, L.G., D. Plucknett, J.V. Pancho and J.P. Herberger (1977). The world's worst weeds: distribution and biology. University Press Hawaii, Honolulu.

Honek, A. and Z. Martinkova (1996). Geograpic variation in seed dormancy among populations of *Echinochloa crus-galli. Oecologia.* 108: 419-423.

Honek, A., Z. Martinkova and V. Jarosik (1999). Annual cycles of germinability and differences between primary and secondary dormancy in buried seeds of *Echinochloa crus-gallis. Weed Res.*, 39: 69-79.

Horng, L.C. and L.S. Leu (1978). The effects of depth of burial on the germination of ten annual weed seeds. *Weed Sci.*, 26: 4-10.

Isikawa, S. (1957). Interaction of temperature and light in the germination of Nigella seeds. *Bot. Mag.* Tokyo, 70:264-275.

Jordan, J.L., L.S. Jordan and C.M. Jordan (1983). Effect of ultrafreezing to – 196°C on *Echinochloa crus-galli* seed coat, secretory epithelial structure and dormancy. *Phyton, Argentina.* 43: 9-19.

Martinkova, Z. and A. Honek (1997). Geographic variation in the rate of seed dormancy termination in barnyard grass, *Echinochloa crus-galli. Ochrana Rostlin.* 29: 251-257.

Martinkova, Z. and A. Honek (2000). Variation in seasonal cycles of germination of buried seeds of barnyardgrass, *Echinochloa crus-galli. J. Plant Dis. Prot. Sonderheft,* 17: 133-138.

Martinkova, Z. and A. Honek (2004). Seed afterripening in barnyardgrass *Echinochloa crus-galli. J. Plant Dis. Prot. Sonderheft.* 19: 100-106.

Martinkova, Z. A. Honek and J. Lukas (2006). Seed age and storage conditions influence germination of barnyard (*Echinochloa crus-galli*). *Weed Science*, 54: 298-304.

Maun, M. A. (1977). Response of seeds to dry heat. *Can. J. Plant Sci.*, 57: 485-490.

Maun, M. A. and S. C. H. Barrett (1986). The biology of Can. weeds, 77: *Echinochloa crus-galli* (L.) Beauv. *Can. J. Plant Sci.*, 66: 739-759.

Miyahara, M. (1975). On dormancy of the seeds of *Echinochloa crus-galli* Beauv. var. *oryzicola* Ohwi, a paddy field weed. *Jpn. Agric. Res. Q.*, 8: 194-198.

Moursi, M. A., T. Y. Rizk and H. R. El-Deepah (1979a). Weed seed germination responses to some chemical treatments. *Egypt. J. Agron.* 2: 197-209.

Moursi, M. A., T. Y. Rizk and H. R. El-Deepah (1979b). Weed seed germination responses to freezing and leaching treatments. *Egypt. J. Agron.* 2: 211-220.

Panggabean, G. (1976). Germination of *Echinochloa crus-galli* (L.) Beauv. seeds. *Weeds in Indonesia.* 5: 43-45.

Rahman, Marlis and Ungar, Irwin A. (1990). The effect of salinity on seed germination and seedling growth of *Echinochloa crusgalli. Ohio Journal of Science.* 90(1): 13-15.

Rahn, E.M., R.D. Sweet, J. Vengris and S. Duhn (1968). Life history studies as related to weed control in the Northeast. 5.-Barnyardgrass. *Agric. Exp. Sta. Univ. Delaware Bull.*, 368: 1-46.

Rizk, T.Y., M.T. Fayed and H.R. El-Deepah (1978). Effect of some promoters on weed seed germination. Faculty Agriculture Ain Shams University, Cairo, Egypt. *Research Bull.* no.818, 30 pp.

Roche, B.F., Jr. and T.J. Muzik (1964). Physiological study of *Echinochloa crus-galli* (L.) Beauv. and the response of its biotypes to sodium 2,2 dichloropropionate. *Agron. J.* 56: 155-160.

Santamaría, L., I. Charalambidou, J. Figuerola and A.J.Y. Green (2002). Effect of passage through duck gut on germination of fennel pondweed seeds. *Arch. Hydrobiol.* 156, 11–22.

Smith, R.J. and W.C. Shaw (1966). Weeds and their control in rice production. U.S. Government Printing Office, Washington, D.C.Agric. Handbook 292, 64 pp.

Sung, S.S., G.R. Leather and M.G. Hale (1987). Development and germination of barnyardgrass (*Echinochloa crus-galli*) seeds. *Weed Science*, 35: 211-215.

Takabayashi, M. and K. Nakayama (1981). The seasonal change in seed dormancy of main upland weeds. *Weed Res. Jpn.*, 26: 249-253.

Taylarson, R.B. and S.B. Hendricks (1972). Interaction of light and a temperature shift on seed germination. *Plant Physiol.*, 49:127-130.

Taylarson, R.B. and L. Dinola (1989). Increased phytochrome responsiveness and a high-temperature transition in barnyardgrass (*Echinochloa crus-galli*) seed dormancy. *Weed Sci.*, 37: 335-338.

Taylorson, R. B (1988). Anaesthetic enhancement of *Echinochloa crus-galli* (L.) Beauv. seed germination: possible membrance involvement. *J. Exper. Bot.*, 39: 50-58.

Toole, E.H., V.K. Toole, H.A. Borthwick and S.B. Hendricks (1955). Interaction of temperature and light in germination of seeds. *Plant Physiol.* 30: 473-478.

Toole, E.H., V.K. Toole, S.B. Hendricks and H.A. Borthwick (1957). Effect of temperature on germination of light-sensitive seeds. *Proc. Intl. Seed Test. Assn.*, 22:1-9.

Torma, M. and L. Hodi. (2002). Reproduction biology of some important monocot weeds in Hungary. *J. Plant Dis. Protect. Sonderheft.* 18: 191-196.

Vengris, J., E.A. Kacperska-Palacz, F.R. Shaw and W.H. Ziener. (1966). Growth and development of barnyardgrass in Massachusetts. *Weeds.* 14: 299-301.

Vicente, M., K. Maoronha, K. Silberschmidt and M. Meneghini. (1968). Successive reversion of the effect of temperature on germination of *Rumex obtusifolius* L. by far-red light. *Phyton* (Buenos Aires), 25: 11-13.

Watanabe, Y. (1977). Ecological studies on the germination and emergence of some summer annual weeds. 6. Effect of light on the germination of seeds wintered in soil. *Weed Res. Jpn.* 22: 80-83.

Watanabe, Y. and F. Hirokawa. (1978). Physiological and ecological studies in upland weeds in Hokkaido. *Res. Bull. Hokkaido National Agric. Exp. Sta.* pp. 17-77.

Wevers, J. D. A. and E. Staas-Ebregt. (1980). The influence of soil moisture, soil pore volume and sowing depth on the emergence of *Echinochloa crus-galli* (L.) P.B. *Proc. 6th Int. Coloquium Weed Ecol. Biol. and Syst.* Columaewrs, Montpellier, France. Vol. 95, p.101.

Wilson, R. (1979). Irrigation weed seed. *Farm, Ranch and Home Quarterly,* 26: 16-18.

Yamasue, Y., K. Sudo and K. Uedi (1977). Physiological studies on seed dormancy of barnyardgrass (*Echinochloa crus-galli* Beauv. var. *oryzicola* Ohwi). *Proc. 6th Asian Pacific Weed Sci. Soc. Conf. Indonesia.* 1: 42-51.

Yoshioka, T., S. Satoh and Y. Yamasue (1998). Effect of increased concentration of soil CO_2 on intermittent flushes of seed germination in *Echinochloa crus-galli* var. *crus-galli. Plant Cell Environ.,* 21: 1301-1306.

JUNGLE RICE (*Echinochloa colona*)

Common name : Awnless barnyard grass, jungle rice

Scientific name : Echinochloa colona

Tamil name : Kudiraivalipul

Hindi name : Jangli, jhangora, Jharwa, Jirio, Shama

Family : Poaceae

Origin and distribution : Jungle rice is of Indian origin. Now, it is widely distributed in tropics and subtropics, including South and Southeast Asia and tropical Africa.

Botany

Habitat : Jungle rice is adapted to full sunlight or partial shade and grows on loam, silt and clay soils. It grows in drains, low-lying grasslands and farmlands in both dry and marshy places. One of the most important weeds of upland rice under moist conditions. It occurs most commonly at low altitudes, but can extend up to about 2000m.

Habit : A slender annual grass, growing up to 60 cm, common in water logged situations, in rice fields and in irrigated lands (Fig. 1).

Stem : Stems are prostrate, ascending or erect and tend to branch.

Leaf : Leaf sheaths can be hairy at the nodes or hairless. Ligules (appendages at the base of the leaf blade) are more or less absent. The blades of the leaves are linear, tapering to the tip, 5-30 x 0.3-1 cm. Leaves are hairless.

Floral biology

Inflorescences : Inflorescences are erect and branched 4-15 cm long with branches up to 2.5 cm long. Spikelets are arranged in 4 irregular rows. The lowest floret of the spikelet has an awn up to 5 cm long. Inflorescence of awnless barnyard grass showing the spikelets arranged in four rows (Fig. 2).

Flower : Flowers bisexual, grouped together in a terminal spike or panicle, sessile, purple or brown, petals not visible (Fig. 3).

Flowering and pollination : Flowering begins 3-4 weeks after emergence occurs (Manidool, 1992) flowers are wind pollinated (Plants for a Future, 2002). *Echinochloa colona* blooms from February to October/November.

Fruit : Caryopsis 1.7-2 mm (0.067 - 0.078 inch) long (Fig. 4).

Fig. 1. Plant

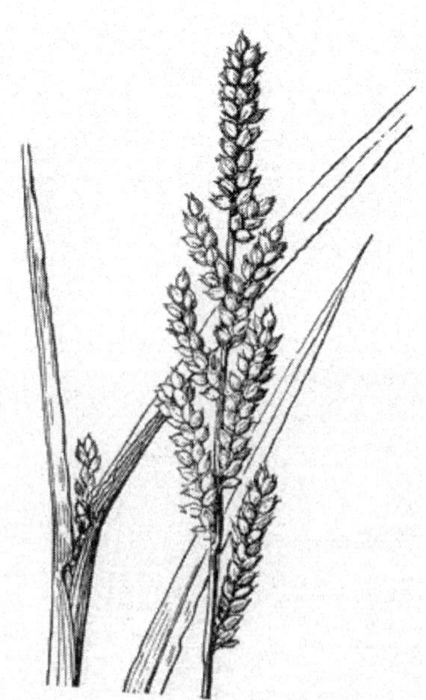

Fig. 2. Inflorescence

Fig. 3. Flower

Fig. 4. Fruit

Seed biology

Seed : Seeds are pale brownish, ovoid with longitudinal ribs, hairy with a long or short awn (Fig. 5).

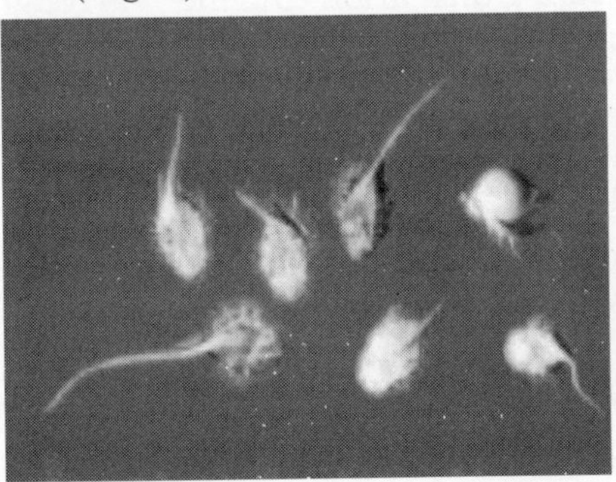

Fig. 5. Seed

Propagation : Echinochloa colona is often propagated by rooted tillers (Holm *et al.*, 1991; Manidool, 1992). It is a fast-growing grass, germinating quickly when rains occur (Manidool, 1992).

Seed dispersal : Echinochloa colona seeds can be dispersed after ingestion by animals, specifically cattle, often germinating in the dung. Also, ducks have been observed to eat the seed of *Echinochloa colona* (Kramer and Euliss, 1986). The seeds also spread by farm machinery, in crop seed or transplants, in irrigation canals, and in mud on the feet, fur, feathers, and skin of rodents, birds and larger animals, including humans (Holm *et al.*, 1991).

Seed dormancy and germination : Freshly harvested seeds of junglerice have been reported to have primary dormancy (Chauhan and Johnson, 2009). Seed dormancy is absent in this weed. *Echinochloa colona* have a short period of dormancy, with the dormancy disappearing in less than 8 weeks of dry storage after harvesting occurs (Holm *et al.*, 1991).

Light is required for *Echinochloa colona* to achieve its best germination (Holm *et al.*, 1991). In Australia, *Echinochloa colona* germination was prevented when rice fields were flooded with 0.4-0.6 in. (1.0-1.5 cm) of water (Holm *et al.*, 1991).

Enhanced germination was noticed due to afterripened at 23°C or treated with high temperatures, acid scarification, or punctured with a scalpel. Puncturing and scarification was not related to removal of a restriction to water entry. The effectiveness of puncturing in overcoming dormancy depended on moisture content of the spikelets at the time of puncturing. Above 18% moisture content, germination percentage was low regardless of the dormancy-releasing treatment.

Life cycle : Mature plants of *E. colona* resemble the rice plant but can grow higher and thus compete for sunlight. They mature easily, producing about 42,000 seeds per plant. According to Chauhan and Johnson (2010), an yield reduction of 84-93% was observed.

Effects : Yield loss from this weed can reach 43%, especially in direct seeded rice. The weed is a host of diseases such as tungro and rice yellow dwarf.

Control measures : Cultural control: Cultivation during early growth can control the weed Manual control is difficult in the early stages.

Biological control: In Japan, the pathogen *Exserohilum monoceras* is being evaluated as a bioherbicide for control of *Echinochloa* species in rice. In the Philippines, *E. monoceras* killed seedlings of *E. colona,* but did not affect rice.

Chemical control: E. colona can be controlled by pre-emergence application of butachlor @ 1.5 kg a.i/ha, anilophos @ 400 g/ha, pretilachlor @ 1.0 kg/ha and pendimethalin @ 1.5 kg/ha.

Seed bank : Great variability in seed bank size with a maximum of 215,000 seeds m^{-2}. Barnyardgrass seed bank was found only in 7% of the cotton fields, while it was 22 and 20%, respectively, for rice and soybean. Barnyardgrass exhibited an extended period of emergence with days to 100% emergence (Muthukumar *et al.*, 2011).

Past work done on Seed Science and Technology

Effect on water stress on production

Germination of junglerice is generally favored by a moist environment (Chauhan and Johnson, 2009). Chauhan and Johnson (2010) reported that seeds (7.400 to 7,800) produced per junglerice plant were statistically similar at 50 to 100% field capacity (FC), indicating possible high water use efficiency in this C_4 species. With further increase in water stress (25 and 12.5% of FC), both plant biomass and seed production were sharply reduced, but junglerice was still able to produce an average of 1,680 seeds plant at 12.5% of FC (Fig. 6).

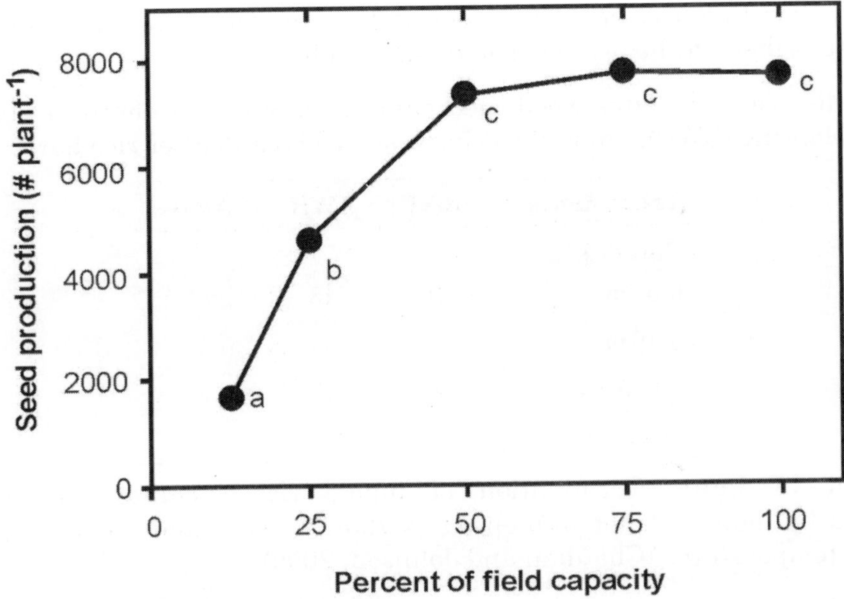

Fig. 6. Effect of soil moisture (percentage of field capacity) on seed production of junglerice (study 1). Means with the same letter are not significantly different based on an LSD at P = 0.05

Germination of junglerice seeds from plants that were water stressed during seed maturation did not differ from that of plants that were not stressed. Irrespective of the treatment and study, germination was greater than 80%. High temperature during seed maturation might have been responsible for the lack of differences in germination between seeds of stressed and non-stressed plants (Fig. 7).

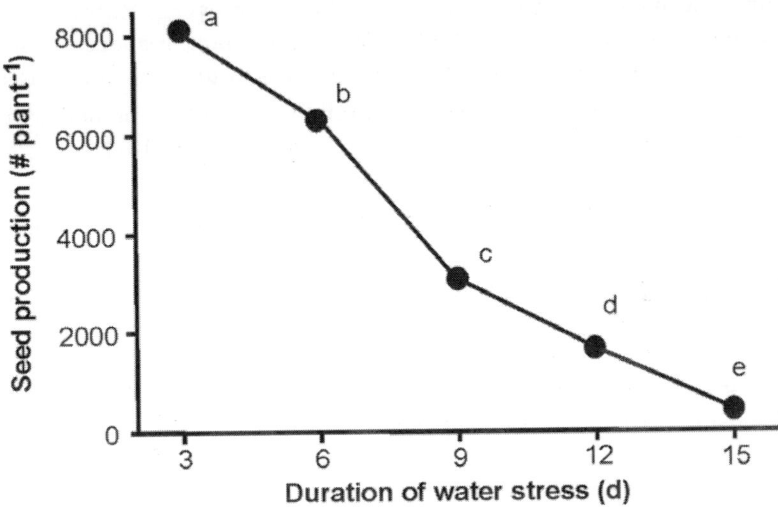

Fig. 7. Effect of duration of water stress (in days) on seed production of junglerice. Means with the same letter are not significantly different based on an LSD at P = 0.05.

Jagat Devi Ranjit *et al.* (2007) showed that the weed seed accumulation of *E. colona* was highest in the 5-10 cm soil depth (Table 1).

Table 1. Response of sedge weed seed bank density to soil depth at before wheat planting (BWP), after wheat harvest (AWH) and after rice harvest (ARH).

Treatments	BWP	AWH	ARH
Soil depth			
0-5 cm	10	43	76
5-10 cm	5	34	1
10-15 cm	9	15	8
15-20 cm	40	4	1

Effect of light: Germination of jungle rice (*Echinochloa colona*) was stimulated by light, but not influenced by (35/25, 30/20 and 25/15°C) alternating day/night temperatures (Chauhan and Johnson, 2009).

REFERENCES

Chauhan B.S. and D.E. Johnson (2009). Seed Germination Ecology of Junglerice (*Echinochloa colona*): A Major Weed of Rice. *Weed Science*: May 2009, 57(3), pp. 235-240.

Chauhan, B.S. and D.E. Johnson (2010). Relative importance of shoot and root competition in dry-seeded rice growing with junglerice (*Echinochloa colona*) and Ludwigia (*Ludwigia hyssopifolia*). *Weed Science*, 58: 295-299.

Holm, L.G., D.L. Plunknett, J.V. Pancho and J.P. Herberger (1991). The world's worst weeds. Distribution and biology. Krieger Publishing Company, Malabar, Florida. 609 pp.

Jagat Devi Ranjit, Rungsit Suwanketnikom, Sombat Chinawong, Sutevee Suprakarn, Isara Sooksathan and Sunanta Juntakool (2007). Weed seed bank response to soil depth, tillage and weed management in the mid hill ecology, *Kasetsart J. (Nat. Sci.),* 41: 17-33.

Kramer, G.W. and N.H. Euliss, Jr. (1986). Winter foods of black-bellied whistling-ducks in northwestern Mexico. *Journal-of-Wildlife-Management* 50(3):413-416.

Manidool, C. (1992). *Echinochloa colona* (L.) Link . Record from Proseabase. Mannetje, L.'t and Jones, R.M. (Eds). PROSEA (Plant Resources of South-East Asia) Foundation, Bogor, Indonesia.

Muthukumar V. Bagavathiannan, Jason K. Norsworthy, Kenneth L. Smith and Nilda Burgos (2011). Seedbank size and emergence pattern of barnyardgrass (*Echinochloa crus-galli*) in Arkansas. *Weed Science*, 59(3):359-365.

Plants for a Future (2002). *Echinochloa colona*. 131 Spencer Place, Leeds, LS7 4DU, England. Website: http://www.ece.leeds.ac.uk/pfaf/index.html.

BHRINGRAJ or BHENGRA (*Eclipta alba* (L))

Common name : Bhringraj or bhengra

Botanical name : Eclipta alba (L.) Hassk. (syn. *Eclipta prostrata* L.)

Tamil : Karisalankanni

Hindi : Bhangara

Family : Compositae

Origin and distribution : Native to U.S. and tropical areas worldwide. It is widely distributed throughout India, China, Thailand and Brazil.

Botany

Habitat : In paddy growing areas of India, it occur as common weed. In many parts of India, it is grown commercially as a medicinal crop.

Habit : An annual herb (Fig. 1).

Stem : 50 cm tall, single from base but with many spreading branches, from fibrous roots, strigose, herbaceous, subsucculent, erect or ascending, often rooting at lowest nodes, terete, purplish in strong sun.

Leaf : Opposite, sessile, lanceolate, shallow serrate, 13 cm long, 3 cm broad, strigose, acuminate.

Floral biology

Flower : Axillary and terminal clusters of 1-3 flower heads. Peduncles 2 cm long, densely antrorse strigose.

Involucral bracts - Typically a single series of 8-10 unequal phyllaries. Phyllaries to 7 mm long, 2.3 mm broad, antrorse strigose externally, glabrous internally, green, acute (Fig. 2).

Ray flowers - Fertile, pistillate. Ligules 2 mm long, 5 mm broad, notched or rounded at apex.

Disc flowers - Disk 4-10 mm broad. Corolla tube 0.9 mm long, 4-lobed, glabrous, white. Lobes acute, 0.3 mm long. Stamens 4. Anthers purple, 0.5-0.6 mm long, included, connate around style. Style whitish at apex (stigmas). Receptacle flat. Chaff to -3 mm long, translucent yellow, antrorsely barbellate (Fig. 3).

Flowering and pollination : Under Indian conditions, the flowering is throughout year and fruiting freely. In Sri Lanka, *Eclipta prostrata* flowers from November to July, in the Philippines all year round.

Fruit : Achene 4-angled, white, 2.5mm long, warty glabrous. Pappus absent to a minute crown. Fruits are many seeded (Fig. 4).

Fig. 1. Plant

Fig. 2. Flower

Fig. 3. Disc Flower

Fig. 4. Fruit

Seed biology

Seed : Seeds are black and resemble cumin seeds.

Propagation : Through seeds

Life cycle : Flowering starts about two months after the seedlings appear and the plant completes its life in about 3 to 4 months. The reproductive capacity value for *E. alba* works out to 564.

Seed dispersal : Ridley (1930) mentions the following method of dispersal of seeds for the species. The surface of the achene is papillose and the fruit is sticky due to mucilagenous exudation. Thus the dispersal is possible by its adherence to the plumage and feet of birds. The human agency is also helping its dispersal either by adhesion to feet or cloths or to importation with other plants. He further mentions of seed drifts in water during floods though not quit satisfactorily as in case of buoyant seeds.

Seed dormancy and germination : According to Dhawan Rupa (2007), *Eclipta alba* seeds germinated immediately after harvest and showed no dormancy. Light is essential for germination. 25°C was seen to be the optimal temperature for germination. Seeds did not germinate beyond a water stress of −0.4 MPa. Low levels of salinity (0–120 mM NaCl) did not affect germination. Acidic pH completely inhibited germination. Only seeds at the surface layers of soil germinated. Those at a depth of even 2.5 cm did not germinate. Seed rain/m^2 was observed to be 13, 63, 534.

Seeds were strongly photoblastic with none germinating in the dark. Seeds germinated in light intensities between 6 and 100% of full light. Seeds germinated 78 to 88% in water potentials of 0.0, -0.1, and -0.2 MPa. In distilled water (pH 5.5) and a buffered solution pH 6 and 7, seeds germinated 85 to 89%; however, some seeds germinated in a buffered solution pH range of 5 to 8.

According to Ferreira *et al.* (2001), the optimum temperature for germination was 30°C and light was must for germination.

Seed germination was completely inhibited in the dark, whereas in the light/dark it was 76, 93, and 87% at 25/15, 30/20, and 35/25°C alternating day/night temperatures, respectively. Germination was greater than 80% up to a temperature of 140°C, when seed were placed in an oven for 5 min followed by incubation at 30/20°C for 14 d, but declined progressively with a further increase in exposure temperature with no germination at 200°C. Seed germination was tolerant of salt stress, but highly sensitive to water stress. Seed germinated (87 to 93%) over a pH range of 4 to 10. Seedling emergence was greatest (83%) for the seed placed on the soil surface, but declined thereafter and no seedlings emerged from a depth of 0.5 cm (Chauhan and Johnson, 2008).

Ramakrishnan (1960) reported that the seeds are found to have no dormancy period, but the germination percentage increases with storage (Table 1).

Table 1. Germination of seeds of *E. alba* after dry storage for different periods

Month of storage	Percentage germination
0	4
2	16
5	62
12	62

The maximum germination of 82 per cent is obtained in continuous light which may be partly due to the warm atmosphere around the electric bulb and also due to the direct effect of light itself. Very poor germination is obtained in the seeds that are kept in continuous darkness. Fresh seeds buried under mud for four months gave a germination of 72 per cent which is greater than that obtained from dry stored seeds.

Uses : It is an old established Hindu medicine. The plant has a better sharp dry taste, good for the complexion of the hair, the eyes, the teeth; cures inflammation, eye diseases, 'kapha' and 'vata', bronchitis, asthma, leucoderma, anemia, diseases of heart and the skin, itching, night blindness, syphilis, used to prevent abortion, miscarriage and uterine pain after delivery.

REFERENCES

Chauhan B.S. and D.E. Johnson (2008). Influence of environmental factors on seed germination and seedling emergence of Eclipta (*Eclipta prostrata*) in a tropical environment. *Weed Science,* 56(3):383-388.

Dhawan Rupa, S. (2007). Germination potential and growth behaviour of *Eclipta alba. Indian Journal of Weed Science,* 39(1&2): 116-119.

Ferreira, A.G., B. Cassol, S.G.T. Da Rossa, T.S. Da Silveira, A.L. Stival and A.A. Silva (2001). Germination of seeds of *Asteraceae* native of Rio Grande do Sul, Brazil. - *Acta Bot. Bras.* 15: 231-242.

Ramakrishnan, P.S. (1960). Ecology of Eclipta alba hassk. *Proc. Nat. Inst. Sci. India,* 26, B(4): 191-204.

Ridley, H.N. (1930). The dispersal of plants throughout the world, L. Reeve and Co., Ltd., Ashford, Kent.

WATER HYACINTH (*Eichhorinia crassipes*)

Common name : Water hyacinth, Floating Water hyacinth, Water-orchid

Scientific name : Eichhorinia crassipes

Tamil : Akasathamarai

Hindi : Jal kumbhi

Family : Pontederiaceae

Origin and distribution : Native to the northern neotropics of South America. It is distributed currently through North America, South America, eastern Africa and Asia.

Botany

Habitat : It is obnoxious aquatic weed distributed all over tropical desert to subtropical or warm temperate desert to rainforest zones. It is floating freely on the surface of free water. Water hyacinths do not grow when the average salinity is greater than 15% that of sea water. In brackish water, its leaves show epinasty and chlorosis, and eventually die.

Habit : Free floating perennial water plant (Fig. 1).

Leaf : Blade is generally oval to round and usually not greater than 10 cm width; leaf stem is somewhat to completely swollen, filled with spongy parenchyma tissue, 1-1.5 in (2-3 cm) diameter, generally longer than the blade.

Inflorescence : The inflorescence is a loose terminal spike with showy light-blue to violet flowers (flowers occasionally white).

Floral biology

Flower : Flowers are 4 to 7 cm across, funnel-shaped, light bluish-purple or dark blue with a yellow centre and have six distinct petals. The upper petal is darker purple with a yellow mark in the centre. Flowers can self-fertilise and are formed on upright stems with between 3 and 35 (but commonly 8) flowers on each spike (Fig. 2).

Fig. 1. Plant

Fig. 2. Flower

Flowering and pollination : Flowering (i.e., sexual reproduction) occurs in mid-summer and early fall. The flower stalks bend back into the water once they are pollinated (it is thought to be self-pollinated) and release the seeds.

Fruit : A capsule, usually surrounded by the remains of dried flowers, has 50 seeds per capsule (Fig. 3).

Seed biology

Seed : Seeds are 1 to 1.5 mm long, roughly egg-shaped with ridges from end to end (Fig. 4).

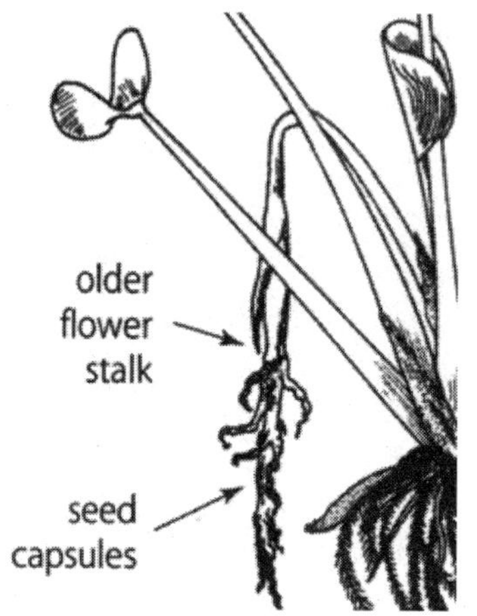

Fig. 3. Fruit

Fig. 4. Seed

Propagation : By seeds and vegetative means. Seeds sink to the bottom and can remain viable in sediments for several years. Upon reflooding, the seeds often germinate and renew the growth cycle.

Vegetative reproduction occurs from late spring through fall. Parts of the stem may break off at the water surface to form independent plants called daughter plants (Penfound and Earle, 1948). These daughter plants are capable of producing additional reproductive stem segments within weeks.

Life cycle: Water hyacinth reproduces from seeds and horizontal stems. Flowers open for only one or two days from mid to late summer before beginning to wither. When all the flowers on a spike have withered, the stalk gradually bends into the water and after two to three weeks the seeds are released and sink.

Autumn and winter frosts cause the leaves to die off but the crowns are able to overwinter. These will commence new growth in the following spring along with the germination of seeds.

Seed dispersal : Water hyacinth spreads through fragmentation of established plants and may resprout from rhizomes or germinate from seeds (Penfound and Earle 1948). Dispersal also occurs by water-borne seeds and by seeds that stick to the feet of birds. Migratory birds may be important in long-distance dispersal.

Seed dormancy and germination : Seeds require warm, shallow water and high light intensity for germination. Wetting, drying and re-wetting gave quickest and complete germination (Obeid and Tag el Seed, 1976).

Seed longevity and burial : Seeds are long lived and may survive in mud for 20 years. After 2 years dry storage, 78% of the seeds were still viable (Obeid and Tag el Seed, 1976). Soil seed density varied between 0 and 2534 seeds/m^2 but did not differ significantly between the type of water body (impoundment vs. river) or the history of control. Average germination was 54.17% with very fast velocities (Vigour index = 36.66) and maximum germination around three days.

Uses : Volatile fatty acids (VFAs) found in water hyacinth can be used as feed supplement in slurry biogas digesters and solid-feed digesters to generate fuels.

Sewage treatment using water hyacinth. Spent water hyacinth biomass for growing mushrooms. Spent mushroom compost for vermicomposting.

The plant is used as a carotene-rich table vegetable in Taiwan. Javanese sometimes cook and eat the green parts and inflorescence.

In Kedah (Java), the flowers are used for medicating the skin of horses. The species is a "tonic."

In East Africa, water hyacinth from Lake Victoria is used to make furniture, handbags and rope.

The plant is also used as animal feed and organic fertilizer although there is controversy stemming from the high alkaline pH value of the fertilizer.

Control measures : Mechanical control

Water hyacinth removal by hand or machine is a practical control method.

Biological control : Weevils, such as *Neochetina eichhorniae* Warner, and *N. bruchi* Hustache (Coleoptera : Curculionidae), and a moth, *Sameodes albiguttalis* Warren (Lepidoptera : Pyralidae) can be used to control this weed. The fungus, *Cercospora rodmanii* Conway (Fungi Imperfecti : Moniliales), can also be used (Conway, 1976).

Herbicide control: Water hyacinth can be controlled using glyphosate as a foliar spray (formulated as RodeoÂ®) and copper complexes used only as a foliar spray.

REFERENCES

Conway, K.E. (1976). *Cercospora rodmanii*; a new pathogen of waterhyacinth with biological control potential. *Can. J. Bot.*, 54: 1079-1083.

Obeid, M. and M. Tag el Seed (1976). Factors affecting dormancy and germination of seeds of *Eichhornia crassipes* (Mart.) Solms. from the Nile. *Weed Res., UK*, 16: 71-80.

Penfound, W.T. and T.T. Earle (1948). The biology of the water hyacinth. Ecol. Monogr. 18:449-72.

HOORAH GRASS (*Fimbristylis miliacea*)

Common name : Grass-like fimbristylis, lesser frimbrstylis, hoorah grass.

Botanical name : Fimbristylis miliacea (L.) Vahl

Tamil name : Poong korai

Hindi name : Ghueen, Dilli

Family : Cyperaceae

Origin and distribution : Native of Tropical America. Distributed all around Asia and Pacific Islands.

Botany

Habitat : It grows in loamy soils, red sandstone, quartzite and sandy soils, generally at low elevations. It appears to prefer open, wet habitats such as the edge of pools and in drainage lines, creek beds, clay pans, damp woodland, wet flats and grassy places.

Habit : Annual tufted erect sedge (Fig. 1).

Fig. 1. Plant

Root : Fibrous root system.

Stem : The culms are slender, 40 to 60 cm tall; their transversal section is four-angled and somewhat flattened.

Leaf : Linear, threadlike and stiff, two ranked, 1.5 to 2.5 mm wide, up to 40cm long, basal leaves are half as long as culm. The leaf bract is shorther than inflorescence.

Floral biology

Flower : Compound umbel rather lax and diffuse, 6 to 10 cm long, globose or subglobose; each umbel 2.5 to 4 mm long, 1.5 to 2 mm wide, round or acute at apex, reddish brown, the lower scales fall early. Stigmas three branched, rarely in a few flowers two branched; anthers yellow; glumes ovate, brown, about 1mm long, spirally arranged, membranous, obtuse, the green midvein or keel broad (Fig. 2).

Fig. 2. Inflorescence, spikelet, flower, glume (Lateral and dorsal view)

Flowering and fruiting : Flowering and fruiting throughout the year.

Fruit : Achene white, yellowish, less than half the length of the glume, three angled, biconvex, broadest above the middle, very finely warty, somewhat sugarcoated (Fig. 3).

Fig. 3. Achene

Seed biology

Seed : The production of empty seeds is common in non-domesticated species of the Cyperaceae. Empty seeds may have a normal seed coat, but an absent or under-developed embryo or limited storage tissue (Fig. 4).

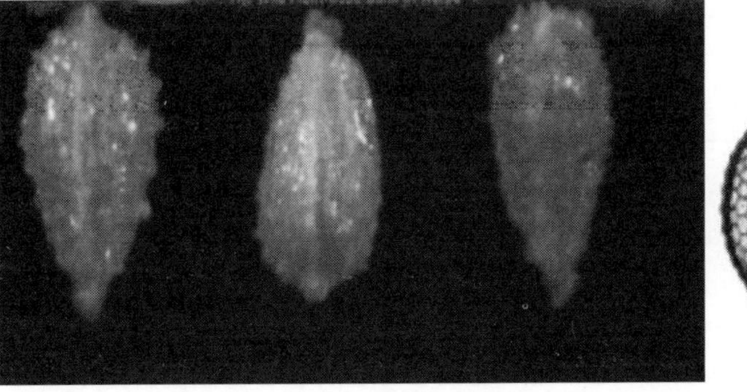

Fig. 4. Seed

Propagation : Through seeds and the seeds tend to germinate throughout the growing season.

Seed production : During the cropping season (4 month period), each *F. miliacea* plant produced on average of 2.3 tillers/plant and a total of 134 inflorescences, with 84 inflorescences/plant ripening within this period. Each inflorescence comprised of 48 spikelets with 511 seeds and matured after 3 weeks of emergence. Total seeds/plant and 1000 seed weight were 42,275 and 0.035 g, respectively Begum (2006).

Seed dispersal : Seeds are dispersed by cattle as the seed survives passage through the gut (Holm *et al.*, 1977). Human mediated dispersal is possible as a result of transport of seeds in contaminated machinery, produce and soil.

Seed dormancy and germination : Seeds sown at 0.5 cm depth had significantly reduced emergence (13%) and no seedlings emerged from 1.0 cm soil depth. For flooding durations of 7, 14 and 21 days, per cent emergence and dry matter production were higher under saturated conditions than at 5 and 10 cm flooding depths. A flooding duration of 14 days or more and in flooding depth up to 10 cm showed a clear trend in reduced emergence and dry matter production of *Fimbristylis miliacea.*

Seed bank : Soil seed bank of *F. miliacea* was the most abundant with 66.07% of total seed reserve (equivalent to 750.84 million seeds/ha), of which 55% emerged in soils incubated in trays (411.48 million seedlings/ha), 8% germinated in incubation tests in petridishes (61.59 million seedlings/ha), while 37% or the equivalent of 277.77 million seeds/ha remained dormant or dead. *Fimbristylis miliacea* exhibited high emergence (44.38%) from surface seeding (Begum, 2006).

The estimated weed seed bank in the top 5 cm of soil ranged from 52.1 to 167 × 10^3 seeds m^{-2} (overall mean of 8.5 × 10^3 seeds m^{-2}) and contained a high proportion (86%) of sedge species, such as *Fimbristylis miliacea* L. and *Cyperus difformis.* No clear difference was found in the weed seed banks between the irrigated fields that were located close to (upstream) and distant from (downstream) the water source or between the irrigated and rain-fed lowland fields, but the weed scores were larger in the rain-fed fields and the downstream fields within the irrigated area (Kamoshita *et al.,* 2010).

Uses : Some species are an important source of fiber for making mats, hats and baskets. The plant roots are given in dysentery by a tribe for central India. The plant is used as cattle fodder.

In the Philippines, it is used as a medicine to treat enlarged spleen. It is also ploughed in as a green manure in rice fields.

Agricultural importance : It is one of the most harmful weeds in rice fields.

Control measures : Cultural control: Keeping a layer of water (15 cm) on rice in Surinam is said to suppress germination of *F. miliacea* completely.

Chemical control : Pre-emergence application of pretilachlor @ 1.0 kg/ha or post-emergence application of 2, 4-D @ 500 /ha or almix @ 4 g/ha.

REFERENCES

Holm, L.G., D.L. Pluknett, J.V. Pancho and J.P. Herberger (1977). The World's worst weeds: Distribution and biology. *The University Press of Hawaii*. Honolulu. p. 273-279.

Begum, M. (2006). Biology and management of *Fimbristylis miliacea* (L.) Vahl. Ph.D. Thesis, Universiti Putra Malaysia, Serdang Darul Ehsan, Malaysia.

Kamoshita, A., H. Ikeda, J. Yamagishi and M. Ouk (2010). Ecophysiological study on weed seed banks and weeds in Cambodian paddy fields with contrasting water availability. *Weed Biol Manag.,* 10: 261–272.

COGON GRASS (*Imperata cylindrica*)

Common name : Cogon grass, satintail, blady grass

Botanical name : *Imperata cylindrica* (L.) Raeusch.

Tamil name : Dharba

Hindi name : Uloo, Sirhu, Dabh, Khans

Family : Gramineae or Poaceae

Origin and distribution : A native to Southeast Asia .It is widely distributed in Africa, Australia, southern Asia, and the Pacific Islands, and less extensively distributed, or a less serious problem, in southern Europe, the Mediterranean, the Middle East, Argentina, Chile, Colombia, the Caribbean, and the southeastern United States. It has been found at latitudes from 45°N (Japan) to 45°S (New Zealand), and from sea level to over 2,000 m elevation (Holm *et al.,* 1977).

Botany

Habitat : It is best suited for wet areas, but invaded a variety of habitats, from highly xeric uplands to fully shaded mesic sites. Sandhills, flatwoods, hardwood hammocks, sand dunes, grasslands, river margins, swamps, scrub, and wet pine savanna. It alters the structure and function of invaded communities (Holm *et al.,* 1977; Lippincott, 1997). It is tolerant of wide variations in soil fertility, organic matter, and moisture, it grows best in relatively acidic soils (pH 4.7) (Hubbard, 1944; Wilcut *et al.,* 1988a).

Habit : Rhizomatous, perennial grass (Fig. 1).

Root : It has tough, branched, whitish, creeping rhizomes. Most rhizomes are found in the upper 40 cm of the soil profile, but they can grow to depths of more than one metre.

Fig. 1. Plant.

Stem : More than 80% of shoots originate from the rhizomes less than 15 cm below the soil surface.

Leaf : It has narrow, rigid leaf-blades. Lower leaf-sheaths bearded at the mouth, upper usually glabrous; blades glabrous or hairy on the lower part, up to 100 cm long, often less, usually 3-10 mm wide, expanded.

Floral biology

Inflorescence : The inflorescence is a cylindrical, spike-like panicle, 3-60 cm long and 0.5-2.5 cm wide, consisting of many spikelets surrounded by 10-15 mm long hairs which give the inflorescence a silky, white appearance (Fig. 2).

Fig. 2. Inflorescence

Flower: It is monoecious (male and female reproductive organs on the same individual) and the flowers are complete (having a pair of female stamens and a pair of male stigma present in each flower), but the species is an obligate out-crosser (Fig. 3).

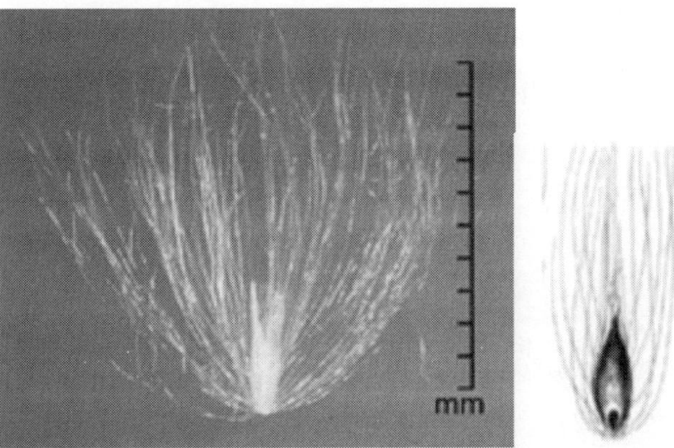

Fig. 3. Flower, spikelet with bracts removed to show caryopsis

Flowering and pollination : It is highly variable and highly productive, but flowering is often sporadic, ranging from none to frequent flowering within and among populations. The flowers production is accelerated by mowing disturbance. Disturbances including nitrogen amendment, slashing, burning, defoliation, and grazing may trigger cogon grass flowering. Cogon grass flowering is not photoperiod-dependent. It is pollinated by wind. Incapable of self-pollination (Gabel, 1982), *I. cylindrica* produces viable seed only when cross-pollinated (McDonald *et al.*, 1996), and the success rate of outcrossing is low (Shilling *et al.*, 1997). It produces as many as 3,000 seeds per plant (Holm *et al.*, 1977).

Shilling *et al.* (1997) determined that cogongrass requires cross-pollination to produce viable caryopses; however, this species is still considered a prolific seed producer capable of yielding an average of 700 seeds per inflorescence with high caryopsis viability (Shilling *et al.* 1997).

Fruit : Fruit are usually seed-like grains (Fig. 4).

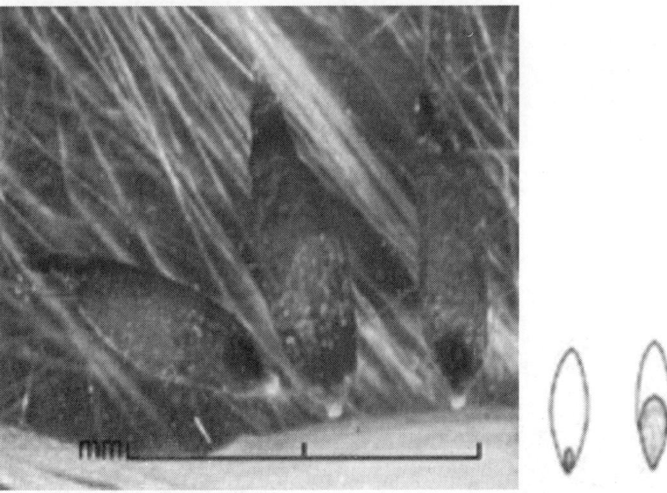

Fig. 4. Caryopsis in ventral view (left and right) and dorsal view (middle)

Seed biology

Seed : The seeds are small (1-1.3 mm long) (Fig. 5).

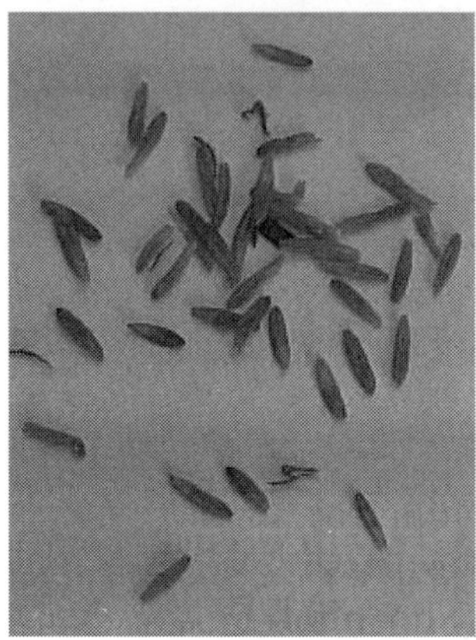

Fig. 5. Seed

Propagation : It reproduces from seed, rhizome expansion and rhizome fragments. However, long-distance dispersal and invasion of new sites is by seed (Shilling *et al.*, 1997).

Seed dispersal : Disperses over long distances into a variety of habitats by wind borne seeds (Bryson and Carter, 1993; in Langeland and Burks 1998); distances dispersed range from an average of 15m to 100m (Holm *et al.*, 1977; Shilling *et al.*, 1997, in Van Loan *et al.*, 2002).

The hairy seeds tend to catch on nearby vegetation. The small seeds are attached to plumes of long hairlike projections to facilitate wind dispersal and as many as 3,000 seeds are produced per plant (FIPR, 1997) (Fig. 6).

The transport of rhizomes and seeds on equipment or in contaminated soil are considered important modes of spread (McDonald, 2004, in Yager and Smith 2009).

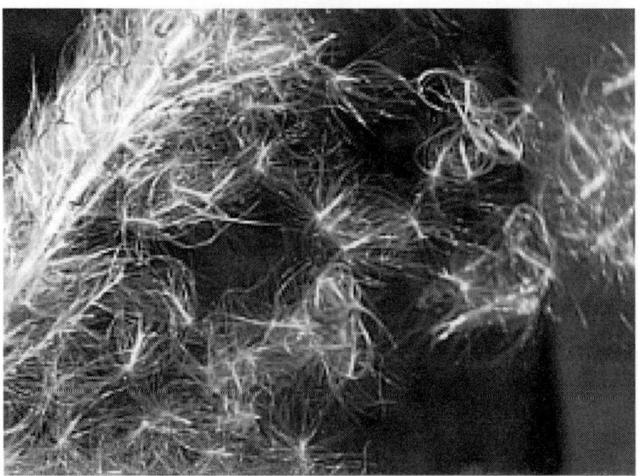

Fig. 6. Wind dispersal

Seed dormancy and germination : Because cogongrass lacks seed dormancy mechanisms (Shilling *et al.*, 1997), suitable site conditions during the period of seed dispersal and early seedling establishment are critical for successful invasion. The seeds are not dormant and do not require stratification. They germinate 1 to 4 weeks after ripening. Seeds began germinating within 7 days of harvest, with 94% germination by day 14. Germination of the seed in the dark increased from 9% at 20°C, to 55% at 30°C, declined somewhat at 35°C, and was about 70% in the light with alternate 12-hour periods at 20 and 30°C. 0.2% KNO_3 solution increased germination in the dark, but not in the light. It is tolerant of burning and rapid regrowth occurs from the rhizomes which are protected by the soil.

Seed longevity and burial : Cogon grass seed is short lived, generally remaining viable in the soil for about 1 year. Viability of seeds stored in a laboratory steadily decreased over 13 months. Field studies in Asia show a maximum seed life of 16 months.

The viable seed banks were small in Imperata grassland sites that had been burnt regularly (200–400 m-2) and very large (2000–3000 m-2) in sites that had not been burnt for five years or more. The seed banks of the burnt and unburnt Imperata grassland sites were composed mostly of agricultural weeds (Hopkins and Graham, 1984).

Economic importance : It negatively affects the growth of teak, cocoa, kola, coffee, cashew, oil palm, coconut, rubber, and *Gmelina arborea* (Komolafe, 1976; Holm *et al.* 1977; Townson, 1991). It caused yield reductions of 51-62 percent in maize when the crop was weeded 2-4 times (Akobundu and Ekeleme, 2000). Higher maize grain yield losses (80-100 percent) have also been reported (Koch *et al.,* 1990; Udensi *et al.,* 1999). Complete crop failure usually occurs when crops are grown in slashed plots without additional weeding. In cassava yield losses of 50-90 percent have been reported (Koch *et al.* 1990; Chikoye *et al.,* 2001). In soybean, Avav (2000), reported yield losses of 29-53 percent in the middle belt of Nigeria.

Medicinal uses : In traditional Asian folk medicine, it is used as a tonic, an emollient, an anti-inflammitory and a fever-reducing agent. A cogon grass extract shows anti-insecticide properties against mosquitoes. It is used as thatch, short-term forage production, soil stabilisation and paper making (Watson and Dallwitz 1992, in Dozier *et al.*, 1998).

Control and management : Physical: In small areas, dig and remove all plants and underground stems.

Chemical: Glyphosate is applied typically @ of 1.8 kg a.e./ha.

Past work done on Seed Science and Technology

Cogon grass can thrive on infertile soils and has highest germination and growth on soils of pH lower than 5 (Wilcut *et al.*, 1988). Sajise (1972) reported that cogon grass seeds require light to germinate and that germination is favored by light quality ratios characteristic of open or partly shaded environments.

In the spring of 1997, 168 plots were each planted with 20 cogon grass seeds, but only 34 plots produced seedlings. This is in sharp contrast to a study conducted in 1998 at the same site, in which 137 of 168 seed plots produced seedlings, with an average germination rate of about 30%. The only obvious difference between these years was rainfall. During 1997, Hurricane Danny deposited over 15 cm of rain on the site three weeks after planting, and research plots remained submersed for a minimum of two weeks following this rainfall event.

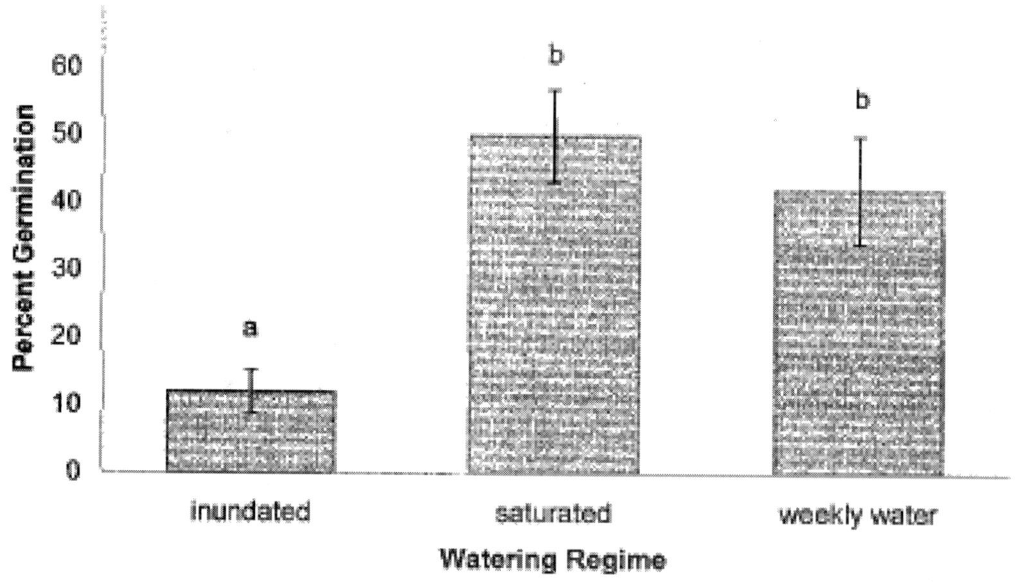

Fig. 7. Mean percent germination (untransformed, $\pm$ 1SE) of cogon grass caryopsis in three watering treatments in the greenhouse. Different letters represent differences among treatment means ($P < 0.05$) as detected by ANOVA and Tukey's test on arcsine-square root transformed data.

These observations suggest that year-to-year variation in moisture can play an important role in determining the success of cogon grass invasion by seed.

King and Grace (2000) reported that mean percent germination differed significantly among watering treatments (Fig. 7). Weekly watering and saturated soil resulted in the highest germination, 42% and 50%, respectively, and results did not differ statistically between the two treatments (Tukey's test, P>0.05). The inundated treatment had significantly lower germination than the other two treatments, having only 12% germination. Final mean seedling height of germinants differed significantly among all three watering treatments (ANOVA, P= 0.0001; Tukey-Kramer, P <0.05). Seedling growth was greatest with weekly watering, moderate in the saturated soil treatment and lowest in the inundated treatment (Fig. 8).

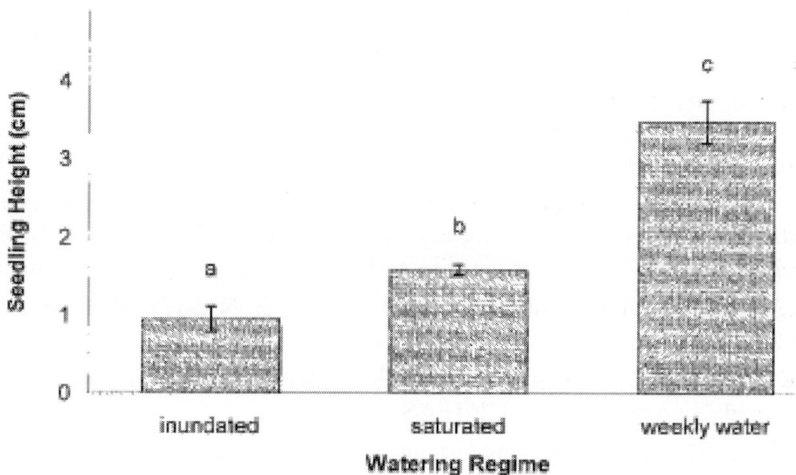

Fig. 8. Mean seedling height (untransformed, ± 1 SE) of cogon grass germinated and grown for 4 weeks in three watering treatments in the greenhouse. Different letters represent differences among treatment means ($P < 0.05$) as determined by ANOVA and the Tukey-Kramer procedure on log transformed data.

REFERENCES

Akobundu, I.O. and F. Ekeleme (2000). Effects of method of *Imperata cylindrica* management on maize grain yield in the derived savanna of south-western Nigeria. *Weed Res.* 40: 335-341

Avav, T. (2000). Control of speargrass (*Imperata cylindrica* (L.) Raeuschel) with glyphosate and fluazifop-butyl for soybean (*Glycine max* (L.) Merr) production in the savanna zone of Nigeria. *J. of the Science of Food and Agriculture,* 80: 193-196.

Bryson, C.T. and R. Carter (1993). Cogongrass, *Imperata cylindrica,* in the United States. *Weed Technology,* 7:1005-1009.

Chikoye, D., F. Ekeleme and E.U. Udensi (2001). *Imperata cylindrica* suppression by intercropping cover crops in *Zea mays/Manihot esculenta* sytems. *Weed Sci.,* 49: 658-667.

Dozier, H., J.F. Gaffney, S.K. McDonald, E.R. R.L. Johnson and D.G. Shilling (1998). Cogongrass in the United States: History, ecology, impacts, and management. *Weed Technology* 12: 737-743.

FIPR. (1997). Ecology, physiology and management of cogongrass (*Imperata cylindrica*). Publication No. 03-107-140, prepared by University of Florida under a grant sponsored by Florida Institute of Phosphate Research (FIPR). 144 p.

Gabel, M.L. (1982). A biosystematic study of the genus Imperata (*Gramineae: Andropogoneae*). Ph.D. dissertation. Iowa State University, Ames, Iowa USA. 94 p.

Holm, L.G., D.L. Plucknett, J.V. Pancho and J.P. Herberger (1977). The World's Worst Weeds: Distribution and Biology, Honolulu, University Press of Hawaii, 609 pp.

Hopkins, M.S. and A.W. Graham (1984). Viable soil seed banks in disturbed lowland tropical rainforest sites in North Queensland. *Australian Journal of Ecology,* 9(2): 71–79.

Hubbard, C.E. (1944). *Imperata cylindrica*: Taxonomy, distribution, economic significance and control. Imperial Agricultural Bureaux Joint Publication No. 7. Oxford, UK: Imperial Forestry Bureau; Aberystwyth, UK: Imperial Bureau of Pastures and Forage Crops. 63 p.

King S.E. and J.B. Grace (2000). The effects of gap size and disturbance type onninvasion of wet pine savanna by cogongrass, *Imperata cylindrica* (poaceae). *American journal of botany*, 87(9): 1279–1286.

Koch, W., F. Großmann, A. Weber, H.J. Lutzeyer and I.O. Akobundu (1990). Weeds as components of maize/cassava cropping systems. pp. 283-298, In *Standortgemaesse landwirtschaft in West Afrika.* Universitaet Hohenheim, Stuttgart, Germany.

Komolafe, D.A. (1976). Weed problems in tree crops in Nigeria. *Pest Articles and News Summary* 22: 250-256.

Langeland, K.A. and K.C. Burks (Eds.). (1998). Identification and biology of non-native plants in Florida's natural areas. UF/IFAS. 165 p.

Lippincott, C.L. (1997). Ecological consequences of *Imperata cylindrica* (cogongrass) invasion in Florida sandhill. Ph.D. Dissertation. University of Florida, Gainesville, FL, USA.

McDonald, S.K., D.G. Shilling, C.A.N. Okoli, T.A. Bewick, D. Gordon, D. Hall and R. Smith. (1996). Population dynamics of cogongrass. *Proc. South. Weed Sci. Soc.* **49**:156.

McDonald, G.E. (2004). Cogongrass (*Imperata cylindrica*)-Biology, ecology and management. *Critical Reviews in Plant Science*, 23:367-380.

Sajise, P.E. (1972). Evaluation of cogon (*Imperata cylindrica* (L.) Beauv.) As a serial stage in Philippine vegetational succession. 1, the cogonal serial stage and plant succession. 2, autecological studies on cogon. Dissertation Abstracts International B (1972). *Weed Abstracts* 3040 – 3041.

Shilling, D.G., T.A. Bewick, J.F. Gaffney, S.K. McDonald, C.A. Chase and E.R. R.L. Johnson. (1997). Ecology, physiology and management of cogongrass (*Imperata cylindrica*). Final report prepared for the Florida Institute of Phosphate Research, Bartow, FL, USA.

Townson, J. K. (1991). *Imperata cylindrica* and its control. *Weed Abstracts*, 40: 457-468.

Udensi, E.U., I.O. Akobundu, A.O. Ayeni and D. Chikoye (1999). Management of congograss (*Imperata cylindrica*) with velveltbean (*Mucuna pruriens* var. *utilis*) and herbicides. *Weed Tech.* 13: 201-208.

Van Loan, A.N., J.R. Meeker and M.C. Minno (2002). Biological control of invasive plants in the Eastern United States. USDA Forest Service Publication FHTET-2002-04,413p.

Watson, L. and M.J. Dallwitz (1992). Grass genera of the World: descriptions, illustrations, identification and information retrieval; including synonyms, morphology, anatomy, physiology, phytochemistry, cytology, classification, pathogens, world and local distribution and references. http://biodiversity.uno.edu/delta/ (accessed August 1999).

Wilcut, J.W., B. Truelove, D.E. Davis and J.C. Williams (1988). Temperature factors limiting the spread of cogongrass (*Imperata cylindrica*) and torpedograss (*Panicum repens*). *Weed Sci.,* 36:49-55.

Yager, L.Y. and M. Smith (2009). Use of GIS to prioritize cogongrass (*Imperata cylindrica*) control on Camp Shelby Joint Forces Training Center, Mississippi. *Invasive Plant Sci. Manag.* 2:74–82.

ROG FRUIT (*Lippia nodiflora*)

Common name : Rog fruit, sawtooth fogfruit, turkey tangle

Botanical name : Lippia nodiflora

Tamil name : Podutalai

Hindi name : Bukkan

Family : Verbenaceae

Origin and distribution : Native to U.S., tropical America, Europe

Botany

Habitat : Lippia grows best on bared ground in periodically moist sites (heavy rainfall, flooding or opening of springs) that also undergo periods of water stress (Richardson, 1994) and with apparently poor soil structure. It is found mostly on high-moisture-holding clay to clayloam soils in floodplains or wetlands.

Habit : Prostrate herb widely creeping and rooting, minutely hairy (Fig. 1).

Fig. 1. Plant

Roots : Roots are produced from leaf axils along stems and consist of a central taproot, usually 50–60 cm in length, but thought to extend up to 2 m (NWWCC 2002). An extensive fibrous root system extends from the taproot. Lippia also produces fibrous roots from some stem nodes.

Stem : Stems lay along ground and root.

Leaf : Green to blue-green; oval, to 3/4inch long; jagged margin.

Floral biology

Inflorescence : The inflorescence consists of a purple-coloured centre encircled by small white-to-pink flowers (Fig. 2). The flower takes on a match-like look, which is why the plant is sometimes called matchweed.

Flower : Bracts broadly obovate, apiculate; calyx membranous 2 to 4 lobed later 2-valved; corolla small salver shaped 2-tiped upper lip, bifid lower 3-lobed; stamens 4, didynamous; ovary bilocular, single ovule in each locule (Fig. 3).

Fig. 2. Inflorscence

Fig. 3. Flower

Flowering and pollination : Lippia flowers anytime in the spring-summer-autumn period given favourable soil moisture conditions (Lucy *et al.*, 1995; Dellow *et al.* 2001). Flowering takes place during May - October. Flowering can be triggered by rain or flooding at any time during this period (Lucy *et al.*, 1995). Lippia can self- and cross-pollinate (McCosker, 1994) resulting in a high percentage seed set and maintenance of sufficient genetic variation to adapt and rapidly colonise new environments (Kumar and Dutt, 1989).

Seed biology

Seed : Dehiscent fruits (1–1.5 mm in diameter) release two tiny brown, oval, flattened seeds at maturity(Fig. 4) (Lucy *et al.*, 1995). Seeds are barely visible to the naked eye and exalbuminous.

Propagation : Through seed. Vegetative reproduction occurs via production of roots at stem nodes, and can occur while the plant grows under water. When inundated with shallow floodwaters, lippia stems grow towards the water surface and become shortened in their stem sections, which then break off as the nodes weaken. The detatched fragments float and can remain dormant before becoming established on suitable soils again when floodwaters recede (McCosker, 1994).

Seed dispersal : Seeds are initially dispersed below the parent plant with secondary dispersal via floodwaters, probably by floating on the water surface, and germinate when they are deposited in silt by the declining waters (McCosker, 1994). Other dispersal agents include ants (McCosker, 1994) and possibly birds (SQDNR

2000). Vegetative dispersal may be assisted via the displacement of plant material (e.g. muddy stem fragments stuck on the feet of animals) by machinery, cattle, sheep and birds, and is especially likely to occur in wet conditions (McCosker, 1994).

Fig. 4. Seed

Seed dormancy and germination : It has the ability to remain dormant when moisture reserves become depleted and can withstand long dry periods. It can also regrow quickly after rain events, allowing it to compete with other species. Germination can occur under a wide range of conditions. Estimations of the viable period of stem fragment dormancy vary from at least six to eight weeks (Lucy *et al.*, 1995) up to three months (Dellow *et al.,* 2001).

Soil seed bank : Viable seed bank remains around a parent plant even after the plant's removal (McCosker, 1994; Lucy et al., 1995). Seeds of lippie may possess an effective drought survival mechanism involving enhanced germination from the seed bank as a result of alternate drying and wetting (McCosker, 1994).

Agricultural importance : Lippia may reduce seed set in commercial crops via pollinator competition. Lippia has allelopathic effects on other plant species (McCosker, 1994). One study showed that lippia leachate inhibited the growth of lettuce seedlings, and the researchers correlated this by the presence of potential allelopathic compounds that have been isolated from the plant (Elakovich and Stevens, 1985).

Uses : Lippia is used as a lawn and garden ornamental in the nursery industry and it is sold as shade-loving lawn replacement species.

Lippia produces many flowers that are very attractive to bees. Bees are able to extract large quantities of nectar from lippia flowers, which is said to produce a pleasant tasting honey. Because of this, and the fact that lippia is very easy to grow, the plant is favoured by honey producers (Richardson 1994).

Control measures : Chemical control: 2,4-D amine, glyphosate and metsul-furon are currently used for the suppression of lippia in pasture and fallow floodplains. Application of herbicides should be carried out during the active growing stage of lippia when plants are starting to flower and there is good soil moisture availability (Dellow *et al.*, 2001; Motley *et al.,* 2001).

Mechanical/physical control: Ploughing is an effective method of lippia control (SQDNR, 2000; Csurhes, 1989).

REFERENCES

Csurhes, S.M. (1989). Occurrence of phyla weed *(Phyla nodiflora)* on the Condamine river, Queensland Lands Department Survey, Allan Fletcher Research Station.

Dellow, J.J., K. Motley, A. Storrie and J. Spenceley (2001). *Lippia*, Agfact P7.6.52, NSW Agriculture.

Department of Natural Resources (SQDNR) (2000), Better management practices manual, Department of Natural Resources, Upper Condamine Floodplain, Natural Heritage Trust and Landcare Queensland, available online at <http://www.nrm.qld.gov.au/community/condaminefp/bmp-manual.htm>, viewed 6 January 2003.

Elakovich, S.D. and K.L. Stevens (1985). Volatile constituents of *Lippia nodiflora*, *Journal of Natural Products*, 48, 504–6.

Kumar, P. and B. Dutt (1989). Cytogenic basis of breeding system in some Verbenaceous species, *Cytologia*, vol. 54, pp. 347–53.

Lucy, M., E. Powell, R. McCosker, G. Inglis and R. Richardson (1995). *Lippia (Phyla canescens):* A review of its economic and environmental impact on floodplain ecosystems in the Murray–Darling Basin, Department of Primary Industries, Brisbane.

McCosker, R.O. (1994). *Lippia* (*Phyla nodiflora*): an invasive plant of floodplain systems in the Murray–Darling Basin, a report on the distribution and ecology of lippia in the lower Gwydir Valley and the Murray–Darling Basin prepared for the Gingham Watercourse Landcare Group, Department of Ecosystem Management, University of New England, Armidale, NSW.

Motley, K., J. Dellow, A. Storrie and J. Spenceley (2001). Using herbicides in a lippia management program, Agnote DPI-384, NSW Agriculture.

North West Weeds County Council (NWWCC). (2002). *Lippia*, available online at <http://www.northwestweeds.nsw.gov.au/Lippia.htm>, viewed 5 December 2003.

Richardson, R. (1994). *Lippia* (*Phyla nodiflora*): its distribution and impact throughout the eastern mainland states of Australia, a postgraduate report, Faculty of Applied Science, The University of Queensland, Gatton College, Lawes.

WATER PRIMROSES (*Ludwigia parviflora*)

Common name : Water-primroses

Botanical name : Ludwigia parviflora

Tamil name : Neer kirmbu

Family : Onagraceae

Origin and distribution : South America

Botany

Habitat : An aquatic vascular plant; characterized by rapid growth; and occurring in transition zones of shallow, slow moving waterways. After establishing in the bank or channel bottom, prostrate stems grow laterally, rooting adventitiously at nodes. Once rooted, secondary shoots grow erect, up to 5 feet in height

Habit : Herbs erect, annual, with taproot (Fig. 1).

Stem : Stems 20–100 cm tall, branched, subglabrous or minutely puberulous on younger parts. Petiole 2–15 mm, winged.

Leaf : Leaf blade narrowly elliptic to lanceolate, 1–11 × 0.3–2.7 cm, lateral veins 6–12 per side, submarginal vein inconspicuous, base narrowly cuneate, apex subacute.

Floral biology

Flower : Sepals 4, rarely 5, deltate, (1.3) 2 –3.5 mm, glabrous or minutely puberulous. Petals yellow, elliptic, 1–3 × 0.7–2 mm. Stamens as many as sepals, or rarely more; filaments 0.3–0.7 mm; anthers 0.5–0.7 mm; pollen in tetrads. Style 0.7–1.5 mm; stigma globose (Fig. 2).

Fig. 1. Plant

Fig. 2. Flower

Fruit : Capsule often nodding, pale brown, oblanceoloid, terete, 3–16 (–19) mm, 2.5–5 mm in diameter, thinly walled, readily and irregularly loculicidal, glabrous or puberulous, sessile or pedicel to 6 mm (Fig. 3).

Flowering and fruiting : July - November

Seed biology

Seed : Seeds in 2 or more rows per locule, free, brown with fine brown lines, 0.3–0.5 mm, raphe very narrow and inconspicuous. The weight of seed is 0.06 mg/seed (Fig. 4).

Propagation : Reproduce primarily from asexual rooting plant fragments and also by seeds.

Life cycle/Seed production : A single plant in rice field may produce up to 75,000 seeds (Pancho, 1964).

Seed dispersal : The seeds and fragments are carried downstream in water, and by waterbirds between rivers as the seeds are palatable to ducks (Parsons and Cuthbertson, 1991).

Seed dormancy and germination : The growth of ludwigia when grown alone and in competition with rice. Ludwigia showed the ability to reduce the effects of rice interference by increasing leaf weight ratio, increasing stem and leaf biomass in upper half of the plant and increasing specific stem length (Chauhan *et al.,* 2011).

Fig. 3. Fruit

Fig. 4. Seed

Soil seed bank : It forms a high soil seed bank for an aquatic (about 65,000/m^2).

Uses : It has attractive flowers and could be used for plantings around large ponds.

Control : Chemical control: Post-emergence application of bensulfuron-methyl 0.05 kg ha^{-1} produced grain yield 5.16t ha^{-1}, which was statistically at par with two hand weeding, pre emergence (PE) trisulfuron 0.009 kg ha^{-1} and bensulfuron-methyl at all the doses.

Integrated weed management:

- Increased seed rate of rice
- Narrow row spacing in rice

REFERENCES

Chauhan, B.S., A.R.P. Pame and D.E. Johnson. (2011). Compensatory growth of Ludwigia (*Ludwigia hyssopifolia*) in response to interference of direct-seeded rice. *Weed Science*, 59: 177-181.

Pancho, J.V. (1964). Seed wizes and production capacities in common weed species of the rice fields of the Philippines. *Philipp. Agric.*, 48: 307-316.

Parsons, W.T and E. Cuthbertson. (1991). Noxious weeds of Australia (Inkata: Melbourne).

AARAI (*Marsilea quadrifolia*)

Common name : Aarai, four leaf clover, water clover, pepperwort and water shamrock.

Botanical name : Marsilea quadrifolia L

Tamil name : Aarai, acanaparni, arai, araikkirai, catuppalli, kurantam, nirarai, niraraikkirai

Hindi name : Caupatiya, Sunsuniya

Family : Marsileaceae

Origin and distribution : It is native to Europe and can be found growing in North America, central and southern Europe, Caucasia, Western Siberia, Afghanistan, southwest India, China and Japan.

Botany

Habitat : An aquatic fern that grows from creeping rhizomes anchored in the muddy bottoms of shallow ponds and lakes. This plant prefers light (sandy) and medium (loamy), acid or neutral or basic (alkaline) soils. It can grow in semi-shade (light woodland) or no shade. It requires moist or wet soil and can grow in water.

Habit : Creeper with submersed, floating and emergent leaves (Fig. 1).

Fig. 1. Plant.

Roots : Roots arising at nodes and 1--3 on internodes.

Stem : The plant's stems will adapt itself to the depth of the water until its leaves float on the water.

Leaf : Leaves that float on the water, grow from the rhizomes. These leaves are divided into 4 parts, resembling four-leaf clover. The flexible leaf stalk is green and slender. The leaflets are produced on the 20 cm tall petiole.

Fig. 2. Flower

Propagation : Reproduces by spores, producing sporocarps that are thick, oval and hairy. The sporocarps will then swell and burst to release the spores. The spores germinate immediately, the highly developed prothallus remains inside the large seed-like spores. The gametophyte generation is completed in 24 hours and the first roots and shoots appear in 2 - 3 days. Mature plants bearing sporocarps can develop in as little as 3 months. These reproductive structures can move the plant downstream to new areas.

Seed dormancy and germination : Most seeds require warm temperature to germinate. When seedlings emerge, thin extras to allow full light around seedlings. The foliage is evergreen but cold water and frosts will send the plant into dormancy.

Seed dispersal : Ducks and other waterfowl also help to move this species between different bodies of water.

Agricultural importance : Parts of *Marsilea drummondii* contain an enzyme which destroys thiamine (vitamin B_1), leading to brain damage in sheep and horses.

Uses : Ornamental : Can be grown in aquaria

Medicinal : A juice made from the leaves is diuretic and febrifuge. It is also used to treat snakebite and applied to abscesses etc. The plant is anti-inflammatory, diuretic, depurative, febrifuge and refrigerant.

Control measures : Preventive control measures

- Use clean seeds
- Keep seed bed weed free
- Keep leaves, bunds and irrigation canal clean
- Keep tools and machinery clean
- Keep livestock out of field
- Prevent weeds from seedlings
- Prevent vegetative reproduction in seed bed

Cultural control measures

- Proper land preparation
- Cultivar selection
- Crop establishment
- Water management
- Control of fertilizer application

Mechanical control measures : Use rotary weeder

Manual weed control : Hand weeding

Biological control method : Weed utilization in rice fields, using mulching effect and allopathic effect.

TOUCH ME NOT (*Mimosa pudica* Linn.)

Common name : Touch me not (Sensible plant)

Scientific name : Mimosa pudica Linn.

Tamil name : Thottal sinungi

Hindi name : Chui-mui, Lajwanti

Family : Marsileaceae

Origin & distribution : Native to south and central America, where it was first documented in Brazil. It's now considered a pan-tropical weed with distribution throughout Africa, Asia, Australia and the Pacific Islands.

Botany

Habitat : Small woody herbs or low-spreading under shrub with hairy and prickly branches, hairs glandular. This herb grows in full sun or light shade in warm climates, because it cannot tolerate frost. It thrives and becomes a ground cover in poor soil because of its ability to convert nitrogen for its own use.

Habit : A creeping annual or perennial herb often grown for its curiosity value (Fig. 1).

Stem : The stem is erect in young plants, but becomes creeping or trailing with age. The stem is slender, branching, and sparsely to densely prickly, growing to a length of 1.5 m (5 ft).

Leaf : Bipinnate, sensitive to touch, pinnae 1-2 pairs, leaflets 10-20 pairs, linear, glabrous.

Floral biology

Flower : Heads small, peduncled, globose, axilalry, pink-purple, calyx campanulate, petals crenate towards base (Fig. 2). The flowers are globe shaped, about 1 inch across.

Flowering and pollination : The plant flowers in mid to late summer. The flowers are pollinated by the wind and insects. In China, the flowering season from March to October, with fructescence from May to November.

Fruit : The fruit consists of clusters of 2-8 pods from 1–2 cm long each, these prickly on the margins. The pods break into 2-5 segments and contain pale brown seeds some 2.5 mm long (Fig. 3).

Seed biology

Seed : The seeds have hard seed coats which restrict germination. 2.4-2.8 (-3.3) mm x 1.7-2 mm; plumply lentiform, fawn-brown, with an open pleurogram. 1000 seed weight is 5.3 g (Fig. 4).

Propagation : By seed.

Seed dispersal : Pods can be spread by attaching to fur of animals, attaching to clothing, by mud on vehicles and by water (Smith, 2002 in Pier, 2005).

Fig. 1. Plant

Fig. 2. Flower

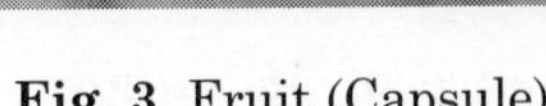

Fig. 3. Fruit (Capsule)

Fig. 4. Seed

Life cycle : M. pudica plants live 1 to 2 years. Seedlings grow slowly for 2 or 3 months and then accelerate, reaching 0.5 to 2m of extension at the end of the first year. Growth of plants that survive into the second year is much slower. Potted and field grown individuals are sensitive to overwatering (Bui, 2001).

Seed dormancy and germination : Seed will germinate in 14 to 21 days at 70° F. The dormancy is induced by a hard seed coat that restricts germination. Scarification released the seeds from dormancy and stimulated germination, though the germination of the scarified seeds was not influenced by light. The germination increased markedly with the exposure to high temperature pretreatment (e.g. 150°C), which was achieved by placing non-scarified seeds in an oven for 5 min followed by incubation at 35/25°C day/night temperatures for 14 days. The germination of the scarified seeds was tolerant of salt and osmotic stress, as some seeds germinated even at 250 mmol L^{-1} NaCl (23%) and at an osmotic potential of −0.8 MPa (5%). The germination of the scarified seeds was >74% over a pH range of 5–10. The seedling emergence of the scarified seeds was 73–88% at depths of 0–2 cm and it gradually decreased with an increasing depth, with no seedling emergence at the 8 cm depth. The rice residue applied to the soil surface at rates of ≤6 t ha^{-1} did not influence the seedling emergence and dry weight. (Chauhan and Johnson, 2009)

The germination of *M. pudica* is independent of light exposure. Optimum germination (percentage and germination speed index) temperature for the species was continuous, at 30°C or alternate temperature of 25-35°C, even though the seeds germinated satisfactorily in continuous temperature above and under 30°C, and in all combinations of alternate temperatures. Comparatively, the germination speed index was more sensitive to the effects of the temperature than the germination percentage.

Uses: According to Ayurveda, root is bitter, acrid, cooling, vulnerary, alexipharmic and used in treatment of biliousness, leprosy, dysentery, vaginal and uterine complaints, inflammations, burning sensation, fatigue, asthma, leucoderma, blood diseases etc. According to the Unani system of medicine, root is resolvent, alternative, useful in diseases arising from blood impurities and bile, bilious fevers, piles, jaundice, leprosy etc. Contains an alkaloid Mimosine.

It is also grown as garden herb, green manuring, fixes nitrogen, used as fodder.

Control measures : Physical: Hand weeding is difficult due to the presence of thorns and a woody root. Repeated burning may in fact encourage the spread of *M. pudica* in pastures (Magda *et al.*, 2006).

Chemical: Very sensitive to dicamba, glyphosate, picloram and triclopyr (Motooka *et al.*, 2002).

Biological: See Waterhouse and Norris (1987) for information on prospective biological control agents (Pier, 2005).

Coir dust can be used as a mulch to suppress *M. pudica.*

Sheep grazing is reported to control the dominance of *M. pudica* in pasture (Magda *et al.*, 2006).

This species has been successfully tested and recommended for erosion control plantings using potted material at a spacing of 60 x 60cm (Coimbra and Magnanini,1953)

Soil Seed bank : Kew's Millennium Seed Bank partnership aims to save plant life world wide, focusing on plants under threat and those of most use in the future. Seeds are dried, packaged and stored at a sub-zero temperature in seed bank vault.

REFERENCES

Bui, L.D. (2001). *Mimosa pudica.* http://bio.maimi.edu/mimosa/mimosa.html. 7 p.

Chauhan, B.S. and D.E. Johnson (2009). Germination, emergence, and dormancy of *Mimosa pudica. Weed Biology and Management,* 9 (1): 38–45.

Coimbra, A.F. and A. Magnanini (1953). Considerations sobre *Mimosa pudica* no combate a arosao superficial. Anucio Brazilero Economico da Floresta 6. Instituto Nacional, Pinho, Brazil. 131-136.

Magda, D., M. Duru, J. Huguenin and B. Gleizes (2006). *Grass and Forage Science.* 61: 89-9.

Motooka, P., Lincoln Ching and Guy Nagai (2002). Herbicidal weed control methods for pastures and natural areas of Hawaii. *Weed Control,* 8: 1-35.

Pacific Island Ecosystems at Risk (Pier) (2005). *Mimosa pudica* Species Information. Francis, J.K. Undated. Sensitive plant fact sheet.

Smith, N.M. (2002). Weeds of the wet/dry tropics of Australia - a field guide. Environment Centre NT, Inc. 112 pp.

Waterhouse, D.F. and K.R. Norris (1987). Biological control: Pacific prospects. Biological control: Pacific prospects, viii + 454pp. [many fig., 26x18 cm]

THYME LEAVED GRATIOLA (*Monniera cuneifolia*)

Common name : Thyme leaved gratiola

Botanical name : Monniera cuneifolia Michx

Tamil name : Nira - Brahmi

Hindi name : Brahmi, safed - ehamni

Family : Scrophulariaceae

Origin and distribution : It is native to India, but has spread throughout the tropics.

Botany

Habitat : It grows on marshy wetlands or shorelines and has the ability to grow in water. It is sometimes grown in aquariums and can even grow in slightly brackish (mildly salty) water.

Habit : Creeping perennial herb with succulent, thick leaves (Fig. 1).

Root : Roots emerge out of the nodules and directly go to the soil.

Stem : Prostrate, (sub) succulent, herbaceous.

Leaf : Decussate, simple, oblong, 1 x 0.4 cm, (sub) succulent, punctate, penninerved, margin entire, apex obtuse, (sub) sessile.

Floral biology

Flower : Axillary, solitary, bracteate (2 bracteoles), linear, pedicel to 0.5 cm, purple in color (Fig. 2). Calyx: 5 lobes (unequal); outer 2 lobes larger, ovate, 7 x 3.5 mm; inner 2 lobes linear, 5.5 x 0.7 mm; median 1 lobe oblong, 5.5 x 2 mm, imbricate, (sub) succulent, punctate, obtuse, acute. Corolla: White with violet and green bands inside the throat, 0.8 cm across, 5 mm tube; 5 lobes, obscurely 2-lipped, 2+3, (sub) equal, obtuse or emarginate. Stamens: 4, didynamous; filament pairs 1 and 2.5 mm anthers oblong, contiguous, 1.5 mm. Ovary: Oblong-globose, 2 mm; style slightly deflexed, 5.5 mm; stigma flat capsule, oblong-globose, 5 x 2.5 mm, septicidal or locilicidal, or 4-valved.

Flowering and fruiting : Flowering occurs in April-June; Fruiting in June-December

Fruit : Capsule - ovate, acute, pointed with the style base, 2-celled, 2-valved, glabrous.

Seed biology

Seed : Seeds many, less than 1 mm, and oblong, striate, pale.

Propagation : By seed, cuttings and also by the root division.

Fig. 1. Plant

Fig. 2. Flower

Seed dormancy and germination : Seeds remain dormant for 6-9 months and due to hard seed coat.

Medicinal uses : Brahmi is used in traditional Indian medicines for centuries. For the treatment of bronchitis, chronic cough, asthma, hoarseness, arthritis, rheumatism, backache, fluid retention, blood cleanser, chronic skin conditions, constipation, hair loss, fevers, digestive problems, depression, mental and physical fatigue and many more. It is used to treat all sorts of skin problems like eczema, psoriasis, abscess and ulceration. It stimulates the growth of skin, hair and nails.

Brahmi posses anticancer activity. It is taken to get relief from stress and anxiety. According to the ayurveda Brahmi has antioxidant properties. It has been reported to reduce oxidation of fats in the blood stream, which is the risk factor for cardiovascular diseases. Brahmi is considered as the main rejuvenating herb for the nerve and brain cells.

Brahmi is rich in Vitamin C and can be used in the salads, soups and sandwiches. Brahmi oil restores and preserves the memory. In India, it is given to the infants to boost memory power intelligence.

PICKERELWEED (*Monochoria vaginalis*)

Common name : Monochoria, oval-leaf monochoria, pickerelweed

Scientific name : Monochoria vaginalis

Tamil name : Neerthamari

Hindi : Neelolpal

Family : Pontederiaceae

Origin and distribution : Native to tropical Asia. It is distributed from Iran to the Philippines and Indonesia, China, Korea, Japan, Taiwan, Northern Australia, Fiji, the Solomon Islands and United States.

Botany

Habitat : An annual or pseudoannual in flooded ricefields, but may grow as a perennial in constantly flooded areas. The plant roots in mud and its upper portions grow above the water (Holm *et al.*, 1977). It is often locally abundant in stagnant backwaters or rivers, in open drains, rice fields, swampy places, ditches, and wet pastures (Smith, 1979)

Habit: Aquatic annual herb (Fig. 1).

Roots : Fibrous roots at the base of petioles.

Stem : Very short rhizome unbranched.

Leaf : The leaves have variable size, 2 to 12.5 cm long, and 0.5 to 10 cm wide. In very young plants without lamina; leaves somewhat older plants with a floating linear or lanceolate blade; leaves of still older plants, ovate-oblong to broadly ovate, sharply acuminate, the base heart-shaped or rounded, shiny, deep green in color, with longitudinal veins; petioles soft, hollow, growing from buds at the base. Leaf sheaths twisted together at the base, slightly reddish when young; crown appears bulbous.

Fig. 1. Plant

Fig. 2. Flower

Floral biology

Inflorescence : Spike like, basally opposite the sheath of the floral, with a large bract arising from a thickened bundle of leaves stalk, about two-thirds of the way up the stalk from the base.

Flower : Flowers 3 to 25, opening simultaneously or in quick succession; on pedicels 4 to 25 mm long; perianth 11 to 15mm long; petals six, violet or lilac blue, spreading at flowering, afterwards spirally contorted; stamens six, one with a lateral obliquely erect tooth; ovary with a long style. After flowering, the inflorescences bend downward (Fig. 2).

Fruit : Capsule of about 1cm diameter, splitting between the partitions into three valves. The fruit may mature below the water surface (Fig. 3).

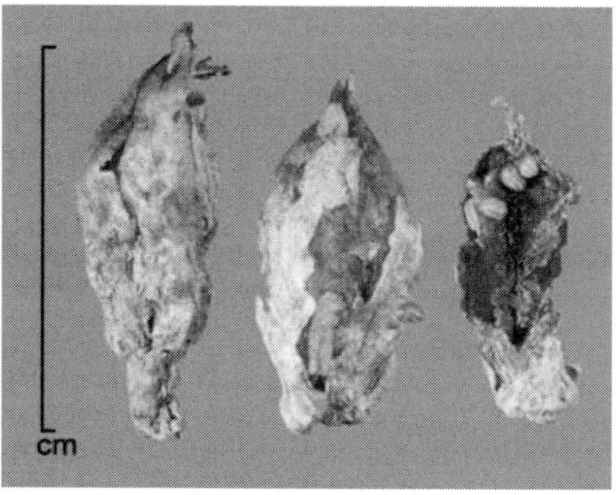

Fig. 3. Fruit

Seed biology

Seed : Numerous seeds longitudinally ribbed. Average length is less than 2mm and the mean seed weight is around 0.17mg (Fig. 4).

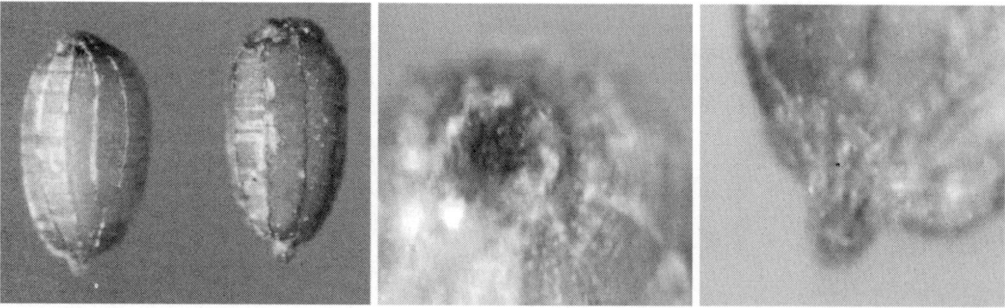

Fig. 4. Seed : Chalazal end (left) and hilar end (right) of seed

Propagation : By seeds

Seed dormancy and germination : The seeds exhumed from the flooded soil were dormant at the beginning of burial and proceeded into a conditional dormancy/non dormancy /conditional dormancy cycle throughout the remaining period of the experiment. The seeds exhumed monthly from the non-flooded soil exhibited an annual dormant cycle, which is dormancy/conditional dormancy/non-dormancy/dormancy.

The seasonal variations in germination and the effects of soil moisture studied in *Monochoria vaginalis*. The greatest number of seedling emerged in the early planted fields. 35 per cent more seedlings emerged from the saturated soil than they did from submerged or dry soil in the early planting. A striking change in germination pattern was shown under submerged conditions. The majority of the seedlings emerged in a very short time, the peak being between 15 and 25 days; whereas in saturated or dried soil there was a tendency toward gradual emergence through the season.

Medicinal uses : The root is used for toothache and the bark is eaten with sugar for asthma. In India, the plant is used as vegetable and the roots are used for medicinal purposes.

Effects and impacts : It reduces rice yield as high as 82%.

Control : Post-emergence application of 2, 4-D @ 500 a.i/ha.

Past work done on Seed Science and Technology

The non-dormant seeds of *M. vaginalis* failed to germinate in the dark. However, the seeds are highly sensitive to dim light. Two days of 1 min. of exposure to light at an intensity of 4 x 10^{-4} mol m^{-2} were sufficient for 50% of the dormant *M. vaginalis* seeds to germinate (Chen and Kuo, 1995).

Chen and Kuo (1999) reported that *M. vaginalis* seeds collected in October 1993 and stored at -20°C for 1 month were dormant at all four temperature regimes. Although still very high after 1 month of burial in water-logged soil (Fig. 5a), the dormancy level was greatly reduced after 2 months of burial. Thereafter, 60-90% of the seeds germinated at 30/25°C, except those exhumed in October 1995. At 30/20°C, the germination was 30-80%. Comparing these two temperat-ure regimes revealed a greater difference in germination during winter than during summer. At 25/20°C, the germination pattern revealed a seasonal fluctuation. Germination increased from winter until early summer and then decreased again until autumn. At 23/13°C germination never exceeded 20% throughout the experiment.

From March to June of the first year of paddy-field burial, 20-40% of the seeds germinated at 30/25°C under the dark condition (Fig. 5b). The lower germination under the dark condition indicated that these seeds were very sensitive to light exposure during exhumation. One year after the burial, a short exposure to light did not stimulate germination.

At 25/20°C, the annual fluctuation in percentage germination of the exhumed seeds buried in upland soil (Fig. 6a) resembled those buried in the paddy fields (Fig. 5a). However, in the upland soil, the seeds became completely dormant during summer, as revealed by a low germination percentage at 30/25°C. The seeds gradually proceeded into conditional dormancy during autumn and winter and became non-dormant in the spring of the next year. According to dark germination tests (Fig. 6b), up to 30% of the seeds lost their light requirement in the first spring. The seeds buried in summer maintained a longer dormancy period (Fig 7 and 8) than those buried in winter (Fig. 5 and 6). Again, the seeds acquired the ability to germinate quicker when buried in the paddy field than in upland soil (Figs. 7b and 8b).

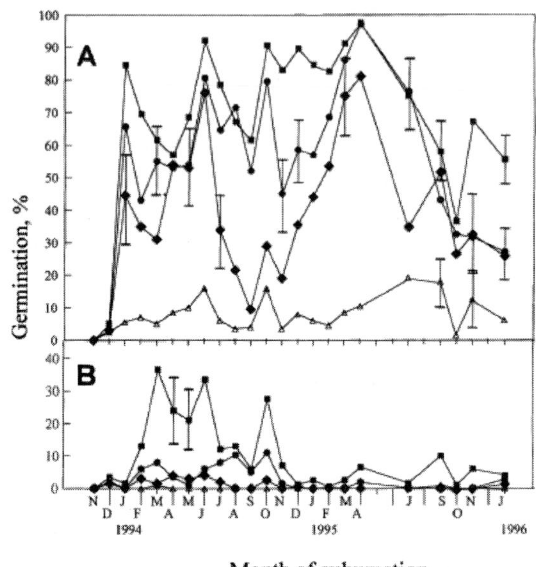

Fig. 5. Percentage germination under light (A) and dark (B) conditions of *Monochoria vaginalis* seeds exhumed monthly from a flooded soil. Vertical bars indicate + standard error, if greater than 5%. Germination tests were performed at 30/25 (■), 30/20 (•), 25/20 (♦) and 23/13°C (Δ).

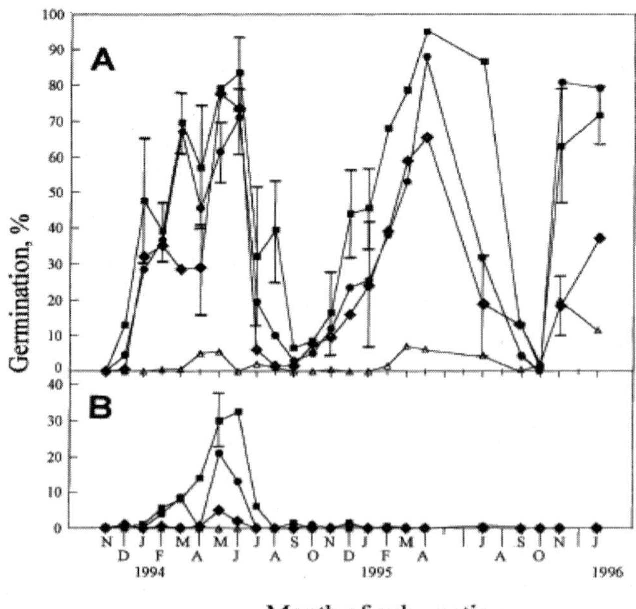

Fig. 6. Percentage germination of *Monochoria vaginalis* seeds exhumed monthly from upland soil

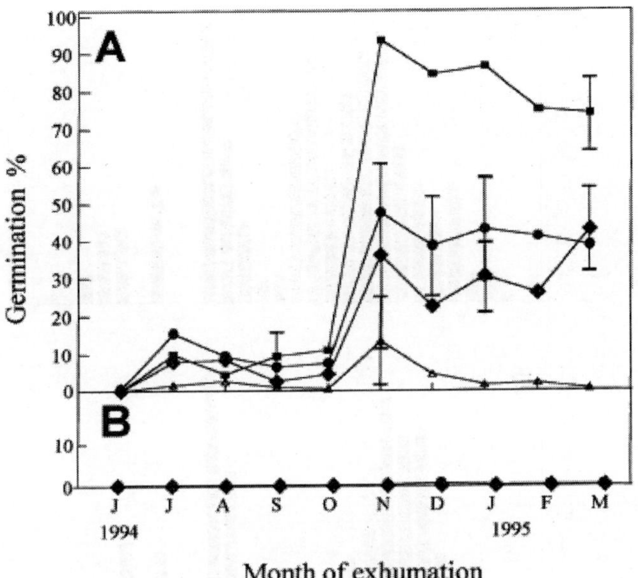

Fig. 7. Percentage germination of *Monochoria vaginalis* seeds exhumed monthly from a flooded soil. The seeds were protected from light during exhumation and the seeds for the dark germination test were prepared under green safe light.

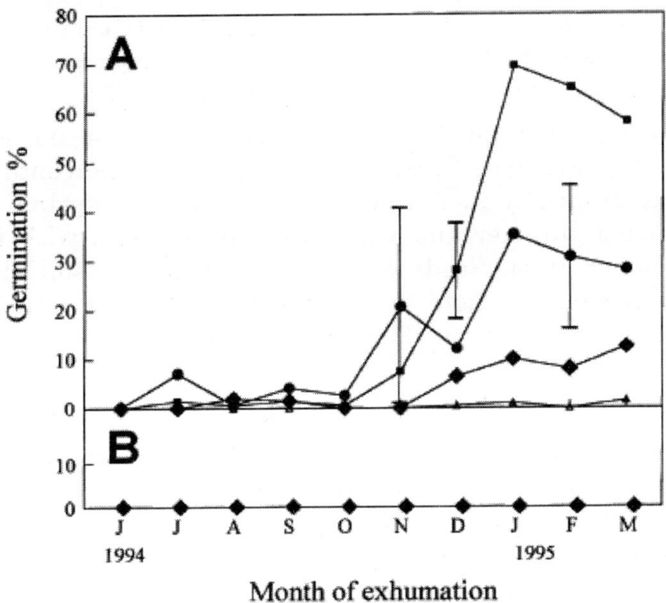

Fig. 8. Percentage germination of *Monochoria vaginalis* seeds exhumed monthly from upland soil. The seeds were kept from light during exhumation and the seeds for the dark germination test were prepared under green safe light.

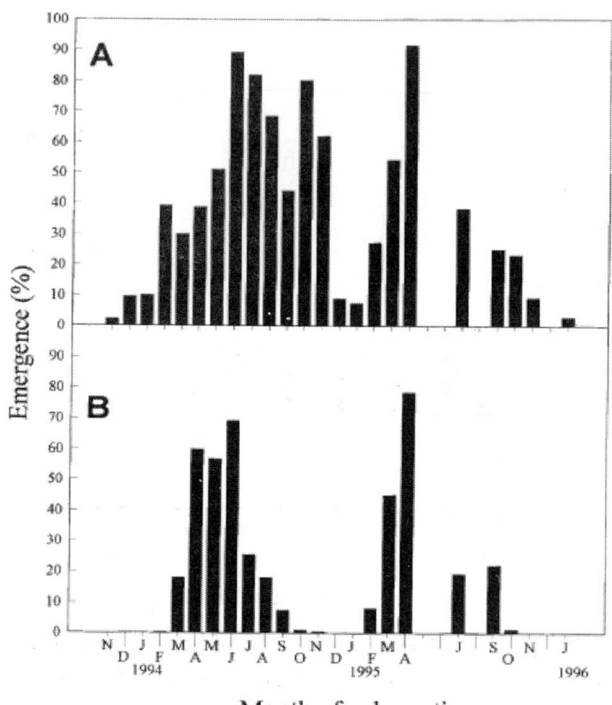

Fig. 9. Percentage emergence of *Monochoria vaginalis* seeds exhumed monthly from a flooded (A) and an upland (B) soil and then replanted into a water-logged field. No exhumation was made in May, June, August and December 1995.

The seeds that were exhumed from water-logged soil did not germinate in the field at the beginning (Fig. 9a). In the 2-year survey, the percentage emergence of the exhumed seeds gradually increased up to 90% during spring. The percentage then decreased during summer and was lowest in winters, with the exception of October and November 1994. Seeds exhumed from upland soil also displayed an annual change in percentage field emergence (Fig. 9b) but were generally more dormant, except from April to June 1994 and in April 1995.

REFERENCES

Chen, P.H. and W.H.J. Kuo (1995). Germination conditions for the non-dormant seeds of *Monochoria vaginallis*. *Taiwania*, 40: 419-432.

Chen, P.H. and W.H.J. Kuo (1999). Seasonal changes in the germination of buried seeds of *Monochoria vaginallis*. *Weed Research*, 39: 107-115.

Holm, L.G., D.L. Plucknett, J.V. Pancho and J.P. Herberger (1977). The world's worst weeds: Distribution and biology. University Press of Hawaii. 609 pp.

Smith, A.C. (1979). Flora Vitiensis nova: a new flora of Fiji. National Tropical Botanical Garden, Lawai, Kauai, Hawaii. Volume 1. 494 pp.

TORPEDO GRASS (*Panicum repens*)

Common name : Torpedo grass

Botanical name : *Panicum repens*

Family : Gramineae

Origin and distribution : Believed native to Europe or Australia and is now widely distributed throughout the tropics and subtropics from approximately 43° North latitude to 35° South latitude (Holm *et al.*, 1977).

Botany

Habitat : Land plant or water plant, rooted in the substrate, partially submerged.

Habit : Herb, less than 50 cm or 50–150 cm, annual or perennial, non-insectivorous, non-aromatic, non-parasitic (Fig. 1).

Stem : Stems not hairy, round, erect or procumbent, hollow, non-succulent.

Leaf : Leaves not hairy on upper surface or on lower surface, hairy on upper surface or on lower surface, sessile, simple, narrow, alternate, evenly distributed on stem, margin entire, apex acute, base rounded, parallel-veined, midrib not-prominent, surface waxy or not-waxy, non-succulent, glands absent, dots absent, hydathodes absent.

Floral biology

Flower : Flowers unisexual or bisexual, grouped together in an inflorescence, terminal. Inflorescence not a spike or a head but other, lax, many (more than 5)-flowered. Flowers clustered, stalked, small (less than 2 cm), yellow or purple or green. Bracts present, not hairy. Spathe absent (Fig. 2).

Seed biology

Fruit : Fruit a nut or nut-like (Fig. 3).

Propagation : Panicum repens can reproduce sexually through flowering and seed production, though the principal means of propagation is reported to be rhizome extension and fragmentation (Holm *et al.*, 1977). Small *P. repens* fragments can give rise to dense clonal stands through vegetative growth and rhizome fragments buried as deep as 50 cm can send up new shoots (Hossain *et al.*, 1999, Brecke *et al.*, 2001). The lack of apical dominance also allows each node the ability to produce axillary buds along entire rhizome lengths (Wilcut *et al.*, 1988). Hossain *et al.* (2001a) reports that a single culm emerging from a single rhizome bud produced approximately 23,000 rhizome buds in a single year.

Fig. 1. Plant

Fig. 2. Flower

Seed dispersal : Deep ploughing favours its spread by breaking up and dispersing its rhizomes.

Seed dormancy and germination : Freshly harvested seeds of *Panicum* spp. can show considerable dormancy and consequently can be difficult to germinate. An indication of the degree of dormancy is provided by the large number of treatments to break dormancy and the observations that substantial periods of after-ripening are required to completely remove dormancy.

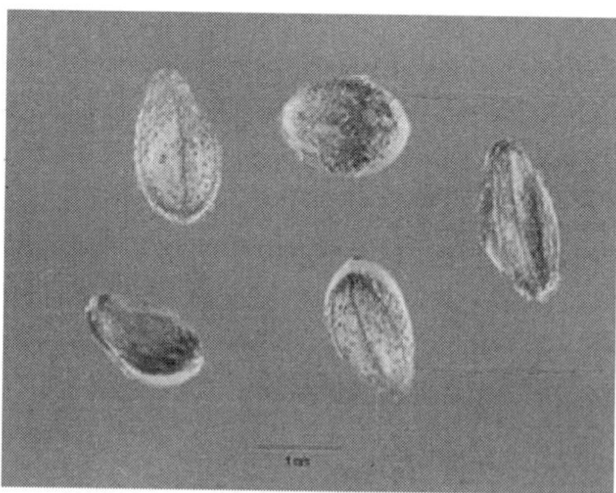

Fig. 3. Fruit/seed

Dormancy breaking treatments

Pre-chill, potassium nitrate (ISTA)

Light, potassium nitrate (AOSA)

Removal of seed covering structures: glumes, plus potassium nitrate, co-applied, 0.2%, at 35°/15°C (16h/8h) in light (6); dehull, germinate at 15°/35°C (20h/4h).

Several treatments combined with sulphuric acid scarification resulted in high, but not full germination. Alternating temperature regimes of 15°/30°C (16h/8h) or 10°/30°C (16h/8h or 20h/4h) in diffuse light applied for 42 to 56 days with or without potassium nitrate co-applied at 10^{-3} M; or an alternating temperature regime of 15°/30°C (20h/4h) in diffuse light with potassium nitrate, co-applied at 10^{-2} M, for 40 days.

Seeds of *P. repens* able to germinate under temperature fluctuation and some of the salinity concentrations (Martínez *et al.*, 1992).

Agricultural importance : Panicum repens is one of the most serious grass weeds because of its rapid rate of spread and the persistence and hardiness of its coarse, enlarged rhizomes. It suppresses other plants by its allelopathic (inhibiting) products.

Uses : Hossain *et al.* (2001b) stated that, *P. repens* is also recognized as a pasture grass, and it could be harvested five to seven times a year in tropical and subtropical areas. A higher amount of rhizomes and roots makes a loose mat-like structure in soil up to 50cm in depth, and indicates that this species could be used for soil erosion control.

Control measures : Mechanical: Continuous tillage will provide good control, but is often impractical in many natural areas. A single tillage operation may help to enhance herbicidal effectiveness, but disking alone may increase infestation levels through rhizome fragmentation. Mowing is only marginally effective.

Chemical: Glyphosate @ 2 to 3% has been the most effective herbicide used to control torpedograss.

Torpedograss is most difficult to control when partially submersed in water. Regardless of habitat conditions, multiple resprays may be required for complete control.

REFERENCES

Brecke, B.J., J.B. Unruh and J.A. Dusky (2001). Torpedograss (*Panicum repens*) control with Quinclorac in bermudagrass (*Cynodon dactylon* x *C. transvaalensis*) turf. *Weed Technology,* 15:732-736.

Holm, L.G., D.L. Plucknett, J.V. Pancho and J.P. Herberger (1977). The world's worst weeds: distribution and biology. Honolulu: University Press of Hawaii. 609 p.

Hossain, M.A., H. Akamine, I. Nakamura, Y.I shimine and H. Kuramochi (2001b). Influence of temperature levels and planting time on the sprouting of rhizome-bud and biomass production of torpedograss (*Panicum repens* L.) in Okinawa Island, southern Japan. *Weed Biology and Management,* 1:164-169.

Hossain, M.A., H. Kuramochi, Y. Ishimine and H. Akamine (2001a). Application timing of asulam for torpedograss (*Panicum repens* L.) control in sugarcane in Okinawa Island. Weed Biology and Management 1:108-114.

Hossain, M.A., Y. Ishimine, H. Akamine, S. Murayama, S.M. Moslem Uddin and K. Kuniyoshi (1999). Effect of burial depth on emergence of *Panicum repens*. *Weed Sci.,* 47: 651-656.

Martinez, M.L., T. Valverde and P. MorenoCasasola (1992). Germination response to temperature, salinity, light & depth of sowing of ten tropical dune species. *Oecologia* 92:343-353.

Wilcut, J.W., R.R. Dute, B. Truelove and D.E. Davis (1988). Factors limiting the distributeon of cogon grass, *Imperata cylindrical*, and torpedo grass, *Panicum repens*. *Weed Sci.*, 36:577-582.

KIZHANEL (*Phyllanthus amarus*)

Common name : Kizhanel

Scientific name : Phyllanthus amarus Schum. & Thonn.

Tamil name : Kizhanel

Hindi name : Jar amla

Family : Euphorbiaceae

Origin and distribution : Native to Asia and has spread as a weed throughout the tropics. It also occurs in tropical Africa and the Indian Ocean islands.

Botany

Habitat : It is found mainly as a weed in waste lands, agricultural lands and riverbanks. It grows abundantly during the rainy season.

Habit : It is an annual, glabrous herb grows up to 15-60 cm height (Fig. 1).

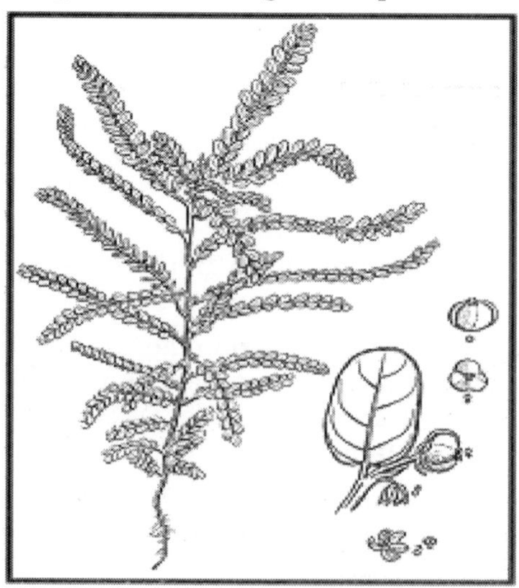

Fig. 1. Plant

Stem : Has an erect stem, naked below and slender and spreading leaf branches above.

Leaf : Alternate, distichous and crowded along lateral branchlets, simple, glabrous, sessile; stipules ovate to lanceolate, long-acuminate, blade oblong to elliptical-oblong, 7–12mm × 3–6 mm, base cuneate to rounded and slightly unequal, apex rounded, often pointed, margins finely toothed.

Floral biology

Flower : Solitary minute flowers unisexual. Female flowers axillaries in the base of twigs, the male flowers regrouped by 2 - 4 in the axils of leaves in the top of twigs. Female flowers in peduncle 2 mm long, with 5 sepals. The disc is flattened, small, with 5 lobes. The ovary is spherical surmounted of a short trifid style. Male flowers in hail peduncle, 0.5 mm long, with 5 elliptic sepals and 3 welded stamens (Fig. 2).

Fig. 2. Flower

Fruit : Capsule dehiscent in 3 cells with 2 seeds each. The capsule is globose, depressed, and smooth, 2 mm in diameter, that appears to the face lower of twigs (Fig. 3).

Seed biology

Seed : Wedge-shaped, one concave and two flat sides, beige, 1mm in height. Seeds on the back longitudinally ribbed, on the sides concentrically ribbed, in between and over the ribs finely transversely ribbed (Fig. 4).

Propagation : Through seeds.

Seed dispersal : The seeds are dispersed by water and animals.

Seed dormancy and germination : Phyllanthus is a prolific seed producer. The seeds can remain dormant for an extended period of time. They germinate massively during the rainy season and the seedlings are very fast growing. Seed requires light to germinate (Unander *et al.*, 1993; Webstar, 1957). A germination rate of up to 80% was observed at temperatures of 25–35°C, but germination was poor at 20°C or 40°C. Germination of seed is also poor under moisture stress conditions (osmotic potential higher than 300 kPa). A 1000-seed weight of 150 mg has been reported for *Phyllanthus amarus.* Stratification significantly improved the germination of the seeds. (Unander *et al.*, 1995).

Fig. 3. Fruit

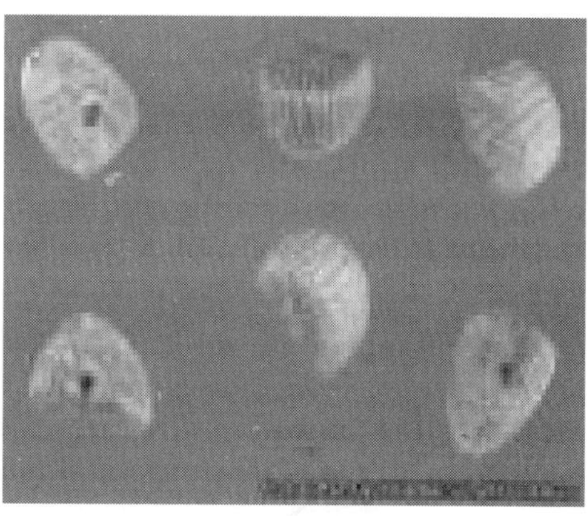

Fig. 4. Seed

Uses : P. amarus is an important plant of Indian ayurvedic system of medicine which is used for stomach problems and for the genitourinary system, liver, kidney and spleen. It possess antiviral activity against hepatitis B and related hepadna viruses, and its wide usage in traditional medicine, suggesting it might possess other biological effects as well (Unander and Blumberg 1991; Unander *et al.*, 1991, 1994)

Control measures : Post-emergence application of 2, 4-D @ 500 a.i/ha or almix @ 4g/ha.

Seed storage : According to Unander *et al.* (1995), the freezing did not significantly affect the germination of dry seeds compared to seeds stored at ambient room temperature or higher, but storage at 10°C was deleterious.

Past work done on Seed Science Technology

Unander *et al.* (1995) reported that seed color was associated with time of dehiscence. All seeds from capsules which dehisced by 30 hours after harvesting plants from the field were dark green. The numbers of light tan seeds increased at subsequent time intervals. Seeds collected at 96 hr were all light tan. Dark-green, early-to-dehisce seeds had greater and faster germination (Table l).

Table 1. Percent germination and a germination index (GI) over seed color in Madras-derived seeds and four time periods of dehiscence of Loiza-derived seeds in *P. amarus*

Madras-derived seeds			Loiza-derived seeds		
Seed color categories	**% germination**	**GI**	**Hours after harvest**	**% germination**	**GI**
Dark (early)	56.6 ± 11.8	0.39 ± 0.07	0-30 hr	22.0 ± 34.4	0.31 ± 0.21
Light (late)	26.5 ± 16.2	0.36 ± 0.11	30-48 hr	13.7 ± 7.9	0.76 ± 0.23
			48-72 hr	7.0 ± 1.6	0.50 ± 0.16
			72-96 hr	6.0 _a	

[a] 95% confidence intervals not applicable because of small seed sample

Seeds kept continuously in the dark did not germinate even after six weeks. For 40, 160, and 480 hours of light exposure, the average germination percentages were 0.3% + 0.66, 11.3% + 4.7, and 33.3% + 14.5, respectively. This indicated that a light requirement exists for seed germination in this species (Unander *et al.*, 1995).

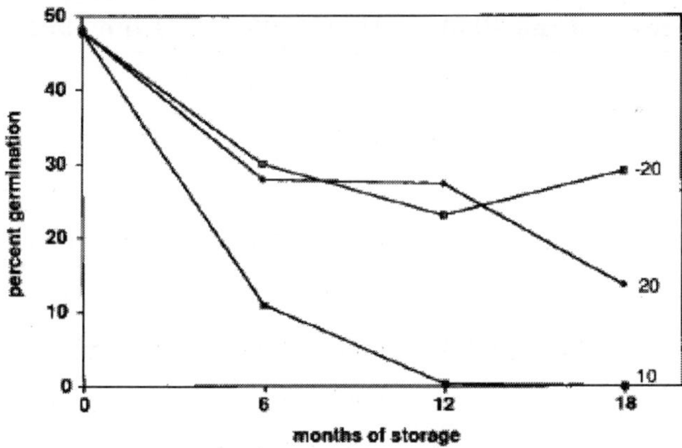

Fig. 5. Mean percent germination of seeds of *Phyllanthus amarus* at harvest and after storage at -20, 10 or 20°C

According to Unander *et al.* (1995), storage temperature significantly affected the per cent germination of 6, 12 and 18 month old seeds, and the GI of 12 and 18 month old seeds. It is concluded that seeds of *P. amarus* are best kept at room temperature or, alteratively, in a frozen state, but should not be stored in a conventional seed storage facility at 10°C. No scarified seeds germinated (Table. 2). These results suggested that percent germination would not be increased in older seeds of *P.amarus* through metabolic elimination of putative germination inhibitors, although a slightly more rapid germination rate with time was reflected in GI values for seed stored at 20°C and -20°C (Fig. 5, 6).

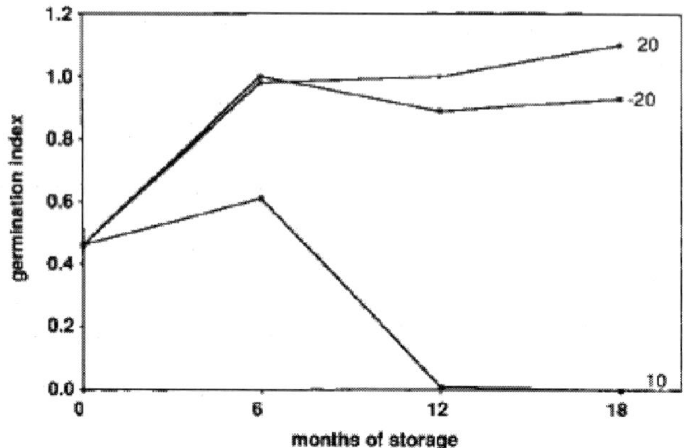

Fig. 6. Mean germination index values of seeds of *Phyllanthus amarus* at harvest and after storage at -20, 10 or 20°C.

Table 2. Means and 95% confidence intervals for percent germination and a germination index (GI) at 0, 6, 12 and 18 months after harvest at three storage temperatures for Madras-derived seeds.

Storage temperatures	June 1990	Dec. 1990	June 1991	Dec. 1991
		Percent germination		
20 C	48.0 + 5.2	28.1 ± 14.6	27.3±5.8	13.7 ± 6.2
10		10.7 ± 8.8	0.3 ± 0.6	0.00
-20		30.2 ± 19.6	23.0 ± 14.7	29.0 ± 5.6
		GI		
20C	0.46 ± 0.03	0.99 ± 0.03	2.27 ± 0.46	1.11 ± 0.06
10		0.62 ± 0.45	0.02 ± 0.03	0.00
-20		1.00 ± 0.00	1.58 ± 0.90	0.93 ± 0.16

REFERENCES

Unander, D.W. and B.S. Blumberg (1991). *In vitro* activity of *Phyllanthus* (Euphorbiaceae) species against the DNA polymerase of hepatitis viruses: Effects of growing environment and inter- and intra-specific differences. *Economic Botany*, 45:225-242.

Unander, D.W. (1991). Callus induction in *Phyllanthus* species and inhibition of viral DNA polymerase and reverse transcriptase by callus extracts. *Plant Cell Reports,* 10: 461-466.

Unander, D.W., G.L. Webster and B.S. Blumberg (1991). A survey of ethnobotanical uses and biological assay results in *Phyllanthus* (Euphorbiaceae). II. The subgenus *Phyllanthus*. *Journal of Ethnopharmacology*, 31:97-133.

Unander, D.W., H.H. Bryan, C.J. Lance and R.T. McMillan, Jr. (1993). Cultivation *of Phyllanthus amarus* and evaluation of variables potentially affecting yield and the inhibition of viral DNA polymerase. *Economic Botany*, 47:79-88.

Unander, D.W., G.L. Webstar and B.S. Blumberg (1994). Usage and bioassays in *Phyllanthus* (Euphorbiaceae): IV. Clustering of antiviral uses and effects and other applications within subgenera. In press for the *Journal of Ethnopharmacology.*

Unander, D.W., H.H. Bryan, C.J. Lance and R.T. McMillan Jr. (1995). Factors affecting germination and stand establishment of *Phyllanthus amarus* (Euphorbiaceae). *Economic Botany* 49(1): 49-55.

Webstar, G.L. (1957). A monographic study of the West Indian species of *Phyllanthus* (continued). *Journal of the Arnold Arboretum*, 38: 51-80; 170-198; 295-373.

INDIAN TOOTHCUP (*Rotala densiflora*)

Common name : Indian toothcup

*Botanical name : Rotala densiflora (*Syn: *Ammannia densiflora* Roth)

Hindi name : Jalmukh

Family : Lythraceae

Origin and distribution : Origin - South East Asia.

Distribution : Tropical and Subtropical Asia, Australia

Botany

Habitat : Rice fields and other wet places.

Habit : Annual herb, erect, 12-22 cm tall, glabrous (Fig. 1).

Stem : Stem usually creeping, sparsely to densely branched, 4-winged, wings not running into leaves.

Leaves : Leaves ovate-lanceolate or linear-lanceolate, rounded at the base, 2-16 mm long, 1-3 mm broad.

Floral biology

Inflorescence : Inflorescence solitary, axillary, in every axil (Fig. 2).

Flowers : Flowers 3 mm long, 5-merous. Bracteoles subulate, ± equalling the flower. Membranous in fruit. Epicalyx subulate-filiform. Petals ovate-oblong, oblong, obovate or elliptic, acute, obtuse or retuse. Ovary ± 3-angled. Style 0.5 mm long (Fig. 3).

Fruit : Capsule 2-3 mm long, 1.5-2 mm broad, 3(-4)-valved (Fig. 4).

Fig. 1. Plant

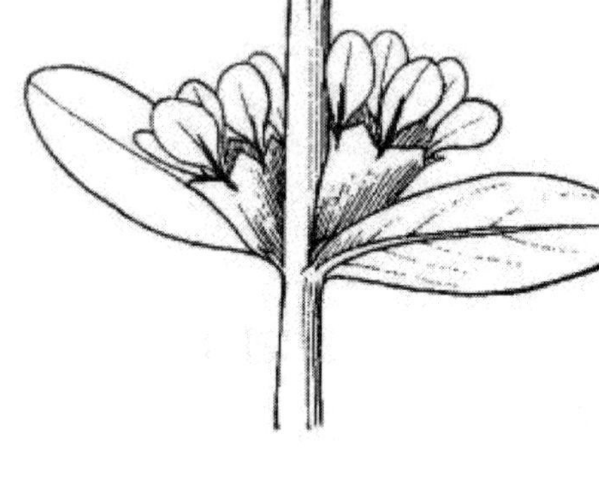

Fig. 2. Inflorescence

Fig. 3. Flower

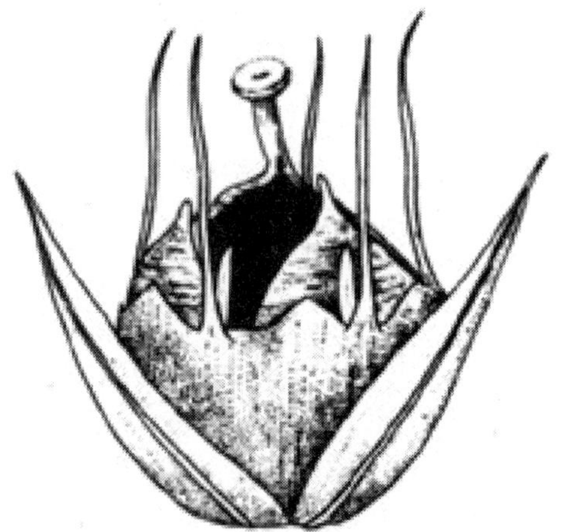

Fig. 4. Fruit

Seed biology

Seed : Seeds plano-convex or ± obovate.

Propagation : Through seeds.

Seed production : A large number of seeds are produced from a single plant (Cook, 1979).

Seed dispersal : The small size and weight of the seeds makes them easily dispersed by wind, gravity and water. The seeds of this species have inverted epidermal hairs that may allow them to attach to the feet of waterfowl.

Seeds may be transported from site to site in mud stuck to the feet of waterfowl. The primary dispersal mechanism is wind. The tiny, virtually weightless seeds could be easily caught by the wind and transported to a new location.

Seed dormancy and germination : Seeds require light and high temperatures to germinate. Darkness appears to be a limiting factor in the germination of this species. Those sown at a depth of greater than 1 mm rarely germinate. Higher day time and evening temperatures seem to increase germination. Under controlled conditions in the lab, day time temperatures of 15°C combined with night time temperatures of 6°C produced no germination. Day time temperatures of 25°C combined with night time temperatures of 15°C resulted in a germination of 32%. The best germination of 99% resulted from day time temperatures of 35°C and night time temperatures of 20°C. Seed germination under cold, moist stratification for 21 days followed by long warm day length yielded the highest germination percentages. Stratification showed an increase in the germination percentages (Fite-Wassilak *et al.*, 2001).

Soil seed bank : Seeds are long-lived in the soil seed bank. (Mattrick, 2001)

Uses : Used as aquarium plants.

REFERENCES

Cook, C.D.K. (1979). A revision of the genus *Rotala ramosior* (Lythraceae). *Boissera*, 29: 3-156.

Fite-Wassilak, A., A. Kane, J. Kelley and A. Leuck (2001). Germination of the rare wetland species *Rotala ramosior (Lythraceae).* Unpublished document. Department of Ecology and Evolutionary Biology (contact Dr. Johanna Schmitt, Box G, Brown University, Providence, Rhode Island, USA).

Mattrick, C. (2001). *Rotala ramosior* (L.) Koehne (Toothcup) conservation and research plan. New England Wild Flower Society, Framingham, Massachusetts, USA.

EAST INDIAN GLOBLE-THISTLE (*Sphaeranthus indicus)*

Common name : East Indian Globle-thistle

Scientific name : Sphaeranthus indicus

Tamil : Kottakaranthai

Hindi : Mundi, gorkhmundi

Family : Asteraceae

Origin and distribution : Native to India distributed throughout India, Sri Lanka, Africa and Australia.

Botany

Habitat : Common rabi weed found in rice fields.

Habit : Annual herb (Fig. 1).

Stem : A much branched, strongly-scented annual with winged stem and the wings toothed.

Leaf : Leaves obovate-oblong, narrowed at the base, dentate and serrate.

Floral biology

Flower : Heads compound, dense globose; bracts papillate, acute, ciliate; outer flowers female, fertile; inner bisexual, fertile or sterile; receptacle small, naked; corolla of the female flowers tubular, pink or purple, minutely 2-3 toothed; of bisexual flowers globosely inflated below, 4-5 lobed; anther bases sagittate; style arms filiform (Fig. 2).

Fig. 1. Plant

Fig. 2. Flower

Fig. 3. Fruit

Seed biology

Fruit : Achene angular, subglabrous; pappus absent (Fig. 3).

Propagation : By seeds.

Seed dispersal : The wind, water and certain animals are the media for the dispersal of seeds.

Uses : Juice styptic, also used in hepatic and gastric disorders. Pulverized seeds and roots used in cough and other chest troubles; credited with antitubercular activities. Flowers tonic and depurative.

Root and seed powder is given orally to kill intestinal worms in children.

Root crushed with *Achyranthus aspera* root and candy is taken thrice daily after each defecation to cure diarrhoea.